# Aircraft Noise



Aircraft noise has adverse impacts on passengers, airport staff and people living near airports, it thus limits the capacity of regional and international airports throughout the world. Reducing perceived noise of aircraft involves reduction of noise at source, along the propagation path and at the receiver.

Effective noise control demands highly skilled and knowledgeable engineers. This book is for them. It shows you how accurate and reliable information about aircraft noise levels can be gained by calculations using appropriate generation and propagation models, or by measurements with effective monitoring systems. It also explains how to allow for atmospheric conditions, natural and artificial topography as well as detailing necessary measurement techniques.

**Oleksandr Zaporozhets** was awarded a D.Sc. for a thesis on the 'Development of models and methods of information provision for environment protection from civil aviation impact' in October 1997 at the Kyiv International University of Civil Aviation and received a Ph.D, for a thesis on 'Optimization of aircraft operational procedures for minimum environment impact' in December 1984 from the Kiev Institute of Civil Aviation Engineers. Jointly with Dr Tokarev, he was awarded a silver medal for achieving successes in the development of the national economy of the USSR in 1987. Currently he is a full Professor at the National Aviation University of the Ukraine.

**Vadim Tokarev** was awarded a D.Sc. in 1990 and a Ph.D. in 1969 at the Kyiv International University of Civil Aviation. Currently he is a full Professor at the National Aviation University of the Ukraine.

**Keith Attenborough** is Research Professor in Acoustics at the Open University, Education Manager of the Institute of Acoustics (UK) and was Editor-in-Chief of *Applied Acoustics* from 2000 to 2010. From 1998 to 2001 he was Head of Department of Engineering at the University of Hull. In 1996 he received the Rayleigh Gold medal from the Institute of Acoustics (UK) for outstanding contributions to acoustics research and teaching. He is a Chartered Engineer, an Honorary fellow of the Institute of Acoustics and a fellow of the Acoustical Society of America.

# Aircraft Noise
## Assessment, prediction and control

Oleksandr Zaporozhets
Vadim Tokarev and
Keith Attenborough

Routledge
Taylor & Francis Group
LONDON AND NEW YORK

First published 2011
by Spon Press

2 Park Square, Milton Park, Abingdon, Oxon OX14 4RN
711 Third Avenue, New York, NY 10017, USA

*Spon Press is an imprint of the Taylor & Francis Group, an informa business*

First issued in paperback 2017



*British Library Cataloguing in Publication Data*
A catalogue record for this book is available from the British Library

*Library of Congress Cataloging in Publication Data*
Attenborough, K. (Keith)
Aircraft noise propagation, exposure & reduction / Oleksandr Zaporozhets, Vadim Tokarev, Keith Attenborough.
p. cm.
Includes bibliographical references and index.
1. Airplanes–Noise. I. Tokarev, V. I. (Vadim Ivanovich)
II. Zaporozhets, Oleksandr. III. Title.
TL671.65.A88 2011
629.132′3–dc22 2010036182

ISBN 13: 978-0-415-24066-6 (hbk)
ISBN 13: 978-1-138-07302-9 (pbk)

Typeset in Sabon by
Glyph International Ltd.

# Contents

# Preface

The motivation to write this book arises from over 40 years of investigations by Oleksandr Zaporozhets and Vadim Tokarev into aviation noise sources and into the technical, ecological, economical and social consequences of their impact on environment. The book also reflects these authors' experience over more than 30 years of teaching undergraduate and graduate courses within the framework of the 'Acoustic Ecology' curriculum at the National Aviation University, Ukraine, including modules on the physical factors that impact the environment, methods of biosphere protection and on environmental noise monitoring. The book contains results of research into aircraft noise modeling (including particular issues relating to aircraft noise propagation), assessment of the efficiency of operational methods of aviation noise reduction, flight planning for minimizing aircraft noise and monitoring of environmental conditions in the vicinities of airports.

The experience of these authors in applied aviation acoustics has been the result of collaborations with many scientific organizations including the State Research Center of the Central Aerohydrodynamic Institute (Moscow), the State Scientific Institute of Civil Aviation (Moscow) and the Aviation Design Offices of Tupolev, Il'ushin (Moscow) and Antonov (Kyiv). Consequently, many of the resulting publications are in Russian and in Ukrainian.

First attempts at writing a systematic overview of the subject of aircraft noise in English were made for a special issue of *Applied Acoustics* published in 1998 and for the final report of the NATO project 'Aircraft noise forecasting' (NATO grant EST.CLG.974767). The latter project also provided the impetus for the subsequent collaborations between the authors based in the Ukraine and Keith Attenborough in the UK. Although the scientific collaboration among the three authors has primarily influenced the contents of Chapters 3 and 6, Attenborough has also contributed by intensive editing of the use of English in the other chapters.

The book places equal emphasis on theory and on practical applications. The authors consider that the text differs in scope from the available texts on same topic [e.g. *Aeroacoustics of Flight Vehicles – Theory and Practice. Vol. 1, Noise Sources, Vol. 2, Noise Control* (1995), edited by H.H. Hubbard, Acoustical Society of America, Woodbury, NY, and *Transportation Noise Reference Book* (1987) edited by P.M. Nelson, Butterworths, UK] in that attention is given to operational and maintenance aspects of aircraft noise assessment and noise reduction methodology. The application of low noise operational procedures provides often neglected opportunities for noise reduction around the airports. This text provides the techniques and scientific basis that will allow for successful modeling and analysis of operational methods for aircraft noise reduction as well as the methods of control at source that are more usually considered.

It is also recognized that noise from aircraft is only part of the noise-associated problem around an airport. Mitigation of airport noise must be investigated as a problem of urban or rural soundscape. The methodology advocated in this text for decreasing the impact of aviation noise is based on a complex approach to a problem of noise reduction around the airports, which is considered as a physical process *and* as a phenomenon of social hygiene, sometimes with economic consequences. The approach to aircraft noise management in the vicinity of an airport used in the book corresponds to the balanced approach advocated by the International Civil Aviation Organization.

An important contribution of the book is to demonstrate how optimization of the control of aircraft noise through operational measures can increase the environmental capacity of the airport, particularly in cases where, otherwise, environmental constraints would reduce the operational and economic capacities of the airport. The basic theme of Chapter 1 of the book follows from the results of research on aviation noise in relation to airport noise capacity. The airport noise capacity is represented by the maximum number of aircraft that can be operated during a given period so that total aircraft noise levels do not exceed a prescribed limitation in critical zones around an airport. The capacity of an airport is a function of many different factors and aspects of airport infrastructure, including airfield layout (the number of runways, the extent of taxiways, apron development), the terminals and landsite facilities, air traffic control procedures, ground handling operations and meteorological conditions.

Aircraft are complex noise sources and a variety of noise protection methods can be employed around airports, including organizational, technical, operational and land-use methods. This is explained in Chapter 1 together with a presentation of the information about the basic noise sources on aircraft necessary for an understanding of the mechanisms of aviation noise generation.

Chapter 2 discusses models used to estimate the acoustical characteristics of the jets, fans, turbines, propellers and elements of the airframe. Parametrical investigations into the fundamental sources enable estimates of the influences of constructional and operational parameters on the overall acoustic fields due to aircraft.

Chapter 3 considers the physical phenomena involved in outdoor sound propagation under various operational conditions. These include atmospheric absorption, propagation over flat ground surfaces, over barriers, through trees, refraction by wind and temperature gradients and propagation through turbulence.

Chapter 4 explores methods for aircraft noise calculation, starting from an acoustic model for an aircraft as a whole. A model for predicting noise under the flight path is essential for operational purposes and for determining low-noise flight procedures. Models for predicting noise levels due to aircraft ground operations are important also for determining total airport noise. Some simplifications are introduced for predicting noise in the vicinity of the airport.

Using the models defined in Chapters 3 and 4, Chapter 5 investigates the influences of operational factors on aircraft noise characteristics at receivers on the ground and under the flight path. The optimal operational procedures for reducing noise impact are deduced for specific situations.

Chapter 6 reviews methods of aircraft noise reduction at source, along the sound propagation path and at the receiver, including the efficiency of acoustic screens for reducing noise from airport ground operations. The selection of optimal features of the operation scenario in the vicinity of the airport informs decision-making procedures for airport noise capacity control.

Chapter 7 introduces monitoring of aircraft noise as an essential tool for noise assessment and control around airports. The reasons for aircraft noise monitoring are operational, technical and economic. Current monitoring systems include powerful instrumentation and software, which besides recording noise levels must control the flight tracks, identify the type of noise source from each particular noise event, register noise complaints and measure meteorological parameters. To achieve effective mitigation of the impact of aviation noise on the environment, the interdependencies and trade-offs between noise and other important environmental factors associated with civil aviation, such as engine emission and third party risk, must be taken into account. It is shown that possible solutions may be reached by informational monitoring systems with the support of specifically predefined Aircraft Design Space, Flight Scenario Design Space and an Aviation Environmental Cost–Benefit Tool.

This book should be of interest to all those concerned with aircraft noise problems. After reading this book, the engineer, consultant or airport designer will be able to implement a balanced approach to airport noise management. This will include use of low noise operational procedures and the results of aircraft noise monitoring. The book should also be useful to those responsible for making or responding to decisions about the requirements for environmental control at airports. Although the book could be used as a reference text, it should be noted that the references listed at the end of the book are far from being exhaustive. Essentially, they contain only the references used in writing the book and reflect the particular questions considered by the authors. Nevertheless, by bringing together their many new scientific and practical results, the authors hope that the book's modern approach to aviation noise assessment and reduction will prove a useful addition to the literature.

Oleksandr Zaporozhets
Vadim Tokarev
Keith Attenborough

# 1 A review of the aircraft noise problem

## 1.1 Environmental impacts of airports

Aviation in the twenty-first century contributes to climate change, noise and air pollution. Together with various social and economic problems, environmental issues have the potential to constrain the operation and growth of airports. Constraints on airport capacity affect the capacity of the air navigation system as a whole. Many international airports are operating at their maximum, and some have already reached their operating limits including those resulting from environmental impact. This situation is expected to become more widespread as air traffic continues to increase. Already aircraft noise is a limiting factor for the capacity of regional and international airports throughout the world.

There are many definitions of airport capacity with regard to various issues: operational, flight safety, economic and environmental. The relative importance of each issue depends on the local, regional and national circumstances of each airport (see Fig. 1.1). Environmental capacity is the extent to which the environment is able to receive, tolerate, assimilate or process the outputs of aviation activity. Local environmental airport capacity can be expressed in terms of the maximum numbers of aircraft, passengers and freight accommodated during a given period under a particular environmental limitation and consistent with flight safety.[1,2] For example, the airport noise capacity is the maximum numbers of aircraft that can be operated during a given period so that total aircraft noise levels do not exceed a prescribed limitation in critical zones around an airport.

*Aircraft noise* is noise associated with the operation and growth of airports that impact upon local communities, in particular the nature and extent of noise exposure arising from aircraft operations. It is the single most significant contemporary environmental constraint, and is likely to become more severe in the future.

*Local air quality* is a capacity issue at some European airports, and is likely to become more widespread in the short to medium term. After aircraft noise, local air quality seems to be the next most significant environmental factor with the potential to constrain airport growth.

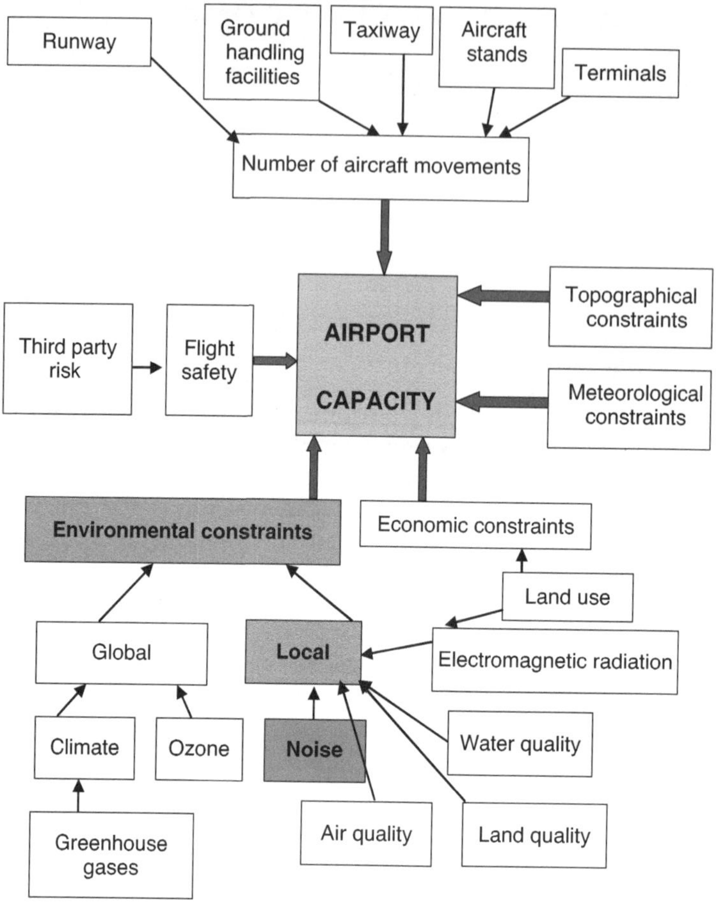


*Figure 1.1* Environmental influences on airport capacity.

*Third party risk* is a potential future constraint for certain larger airports located close to built-up areas. The communities surrounding such airports are exposed to the small risk of an aircraft crash.

*Water usage/pollution* is both an existing and a potential constraint at certain European airports.

*Surrounding land use and habitat value* are both existing and potential constraints at a number of European airports.

*Greenhouse gas emissions* pose a potential constraint in the long term.

The capacity of an airport is a function of many different factors and the airport infrastructure, including the airfield layout (the number of runways, the extent of the taxiway, apron development), the terminals and landsite facilities, air traffic control procedures, ground handling operations and meteorological conditions. An individual airport capacity depends on the time between an aircraft landing and its leaving the airport, the ability of the airport to accept aircraft within a specified delay, the airport air traffic control system and its runway approach facilities.

In 2001, the International Civil Aviation Organization (ICAO) developed a balanced approach to noise management at airports. The balanced approach includes four elements: reduction at source, land-use planning and management, operational procedures for noise abatement and aircraft operational restrictions. The balanced approach has been applied to European airports by means of EU Directive 2002/30/EC concerning rules and procedures for introducing noise-related practices at airports. The noise mitigation measures should take into account specific features of the particular airport and the maximum achievable efficiency of suggested methods.

The potential to reduce noise at source is limited and land-use measures are difficult to implement in densely populated zones. Operational procedures which depend on pilot behavior may lead to a reduction in the level of flight safety. The growth of air traffic is faster than developments in new technologies and methods of noise reduction.

At present, only 2 per cent of the population is exposed to aircraft noise. This proportion should be compared with, for example, the 45 per cent of the population exposed to noise of road traffic and the 30 per cent to industrial noise. Nevertheless, ICAO analysis has suggested that there will be a 42 per cent increase in the number of people affected by aircraft noise in Europe by the year 2020.[3]

The noise produced by aircraft during operations in the areas around airports represents a serious social, ecological, technical and economic problem. Substantial levels of noise emission can bring about worsening of people's health, lowering their quality of life and lessening their productivity at work, through speech interference for example. In the areas around airports, aircraft noise has adverse influences on ground, maintenance and flight operations personnel, on passengers and on the local residential population. In abating aircraft noise, it is necessary to consider several criteria: ecological, technical, economic and social.

Methods of reducing aircraft noise have to take into account many requirements as follows:

1 Noise sources must be placed as far away as possible from built up areas.
2 Noise should be reduced to the lowest level achievable in a given case.
3 Noise abatement of aircraft involves several acoustic sources: jet exhaust stream, engine fan, turbine, combustion chamber, propellers (including the number of rotors and the tail rotor on a helicopter) and the airframe.
4 Since there are different types of aircraft in operation at an airport, the aircraft noise in the vicinity of the airport depends on the type of aircraft

in service, the number of flights by each type, the times of day and the meteorological conditions.

5 Propagation of sound from aircraft to a receiver involves direct transmission through air, reflection, diffraction and scattering from the surface of the Earth, screens and buildings, and through a turbulent and inhomogeneous atmosphere.

6 Apart from dwellings, there might be particularly noise sensitive receiver locations such as in laboratories, schools and hospitals. In developing measures for reducing noise around airports, it is necessary to take into account the short- and long-term forecasts of airport development.

7 There is a need for a balanced approach to engineering noise abatement practice from complex sources taking account not only noise levels but also the spectral characteristics at the receiver.

8 Noise abatement on aircraft can be realized at various stages including their design, manufacture, operation and repair. During operation, noise-reducing activities include reduction at the source, along the propagation path and at the receiver. The most cost-effective is to reduce noise at the source or at the design stage.[4]

9 Noise abatement requires identification of the noise sources, assessment of their contributions to the overall acoustic field and acquaintance with the accumulated knowledge of the effectiveness of available noise abatement methods.

10 The full costs associated with noise pollution (monitoring, management, lowering levels and supervision) should be met by those responsible for the noise.

Although aircraft are not the only sources of environmental noise around airports, they are the main ones. The working cycle of aircraft can be subdivided into starting engine operation, preflight engine run, taxiing to lineup, acceleration on the runway with full or reduced throttle, takeoff and roll-on, flight path, landing, run-on operation and engine run-up. The maximum noise levels are made during the acceleration on the runway, takeoff and roll-on. But these stages are of relatively short duration. Other periods of aircraft noise generation around an airport occur during engine testing, maintenance work, temporary repair and engine replacement after the end of their service life. Maintenance operations and engine run-ups have a long duration and take place at comparatively short distances in relation to surrounding residential zones, passengers and technical staff. So, although they involve lower levels than those from moving aircraft, noise from these ground operations must be considered.

The historical changes in priorities among the various operational factors during the development of civil aviation are indicated in Table 1.1.[5]

Although flight safety remains paramount in importance, currently the problems of flight operation of aircraft and environmental protection,

*Table 1.1* Changes in priorities for civil aviation

| *1950–1970* | *1970–1990* | *1990–2020* |
|---|---|---|
| Flight safety | Flight safety | Flight safety |
| Speed | Economic indexes | Environmental protection (including noise) |
| Range | Noise around airports | Resources |
| Economic indices | Regularity of operation | Regularity of operation |
| Comfort | Comfort | Economic indices |
| Regularity of operations | Speed | Comfort |
| Noise around airports | Range | Speed and range |

including noise abatement, are combined. Noise abatement by operational measures involves additional pilot workloads for pilots and air traffic control and can result in additional operational costs for the aircraft operator. Aviation safety will always have priority over noise abatement operating measures. The pilot-in-charge will make the decision not to use low-noise flight procedures if it prejudices flight safety. For example, the pilot will ignore the demands of minimum noise impact under any kind of failure or shut-down of an engine, equipment failure or any other apparent loss of performance at any stage of takeoff. Noise abatement procedures in the form of reduced power takeoff should not be required in adverse operational conditions such as when the runway is not clear and dry, when horizontal visibility is less than 1.9 km, when a cross-wind component, including gusts, exceeds 28 km/h, when a tail-wind component, including gusts, exceeds 9 km/h, when wind shear has been reported or forecast or when thunderstorms are expected to affect approach or departure.

## 1.2 Description of aircraft noise

Aircraft are complex noise sources (see Fig. 1.2). So a variety of noise protection methods are employed around airports; including organizational, technical, operational and zoning methods. The main noise sources on an aircraft in flight are the power unit and the aerodynamic noise. Aerodynamic noise becomes particularly noticeable during the landing approach of heavy jet aircraft, when the engines are at comparatively low thrust.

The scientific basis for abating noise from aircraft relies on advances that have been made in aeroacoustics. Unlike classical acoustics (which is concerned mainly with the sound caused by oscillating surfaces), aeroacoustics investigates aerodynamic noise conditioned by turbulent non-stationary flow. Typically, jet aircraft noise sources include: jet noise, core noise, inlet and aft fan noise, turbine noise and airframe noise. Table 1.2 shows a classification of aircraft noise sources.

Usually third-octave band spectra are used for noise assessment of any type of aircraft in any mode of flight or during maintenance activities in the

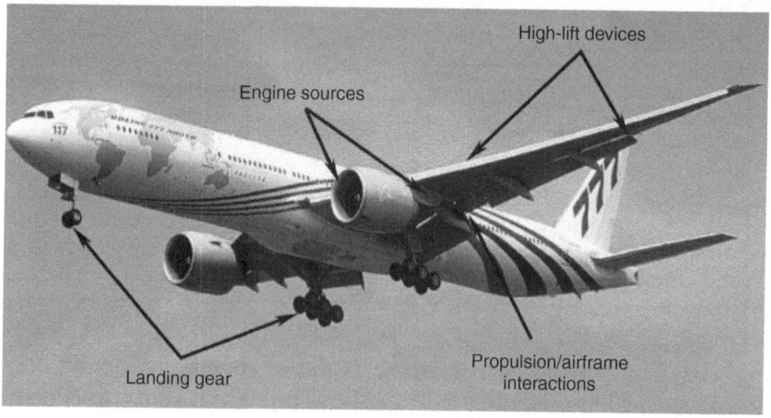


*Figure 1.2* Aircraft noise sources.

*Table 1.2* A classification of noise sources on aircraft

| *Aircraft class* | | *Main sources of noise* | |
|---|---|---|---|
| | | *Power-unit* | *Airframe* |
| Aircraft – ordinary takeoff and landing | Turbojet | Jet, fan, core noise | Flap and wing trailing edges, flap side edges, slats, gear sources, fuselage and wing turbulent boundary layers |
| | Turboprop | Propeller, propfan, engine exhaust | |
| Aircraft – short takeoff and landing | Turbojet | Fan, engine exhaust | Interaction jet with flap |
| | Turboprop | Propeller | |
| Supersonic aircraft | | Jet | Interaction of flow with frame |
| Helicopters | | Blades of main rotor, engine exhaust | Not important |
| Aircraft of general aviation | Turbojet | Jet, fan | Not important |
| | Turboprop | Propeller, engine exhaust | |

vicinity of the airport. In this case, the common computational procedure for the prediction of the aircraft noise under the flight path or around the aircraft on a ground (run-ups, taxiing, waiting for the takeoff along the runway) is based on the assumption that sound waves are spreading along the shortest distance between the aircraft and the point of noise control.

From measurement experience it can be argued that the acoustic field produced by an aircraft moving at constant altitude, speed, attitude and engine power setting through a uniform atmosphere represents a stationary random process. The acoustic signal received from a moving aircraft at a fixed microphone location, however, is clearly non-stationary. The characteristics of the spectrum of the received signal change because of the directionality of the source, spherical spreading, atmospheric absorption and refraction, Doppler effect and ground reflection and attenuation. The received acoustic signal can be assumed to be weakly stationary only over some sufficiently small time interval. However, use of too small analysis time intervals results in too few statistical degrees of freedom and poor confidence in the sound pressure level.

Any type of aircraft noise criterion or index is estimated from a set of noise spectra (in third-octave frequency bands from 50 to 10,000 Hz) and sample duration 0.5 s, that vary during the particular noise event or during any kind of noise exposure. Several methods of sound pressure filtering in the frequency domain are used. The most appropriate for aircraft noise analysis are the A-weighting correction, which gives a measure of the loudness, and the perceived noise calculation scheme, which gives a measure of the noisiness.

The jet and the fan are the main noise sources in a jet engine. Bypass engines have inner and outer contours. The bypass ratio ($m$) represents the ratio of air masses flowing through the outer and inner contours of the engine. On an engine with a high bypass ratio ($m > 3$), used typically for contemporary subsonic heavy aircraft, the fan is the predominant source of noise, spreading forward over the engine inlet and backwards over the fan exhaust system. On engines with a low bypass ratio ($m < 2$), such as those used on first-generation supersonic transport, jet noise is predominant. Increase in the bypass ratio lowers the contribution of jet noise to the total acoustic field of the engine and increases the contributions of fan and turbine noise.

A third-octave frequency band spectrum and an overall sound pressure level (OASPL) for an aircraft with a low bypass engine ($m = 1$) during takeoff engine mode measured at the lateral noise monitoring point 1 (450 m from the runway axis) are shown in Fig. 1.3. Figure 1.4 shows the measured noise characteristics of the same aircraft at the flyover noise monitoring measurement point 2 (6500 m from aircraft gear release on runway during takeoff). The engine mode is nominal. For the same aircraft, Fig. 1.5 shows the landing noise characteristics measured at the approach noise monitoring point 3, located 2000 m before runway edge. The engine mode is around 60 per cent of nominal thrust.

During takeoff (as measured at monitoring points 1, 2) the dominant aircraft engine noise source is the jet. During landing (at monitoring point 3) the dominant engine source is the fan in the high-frequency range and the jet and airframe (noise from flaps, gears, other airframe components) are dominant in the low-frequency range.

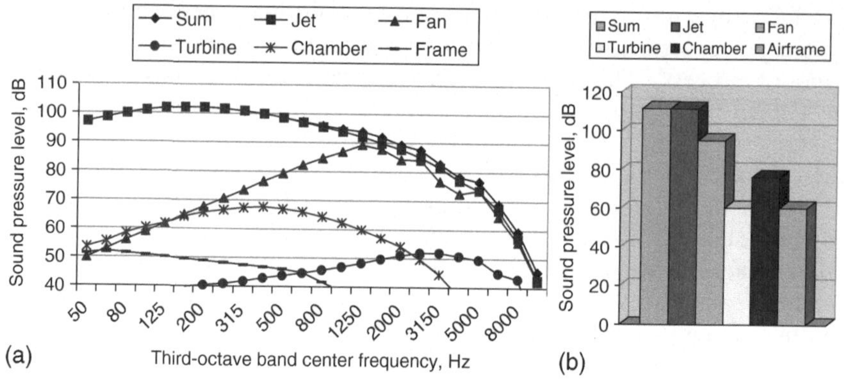


*Figure 1.3* Noise source contributions for aircraft with low bypass ratio engines (bypass engine ratio, $m = 1$) at control point No. 1 (takeoff mass 160 t, distance 450 m, engine mode at maximum thrust, 'lateral attenuation' neglected): (a) spectra; (b) overall sound level.

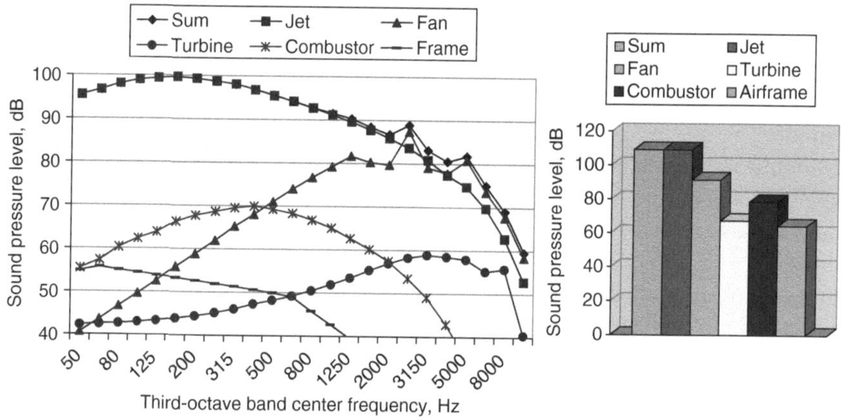


*Figure 1.4* Noise source contributions for aircraft with low bypass ratio engines (bypass engine ratio, $m = 1$) at control point No. 2 (takeoff mass 160 t, distance 450 m, engine mode at maximum thrust, 'lateral attenuation' neglected): (a) spectra; (b) overall sound level.

Noise characteristics of an aircraft with middle bypass ratio ($m \sim 2.5$) engines are shown in Figs 1.6–1.8 for noise monitoring points 1, 2 and 3 respectively. At takeoff (points 1, 2; Figs 1.6 and 1.7) the dominant noise sources of the aircraft are the jets (in the low-frequency range) and the fans (in the high-frequency range). During landing (monitoring point 3; Fig. 1.8) the dominant sources are the fans and the airframe.

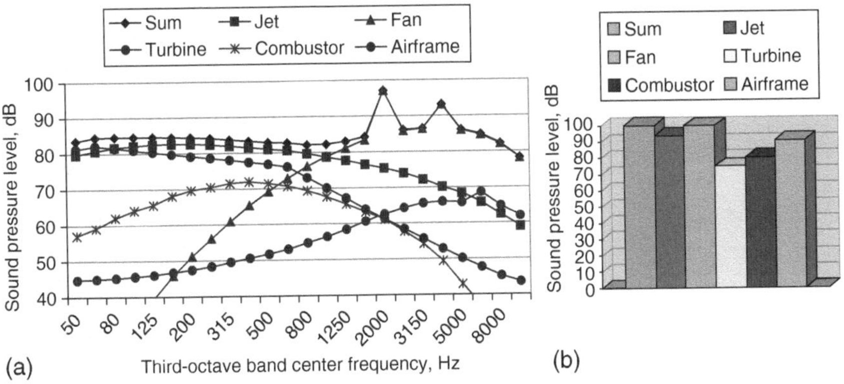


*Figure 1.5* Noise source contributions for aircraft with low bypass ratio engines (bypass engine ratio, $m = 1$) at control point No. 3 (takeoff mass 160 t, distance 450 m, engine mode at maximum thrust, 'lateral attenuation' neglected): (a) spectra; (b) overall sound level.

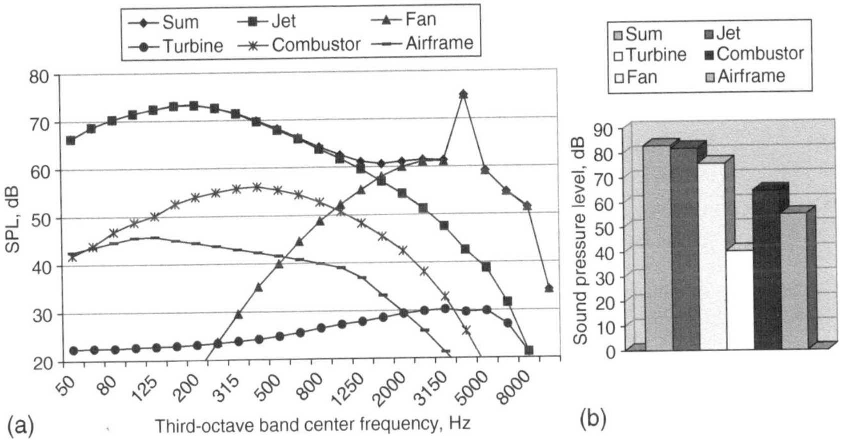


*Figure 1.6* Noise source contributions for aircraft with intermediate bypass ratio engines (bypass ratio, $m = 2.5$) at control point No. 1 (take-off mass 160 t, distance 450 m, engine mode at maximum thrust, 'lateral attenuation' neglected): (a) spectra; (b) Overall sound level

The noise characteristics of the aircraft with high bypass ratio ($m = 6$) engines are shown in Figs 1.9–1.11 for noise monitoring points 1, 2 and 3, respectively. During takeoff (points 1 and 2; Figs 1.9 and 1.10) the dominant noise sources (in the high-frequency range) on the aircraft are the fans, the combustion chambers of the engines and the airframe. During landing (point 3; Fig. 1.11) the dominant sources are the fans (in the high-frequency range), the airframe and the combustion chambers.

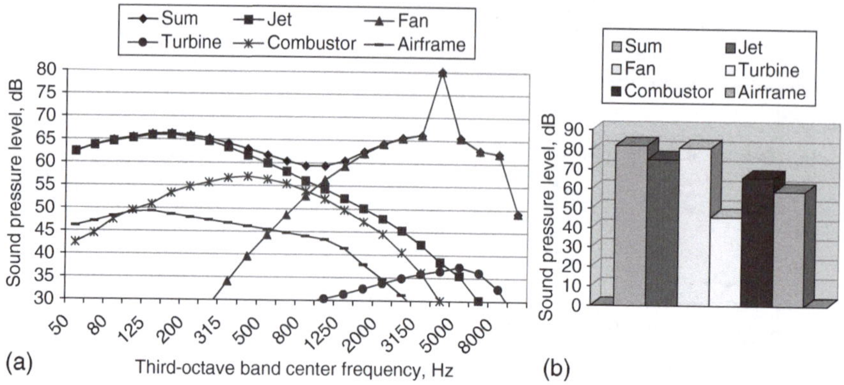


*Figure 1.7* Noise source contributions for aircraft with intermediate bypass ratio engines (bypass ratio, $m = 2.5$) at control point No. 2 (takeoff mass 160 t, distance 450 m, engine mode at maximum thrust, 'lateral attenuation' neglected): (a) spectra; (b) overall sound level.

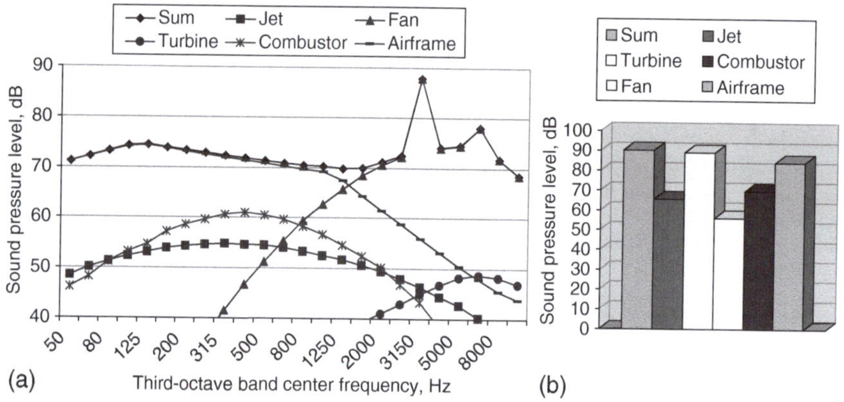


*Figure 1.8* Noise source contributions for aircraft with intermediate bypass ratio engines (bypass ratio, $m = 2.5$) at control point No. 3 (takeoff mass 160 t, distance 450 m, engine mode at maximum thrust, 'lateral attenuation' neglected): (a) spectra; (b) overall sound level.

At present, attention is focused mainly on the noise reduction of engines with high bypass ratios ($m \geq 6$), since they are widely used. Consideration is given to possible design methods: optimization of fan, gas-dynamic and operation parameters on the basis of integrated aeroacoustic design and installation of intake and exhaust silencers.

The noise characteristics of an aircraft with turboprop engines are shown in Figs 1.12 and 1.13 corresponding to noise monitoring points 2 (Fig. 1.12) and 3 (Fig. 1.13). The use of third-octave frequency bands means that the

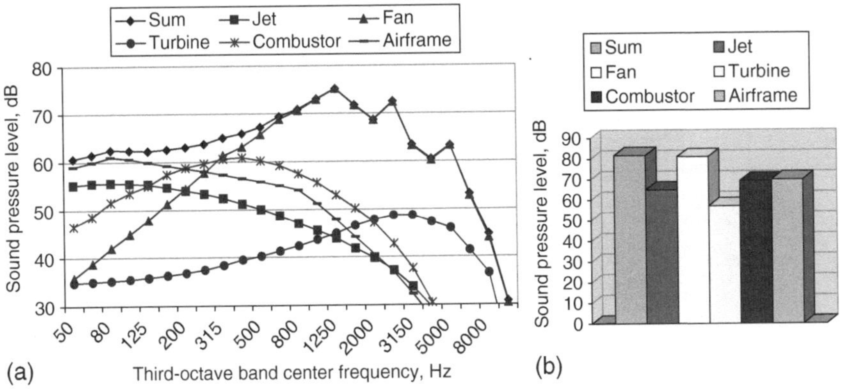


*Figure 1.9* Noise source contributions for aircraft with high bypass ratio engines (bypass ratio, $m = 6$) at control point No. 1 (takeoff mass 160 t, distance 450 m, engine mode at maximum thrust, 'lateral attenuation' neglected): (a) spectra; (b) overall sound level.

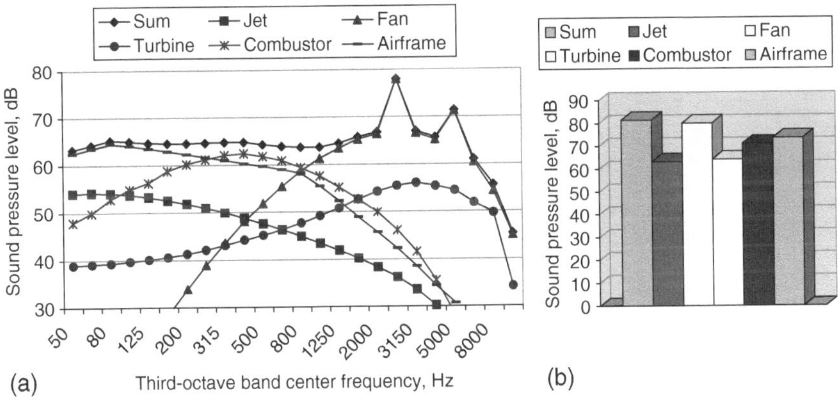


*Figure 1.10* Noise source contributions for aircraft with high bypass ratio engines (bypass ratio, $m = 6$) at control point No. 2 (takeoff mass 160 t, distance 450 m, engine mode at maximum thrust, 'lateral attenuation' neglected): (a) spectra; (b) overall sound level.

broad band noise emission masks the discrete harmonics. During takeoff and landing the dominant noise sources on such aircraft are the propellers. Their noise levels exceed those from other sources by more than 10 dB.

Figure 1.14 shows the stages in the procedure for reducing aircraft noise at its source.

Turbulent airflow over the airfoil (corresponding to high speed and Reynolds number) results in radiation of aerodynamic noise. In turbulent flow one can distinguish the disturbances due to vorticity, entropy and

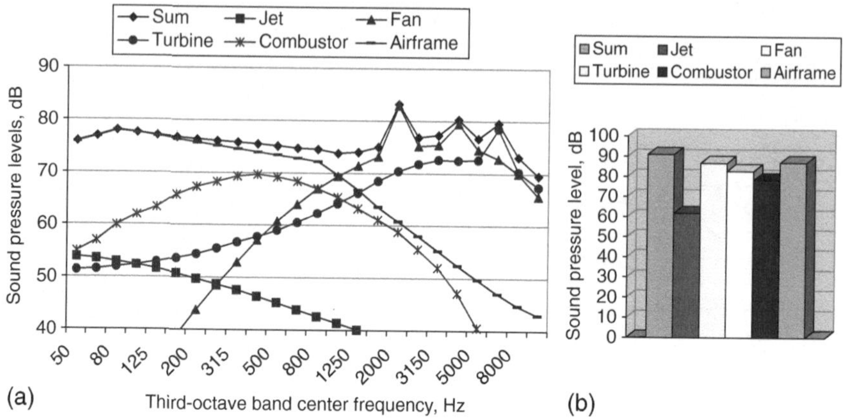


*Figure 1.11* Noise source contributions for aircraft with high bypass ratio engines (bypass ratio, $m = 6$) at control point No. 3 (takeoff mass 160 t, distance 450 m, engine mode at maximum thrust, 'lateral attenuation' neglected): (a) spectra; (b) overall sound level.

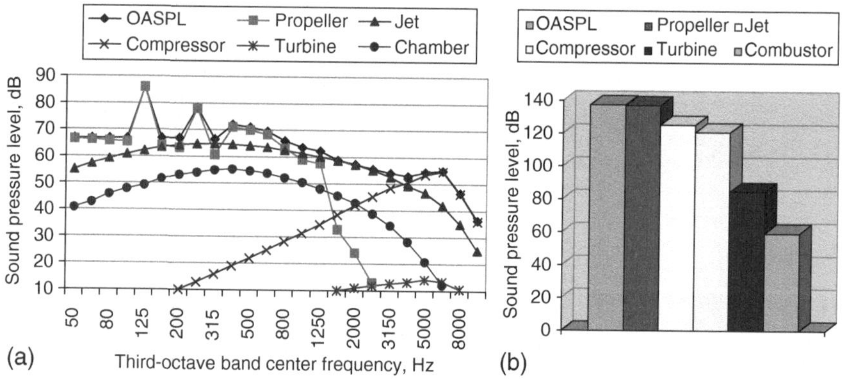


*Figure 1.12* Noise source contributions for turboprop aircraft at control point No. 2 (takeoff mass 9.8 t, distance 300 m, engine mode – maximum, 'lateral attenuation' neglected): (a) spectra; (b) overall sound level.

sound. Interaction between these disturbances, described mathematically by non-linear equations, is determined by the turbulent flow structure and the acoustical field characteristics.

Radiation of the sound usually results from non-stationary flows, and separated flows associated with elements of the aircraft with imperfect aerodynamics. These destabilize the flow and a large part of the kinetic energy of the flow turns into energy of acoustic radiation. Table 1.3 lists some values of the acoustic efficiency $\eta_a$, which is the ratio of acoustic power

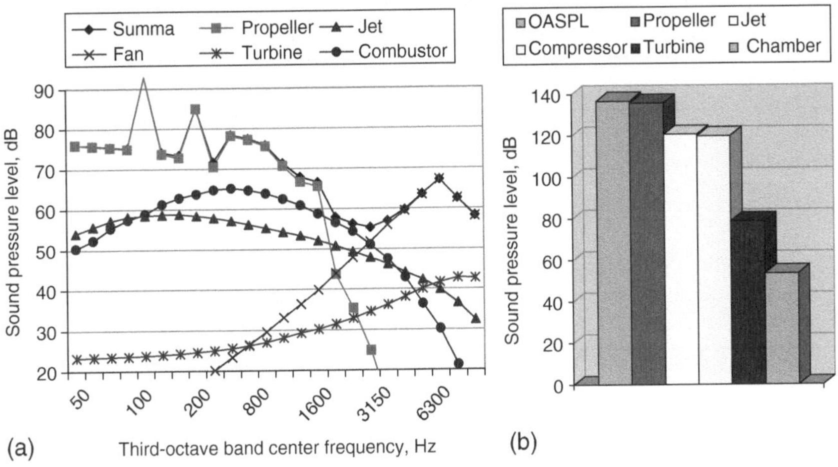


*Figure 1.13* Noise source contributions for turboprop aircraft at control point No. 3 (landing mass 9.8 t, distance 100 m, engine mode – 0.6 nominal, 'lateral attenuation' neglected).

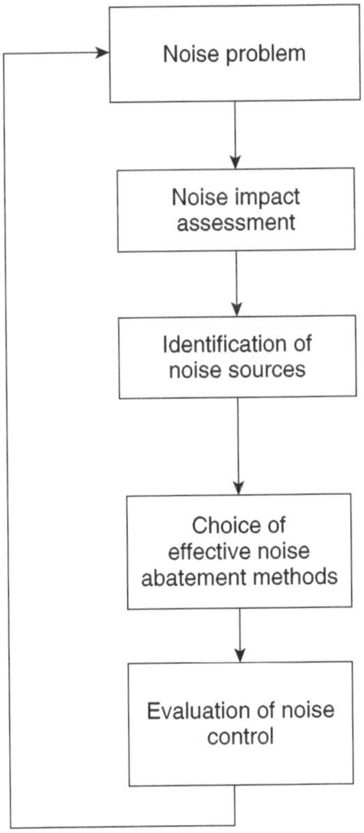


*Figure 1.14* An algorithm for noise management.

*Table 1.3* A comparison of acoustic efficiency coefficient ($\eta_a$) values

| *Source type* | *Coefficient* $\eta_a$ |
|---|---|
| Human voice | $5 \times 10^{-4}$ |
| Noise of jet aircraft engine | $5 \times 10^{-4} M^5$ for $M \leq 0.7$ |
| | $10^{-4} M^5$ for $0.7 \leq M \leq 1.6$ |
| | $2 \times 10^{-3}$ for $M \geq 2$ |
| Separated flow in regulator of airborne air-conditioning system | $10^{-3}$ for $M \leq 1.3$ |
| Siren | 0.5 |

to the strength of the flow for particular sources. The flow Mach number $M$ is the ratio of typical flow velocity $V$ and ambient sound velocity $a_0$, $M = V/a_0$.

The transformation of kinetic energy of the flux into acoustic power can be described using three types of noise sources: the monopole (representing a volume source of gas mass changing in time), the dipole (representing two monopole sources at a small distance from one another in comparison with sound wave length and pulsating in opposite phase) and the quadrupole (representing the superposition of four equal monopole sources in phase opposition to each other in pairs and at small distances from one another in comparison with sound wavelength). The acoustic efficiency diminishes from monopole to dipole and then to quadrupole.

In turbulent flow, a typical eddy length scale $L$ is used. For a sound wave, a typical scale is the wavelength, $\lambda$. If the source distribution for subsonic flows ($M = V/a_0 < 1$) is assumed to be compact and proportional to $V/L$, then the wavelength is given by $\lambda = LM^{-1}$. If M $<< 1$, the wavelength is larger than the scale of the turbulent flow: $\lambda >> L$. The noise of turbulent flow has a multipole nature. Table 1.4 gives the parameters of density fluctuation $\rho'(x, t)$(relative to ambient density) and acoustic power $W(W = 4\pi |x|^2 a_0^3 \rho_0^{-1} < \rho'^2 >)$ for monopole, dipole and quadrupole in compact and non-compact sources. The symbol $<>$ indicates the mean square average of the density fluctuation.

The effective transformation mechanical energy of flow into acoustical energy for compact sound sources of monopole, dipole and quadrupole nature are proportional to $M, M^3$ and $M^5$, respectively. The decrease of the efficiency with increase in multipole order ($M < 1$) is the result of partial suppression of radiation sources, located at a small distance (in comparison with $\lambda$) from one another. With increasing Mach number of flow (for example, in the case of supersonic flow), sound sources become non-compact. For these non-compact sound sources, the radiation of separate sources is prevalent, and the dependence on the multipole structure of acoustical sources is insignificant.

*Table 1.4* The characteristics of compact and non-compact acoustic radiators

| *Acoustic radiator* | *Compact sources of sound* | | *Non-compact sources of sound* | |
|---|---|---|---|---|
| | $\rho'(x,t)$ | W | $\rho'(x,t)$ | W |
| Monopole | $\rho_0 \frac{L}{\lvert x\rvert} M^2$ | $\rho_0 V^2 L^2 M$ | | |
| Dipole | $\rho_0 \frac{L}{\lvert x\rvert} M^3$ | $\rho_0 V^2 L^2 M^3$ | $\rho_0 \frac{L}{\lvert x\rvert} M$ | $\rho_0 V^3 L^2 M^{-1}$ |
| Quadrupole | $\rho_0 \frac{L}{\lvert x\rvert} M^4$ | $\rho_0 V^3 L^2 M^5$ | | |

The analysis of acoustic sources given above is based on qualitative investigations of the radiation. Only solutions of the basic continuum equations will allow descriptive relationships between the parameters of noise radiation and turbulent flow to be obtained.

## 1.3 Basic equations

The propagation of acoustic waves in a medium depends on its properties. If the airflow is homogeneous and in thermodynamic equilibrium, the airflow and sound field are described by differential equations, which are based on conservation of flow mass, momentum and energy.

$$\begin{aligned}
&\frac{\partial \rho}{\partial t} + \frac{\partial \rho v_j}{\partial x_j} = 0 \\
&\frac{\partial \rho v_i}{\partial t} + \frac{\partial \rho v_i v_j}{\partial x_j} = -\delta_{ij}\frac{\partial p}{\partial x_j} + \frac{\partial \tau_{ij}}{\partial x_j} \\
&\rho\left(\frac{\partial s}{\partial t} + v_j \frac{\partial s}{\partial x_j}\right) = U,
\end{aligned} \tag{1.1}$$

where $x_i$ are Cartesian coordinates, $p$ is pressure, $\rho$ is density, $v_i$ are the velocity vector components, $U = \frac{\partial}{\partial x_j}\left(\frac{Q_j}{T}\right) + \frac{\rho q_0}{T} + \sigma$, $T$ is the temperature, $Q_j = \chi \frac{\partial T}{\partial x_j}$, $\chi$ is the thermal conductivity, $\tau_{ij} = \mu\left(\frac{\partial v_i}{\partial x_j} + \frac{\partial v_j}{\partial x_i} - \frac{2}{3}\delta_{ij}\frac{\partial v_l}{\partial x_l}\right) + \varsigma_B \delta_{ij}\frac{\partial v_l}{\partial x_l}$ is the viscous stress tensor, $\sigma = T^{-2}\chi\delta_{ij}\frac{\partial T}{\partial x_i}\frac{\partial T}{\partial x_j} + \frac{\mu}{2T}\left(\frac{\partial v_i}{\partial x_j} + \frac{\partial v_j}{\partial x_i} - \frac{2}{3}\delta_{ij}\frac{\partial v_l}{\partial x_l}\right)^2 + \varsigma_B T^{-1}\frac{\partial v_i}{\partial x_i}\frac{\partial v_j}{\partial x_j}$, $\mu, \varsigma_B$ are the coefficients of dynamic and bulk viscosity, respectively, $s$ is the specific entropy, $i, j, l = 1, 2, 3$, $\delta_{ij} = 0$ if $i \neq j$, $\delta_{ij} = 1$, if $i = j$, $q_0$ is amount of heat. Repeated indices imply summation.

In general, the entropy change for finite volume of gas is described by

$$\frac{dS}{dt} = \frac{d_eS}{dt} + \frac{d_iS}{dt}. \tag{1.2}$$

The first item in (1.2) is an entropy flux which determines the entropy change due to interaction with ambient medium (this change can have any sign). The second component of the equation (1.2) represents the production of the entropy and determines the entropy flux for irreversible processes $\left(\frac{d_iS}{dt} \geq 0\right)$. Making use of some continuous function $F(s)$, one can rewrite the third equation of the system (1.1) in the form

$$\rho\left(\frac{\partial F}{\partial t} + v_j\frac{\partial F}{\partial x_j}\right) = F_s U, \tag{1.3}$$

where $F_s = dF/ds$. After multiplying the first equation of system (1.1) by $Fv_i$, the second by $F$ and equation (1.3) by $v_i$, and summing the results, the following result is obtained:

$$\frac{\partial(\rho F v_i)}{\partial t} + \frac{\partial(\rho F v_i v_j)}{\partial x_j} = -\delta_{ij}F\frac{\partial p}{\partial xi} + F\frac{\partial \tau_{ij}}{\partial x_j} + F_s v_i U. \tag{1.4}$$

From the first equation of the system (1.1) and equation (1.3)

$$\frac{\partial(\rho F)}{\partial t} + \frac{\partial(\rho F v_j)}{\partial t} = F_s U. \tag{1.5}$$

After transformations of the equations (1.4) and (1.5) and adding the expression

$$-\frac{\partial}{\partial x_i}a^2\frac{\partial(\rho F)}{\partial x_i}$$

to both parts of equation, it is found that

$$\frac{\partial^2 \rho F}{\partial t^2} - \frac{\partial}{\partial x_i}a^2\frac{\partial \rho F}{\partial x_i} = \frac{\partial^2 \rho F v_i v_j}{\partial x_i \partial x_j} + \Phi, \tag{1.6}$$

where $a^2 = (\partial p/\partial \rho)_s$, $a$ is the speed of sound and

$$\Phi = \frac{\partial UF}{\partial t} - \frac{\partial}{\partial x_i}\left(F\frac{\partial \tau_{ij}}{\partial x_j} + v_i F_s U\right).$$

The parameters of the gas are connected by the equation of state

$$\rho F(s)A(p) = constant \tag{1.7}$$

For an ideal gas ($\frac{d_e S}{dt} = 0$), $F(s) = \exp(s/c_p)$ and $A(p) = p^{-\frac{1}{\gamma}}$, therefore equation (1.6) can be written as

$$\frac{\partial^2 p^{\frac{1}{\gamma}}}{\partial t^2} - \frac{\partial}{\partial x_i} a^2 \frac{\partial p^{\frac{1}{\gamma}}}{\partial x_i} = \frac{\partial^2 (p^{\frac{1}{\gamma}} v_i v_j)}{\partial x_i \partial x_j} + \Phi, \tag{1.8}$$

where $\gamma$ is the specific heat ratio and $c_p$ is the specific heat of the gas at constant pressure. Equation (1.8) has the form of a wave equation. The terms on the right-hand side are determined by the aerodynamic noise sources connected with speed, entropy and viscous stress. Equation (1.8) is an exact consequence of conservation of mass, momentum and energy of flow, since it is derived from equations (1.1). It is necessary to make supplementary hypotheses for practical application of equation (1.8).

Suppose that the entropy per unit of mass of any given flow particle remains constant, then equation (1.7) yields

$$\frac{p}{\rho^\gamma} = constant \quad \text{or} \quad \frac{p^{\frac{1}{\gamma}}}{\rho} = constant$$

In this case, equation (1.8) is transformed to

$$\frac{\partial^2 \rho}{\partial t^2} - \frac{\partial}{\partial x_i} a^2 \frac{\partial \rho}{\partial x_i} = \frac{\partial^2 (\rho v_i v_j)}{\partial x_i \partial x_j} + \Phi. \tag{1.9}$$

At flow with high Reynolds number the viscous contribution terms in equation (1.9) can be neglected, and if there are no heat transfer effects, then

$$\frac{\partial^2 \rho}{\partial t^2} - \frac{\partial}{\partial x_i} a^2 \frac{\partial \rho}{\partial x_i} = \frac{\partial^2 (\rho v_i v_j)}{\partial x_i \partial x_j} + \frac{\partial}{\partial x_i}(a^2 - a_0^2)\frac{\partial \rho}{\partial x_i}. \tag{1.10}$$

In orthogonal curvilinear coordinates $q_i$, equation (1.10) (for which the viscous contribution was neglected) becomes

$$\frac{\partial^2 \rho'}{\partial t^2} - \frac{a_0^2}{h_1 h_2 h_3}\frac{\partial}{\partial q_i}\frac{h_1 h_2 h_3}{h_i^2}\frac{\partial \rho'}{\partial q_i} = \frac{1}{h_1 h_2 h_3}\frac{\partial}{\partial q_i}\frac{1}{h_i}\frac{\partial}{\partial q_j}\frac{(\rho v_i v_j h_1 h_2 h_3)}{h_j} + \frac{1}{h_1 h_2 h_3}\frac{\partial}{\partial q_i}(a^2 - a_0^2)\frac{h_1 h_2 h_3}{h_i^2}\frac{\partial \rho'}{\partial q_i}, \tag{1.11}$$

where $h_1, h_2, h_3$ are Lame's coefficients.

Sound is a consequence of fluctuations of the variables that describe the flow with typical wavelength $\lambda$ and time scale $T = 1/f$ at the oscillation frequency $f$. The total values of variables are the sum of the variable values for the ambient medium and their fluctuations. The fluctuations are represented by primes on the symbols: $v' = v - V_0$ (velocity), $\rho' = \rho - \rho_0$ (density), $p' = p - p_0$ (pressure), $a'^2 = a^2 - a_0^2$ (square of the adiabatic sound speed – $s = const$). The perturbation terms due to a sound wave are small ($\varepsilon << 1$):

$$\frac{v'}{< V_0 >} = O(\varepsilon)$$

$$\frac{\rho'}{< \rho_0 >} = O(\varepsilon)$$

$$\frac{p'}{< p_0 >} = O(\varepsilon). \quad (1.12)$$

If it is assumed that there are no heat transfer effects in the flow and that entropy is homogeneous throughout the ambient medium, then from equation (1.9) one obtains Lighthill's equation for sound generated by free turbulence.[6,7]

$$\frac{\partial^2 \rho'}{\partial t^2} - a_0^2 \frac{\partial^2 \rho'}{\partial x_i^2} = A(\vec{x}, t) = \frac{\partial^2 T_{ij}}{\partial x_i \partial x_j}, \quad (1.13)$$

where $T_{ij} = \rho v_i v_j - \tau_{ij} + p\delta_{ij} - a_0^2 \rho \delta_{ij}$.

Equation (1.13) can be rewritten as an integro-differential equation for the fluctuation in density:

$$\rho' = \frac{1}{4\pi a_0^2} \int_{V(\vec{y})} A(\vec{y}, \tau) \frac{dV(\vec{y})}{r}, \quad (1.14)$$

where $V(\vec{y})$ is the domain of turbulent flow, $r = |\vec{x} - \vec{y}|$, $\vec{y}, \vec{x}$ are coordinates, respectively, of the sources in domain of turbulent flow and the observation point, $\tau = t - \frac{r}{a_0}$ is a retarded time.

If the velocity due to the noise sources is denoted by $\vec{V}$, use of the new variable $\vec{\eta} = \vec{y} - a_0 \vec{M} \tau$, enables the solution of Lighthill's equation (1.13) to be written

$$\rho'(\vec{x}, t) = \frac{1}{4\pi a_0^2} \iiint \frac{A(\vec{\eta}, \tau)}{|\vec{x} - \vec{y}| - \vec{M}(\vec{x} - \vec{y})} dV(\vec{\eta}), \quad (1.15)$$

where $\vec{M} = \frac{\vec{V}}{a_0}$, $\tau = t - \frac{|\vec{x} - \vec{y}|}{a_0}$. Suppose also that the function $A(\vec{y}, \tau)$ decreases sufficiently rapidly and the receiver is sufficiently far from source. In the far field, the density perturbation is approximately given by

$$\rho'(\vec{x},t) \approx \frac{1}{4\pi a_0^4 |\vec{x}|} \iiint \frac{\partial^2 T_0(\vec{y},\tau)}{\partial \tau^2} dV(\vec{y}), \tag{1.16}$$

where $T_0 = \frac{x_i x_j}{|\vec{x}|^2} T_{ij}$.

For a subsonic turbulent jet ($0.3 \leq M \leq 1$), the equation (1.16) yields an expression for sound power

$$W_j = \frac{K \rho_j^2 S_j V_j^8}{\rho_0 a_0^5}, \tag{1.17}$$

where $K \approx 10^{-5}$ is an empirical constant and $\rho_j$, $V_j$, $S_j$, are respectively the density, velocity and area of the jet nozzle. The ratio of the sound power to the mechanical power of the jet is given by $\frac{W_a}{W_j} = K\left(\frac{\rho_j}{\rho_0}\right)\left(\frac{V_j}{a_0}\right)^5$. For subsonic flow, only a small part of the mechanical power of the jet is transformed into acoustic energy. On the other hand, the turbulent structure of the jet produces a powerful sound.

Neglecting the viscous contribution and supposing $v_i = 0$, then equation (1.10) becomes a homogeneous wave equation

$$\frac{\partial^2 \rho'}{\partial t^2} - a_0^2 \Delta \rho' = 0, \tag{1.18}$$

where $\Delta$ is the Laplacian.

The acoustic equation is determined neglecting the second and higher-order terms in the non-linear equations of continuum mechanics and retaining only the first order terms. Taking into account (1.12), then, after neglecting the quadratic and higher-order terms in the expansion of the second equation in (1.1), one obtains

$$\rho_0 \frac{\partial v_i'}{\partial t} + \frac{\partial p'}{\partial x_i} = 0. \tag{1.19}$$

Integrating equation (1.19) over time yields

$$v_i' = -\frac{1}{\rho_0} \frac{\partial}{\partial x_i} \int p' dt. \tag{1.20}$$

So, the acoustical field is irrotational and can be described in terms of a velocity potential $\varphi$, given by

$$\varphi = -\frac{1}{\rho_0} \int p' dt. \tag{1.21}$$

The perturbations in pressure, density and velocity in the sound wave are

$$p' = -\rho_0 \frac{\partial \varphi}{\partial t}, \quad \rho' = \frac{p'}{a_0^2}, \quad v_i' = \frac{\partial \varphi}{\partial x_i}. \tag{1.22}$$

From equations (1.22) and (1.18), it follows that the perturbations of pressure, density and velocity potential satisfy homogeneous wave equations:

$$\begin{aligned} &\frac{\partial^2 \rho'}{\partial t^2} - a_0^2 \Delta \rho' = 0 \\ &\frac{\partial^2 p'}{\partial t^2} - a_0^2 \Delta p' = 0 \\ &\frac{\partial^2 \varphi}{\partial t^2} - a_0^2 \Delta \varphi = 0. \end{aligned} \tag{1.23}$$

If we consider harmonic time dependence $\exp(-i\omega t)$ for the pressure perturbation

$$p'(\vec{x}, t) = p(\vec{x}) \exp(-i\omega t), \tag{1.24}$$

where $\omega$ is the angular frequency ($\omega = 2\pi f$) and $p(\vec{x})$ is the amplitude of the complex sound pressure, then the acoustic equation in (1.23) reduces to the Helmholtz equation

$$\Delta p + k^2 p = 0, \tag{1.25}$$

where $k = \omega / a_0$ is wave number ($k = 2\pi / \lambda$), $\Delta = \frac{\partial^2}{\partial x^2} + \frac{\partial^2}{\partial y^2} + \frac{\partial^2}{\partial z^2}$ is the Laplacian, and $x = x_1$, $y = x_2$, $z = x_3$ are Cartesian coordinates.

The boundary conditions for the acoustic field equation are determined by the situation to be modelled. When modelling the radiation, reflection and diffraction sound in flow without viscosity and with thermal conduction at the surfaces, it is usual to specify:

(a) the normal component of acoustic velocity on surface $S$ (for harmonic waves)

$$V_S = \left( \frac{\partial \varphi}{\partial n} \right)_S = f_1(S), \tag{1.26}$$

where $n$ is the normal vector pointing out of the surface into the flow;

(b) the sound pressure on surface $S$ is (for harmonic waves)

$$(\varphi)_S = f_2(S); \tag{1.27}$$

(c) a mixed boundary condition on the surface (for example, for a velocity potential)

$$\frac{1}{a_0}\frac{\partial\varphi}{\partial t} - \frac{1}{\beta}\frac{\partial\varphi}{\partial z} = 0, \tag{1.28}$$

where $\beta$ is the normalized admittance of surface and $z$ is the coordinate pointing out to normal to surface into the flow.

For $f_1(S) = 0$, the boundary value problem (1.26) represents sound reflection on an absolutely hard surface. If $f_2(S) = 0$, (1.27) represents sound reflection on an absolutely soft surface. The relation (1.28) represents sound reflection from an impedance boundary. In reflection problems, usually the total acoustic field $\varphi_t = \varphi_i + \varphi$ is the sum of an incident field $\varphi_i$ and a scattered field $\varphi$. The remaining condition is Sommerfeld's radiation condition for outgoing waves. For a three-dimensional pressure perturbation, this is written as:

$$|rp| < C,$$

$$r\left(\frac{\partial p}{\partial r} + ikp\right) \to 0$$

uniformly with respect to direction as $r \to \infty$, where C is some finite constant.

For medium that is at rest, then equations (1.23) of linear acoustics yield the principle of superposition of acoustic waves. In a linear ambient medium, free waves propagate irrespective of other waves, and a sound field is a sum of separate free waves. For scalar variables (for example, pressure), the summation is algebraic. For vector variables (for example, velocities), the summation is vectorial.

Consider some domain $V$, enclosed by surface $S$. In terms of the velocity potential $\varphi$ and its normal derivative $\partial\varphi/\partial n$ on the surface, Kirchhoff's solution yields

$$\varphi(R) = \frac{1}{4\pi}\iint_S \left\{\frac{\partial\varphi}{\partial n}\frac{\exp(ikr)}{r} - \varphi\frac{\partial}{\partial n}\left[\frac{\exp(ikr)}{r}\right]\right\} dS, \tag{1.29}$$

where $R$ is the radial vector of the observer, $r$ is the radial vector between the observer point and radiation point in domain $V$ and $n$ is the vector normal pointing into the surface. The form of the solution (1.29) represents the sound field as the sum of spherical and dipole sources on the surface.

Sound radiation by a flat surface is given by Huygen's formulas. The first Huygen's formula gives the field over a perfectly hard surface as

$$\varphi(R) = \frac{1}{2\pi} \iint_S \frac{\partial \varphi}{\partial n} \frac{\exp(ikr)}{r} dS. \tag{1.30}$$

The acoustical field over a perfectly soft surface is determined by the second Huygen's formula as

$$\varphi(R) = -\frac{1}{2\pi} \iint_S \varphi \frac{\partial}{\partial n} \left[ \frac{\exp(ikr)}{r} \right] dS. \tag{1.31}$$

For the inhomogeneous Helmholtz equation (1.23), the solution for pressure can also be represented as the sum of an incident pressure and a secondary pressure

$$p_t = \frac{1}{(2\pi)^{\frac{3}{2}}} \int_{-\infty}^{\infty} \int_{-\infty}^{\infty} d\beta_x d\beta_y \int_{-\infty}^{\infty} \frac{\Gamma(\beta_x, \beta_y, \beta_z) \exp[-i(\beta_x x + \beta_y y + \beta_z z)]}{k^2 - \beta_x^2 - \beta_y^2 - \beta_z^2} d\beta_z$$

$$+ \frac{1}{2\pi} \int_{-\infty}^{\infty} \int_{-\infty}^{\infty} A(\alpha, \delta) \exp(-\gamma z - i\alpha x - i\delta y) d\alpha d\delta, \tag{1.32}$$

where $\gamma = \sqrt{\alpha^2 + \delta^2 - k^2}$, function $\Gamma(\beta_x, \beta_y, \beta_z)$ is a Fourier transformation of the multipole source term $\Gamma(x, y, z)$, $\alpha, \delta, \beta_x, \beta_y, \beta_z$ are complex variables, $k$ is the wave number and $A(\alpha, \beta)$ is an unknown function defined by the solution of a boundary value problem.

Many acoustical models have been developed following the classical work of Lighthill on 'sound-generated aerodynamically'.[6–15] Lighthill's theory provides the basic theory of free jet noise. The sound generated by free turbulence is given by equation (1.13). According to Lighthill's acoustic analogy, equation (1.13) describes the generation of sound waves by quadrupole sources. In Lighthill's acoustic analogy, the sound sources are in the domain of turbulent flow and embedded in a medium at rest (with density and sound velocity, respectively, $\rho_0, a_0$). If there are no heat transfer and viscosity effects, then Lighthill's stress tensor reduces to $T_{ij} \approx \rho_0 v_i v_j$ (neglecting also the fluctuations of density at source, i.e. $\rho \approx \rho_0$).

A circular turbulent jet may be subdivided into the initial mixing region (extending about four diameters from the jet exit), the intermediate downstream region and the main extensive mixing region (reaching to between 16 and 18 diameters from the jet exit). The initial region includes a mixing layer with ambient and potential core. The initial mixing region and the extensive mixing region have a self-preserving structure. In the

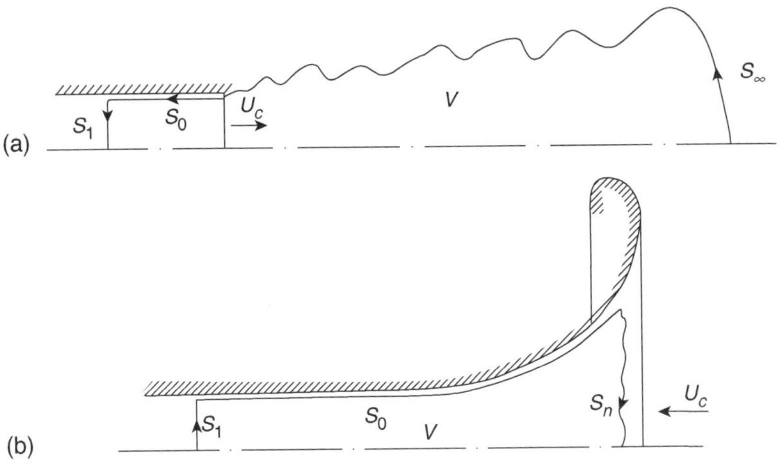


*Figure 1.15* Schematics of an air stream: (à) a free jet; (b) jet suction.

intermediate region, the turbulent structure transforms from the self-preserving structure of the initial mixing region to the new structure of the extensive mixing region. The end of the initial mixing region and the intermediate region generate the most acoustic power. The turbulent mixing region of a circular jet separates from the ambient irrotational flow, which is the inflow into the jet. The thickness of the separation zone is small in comparison with the typical turbulence scale. Therefore, the separation zone is considered as a geometrically random surface, distorted by the instability of the vortex sheet (see Fig. 1.15a).

At the separation surface, there is a jump in vorticity because outside the turbulent volume $V$, the flow is potential (the gas velocity of inflow into the jet over the separation is continuous).

We suppose that a subsonic jet contains compact noise sources. To calculate the parameters of turbulent flow, we introduce non-dimensional inner variables

$$\overline{x}_i = \frac{x_i}{L}, \overline{v}_i = \frac{v_i}{V_j}, \overline{\tau} = \frac{tV_j}{L}, \overline{a} = \frac{a}{a_0},$$

where $L$, $V_j$ define jet turbulence length and velocity scales, respectively. Equation (1.10) can be rewritten in non-dimensional inner variables

$$M^2 \frac{\partial^2 h}{\partial \overline{\tau}^2} - \Delta h = M^2 \frac{\partial^2 h \overline{v}_i \overline{v}_j}{\partial \overline{x}_i \partial \overline{x}_j} + \frac{\partial}{\partial \overline{x}_i} (\overline{a}^2 - 1) \frac{\partial h}{\partial \overline{x}_i}, \tag{1.33}$$

where $h = \rho/\rho_0$. The asymptotic expansion of the inner solution is

$$h(\overline{x}, \overline{\tau}, M) = 1 + \delta_1(M)\sigma_1(\overline{x}, \overline{\tau}) + \delta_2(M)\sigma_2(\overline{x}, \overline{\tau}) + \ldots,$$

$$\overline{a}^2 = 1 + \delta_1(M)\overline{a}_1^2(\overline{x}, \overline{\tau}) + \ldots,$$

$$\overline{v}_i = \overline{v}_i^{(0)}(\overline{x}, \overline{\tau}) + \delta_1(M)\overline{v}_i^{(1)} \ldots,$$

where for $M \to 0$, the variables $\overline{x}, \overline{\tau}$ are fixed. The solution for $\sigma_i(\overline{x}, \overline{\tau})$ satisfies the equations

$$\delta_1(M) = M^2;\ \Delta\sigma_1 = -\frac{\partial^2 \overline{v}_i^{(0)} \overline{v}_j^{(0)}}{\partial \overline{x}_i \partial \overline{x}_j}; \tag{1.34}$$

$$\delta_2(M) = M^4;$$

$$\Delta\sigma_2 = \frac{\partial^2 \sigma_1}{\partial \overline{\tau}^2} - \frac{\partial^2}{\partial \overline{x}_i \partial \overline{x}_j}\left(\overline{v}_i^{(1)}\overline{v}_j^{(0)} + \overline{v}_i^{(0)}\overline{v}_j^{(1)} + \sigma_1 \overline{v}_i^{(0)}\overline{v}_j^{(0)}\right) - 2\frac{\partial}{\partial \overline{x}_i}\overline{a}_1^2 \frac{\partial \sigma_1}{\partial \overline{x}_i}.$$

The solution of the first equation (1.34) is

$$\sigma_1(\overline{x}, \overline{\tau}) = \frac{1}{4\pi}\int_V \frac{\partial^2 \overline{v}_i^{(0)} \overline{v}_j^{(0)}}{\partial \xi_i \partial \xi_j} \frac{d\xi}{|\overline{x} - \xi|}$$
$$+ \frac{1}{4\pi}\int_S \left[\frac{1}{|\overline{x} - \xi|}\frac{\partial \sigma_1(\xi)}{\partial n} - \sigma_1(\xi)\frac{\partial}{\partial n}\frac{1}{|\overline{x} - \xi|}\right] dS, \tag{1.35}$$

where integration has been performed on the volume of turbulent flow $V(\xi)$, and over bounding surfaces $S = S_1 + S_0 + S_\infty$ (Fig. 1.15a): $S_1$ is the surface on the nozzle to a distance on the order of typical sound wavelength, $S_0$ is determined by the jet nozzle surface, $S_\infty$ is the part of the surface sufficiently far from jet exit and separation surface.

The integral along the surface $S$ in equation (1.35) relates to noise sources along surfaces $S_0, S_1$. Some noise sources exist outside the separation surface in the surrounding non-turbulent ambient medium. The solution of the first equation (1.34) is determined by integrating over $V(\xi)$. Using the result from differentiation of $f(\xi)$,

$$\frac{1}{|\overline{x} - \xi|}\frac{\partial^2 f}{\partial \xi_i \partial \xi_j} = \frac{\partial^2}{\partial \overline{x}_i \partial \overline{x}_j}\frac{f}{|\overline{x} - \xi|} + \frac{\partial^2}{\partial \overline{x}_i \partial \xi_j}\frac{f}{|\overline{x} - \xi|} + \frac{\partial}{\partial \xi_i}\frac{\frac{\partial f}{\partial \xi_j}}{|\overline{x} - \xi|},$$

in equation (1.35),

$$\sigma_1(\overline{x},\overline{\tau}) = \frac{1}{4\pi}\left\{\frac{\partial^2}{\partial\overline{x}_i\partial\overline{x}_j}\int\limits_V \frac{\overline{v}_i^{(0)}\overline{v}_j^{(0)}}{|\overline{x}-\xi|}d\xi + \frac{\partial}{\partial\overline{x}_i}\int\limits_V \frac{\partial}{\partial\xi_j}\frac{\overline{v}_i^{(0)}\overline{v}_j^{(0)}}{|\overline{x}-\xi|}d\xi\right.$$

$$\left. + \int\limits_V \frac{\partial}{\partial\xi_j}\frac{\frac{\partial\overline{v}_i^{(0)}\overline{v}_j^{(0)}}{\partial\xi_j}}{|\overline{x}-\xi|}d\xi + \int\limits_S\left[\frac{1}{|\overline{x}-\xi|}\frac{\partial}{\partial n} - \sigma_1(\xi)\frac{\partial}{\partial n}\frac{1}{|\overline{x}-\xi|}\right]dS.\right\} \tag{1.36}$$

After some transformations in equation (1.36) one obtains

$$\sigma_1(\overline{x},\overline{\tau}) = \frac{1}{4\pi}\left\{\frac{\partial^2}{\partial\overline{x}_i\partial\overline{x}_j}\int\limits_V \frac{\overline{v}_i^{(0)}\overline{v}_j^{(0)}}{|\overline{x}-\xi|}d\xi + \frac{\partial}{\partial\overline{x}_i}\int\limits_S l_j\frac{\overline{v}_i^{(0)}\overline{v}_j^{(0)}+\sigma_1(\xi)\delta_{ij}}{|\overline{x}-\xi|}dS\right. \tag{1.37}$$

$$\left. + \int\limits_S l_i\frac{1}{|\overline{x}-\xi|}\frac{\partial}{\partial\xi_j}\left[\overline{v}_i^{(0)}\overline{v}_j^{(0)}+\sigma_1(\xi)\delta_{ij}\right]dS\right\}.$$

The equation of conservation of momentum (1.1) in the approach considered has the form

$$\frac{\partial\overline{v}_i^{(0)}}{\partial\overline{\tau}} + \frac{\partial}{\partial\xi_j}[\overline{v}_i^{(0)}\overline{v}_j^{(0)} + \sigma_1(\xi)] = 0. \tag{1.38}$$

Therefore, after taking into consideration equation (1.35), equation (1.37) becomes

$$\sigma_1(\overline{x},\overline{\tau}) = \frac{1}{4\pi}\left[\frac{\partial^2}{\partial\overline{x}_i\partial\overline{x}_j}\int\limits_V \frac{\overline{v}_i^{(0)}\overline{v}_j^{(0)}}{|\overline{x}-\xi|}d\xi + \frac{\partial}{\partial\overline{x}_i}\int\limits_S l_j\frac{\overline{v}_i^{(0)}\overline{v}_j^{(0)}+\sigma_1(\xi)\delta_{ij}}{|\overline{x}-\xi|}dS\right.$$

$$\left. - \frac{\partial}{\partial\overline{\tau}}\int\limits_S l_i\frac{\overline{v}_i^{(0)}}{|\overline{x}-\xi|}dS\right]. \tag{1.39}$$

For a subsonic jet, ($M = \frac{V_j}{a_0} < 1$) and if the source distribution is assumed compact, then $L = M\lambda$ and we can introduce non-dimensional outer variables

$$\tilde{x}_i = \frac{x_i}{\lambda},\ \tilde{v}_i = \frac{v_i}{a_0},\ \tilde{\tau} = \overline{\tau} = \frac{tc_0}{\lambda},\ \tilde{a} = \frac{a}{a_0},\ M < 1.$$

Equation (1.10) can be rewritten in non-dimensional outer variables

$$\frac{\partial^2 h}{\partial \tilde{\tau}^2} - \Delta h = \frac{\partial^2 h \tilde{v}_i^{(0)} \tilde{v}_j^{(0)}}{\partial \tilde{x}_i \partial \tilde{x}_j} + \frac{\partial}{\partial \tilde{x}_i}(\tilde{a}^2 - 1)\frac{\partial h}{\partial \tilde{x}_i}. \tag{1.41}$$

The asymptotic expansion of the outer solution is

$$\begin{aligned} h(\tilde{x}, \tilde{\tau}, M) &= 1 + \Delta_1(M) h_1(\tilde{x}, \tilde{\tau}) + \Delta_2(M) h_2(\tilde{x}, \tilde{\tau}) + \ldots, \\ \tilde{v}_i(\tilde{x}, \tilde{\tau}, M) &= \Delta_1(M)\tilde{v}_i^{(1)} + \Delta_2(M)\tilde{v}_i^{(2)} + \ldots, \\ \tilde{a}^2(\tilde{x}, \tilde{\tau}, M) &= 1 + \Delta_1(M)\tilde{a}_1^2 + \ldots, \end{aligned} \tag{1.42}$$

where, for $M \to 0$, variables $\tilde{x}, \tilde{\tau}$ are fixed. The function $h_1(\overline{x}, \overline{\tau})$ satisfies the homogeneous wave equation

$$\frac{\partial^2 h_1}{\partial \tilde{\tau}^2} - \Delta h_1 = 0.$$

In general, the solution of homogeneous wave equation in far field ($|\tilde{x}| >> 1$) is a spherical symmetric wave spreading out from the source in the ambient medium

$$h_1(\tilde{x}, \tilde{\tau}) = \frac{1}{4\pi\,|\tilde{x}|} H(\tilde{\tau} - |\tilde{x}|), \tag{1.43}$$

where $H(\tilde{\tau} - |\tilde{x}|)$ is any twice differentiable function.

To carry out matching of inner and outer expansions, we rewrite the first term of the inner expansion (1.39) in outer variables ($\tilde{x} = M\overline{x}$)

$$\begin{aligned} M^2 \sigma_1(\tilde{x}, \tilde{\tau}) = \frac{M^2}{4\pi} \Bigg[ & M^2 \frac{\partial^2}{\partial \tilde{x}_i \partial \tilde{x}_j} \int_V \frac{\overline{v}_i^{(0)} \overline{v}_j^{(0)}}{|\tilde{x}M^{-1} - \xi|} d\xi \\ & + M \frac{\partial}{\partial \tilde{x}_i} \int_S l_j \frac{\overline{v}_i^{(0)} \overline{v}_j^{(0)}}{|\tilde{x}M^{-1} - \xi|} dS - \frac{\partial}{\partial \tilde{\tau}} \int_S \frac{l_i \overline{v}_i^{(0)}}{|\tilde{x}M^{-1} - \xi|} dS \Bigg]. \end{aligned} \tag{1.44}$$

For $M < 1$ equation (1.44) simplifies to

$$\begin{aligned} M^2 \sigma_1(\tilde{x}, \tilde{\tau}) = \frac{1}{4\pi} \Bigg[ & M^5 \frac{\partial^2}{\partial \tilde{x}_i \partial \tilde{x}_j} \frac{1}{|\tilde{x}|} \int_V \overline{v}_i^{(0)} \overline{v}_j^{(0)} d\xi - M^4 \frac{\partial}{\partial \tilde{x}_i} \frac{1}{|\tilde{x}|} \frac{\partial}{\partial \tilde{\tau}} \int_S \overline{v}_i^{(0)} dS \\ & - \frac{M^3}{|\tilde{x}|} \frac{\partial}{\partial \tilde{\tau}} \int_S l_i \overline{v}_i^{(0)} dS \Bigg]. \end{aligned} \tag{1.45}$$

Matching three terms of the inner expansion [orders $M^3$, $M^4$, $M^5$in equation (1.45)] with the outer solution [equation (1.43)] gives

$$\Delta_1(M)h_1(\tilde{x},\tilde{\tau}) = \frac{M^5}{4\pi}H^{(1)} + \frac{M^4}{4\pi}H^{(2)} + \frac{M^3}{4\pi}H^{(3)}, \tag{1.46}$$

where

$$H^{(1)} = \frac{\partial^2}{\partial\tilde{x}_i\partial\tilde{x}_j}\frac{1}{|\tilde{x}|}\int\limits_V [\overline{v}_i^{(0)}\overline{v}_j^{(0)}]_{\bullet}d\xi,$$

$$H^{(2)} = \frac{\partial}{\partial\tilde{x}_i}\frac{1}{|\tilde{x}|}\int\limits_S l_j[\overline{v}_i^{(0)}\overline{v}_j^{(0)}]_{\bullet}dS,$$

$$H^{(3)} = -\frac{1}{|\tilde{x}|}\frac{\partial}{\partial\tilde{\tau}}\int\limits_S l_i[\overline{v}_i^{(0)}]_{\bullet}dS,$$

The symbol '[ ]•' signifies retarded time. In dimensional variables, the solution is

$$\frac{\rho-\rho_0}{\rho_0} = \frac{1}{4\pi a_0^2}\left\{\frac{\partial^2}{\partial x_i\partial x_j}\frac{1}{|x|}\int\limits_V \left[v_i^{(0)}v_j^{(0)}\right]_{\bullet}d\xi + \frac{\partial}{\partial x_i}\frac{1}{|x|}\int\limits_S l_j\left[v_i^{(0)}v_j^{(0)}\right]_{\bullet}dS\right.$$
$$\left. -\frac{1}{|x|}\frac{\partial}{\partial t}\int\limits_S l_i\left[v_i^{(0)}\right]_{\bullet}dS\right\} \tag{1.47}$$

In the far field, equation (1.47) simplifies to

$$\frac{\rho-\rho_0}{\rho_0} \approx \frac{1}{4\pi a_0^4}\frac{x_ix_j}{|x|^3}\int\limits_V \frac{\partial^2 v_i^{(0)}v_j^{(0)}}{\partial\tau^2}d\xi + \frac{1}{4\pi a_0^3}\frac{x_i}{|x|^2}\int\limits_S l_j\frac{\partial}{\partial\tau}\left(v_i^{(0)}v_j^{(0)} + \sigma_1\delta_{ij}\right)dS$$
$$-\frac{1}{4\pi a_0^2|x|}\int\limits_S l_i\frac{\partial v_i^{(0)}}{\partial\tau}dS. \tag{1.48}$$

In expression (1.48), the first term of the asymptotic expansion represents a quadrupole source term. The second and third terms represent the effects of interaction of the jet with the external medium including elements of the jet nozzle surface and elements of any jet noise suppression devices.

Composite expansions can be obtained by using standard methods of matching the inner and outer expansions.[16]

In accordance with equation (1.10) for an unbounded flow (neglecting the effects of heat conductivity, viscosity and solid boundaries on $S$), the noise of an isothermal jet source is quadrupole. If $\frac{\partial^2 v_i^{(0)} v_j^{(0)}}{\partial \overline{x}_i \partial \overline{x}_j} = 0$ from equation (1.11), it follows that aerodynamic acoustical sources in the turbulent domain $V(\xi)$ will be less effective radiators of sound than acoustic sources defined by the term $M^2\sigma_1$. The aerodynamic acoustical source is multipole, the effectiveness of which diminishes with source-order growth. Besides the trivial solution, $\overline{v}_i^{(0)} = 0$, there are several aerodynamic fields with the following properties in Cartesian coordinates $(x, y, z)$:

1 $v_x^{(0)}(t), v_y^{(0)}(x,t), v_z^{(0)}(x,y,t)$;
2 $v_x^{(0)}(t), v_y^{(0)}(x,z,t), v_z^{(0)}(x,t)$;
3 $v_x^{(0)}(y,z,t), v_y^{(0)}(t), v_z^{(0)}(x,y,t)$;
4 $v_x^{(0)}(y,t), v_y^{(0)}(t), v_z^{(0)}(x,y,t)$;
5 $v_x^{(0)}(z,t), v_y^{(0)}(x,z,t), v_z^{(0)}(t)$;
6 $v_x^{(0)}(y,z,t), v_y^{(0)}(z,t), v_z^{(0)}(t)$.

Cylindrical polar coordinates $(r, \varphi, z)$:

7 $v_r^{(0)} = 0, v_\varphi^{(0)}(r,t), v_z^{(0)}(r,\varphi,t)$;
8 $v_r^{(0)} = 0, v_\varphi^{(0)}(r,z,t), v_z^{(0)}(r,t)$.

Toroidal coordinates $\{x = r\sin(\theta),\ y = [l + r\cos(\theta)]\cos(\psi),\ z = [l + r\cos(\theta)]\sin(\psi)\}$:

9 $v_r^{(0)} = v_\theta^{(0)} = 0, v_\psi^{(0)}(r,\theta,t)$.

Paraboloidal coordinates $\{x = sp\cos(\varphi), y = sp\sin(\varphi), z = \frac{1}{2}(s^2 - p^2)\}$:

10 $v_s^{(0)} = v_p^{(0)} = 0, v_\varphi^{(0)}(s,p,t)$.

Prolate spheroidal coordinates: $x = \alpha\cos(s)sh(p)\cos(\varphi)$, $y = \alpha\cos(s)sh(p)\sin(\varphi)$, $z = \alpha\sin(s)ch(p)$.

11 $v_s^{(0)} = v_p^{(0)} = 0, v_\varphi^{(0)}(s,p,t)$.

Oblate spheroidal coordinates: $x = \alpha\sin(s)sh(p)\cos(\varphi)$, $y = \alpha\sin(s)ch(p)\sin(\varphi)$, $z = \alpha\cos(s)sh(p)$.

12 $v_s^{(0)} = v_p^{(0)} = 0, v_\varphi^{(0)}(s,p,t)$.

The flows forming such types of jet enable lowering of acoustic radiation of free jet by modification of the shape of the nozzle exit, for example, with corrugated nozzles and screw jets.

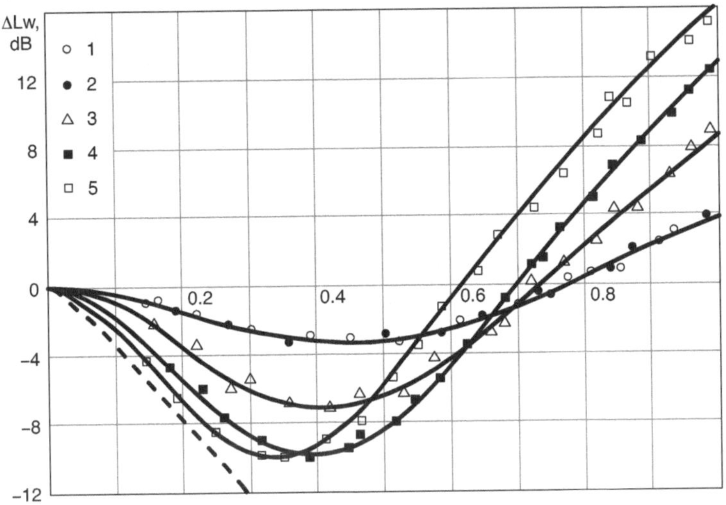


*Figure 1.16* Dependence of surplus noise output of a coaxial jet (in comparison with that of a single jet) on relative velocity $\overline{u}$. The coaxial jet is specified by the primary diameter $d_1$ and the secondary diameter $d_2$: 1 – $d_1$ = 30 mm, $d_2$ = 50 mm; 2 – $d_1$ = 18 mm, $d_2$ = 30 mm; 3 – $d_1$ = 18 mm, $d_2$ = 50 mm; 4 – $d_1$ = 12 mm, $d_2$ = 50 mm; 5 – $d_1$ = 8 mm, $d_2$ = 50 mm. The broken line represents $80\lg(1 - \overline{u})$.

Jet noise reduction techniques are based on reducing noise emission with minimal loss of jet thrust (less than between 3 and 5 per cent). Corrugated nozzles have been used on many civil aircraft flying between 1955 and 1980. Derivatives of such nozzles are used on modern aircraft.[12] Substantial noise reduction could be obtained with the coaxial airstream (used in bypass jets and in the turbofan engine).

The application of coaxial jets changes the flow structure of the jets, slows down the jet velocity and reduces the velocity gradient, which results in a decrease in jet noise. Figure 1.16 shows the dependence of the ratio of the noise power level of coaxial jets ($W_{cj}$) and the power of the jet ($W_1$) on the dimensionless parameters ($\overline{d} = d_2/d_1, \overline{u} = u_2/u_1$)

$$\Delta L_W = 10\lg\frac{W_{cj}}{W_1}.$$

The noise power $W_{cj}$ for $\overline{u}$=1 is given by equation (1.17) (the sources are quadrupoles). The acoustical radiation of the coaxial jets for $\overline{d} \geq 4$ and $\overline{u} \leq 0.3$ is determined by the interaction between the primary and secondary jet. The dashed line in Fig. 1.16 is determined by the formula $\Delta L_W = 80\lg(1 - \overline{u})$. Substantial noise reduction of 10 dB is achieved for $\overline{d} \approx 5$ and $\overline{u} = 0.3$ to 0.4.

*Table 1.5* Value of approximating coefficients $a_{ij}$

| $j$ | $i$ = 0 | 1 | 2 | 3 | 4 | 5 |
|---|---|---|---|---|---|---|
| 0 | 0.1667 | −13.416 | 246.8 | −673.0 | 648.213 | −214.3 |
| 1 | −0.1638 | 13.052 | −234.69 | 590.416 | −516.959 | 154.714 |
| 2 | 0.016 | −1.5114 | 19.774 | −39.586 | 24.837 | −3.995 |

The sound power level ($L_{Wcj}$) of cold coaxial jets can be defined as

$$L_{Wcj} = L_{W_1} + \Delta L_{Wcj},\ \Delta L_{Wcj} = \sum_{i=0}^{i=5}\sum_{j=0}^{j=2} a_{ij}\bar{u}^i\bar{d}^j,$$

where $L_{W_1}$ is the sound power level of the primary jet and $a_{ij}$ are coefficients (see Table 1.5); $0 \le \bar{u} \le 1$; $1.7 \le \bar{d} \le 6.3$.

To investigate the effect of a suction jet, we have used various nozzle shapes on a model in an anechoic chamber (Fig. 1.15b). The sound power level due to air nozzle suction can be determined from

$$W_s = K_s(1 - M_s)^2 u_s^6 S_1 \rho c^{-3},$$

where $u_s$, $M_s$ are respectively the velocity and Mach number in the most narrow cross-section of the nozzle, $(1 - M_s)^2$ is a factor taking into account the sound convection effect in the axial direction, $S_1$ is the area of the suction nozzle in its most narrow cross-section ($d_s$ is the diameter of the cross-section) and $K_s = 1.8958 \times 10^{-5}$. The sound power level spectra are determined from

$$L_W(f) = 10 \lg W_s + \Delta L_W(f) + 120,$$

where $\Delta L_W(f)$ is a spectrum amendment defined in Fig. 1.17 (the abscissa of which uses a parameter $Sh \cdot M = fd_s/c$, where $Sh$ is the Strouhal number).

The sound of supersonic jets is generated by different mechanisms: vortex–vortex interaction, interaction of the vortex with the shock wave and instability waves. The noise spectrum of a supersonic jet may include both broadband and discrete components. For an imperfectly expanded supersonic jet, the essential characteristic is noise emission from the shock cell structure. In this case screech tones are observed. For a perfectly expanded jet, the screech tone disappears and broadband noise is generated by turbulent mixing of air within the jet. At high Strouhal numbers, the broadband noise is due the interaction of disturbances in the jet with shock waves. In experiments on supersonic jets at low to moderate Reynolds numbers, the large coherent structures in the flows were found to influence the emitted sound.

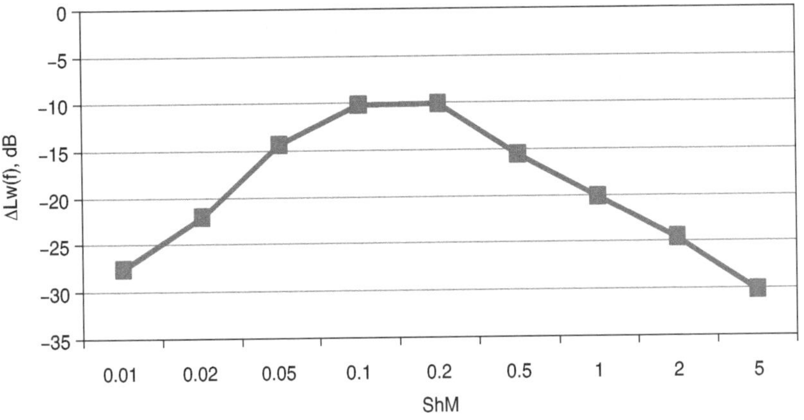


*Figure 1.17* Sound power level spectrum of jet suction.

For predicting propeller noise, propfan noise, rotor noise, turbomachinery noise and airframe noise, the Ffowcs-Williams and Hawkings equation for a moving surfaces can be used.[17]

$$\frac{1}{a^2}\frac{\partial^2 p}{\partial t^2} - \Delta p = \frac{\partial}{\partial t}[\rho_0 v_n \delta(f)] - \frac{\partial}{\partial x_i}[l_i \delta(f)] + \frac{\partial^2}{\partial x_i \partial x_j}[T_{ij} H(f)], \tag{1.49}$$

where $\delta(f)$ is the Dirac delta function, $f=0$ gives the equation of the moving surface, $l_i$ is the $i$th component of the surface force and $T_{ij}$ is Lighthill's stress tensor. The right side of equation (1.49) contains the following source terms: the first is the thickness source (monopole), the second is the loading source (dipole) and the third is the quadrupole source. The integral formulation equation (1.49) for subsonic motion can be represented as in Farassat's formula (neglecting the quadrupole contribution).[18]

$$\begin{aligned} 4\pi p = {} & \frac{1}{a}\int\limits_{f=0}\left[\frac{\dot{l}_i r_i}{r(1-M_r)^2}\right]_{\bullet} dS + \int\limits_{f=0}\left[\frac{l_r - l_i M_i}{r^2(1-M_r)^2}\right]_{\bullet} dS \\ & + \frac{1}{a}\int\limits_{f=0}\left[\frac{l_r(r\dot{M}_i r_i + aM_r - aM^2)}{r^2(1-M_r)^3}\right]_{\bullet} dS \\ & + \int\limits_{f=0}\left[\frac{\rho_0 v_n(r\dot{M}_i r_i + aM_r - aM^2)}{r^2(1-M_r)^3}\right]_{\bullet} dS, \end{aligned} \tag{1.50}$$

where dots on the Mach number $M_i = \frac{v_i}{a}$ and $l_i$ denote derivatives with respect to source time, $v_i$ is the local velocity of moving surface and $M_r$ is the Mach number of the source in the direction of the observer.

The propeller is the main noise source on a turboprop engine. Propeller noise arises as a result of the periodic displacement of the air by the volume of a passing blade (thickness noise). The pressure fluctuation due to lift and draft disturbance gives the loading noise of the propeller. The propeller noise spectrum contains broadband and harmonic noise. The harmonic noise has frequencies given by $f_k = nzk$, where $k = 1, 2, \ldots, n$ is rotational speed and $z$ is the number of blades. If the propeller has a small number of blades, then for subsonic blade section speeds, the noise is determined in the main by the first two harmonics. For such a propeller, the level of broadband noise is 10 dB less than that of the first harmonic. Because the propeller blade tip speed is one of the important parameters, reducing the tip speed results in lowering of the noise. Increasing the number of blades also reduces propeller noise for small blade rotation speeds (less than 240 m/s). On a propeller, a small thickness of blades reduces thickness noise, a large diameter with many blades diminishes loading noise and a large blade sweep reduces quadrupole noise. Increasing the propeller diameter for a given thrust combined with a relatively small tip speed reduces loading noise. On an aircraft with few propellers, noise levels in the aircraft cabin can be decreased by anti-phasing the rotation of the propellers on opposite sides of the fuselage (so-called synchrophasing). In this case, use is being made of noise cancellation through the opposite phases of waves.

On a helicopter, the main noise sources are aerodynamic and mechanical (see Fig. 1.18). The aerodynamic noise sources include the main rotor, tail rotor and the gas turbine of the engine. The mechanical noise sources include the gearbox and transmission. The mechanical noise sources radiate high-frequency sound, which is more rapidly attenuated by the atmosphere. The main rotor is a source of impulsive noise.

The acoustic spectrum of a helicopter (for example, with a single main rotor) includes discrete and broadband noise. The main rotor and tail rotor form a discrete frequency rotational noise spectrum. Broadband noise originates from the interaction of the blades with atmosphere turbulence

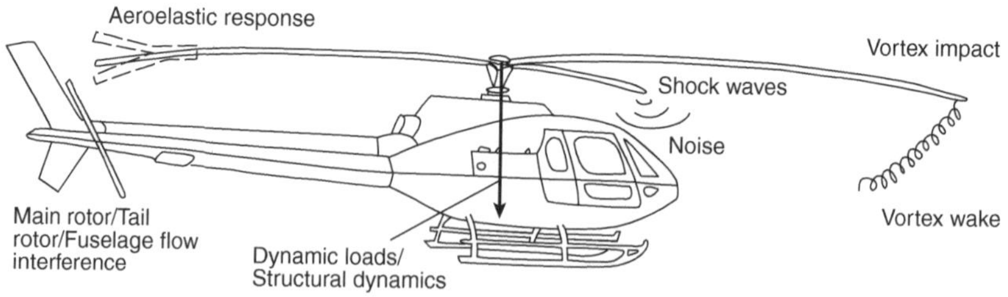


*Figure 1.18* Helicopter noise sources.

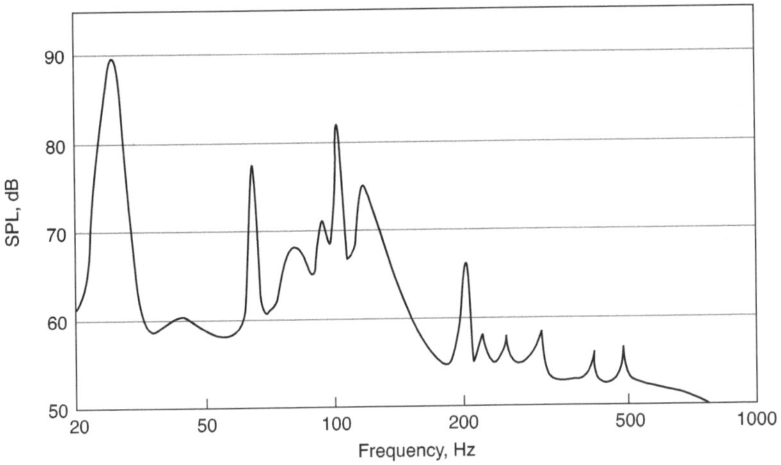


*Figure 1.19* Typical form of helicopter noise spectrum.

and the vortex following the blade itself. The typical spectrum of helicopter noise is shown schematically in Fig. 1.19.

The main parameter determining helicopter rotor noise is the blade tip speed. To reduce noise emission, tip speeds can be reduced. In helicopters with a single main rotor, a reduction in the intensity of impulsive noise at subsonic speeds of the main rotor can be achieved by adding special blade tips (thinned, tapered or swept). Reduction in the noise of the tail rotor for a single main rotor can be achieved by increasing the number of blades, tapering the tip of the rotor and placing the tail rotor in an annulus.

## 1.4 Criteria and methods of aircraft noise assessment

To evaluate the effectiveness of aircraft noise abatement during airport operations, several different criteria may be employed:[19–24]

- the number of people within specified noise contours;
- the physical extent of specified noise contours;
- the number of noise complaints received;
- the amount of time communities are exposed to noise above a predetermined level;
- the number of noise events above a predetermined level to which communities are exposed.

Minimization of aircraft noise impact involves a systems approach and assessment of various technical, ecological, economic and social factors. The 'minimum noise impact' vector of criteria is large but the determination of its structure is an important task. For example, certain optimization criteria determine the parameters for low noise flight procedures.

The effects of aircraft noise include speech and sleep interference, and physiological, psychological and social problems. Annoyance from aircraft noise depends on both acoustical and non-acoustical factors. The latter include fear of aircraft crashing, potential benefits, housing availability, possibilities for compensation, sensitivity to noise and so on.

For aircraft noise impact assessment, in addition to choosing the relevant aircraft noise index, it is necessary to consider aircraft noise source descriptions, the flight procedures, details of population affected by noise and the airport staff and passengers' reaction to aircraft noise. Important roles are played by the requirements of health norms, flight safety, economic efficiency of aircraft operation and maintenance.

Aircraft noise annoyance is determined by several factors including the intensity, frequency content, duration and repetition, individual noise sensitivity and time of the day. The various criteria for aircraft noise assessment differ in the extent to which they account for these factors, and also in their mathematical structure. At present several criteria are used in the assessment of aircraft noise. Table 1.6 lists the most frequently used aircraft noise indices.

Models and methods used for assessing environmental noise problems must be based on the noise exposure ratings used by the relevant national and international noise control regulations and standards. These vary greatly one from one another both in their structure, and in the basic approaches used for their definitions (see Table 1.7).

According to ICAO, the effective perceived noise level (EPNL) should be used for evaluating the acoustical performance of aircraft. On the other hand, it is well known that $L_{Aeq}$ correlates well with the effects of noise on any kind of human activity. Moreover, the percentage of highly noise-annoyed people $p$ living in zones of significant aircraft noise impact is one of the best descriptors of total noise impact. This criterion is used as the regulatory basis in some countries and is based on the well-known relationship by Shultz.[5] The latter approach is convenient for the efficient comparison of alternative approaches to the noise problem in particular cases. For example, the value of the unit cost of noise protection for a population (UCNPP) rises with growth in noise levels in the protection zone.

The total cost for any probable type of noise protection $CNP$ is given by:

$$CNP = \sum_k UCNPP_k(L_{Aeq})p(L_{Aeq})P_k S_k, \tag{1.51}$$

where $P_k$ is the number of people living in the $k$th zone of noise control with area $S_k$ and the boundaries of the zones are defined by values of the noise control criterion, for example, by values of $L_{Aeq}$.

*Table 1.6* Aircraft noise criteria

| *Criterion symbol* | *Name* | *Basic unit* | *Tone corrections* | *Duration factor* | *Event repetition factor* | *Duration and/or time periods* |
|---|---|---|---|---|---|---|
| $L_A$ | A-frequency weighted level | dBA | | | | |
| *PNL* | Perceived noise level | PNdB | | | | |
| *PNLT* | Tone corrected perceived noise level | TPNdB | • | | | |
| *EPNL* | Effective perceived noise level | EPNdB | • | • | | Aircraft pass by |
| $L_{Aeq,24h}$ | A-weighted average sound level | dBA | | • | $10\lg N$ | 24 hours |
| *DNL* | Day–night sound level | dBA | | • | $10\lg N$ | 2 periods |
| *NEF* | Noise exposure forecast | PNdB | • | • | $10\lg N$ | 2 periods |
| *NNI* | Noise and number index | PNdB | | | $15\lg N$ | Depends on aim |
| *WECPNL* | Weighted equivalent continuous perceived noise level | EPNdB | • | • | $10\lg N$ | 2 periods |
| *CNR* | Composite noise rating | PNdB | | | $10\lg N$ | 2 periods |
| *CNEL* | Community noise equivalent level | dBA | | • | $10\lg N$ | 3 periods |
| *B* | Noise load in Kosten units | dBA | | | $20\lg N$ | Aircraft pass by |
| *DENL* | European reporting data noise exposure | dB | | • | $10\lg N$ | 3 periods |
| $L_{eqnight}$ | Overall night-time noise exposure | dB | | • | $10\lg N$ | 8- hour night-time |

*Table 1.7* Noise ratings used for aircraft noise impact assessment

| *Aircraft noise index* | *Loudness scale approach A-weighted noise descriptors* | *Nuisance approach perceived noise descriptors* |
|---|---|---|
| Maximum value of time-varying levels | $L_{Amax}$ | $PNL_{max}$ (PNLTM) |
| Effective levels | LAX, SEL | EPNL |
| Equivalent levels | $L_{Aeq}$ | ECPNL |
| Time-of-day weighted levels | DNL | WECPNL |
| Noise exposure indices | TNI, NII | NEF, NNI |
| Number or percentage of population annoyed by noise | $p$[%]-relationship by Shultz | $\pi$-function of ICAO |

In addition, the area of noise contour $S$ defined by a noise level of particular significance (for example, that specified by the national regulations) is given by

$$S = \sum_k S_k, \; k = 1, N$$

EPNL values of 100 or 90 EPNdB are used for noise contour analysis in current research and they are correlated with $L_{Aeq}$ values of 75 and 65 dBA, respectively. The correlation between the values of noise contour areas $S$ and the values of EPNL at the noise monitoring points are also good.

CAEP/5 suggests use of Day–Night Sound Level to forecast the population annoyed in the area with the prescribed value of noise level $L_{dn}$. Two thresholds are specified: 'significant' exposure is defined by $L_{dn} = 55$ dB or higher and 'high' exposure is defined by $L_{dn} = 65$ dB or higher.

Harmonization of noise indicators (*NI*) is an essential component of European strategy to reduce noise. The Commission of the European Communities has suggested use of the following criteria for *NI*:[1,5]

- validity (relationship with effects);
- practical applicability (easy to calculate using available data, or to measure using available equipment);
- transparency (easy to explain, the relationship with physical units);
- enforceability (use of indicator in assessing changes or when set limits are exceeded);
- consistency.

Since the purpose of *NI* is to reduce a large volume of information to a single noise metric value, information about the individual contributions of noise sources can be lost. Five steps are considered when designing a method for estimating reaction to noise. The first step is to reduce the frequency spectrum to a single number (using A-, B-, C-, D-weightings, PNdB or Zwicker/Stevens phons). The second step produces a single value per event by means of energetic summation (for example, $L_{AX}$) or the maximum level per event ($L_{A\max}$). In the third step, the number of events per day (day, evening, night) is incorporated by means of energetic summation over the period $T$ of interest (for example, $L_{Aeq,T}$). The fourth step summation with a specified weighting factor for different periods. In the simplest form of the fourth step, the day, evening or night periods are summed to give a 24-hour value. In some cases either evening/night corrections are used. Finally, as the fifth step, a long-term value is obtained by means of energetic summation and averaging.

The following types of indices or rating schemes result:[23]

- noise exposure levels, which take into account human noise response (for example, overall levels in A-, B-, C-, D-weighted decibel, *PNL*);
- effective noise levels, which take into account noise exposure levels and the duration of the event (for example, EPNL);
- noise indices, which take into account the variation of noise levels with time and the number of flights per specified time period (for example, $L_{Aeq}$, $L_{DN}$, *NEF*, *NNI*, *N*, *CNEL*, *WECPNL*);
- noise criteria, which indicate simultaneously the change of noise levels over time, the number of flights per specified time period and population reaction to aircraft noise (for example, number or percentage of the population annoyed by noise).

Another method is to use noise impact scaling as shown in Fig. 1.20. On an axis corresponding to a particular noise index or rating, it is necessary to decide upon the border (*O* indicates the border point) between unacceptable and acceptable levels of annoyance. Such scaling allows a balanced approach in aircraft noise abatement.

Some tasks, for example, the allocation of the aircraft fleet in the vicinity of an airport to minimize aircraft noise can be turned into two optimization tasks by the method of decomposition. In the first stage, it is necessary to determine a low noise takeoff and approach flight procedure for each separate aircraft using, for example, EPNL. The second optimization task involves a higher-level criterion (for example, the area restricted by a given noise contour EPNL = *constant*).

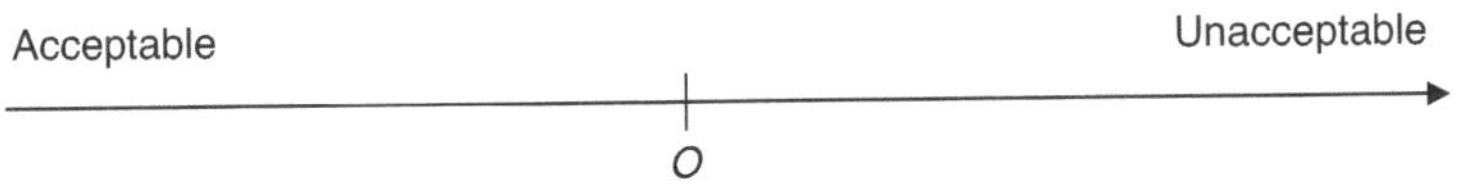


*Figure 1.20* Structure for one-dimensional scaling of annoyance.

## 1.5 Control of noise impact

The aircraft noise problem can be addressed in several ways: by government policy through aircraft noise control legislation; by the aircraft industry through the creation of low noise aeronautical equipment; by regional zoning for noise; and by local application of low noise flight procedures around an airport. The flight procedures for noise abatement are developed by the airline in consultation with the aircraft manufacturer. The authority for approving the introduction of any noise abatement operating procedure may be the national Civil Aviation Authority and/or the national authority for civil aviation safety. Low noise flight procedures for any particular flight must conform to the requirements of the state or country in which the airport is located.[16] Consequently, aviation noise reduction through operating procedures depends on communication between airlines, airports and the Civil Aviation Authority.

The abatement of aircraft noise involves limiting the noise at the source, noise control along the sound transmission path, low noise takeoff and approach flight procedures, optimal distribution of aircraft between the arrival and departure routes and land-use planning. Methods of noise abatement can be realized at all stages of the 'life cycle' of aircraft, from designing to aircraft phase-out.

Increasing stringency of the noise certification limits for subsonic airplanes, the phasing out of Chapter 2 aircraft, noise-based operational restrictions, reduction of noise at source, land-use planning and low noise operational procedures are all elements of the ICAO program on noise reduction. International organizations, for example, the ICAO and the European Civil Aviation Conference (ECAC), and the aviation industry, work together to formulate and implement environmental regulations.

According to the World Health Organization (WHO) data, recent years have seen an uncontrolled increase in environmental noise. The European Union (EU) has a strategic goal – to prevent an increase in the number of people affected by noise within the next few years. Indeed, the aim of the long-term EU program inside the VIth Action Plan on Environment Protection is to decrease the number of people exposed to high noise levels.

In Europe, for example, programs X-NOISE, SILENCE(R), the Fifth EU Framework Program has an objective to reduce noise impact by 10 EPNdB within 10 years. The 'critical technology' for source noise reduction embraces the engine (fan, compressor, turbine, core, jet noise reduction, progress in nacelle design), the airframe and the use of favourable installation effects. Large-scale integration programs for reduction of aircraft noise impact include source noise reduction and low noise operational procedures.

Aircraft noise management includes noise exposure simulation, environmental regulations, land-use planning, noise monitoring and air traffic control. This approach to the noise problem includes the following steps:

- reduction of aircraft noise at source by means of new technologies to mitigate noise impacts (propulsion system noise reduction using higher bypass ratios and turbomachinery noise reduction);
- special operational measures (for example, throttling back the engine on takeoff, low-power low-drag approaches, continuous descent approaches, and delayed flap and landing-gear extension);
- rational distribution of aircraft in the zone of the airport (preferred runway operation and flight tracks/corridors, use of less noisy aircraft, particularly at night, and fewer night flights);
- restricted building in high noise-level zones around airports and the introduction of noise-mitigation measures;
- noise monitoring systems in the vicinity of the airport and effective policing of them.

The essential elements of air traffic noise management are airport noise prediction and forecasting; noise exposure simulation; elaboration of noise-reducing strategies; noise certification of aircraft, accounting for noise propagation under various operational conditions; local environmental adjustments at the airport and the monitoring of aircraft noise. Airport noise forecasting needs information about the traffic pattern, the structure of the aircraft fleet, aircraft noise characteristics, aircraft weight and flight path, the number of aircraft operating on the flight path, their schedule and operational measures, the atmospheric parameters, sound propagation in atmosphere and the ground surfaces and topography in the vicinity of airport.

Environmental adjustments at an airport include some restrictions: the limitation of number of operations; bans on the operation of noisy aircraft; special operational measures and the rational distribution of aircraft in the zone of the airport. Table 1.8 summarizes aircraft noise abatement measures.

It is important to distinguish between the notions of imission and emission with regard to noise abatement. Imission is the influence of noise received by a receiver in a noise source action zone. Emission describes the radiation of noise from the source. Permissible emission is related to standardized imissions after taking into account the noise propagation from source to receiver. International and national experience allows the derivation of normative requirements on noise in the working zone, to define noise control methods and to establish normative noise levels for individual acoustic sources. There are several groups of normative documents on noise. The first group determines the terms, system of units and the permissible noise level norms (for example, in dwellings and public buildings in the vicinity of the airport). The second group determines methods for measuring noise in the working zone. The third group of normative documents establishes methods for measuring the noise of the sources (for example, of engines and equipment). The fourth group of normative documents establishes procedures for verifying the effectiveness of noise suppressors (for example,

*Table 1.8* A classification of noise abatement methods

| *Noise abatement measures* | *Procedure* | *Aircraft and equipment operation* | | | | | | |
|---|---|---|---|---|---|---|---|---|
| | | *Taxiing* | *Takeoff* | *Landing* | *Roll-on* | *Training flight* | *Run-up* | *Ground equipment* |
| Aircraft noise emissions | Standard and recommended practices of ICAO, Annex 16, volume 1, National standards | | • | • | | | | |
| Aircraft noise transmission and imissions | National imission standards, noise monitoring | • | • | • | • | | • | • |
| Planning of airport | Change of direction runway, length of runway | • | • | • | • | • | | |
| | Displacement of runway threshold | | | • | | • | | |
| | Building of high speed taxiway | • | | | • | | | |
| | Reconstruction of terminal building | • | | | | | • | • |
| | Use of noise mufflers or screens for isolation of zones run-up of engines | • | | | | | • | • |
| Rational use of a zone of the airport and airspace near to airport | Preferred runway operations | • | • | • | • | • | | |
| | Preferred on noise flight tracks operations or change of flight procedures of takeoff and landing | | • | • | • | • | | |
| | Restriction of aircraft taxing | • | | | | | | |
| | Restriction of run-up of the engine | | | | | | • | • |
| | Restriction of intensity aircraft operation of separate types | • | • | • | • | • | • | • |

*Table 1.8* Cont'd

| *Noise abatement measures* | *Procedure* | *Aircraft and equipment operation* | | | | | | |
|---|---|---|---|---|---|---|---|---|
| | | *Taxiing* | *Takeoff* | *Landing* | *Roll-on* | *Training flight* | *Run-up* | *Ground equipment* |
| | Ban of noisy aircraft, limiting night flights | • | • | • | • | • | • | • |
| | Augmentation of inclination of glidepath or entrance height in glidepath | | | • | | • | | |
| | Descent of aircraft in configuration with low drag | | | • | | • | | |
| Operation of aircraft | Engine, flaps deployment and flight speed control | | • | • | | • | | |
| | Reverse thrust limitation | | | | • | | | |
| Land-use planing | Alienation of land-use in the vicinity of the airport | • | • | • | • | • | • | • |
| | Purposeful airport development | • | • | • | • | • | • | • |
| | Zoning territory in the vicinity of the airport | • | • | • | • | • | • | • |
| | Observance of national building norms and sound insolating of dwelling | • | • | • | • | • | • | • |
| Development of noise abatement programs | Landing charge with noise factor calculation | • | • | • | • | • | | |
| | Research of population complaints on noise | • | • | • | • | • | • | • |
| | Inculcation of automation noise monitoring systems | • | • | • | • | • | • | • |

mufflers and screens). The fifth group establishes the requirements on the performance of noise suppressors and sound-proofing materials.

Normative documents are made by the International Organization on Standardization (ISO), the International Electrotechnical Commission (IEC), ICAO, the WHO and other organizations. The approaches to noise normalization can be categorized as sanitary or technical. Sanitary noise normalization establishes the limitations on noise under conditions of insignificant harmful influence on man. Technical noise normalization establishes the maximum noise levels with regard to technically achievable methods of noise reduction for given acoustic sources. The sanitary norms determine the necessary noise attenuation and the technical norms specify the attainable noise levels of the equipment.

The government policy on the control, abatement and mitigation of aircraft noise involves a balance between the needs of an efficient aviation industry and the need to minimize the impact of noise around airports. An important improvement in aircraft certification noise levels has been achieved over the past 25 years. It is necessary to take into account ICAO long-term strategy on the stringency of noise standards. In January 2001, the Committee on Aviation Environmental Protection (CAEP-5) recommended new, stricter noise certification standards, which was forwarded to the ICAO Council for a final decision. The new standards that the sum of noise levels measured at three measurement points (under the takeoff, under the arrival, and along the side of the runway) is 10 EPNdB lower than current Chapter 3 standards for noise reduction. At the federal level, noise control tools can include general environmental legislation, a noise quota system and quota limits, optimization of flight procedures, land-use planning, legislation to avoid construction of noise sensitive buildings and a noise monitoring system.

At the regional level, tools for noise control can include noise limits for aircraft flying over a region, noise emission limits, restrictions on the number of inhabitants within certain noise contours and an environmental audit.

In the vicinity of an airport, the noise control tools can include zoning and land use with respect to aircraft noise, limitations on the number of night movements, noise charges, noise minimization operations through optimization of airplane movements, layout of parking areas, construction of acoustic screens for reducing aircraft noise impact, construction of engine test places and actions to take care of ecological aspects around the airport.

Noise abatement tools for the airline can include the process of aircraft fleet formation and training the flight crew on low noise operational procedures.

## 1.6 Regulations and standards for aircraft noise

Since noise is the main source of environmental disturbance caused by air transport, the dominant environmental issue for air transport and a major constraint to growth, noise issues have received more regulatory and technological attention than any other aviation environmental problem. The most important effect in terms of the number of affected people is so-called *annoyance*, which can be determined from structured field surveys. Noise annoyance is strongly connected with specific effects, such as the necessity to close windows in order to enable sleep, disturbance or interference with communication, listening to the TV, the radio or music. Additionally, a number of serious medical effects may arise, such as high blood pressure, mental stress, heart attacks and hearing damage, although the latter concerns a smaller part of the population. Furthermore, there are negative effects on the learning capabilities of children. It is evident that people reporting noise-induced annoyance experience a reduced quality of life. This is a reality for at least 25 per cent of the EU population who are exposed to A-weighted transportation noise levels exceeding 65 dB, which many countries consider to be unacceptable. Between 5 and 15 per cent of the EU population suffers serious noise-induced sleep disturbance.[20] Even though the uncertainty of these estimates is very large, there is no doubt about the high prevalence of noise annoyance in the EU.

Aircraft, road traffic and railways are the most important sources of environmental noise in Europe.[21,22] However, locally, the noise situation can be dominated by other types of sources, for example, by noise from industrial sources or from residential and leisure areas. Important environmental noise effects are perceived in the domestic environment – in and near the home, in public parks, in schools. Current economic estimates of the annual damage in the EU due to environmental noise range from EUR 13 billion to 38 billion.[25] Elements that contribute are a reduction of housing prices, medical costs, reduced possibilities of land use and cost of lost labor days. In spite of some uncertainties, it seems likely that the costs of noise involve tens of billions of euros per year.

Public concern about exposure to noise pollution remains high, for example, in the EU, in spite of existing legislation. Noise emission from products is covered by Council Directive 86/188/EEC of 12 May 1986 on the protection of workers from risk related to exposure to noise at work, as amended by Directive 98/24/EC, and noise insulation between dwellings is converted by Council Directive 89/106/EEC of 21 December 1988 through regulations and administrative provisions of the Member States relating to construction products, as amended by Directive 93/68/EEC.

Legislation on environmental noise is divided into two major categories namely, legislation on noise emission by products (cars, trucks, aircraft and industrial equipment) and on allowable noise levels in the domestic environment. Emission standards consist of emission limit values applicable to individual sources and included in type approval procedures to ensure that new products comply with the noise limits at the time of manufacture. Emission standards must be grounded on the best available technology, that is, the technology that minimizes noise emissions for a given source. However, because the technology is sometimes too expensive for the industry to adopt without operating at a loss (considering civil aviation, for example), a determination is made of best practicable technology, which means taking socio-economic, as well as engineering factors into account. The regulations may be stricter for new sources than for existing ones. The aircraft noise norms and certification procedures declared by ICAO Annex 16,[26] are good examples of this approach.

Immission standards are based on noise quality criteria or guideline values for noise exposure[24,27] to be applied to specific locations and are generally built into planning procedures. Noise is associated with several negative feelings: annoyance, disturbance, bother and intrusion. In discussing noise annoyance, Langdon[28] considers the noise itself, its source, meaning, perception and the degree of interference as external factors. Except at higher sound levels, increases in annoyance are not always related to increases in sound, but some degree of relatedness remains.

Where sound pressure levels vary quite substantially and rapidly, such as during a low-level jet aircraft pass by, one might also want to consider the rate of change of sound pressure levels (the onset rate, for example). At the same time, the frequency content of each noise will also determine its effect on people, as will the number of events when there are relatively small numbers of discrete noisy events. Combinations of these characteristics determine how each type of environmental noise affects people.

Noise limits have been fixed for aircraft noise to ensure that rules are followed when building new dwellings and other noise sensitive installations close to existing airports and are taken into consideration for airport capacity expansion. Zones are generally designed to separate land uses. This is done by mapping noise contours and relating permissible land use to ambient noise levels.

The EU Green Paper on Future Noise Policy[21] and underlying studies analyzed the characteristics and impact of the EU and Member State approaches. It concluded that, to date, the total effect is unsatisfactory.

Directive 92/14/EEC, which came into force in April 1995, is the latest in a series of legislative measures in the EU begun in 1979 (Directives 80/51/EEC and 89/629/EEC) aimed at limitation of aircraft noise emission. These directives, like broadly similar legislation in other 'noise restrictive states' (most of non-EU Europe, Japan, Australia and New Zealand and the USA), use the benchmark standards specified by the ICAO in the Environmental Protection Annex 16, Volume I,[26] to the Chicago Convention, to which most countries in the world adhere. The limit values for individual aircraft types during takeoff and landing are specified in the terms of EPNL in EPN dB, and depend on the aircraft weight and number of engines. The oldest, noisier jet transport aircraft are 'non-noise certificated' (NNC), the second generation's characteristics are reflected in Chapter 2 of Annex 16 and the most modern, quieter aircraft meet the standards in Chapter 4.

Air transport was among the first of the world's transport industries to meet internationally accepted noise regulations. People living around airports often feel that air transport is a heavy strain on the local environment particularly with regard to noise. Their complaints cause airport authorities to introduce their own noise rules, restricting airline operations. In turn, the airlines look to manufacturers for quieter aircraft.

Quieter aircraft have significantly reduced the number of people affected by aircraft noise. Environmental performance of the product is only one of many considerations taken into account in the design process. The aeroplane manufacturer must carefully weigh and re-weigh each objective against all others to establish a design eventually that meets all regulatory and market requirements for the product. Noise reduction as a design metric must be balanced with respect to other customer needs. Anyway, the aircraft noise reduction that has been achieved is quite impressive. For example, on takeoff, a Boeing 727 (a 1960s aircraft) created an intrusive noise 'footprint' (defined as the area bounded by a contour specified by a normative value of noise index, for example, in terms of $L_{Amax}$, EPNL or $DNL$), which covered an area exceeding 14 km$^2$. In contrast, a modern commercial jet of similar capacity, but with greater takeoff power, such as the Airbus A-320, creates a 'noise footprint' covering only 1.5 km$^2$. This represents an area reduction by nearly a factor of 9 (Fig. 1.21). In the turboprop segment, the takeoff noise footprint area of the Fokker 50 (1987) compared to the Fokker F27 MK500 (1968) shows a reduction of the 80 dB contour from 3.77 km$^2$ to 0.84 km$^2$, that is, an area reduction factor of 4.5.

As a result, in the United States and Europe, the number of people directly affected by aircraft noise is now only about 5 per cent of what it was with 1970s technology aircraft. The 19 million inhabitants of the USA and Europe affected by aircraft noise in 1970s have been reduced to only 0.8 millions in 1990s owing to improvements in engine and

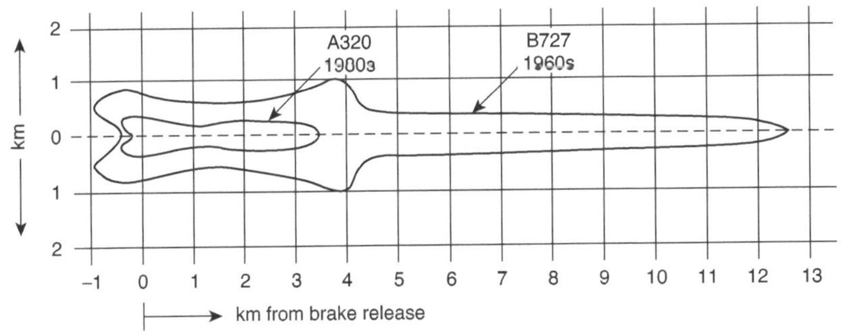


*Figure 1.21* Noise impact at takeoff [noise contour for 85 dBA]: A320–200 with CFM-56-5 engines and TO weight 67.5 t (noise footprint = 1.55 km$^2$); B727–200 with JT8D-15 engines and TO weight 76.5 t (noise footprint = 14.25 km$^2$).

aircraft technologies. These data are derived from the Nationwide Airport Noise Impact Model developed for the Federal Aviation Administration (FAA) Nationwide Airport Noise Impact Model (NANIM, 1993). NANIM calculated the regional and national totals of the number of people, the land area and the number of housing units exposed to a *DNL* of 65 dB or higher.

The MAGENTA (Model for Assessing Global Exposure form Noise of Transport Airplanes) study[29] of the ICAO has been used to help in the identification of noise problems around airports and assess the relevance of the noise certification scheme to these problems.

Based on the initial success of the FAA activity, the fourth meeting of CAEP (CAEP-4) recommended that a task group be formed to complete the development of the same tool for CAEP analysis. The model calculates noise contours for major airports and overlays them on population maps. Global noise exposure calculations were made for a number of 'snapshot' years between 1998 and 2020, taking account of increases in traffic. The airline fleets, aircraft noise and performance characteristics and airport operations by type were defined by CAEP.

From the 1998 starting point to 2002, noise exposure (levels of *DNL* 55 and *DNL* 65 were taken as the thresholds of significant and high noise impact, respectively) generally falls as a result of the ongoing phase-out of Chapter 2 aircraft. Beyond 2002, airport capacity is a key factor. The extent to which global noise exposure goes up or down will depend very much on how airport operators accommodate the forecasted doubling of passenger traffic over the next 20 years. Around 25 million inhabitants, living in the vicinity of the airports, are likely to be impacted significantly worldwide.[29]

Many of the adverse environmental effects of civil aviation activity can be reduced by the application of integrated measures embracing technological improvements, appropriate operating procedures, proper organization of air traffic and the appropriate use of airport planning and land-use control mechanisms. States and international organizations recognize the leading role of ICAO in dealing with the problems of aircraft noise and they keep the Council informed of their policies and programs to alleviate the problem of aircraft noise in international civil aviation.

The ICAO is becoming involved in activities relating to environmental policies affecting air transport and it strives to achieve a balance between the benefit accruing to the world community through civil aviation and the harm caused to the environment. The ICAO Assembly considers that improvements in the noise climate achieved at many airports by the introduction or revision of aircraft-related measures (for example, the phase-out of Chapter 2 aircraft) should be safeguarded by taking into account the sustainability of future growth and should not be eroded by incompatible urban encroachment in areas where reductions in noise levels have been achieved.

*Guidance on the Balanced Approach to Aircraft Noise Management* was published by ICAO in 2004. The Balanced Approach encompasses four principal elements: reduction of noise at source, land-use planning and management, operational procedures for noise abatement and operating restrictions on aircraft. The process of implementing the Balanced Approach would typically consist of an assessment of the noise situation at an individual airport, definition of the objective, provision for consultation, identification of measures available to reduce the noise impact, evaluation of the relative cost-effectiveness of the measures, selection of measures, adequate public notification of intended actions, implementation of measures and a provision for dispute resolution available to stakeholders.

Under the Balanced Approach, reduction of noise at source is limited to noise reduction through the adoption and implementation of noise certification standards set by ICAO and is not within the control of individual airports. The prime purpose of noise certification is to ensure that the latest available noise reduction technology is incorporated into aircraft design demonstrated by procedures which are relevant to day-to-day operations, to ensure that noise reduction offered by technology is reflected in reductions around airports.

Standards for aircraft noise were first adopted by the Council of ICAO on 2 April 1971 pursuant to the provisions of Article 37 of the Convention and designated as Annex 16 to the Convention. All of the standards and recommended practice for aircraft noise are included in Volume 1 of the Annex 16 'Aircraft noise', and all the questions concerning engine emissions are in Volume 2 – 'Aircraft engine emission'. Part II of the Volume 1 contains standards and guidelines for noise certification applicable to the classification of aircraft specified in individual chapters of this part. Chapter 2 contains the oldest version of the standards for the most popular types of the subsonic jet aircraft that have not been in production since the 1980s. More stringent standards were implemented and applied to new airplane designs in the late 1970s. These new standards were included in Chapter 3, Annex 16 of Volume 1 to the ICAO Convention and applied to new jet aircraft types starting on 6 October 1977. ICAO has recently confirmed the Chapter 4 noise certification standards. Effectiveness and reliability of certification schemes from the viewpoint of technical feasibility, economic reasonableness and environmental benefit need to be achieved in all the standards.

Several steps are involved in developing a noise reduction concept, beginning with the initial idea, evaluating the feasibility of the idea, evaluating the merits of the idea and establishing and executing a development plan to carry the idea through to a practical, useful concept.

Two types of cost assessments were made before introduction of the new Chapter 4 noise certification standards.[30]

(1) First, it was used to determine the amount of noise reduction for each noise certification reference condition required to enable each current or near-term production aircraft type10 to meet the noise stringency options under study, in this case Chapter 3 limits minus 8, 11 and 14 EPNdB. The incremental operating and capital costs associated with the required insertion of noise reduction technology for aircraft failing each of the proposed noise stringency levels was also determined in this step.

(2) For the noise policy options that included phasing out the noise non-compliant aircraft among the in-service fleet, the Aircraft Noise Design Effects Study (ANDES) model was used to estimate the combined incremental capital and operating costs resulting from the insertion of the required noise reduction technology that would bring into compliance aircraft that did not meet the proposed noise stringency levels (Fig. 1.22). It was also used as a screening tool to determine which in-service aircraft could be potential candidates for re-certification (that is, modification of the airplane with retrofit noise reduction technology packages).

Noise reduction features can be divided into three classes:

- features that attack those sources which need reduction to meet requirements;
- features with a higher technology readiness level in order to minimize risk of failure and the need for a redesign; and
- features offering the required benefit at the minimum cost.

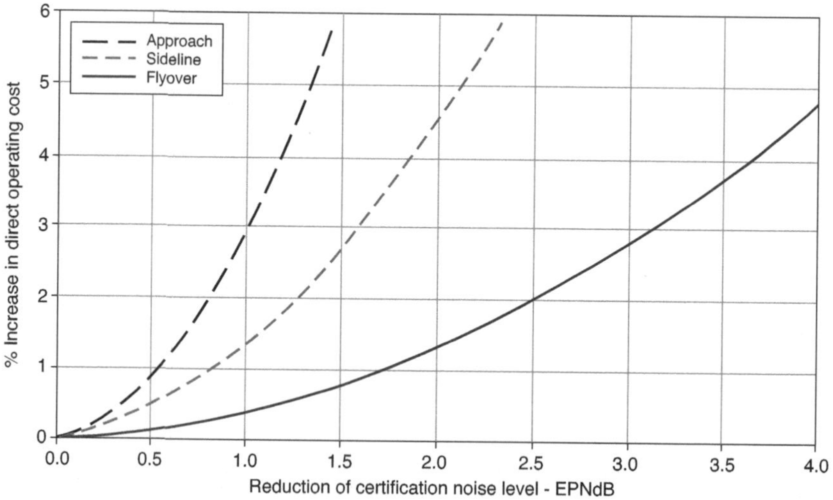


*Figure 1.22* Combined incremental capital and operating costs resulting from the application of the required noise reduction technology.

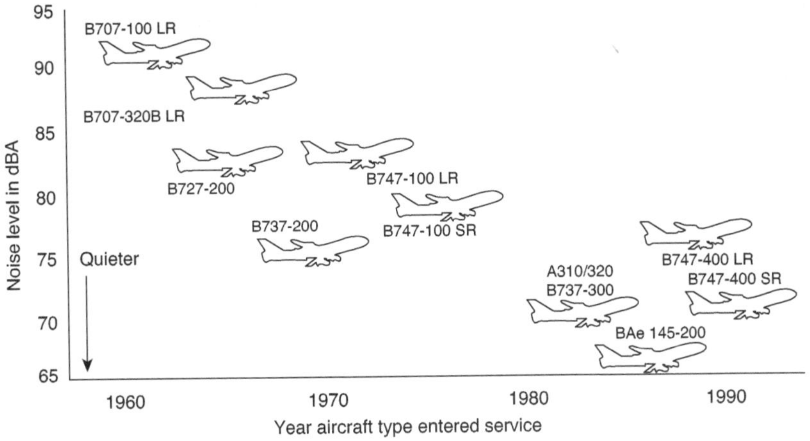


*Figure 1.23* Progress in aircraft noise reduction due to aircraft and engine development.

Since the 1970s, the limits set by legislation and certification have been tightened and distributed over the whole spectrum of possible airplane types. Reduction in engine noise has contributed most to the dramatic reduction of the overall aircraft noise. Figure 1.23 gives an overview of the progress in aircraft noise reduction since 1960.

CAEP-5 has defined Chapter 4 limits as giving rise to a cumulative 10 dB below the Chapter 3 standards. However, there is no requirement for a minimum improvement at each of the three reference noise measurement points. Given that aircraft noise affects communities on either side of airports as well as under the approach and departure routes, Chapter 4 does not ensure that ICAO's goal to limit or reduce the number of people affected by noise will be met, as it does not guarantee an improvement at all of the reference noise measurement points.

When Chapter 4 was decided in 2001, the cumulative decrease of 10 dB was already being met by virtually all aircraft in production. This means that aircraft certificated after the commencement date of 2006 are not required to perform better than the majority of aircraft already in production in 2001. A substantial reduction in the noise impact around airports is therefore not expected from the Chapter 4 standard.

Table 1.9 illustrates some of the major parameters that are currently considered in the course of an aircraft 'design-to-noise-objectives' process. The table shows the trend of each parameter's potential impact on certification noise levels. The sign indicates the direction of change required for the parameter to achieve a noise reduction (+, increase; –, decrease; = no direct influence). Many of these parameters result from general trade-off studies

*Table 1.9* Useful parameters for an aircraft 'design-to-noise-objectives' process

| *Design parameters* | *Sideline* | *Takeoff flyover* | *Approach* |
|---|---|---|---|
| *Aircraft performance* | | | |
| Weight | − | − | − |
| Thrust rating | − | + | = |
| Wing L/D | + | + | + |
| *Engine* | | | |
| Bypass ratio | + | + | + |
| Fan pressure ratio | − | − | − |
| *Nacelle* | | | |
| Intake liner efficiency | + | + | + |
| Fan duct liner efficiency | + | + | + |
| Intake and/or fan duct length | + | + | + |
| *Airframe* | | | |
| Low noise high lift devices | + | + | + |
| Low noise landing gear | = | = | + |

leading to the detailed design of the aircraft and are not fully independent of each other.

Their individual impact on other aircraft requirements can also be significant. Nevertheless, it shows that, with the exception of engine thrust rating that can have an opposite effect on the takeoff, flyover and sideline certification points (see later), any variation causing a noise reduction at one point will cause either no change or a corresponding noise reduction at any other point. Whether this remains true of all design parameters has to be evaluated on a case-by-case basis. The consequence is that designing to a 'cumulative' noise objective most likely results in improving preferentially one or two out of the three certification points, without significantly negatively impacting the third one, but possibly minimizing the impact on other aircraft requirements. On the other hand, designing to one challenging individual noise level could lead to a detrimental impact on non-noise aircraft design requirements, could potentially yield limited benefit to the other two noise point levels, and result in poor cumulative performance.

For example, for subsonic transport category large airplanes and subsonic turbojet-powered airplanes, compliance with the requirements of the Chapter 2 must be shown with noise levels measured and evaluated as prescribed in this chapter and demonstrated at the measuring points and in accordance with the flight test conditions prescribed under this chapter.

Compliance with the noise level standards must be shown at the three measuring points:

(a) For takeoff, at a point 6500 m from the start of the takeoff roll on the extended centerline of the runway.
(b) For approach, at a point 2000 m from the threshold on the extended centerline of the runway.
(c) For the sideline, at the point, on a line parallel to and 450 m from the extended centerline of the runway, where the noise level after liftoff is greatest.

It must be shown by a flight test that the noise levels of the airplane, at these three monitoring points, do not exceed the following limits (with appropriate interpolation between weights):

(1) Stage 2 noise limits for airplanes, regardless of the number of engines, are as follows:

   (i) For takeoff: 108 EPNdB for maximum weights of 272 t or more, reduced by 5 EPNdB per halving of the 272 t maximum weight down to 93 EPNdB for maximum weights of 34 t and less.
   (ii) For sideline and approach: 108 EPNdB for maximum weights of 272 t or more, reduced by 2 EPNdB per halving of the 272 t maximum weight down to 102 EPNdB for maximum weights of 34 t and less.

(2) Stage 3 noise limits are as follows:

   (i) For takeoff:

      (A) For airplanes with more than three engines: 106 EPNdB for maximum weights of 385 t or more, reduced by 4 EPNdB per halving of the 385 t maximum weight down to 89 EPNdB after which the limit is constant (under maximum weight of 20.2 t).
      (B) For airplanes with three engines: 104 EPNdB for maximum weights of 385 t or more, reduced by 4 EPNdB per halving of the 385 t maximum weight down to 89 EPNdB after which the limit is constant (under a maximum weight of 28.6 t).
      (C) For airplanes with fewer than three engines: 101 EPNdB for maximum weights of 385 t or more, reducing by 4 EPNdB per halving of the 385 t maximum weight down to 89 EPNdB after which the limit is constant (under a maximum weight of 48.1 t).

   The various noise limits as a function of takeoff weight are shown in Fig. 1.24.

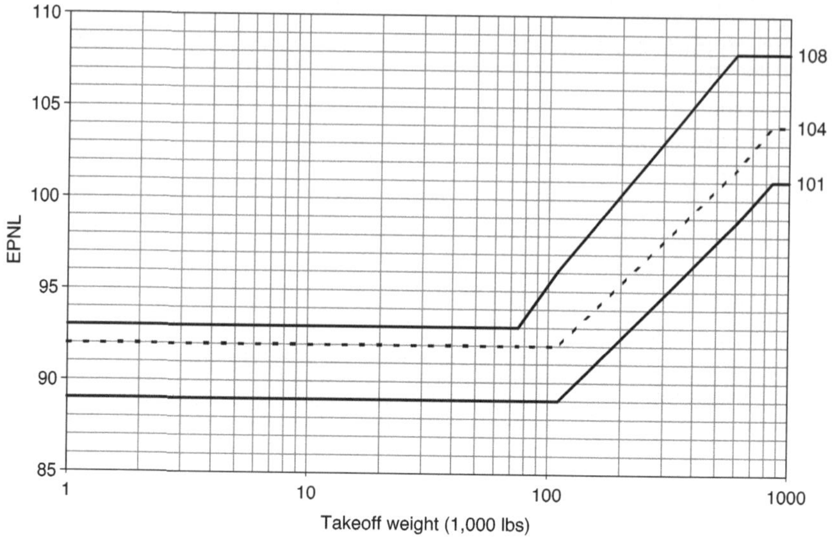


*Figure 1.24* Stage 3 aircraft noise limits.

(ii) For the sideline, regardless of the number of engines: 103 EPNdB for maximum weights of 400 t or more, reduced by 2.56 EPNdB per halving of the 400 t maximum weight down to 94 EPNdB after which the limit is constant (under the maximum weights of 35 t).

(iii) For the approach, regardless of the number of engines: 105 EPNdB for maximum weights of 280 t or more, reduced by 2.33 EPNdB per halving of the 280 t weight down to 98 EPNdB after which the limit is constant (under maximum weights of 35 t).

All of the limits can be calculated from equations of noise levels as a function of takeoff mass $M$ (t). They are shown in Tables 1.10–1.14 for the particular types of airplanes considered in Annex 16.

*Table 1.10* Noise and mass relationships for Chapter 2 aircraft requirements (before 6 October 1977)

| | | | |
|---|---|---|---|
| Maximum takeoff mass, $M$ (t) | 0 | 34 | 272 |
| Lateral noise level (EPNdB) | 102 | $91.83 + 6.64 \lg M$ | 108 |
| Approach noise level (EPNdB) | 102 | $91.83 + 6.64 \lg M$ | 108 |
| Flyover noise level (EPNdB) | 93 | $67.56 + 16.61 \lg M$ | 108 |

*Table 1.11* Noise and mass relationships for Chapter 2 aircraft requirements (after 26 October 1981)

| | | | | | | | | | |
|---|---|---|---|---|---|---|---|---|---|
| Maximum takeoff mass $M$ (t) | 0 | 34 | 35 | 48.3 | 66.7 | 133.45 | 280 | 325 | 400 |
| Lateral noise level (EPNdB) | 97 | | $83.87 + 8.51 \lg M$ | | | | | | 106 |
| Approach noise level (EPNdB) | 101 | | $89.03 + 7.75 \lg M$ | | | | 108 | | |
| Flyover noise level, 2 engines (EPNdB) | 93 | | | $70.62 + 13.29 \lg M$ | | | | 104 | |
| Flyover noise level, 3 engines (EPNdB) | 93 | $67.56 + 16.61 \lg M$ | | | $73.62 + 13.29 \lg M$ | | | 107 | |
| Flyover noise level, 4 engines (EPNdB) | 93 | $67.56 + 16.61 \lg M$ | | | | $74.62 + 13.29 \lg M$ | | 108 | |

*Table 1.12* Noise and mass relationships for Chapter 3 aircraft requirements

| | | | | | | | | |
|---|---|---|---|---|---|---|---|---|
| Maximum takeoff mass $M$ (t) | 0 | 20.2 | 28.6 | 35 | 48.1 | 280 | 385 | 400 |
| Lateral noise level (EPNdB) | 94 | | | $80.87 + 8.51 \lg M$ | | | | 103 |
| Approach noise level (EPNdB) | 98 | | | $86.03 + 7.75 \lg M$ | | 105 | | |
| Flyover noise level, 2 engines (EPNdB) | 89 | | | | $66.65 + 13.29 \lg M$ | | 101 | |
| Flyover noise level, 3 engines (EPNdB) | 89 | $69.65 + 13.29 \lg M$ | | | | | 104 | |
| Flyover noise level, 4 engines (EPNdB) | 89 | $69.65 + 13.29 \lg M$ | | | | | 108 | |

*Table 1.13* Noise and mass relationships for Chapter 5 aircraft requirements

| | | | | |
|---|---|---|---|---|
| Maximum takeoff mass $M$ (t) | 0 | 34 | 358.9 | 384.7 |
| Lateral noise level (EPNdB) | 96 | $85.83 + 6.64 \lg M$ | | 103 |
| Approach noise level (EPNdB) | 98 | $87.83 + 6.64 \lg M$ | | 105 |
| Flyover noise level (EPNdB) | 89 | $63.56 + 16.61 \lg M$ | 106 | |

*Table 1.14* Noise and mass relationships for Chapter 8 aircraft requirements

| | | | |
|---|---|---|---|
| Maximum takeoff mass $M$ (t) | 0 | 0.788 | 80 |
| Lateral noise level (EPNdB) | 86 | $87.03 + 9.97 \lg M$ | 106 |
| Approach noise level (EPNdB) | 87 | $88.03 + 9.97 \lg M$ | 107 |
| Flyover noise level (EPNdB) | 85 | $86.03 + 9.97 \lg M$ | 105 |

Tests for compliance with the required noise limits must be conducted under the following conditions:

(a) Takeoff power or thrust must be used from the start of takeoff roll to at least the specified altitude above the runway:

(1) ICAO Annex 16: for Stage 2 airplanes (and in accordance with FAR-36 for Stage 1 airplanes and for Stage 2 airplanes that do not have turbojet engines with a bypass ratio of 2 or more), the following apply:

(i) For all types of the airplanes: an altitude of 210 m.
(ii) For airplanes with more than three turbojet engines: an altitude of 214 m.
(iii) For all other airplanes: an altitude of 305 m.

(2) For Stage 2 airplanes that have turbojet engines with a bypass ratio of 2 or more (FAR-36) and for Stage 3 airplane (FAR-36 and ICAO Annex 16), the following altitudes apply:

(i) for airplanes with more than three turbojet engines: 210 m.
(ii) for airplanes with three turbojet engines: 260 m.
(iii) for airplanes with fewer than three turbojet engines: 300 m.
(iv) for airplanes not powered by turbojet engines (only in accordance with FAR-36): 305 m.

(b) Upon reaching the altitude specified in paragraph (a) of this section, the power or thrust may not be reduced below that needed to maintain level flight with one engine inoperative, or to maintain a 4 per cent climb gradient, whichever power or thrust is greater.
(c) A constant takeoff configuration, selected by the applicant, must be maintained throughout the takeoff noise test, except that the landing gear may be retracted.
(d) For applications made for subsonic airplanes after September 17, 1971, and for Concorde airplanes, the following requirements on speeds apply:

(1) For subsonic airplanes, the test day speeds and the acoustic day reference speed must be the minimum approved value of $V_2 +$ 19 km/h ($V_2$ is takeoff safety speed), or the all-engines-operating speed at 35 feet (for turbine engine powered airplanes) or 50 feet (for reciprocating engine-powered airplanes), whichever speed is greater as determined under the regulations constituting the type certification basis of the airplane. These tests must be conducted at the test day speeds $\pm 3$ knots. Noise values measured at the test day speeds must be corrected to the acoustic day reference speed.

In all cases, compliance must be tested against the defined noise limits for approach.

(a) The airplane's configuration must be that used in showing compliance with the landing requirements in the airworthiness regulations constituting the type certification basis of the airplane. In accordance with ICAO Annex 16 Chapter 2, the configuration shall be with the maximum allowable landing flap setting. In accordance with FAR-36 and ICAO Annex 16 Chapter 3, if more than one configuration is used in showing compliance with the landing requirements in the airworthiness regulations constituting the type certification basis of the airplane, the configuration that is most critical from a noise standpoint must be used.
(b) The approaches must be conducted with a steady glide angle of 3 degrees +0.5 degrees and must be continued to a normal touchdown with no airframe configuration change.
(c) All engines must be operating at approximately the same power or thrust (FAR-36 requirement).
(d) For applications made for subsonic airplanes after September 17, 1971, and for Concorde aircraft, the following apply:

    (1) For subsonic airplanes, a steady approach speed, that is, either $1.30V_s$ + 19 km/h ($1.30V_s$+ 10 knots; $V_s$ is stall speed) or the speed used in establishing the approved landing distance under the airworthiness regulations constituting the type certification basis of the airplane, whichever speed is greater, must be established and maintained over the approach measuring point.
    (2) In accordance with FAR-36 for Concorde aircraft, a steady approach speed, that is, either the landing reference speed +10 knots or the speed used in establishing the approved landing distance under the airworthiness regulations constituting the type certification basis of the airplane, whichever speed is greater, must be established and maintained over the approach measuring point.
    (3) A tolerance of +3 knots may be used throughout the approach noise testing.

Besides the requirements for jet aircraft, the ICAO Annex 16 includes the standards for noise levels of propeller-driven aircraft (Chapter 5 for weights over 5700 kg, Table 1.13;[26] Chapter 6 for weights not exceeding 5700 kg), propeller-driven STOL aircraft (Attachment C to Annex 16, which is only a guideline, not an enforceable standard), helicopters (Chapter 8 and Chapter 11; see Table 1.14). Noise limits and guidelines are also defined for auxiliary power units that are installed on board the aircraft (Attachment D to Annex 16). Some standards use indices such as $L_{Amax}$ or *SEL* for the limits that differ greatly from EPNL. As a rule, test circumstances are different for these sources too. Nevertheless, the aircraft noise emission limits take into account not only the acoustical properties of the engines as a main contributor to the total noise levels under the flight path aside the aircraft, but also their installation effects, the aerodynamic properties of the airplanes, possibilities for effective flight procedures and so on. That is why the noise limits must be considered as a complex measure for acoustic efficiency of the particular type of airplane as a whole.

Chapter 2 aircraft have not been in production since the late 1980s and the retirement of noisier aircraft has been underway for many years. Manufacturers estimate that, as of the year-end 1991, nearly 2000 jet aircraft had been permanently retired. The peak of retirements, which occurred in the early 1980s, coincided with an economic recession and mainly involved the noisiest aircraft banned by noise regulations. Further retirements are now taking place and there will be a substantial need to replace large numbers of aircraft with newer models. Currently, the standards are used only for new designs of aircraft and not applied to aircraft in operation. In resolution A29-12 (1995), the 29th ICAO Assembly proposed in resolution A29-12 to phase out from operation the oldest, noisiest jet transport aircraft – those that are 'non-noise certificated' and have second-generation characteristics (reflected in Chapter 2 of Annex 16). The recommendations of the ICAO were reflected, for example, in USA FAR-91 and EU Directive 92/14. By 2002, aircraft which do not meet the latest ICAO noise certification standards – the so-called Chapter 2 aircraft – have to be phased out of commercial airline operations in most countries, or modified to meet these standards.

Subsonic non-noise certificated aircraft have been excluded from airports for several years and, under the terms of Directive 92/14 Chapter 2, aircraft over 25 years old have been banned from European Community airports since April 1995 unless granted exemptions designed to avoid unreasonable economic hardship to the airlines of developing nations, for instance. Chapter 2 aircraft were systematically phased out over the 1995–2002 period, and as of 1 April 2002, only Chapter 3 aircraft were allowed to use Community airports. FAR-91 considers more stringent terms, from 1995 up to 2000 only. Meanwhile, increased stringency is being considered by ICAO's Committee on Aviation Environmental Protection (CAEP) and the European Civil Aviation Conference (ECAC).

Aircraft manufacturers estimate that over the next 20 years, the world's airlines are expected to take delivery of up to 12,000 aircraft, worth an estimated US$ 857 billion (1992 dollars). While the three major aircraft manufacturers differ in their forecasts of the size and total number of the future aircraft fleet, all predict delivery of around 600 new planes a year for the next 15 years. In addition to significant enhancements in performance, these new aircraft will also bring major improvements in airline utilization, thereby reducing emissions and aircraft noise. So, wherever possible, airlines are:

- buying or leasing aircraft that satisfy the strictest international environmental requirements;
- replacing aircraft that do not meet these requirements as soon as operations and economics allow; and
- actively encouraging manufacturers to develop new, quieter, cleaner and more energy-efficient aircraft.

These developments, coupled with the high growth in the past and projected high growth for the future, may mean that only short to medium term benefit will be gained from the phase out of Chapter 2 aircraft and that after 2002, the overall noise emissions and consequently, the overall noise footprints may not be contained within the reduced boundaries expected to be achieved by that date. Further improvements in the noise impact around airports will require a combination of measures, including changes in operating procedures and better land-use planning.

Land-use planning procedures to ensure separation of dwellings and other noise sensitive buildings from noise sources are one of the means of putting immissions regulations into practice and are a key tool for noise abatement. It is a principal element of the Balanced Approach. The airport authority should work closely with local and regional authorities responsible for land-use management aiming to implement all noise-control measures around airports with regard to the noise impact of aviation operations. It is a planning, mitigating and financial instrument at the same time.

Airport planning must be recognized as an integral part of an area-wide comprehensive planning program. The location, size and configuration of the airport need to be coordinated with patterns of residential, industrial, commercial, agricultural and other land uses of the area, taking into account the effects of the airport on people, flora, fauna, the atmosphere, water courses, air quality, soil pollution and other facets of the environment. The social and economic impact, together with the environmental effects of the airport, can then be evaluated.

The need for some public control of land in the vicinity of an airport was recognized in the early history of civil aviation. In general, these early actions were usually concerned with height control of possible hazards or obstacles to flight into or out of airports. The compatibility of land use with noise exposure in the vicinity of airports did not become a major consideration until the early 1960s, a few years after the widespread introduction of commercial turbojet aircraft operations, although litigation regarding aircraft noise was not infrequent before that time.

Today, aircraft noise is probably the most significant influence on land-use planning in the vicinity of airports. Over the long-term it is one of the most efficient ways of reducing noise impact, as it can be used to prevent new problems occurring. In particular, noise abatement through land-use planning can include:[31] restricting the use of land that is already subject to high levels of noise, restricting the siting of new noise generators, such as traffic routes or industrial installations, in order to protect existing developments and encouraging noise-generating activities to cluster together in order to preserve other low noise areas. Noise is one of the considerations to be dealt with in environmental statements for developments requiring an environmental impact assessment.

The first noise limit of 112 PNdB for the new family of jet-powered aircraft, the Boeing 707 and McDonnell–Douglas DC-8 was set in 1959 at New York airport. This placed them in the same noise bracket as the quietest 75 per cent of propeller-driven airplanes. The first national noise standards for commercial aircraft were implemented in the United States in 1969. Now, all of the existing aircraft noise standards are included in US Federal Aviation Rules FAR-36 and they are equivalent to the ICAO Annex 16 standards due to the practice of worldwide harmonization of all rules that have an influence on civil aviation. The technical standards mentioned are examples of emission standards and relate to technological possibilities of the manufacturer to design a quiet aircraft.

Noise limits around the airports are considered to be ecological immission standards that must mitigate the noise impact on the population in the vicinity of the airports due to the environmental requirements and the possibilities of fulfilling these requirements by technical and financial efforts. Immision standards are included in special rules, for example, in the USA in the FAR-150 'Airport Noise Compatibility Programs'.

As a rule, noise immision standards are very specific for a particular nation. Work done under the ICAO coordination or in the EU community for their harmonization is only at a very early stage. Two basic approaches are generally followed. One uses the $L_{Aeq}$ as for road and rail, the other uses indices that consider the number of aircraft movements and the peak noise level of each movement, with weightings for different periods of the day. In view of the diversity of the indices, it is difficult to compare immission limits.

In the USA, the Aviation Safety and Noise Abatement Act of 1979 directed the FAA to:

- establish a single system of measuring noise for which there is a highly reliable relationship between projected noise exposure and surveyed reactions of people to noise, to be uniformly applied in measuring the noise at airports and the areas surrounding airports;
- establish a single system for determining the exposure of individuals to noise that results from airport operations, which includes consideration of the noise intensity, duration, number of occurrences and time of occurrence;
- identify land uses that are normally compatible with various exposures of individuals to noise;

- establish a program for airport operators voluntarily to develop and submit to the FAA (1) noise exposure maps showing present and future non-compatible land uses around an airport, and (2) a noise compatibility program setting forth measures to reduce existing incompatible land uses and to prevent the introduction of additional incompatible land uses around an airport; and
- make Federal funding available for preparing a noise compatibility program and for projects to carry out a noise compatibility program.

The FAA implemented the Aviation Safety and Noise Abatement Act through the issuance of Part 150 of the Federal Aviation Regulations (FAR 150 or Part 150) (DOT, FAA 1989)[36]. Part 150 prescribes the procedures, standards and methodology governing the development, submission and review of airport noise exposure maps and airport noise compatibility programs, including the process for evaluating and approving or disapproving of those programs. It prescribes single systems for: (a) measuring noise at airports and surrounding areas that generally provide a highly reliable relationship between projected noise exposure and surveyed reaction of people to noise; and (b) determining exposure of individuals to noise, which results from operations.

In Part 150, the FAA designated *DNL* as the noise metric to be used, and required noise exposure maps to include *DNL* contours of 65 dB, 70 dB, 75 dB. The FAA has received noise exposure maps that include the *DNL* 60 dB contour and the *DNL* 55 dB contour. Part 150 includes the 1980 FICUN land-use compatibility criteria. These criteria are guidelines only, and Part 150 specifically allows local discretion in using the criteria. The FAA has received noise exposure maps and noise compatibility programs that include variations to these criteria, usually identifying land uses as incompatible at levels lower than *DNL* 65 dB.

An airport operator that undertakes airport noise compatibility planning under Part 150 is required to: develop present and future noise exposure maps; examine the airport's current and forecast future noise problems based on these maps; consider ways to reduce the exposure of noise sensitive land uses to levels of aircraft noise that are not compatible with those land uses; recommend noise reduction/land-use compatibility measures to be implemented; and submit its maps and recommended program to the FAA.

The Act directs the FAA to approve an airport operator's noise compatibility program if it meets specified standards. These standards require the program to:

- provide for reduction of existing incompatible land uses and prevention of the establishment of additional incompatible land uses;
- impose no undue burden on interstate or foreign commerce;
- not unjustly discriminate among users;

- not result in derogation of safety or adversely affect the safe and efficient use of airspace;
- meet both the local needs and the needs of the national air transportation system, to the extent practicable, considering trade-offs between economic benefits derived from the airport and the noise impact;
- be capable of implementation in a manner consistent with all of the powers and duties of the FAA Administrator; and
- provide for program revision, if necessary.

A large degree of international consensus has emerged over the years as to what constitutes unacceptable levels of noise exposure and what should be the maximum levels of exposure for certain specific situations. At the international level, the WHO together with the Organization for Economic Cooperation and Development (OECD) are the main bodies that have collected data and developed their own assessments on the effects of exposure to environmental noise. On the basis of these assessments, guideline values for different time periods and situations have been suggested.

In the mid-1980s the OECD[32] reported the thresholds for noise nuisance as follows (in day-time $L_{Aeq}$):

- at 55–60 dBA noise creates annoyance;
- at 60–65 dBA annoyance increases considerably;
- above 65 dBA constrained behavior patterns, symptomatic of serious damage caused by noise arise.

The World Health Organization (WHO)[33] has suggested a standard guideline value for average outdoor noise levels of 55 dBA, applied during normal daytime in order to prevent significant interference with the normal activities of local communities. Additional guideline values suggested for specific environments (WHO, all figures are in $L_{Aeq}$) are shown in Table 1.15.

The Fifth Environmental Action Program established a number of broad targets on which to base action up to the year 2000 for night time ($L_{Aeq}$):

- to phase out average exposure above 65 dBA;
- to ensure that at no point in time a level of 85 dBA should be exceeded coupled with the aim of ensuring that the proportions of the population exposed to average levels between 55 and 65 dBA should not increase; and
- exposure in quiet areas should not increase beyond 55 dBA.

A survey of the situation in Community countries has shown that most Member States have adopted legislation or recommendations aiming for immission limits in noise sensitive areas similar to these guideline values.[25] The national regulations were initially developed in the 1970s and

*Table 1.15* World Health Organization guideline values ($L_{Aeq}$ dB) for specific environments

| | *Day* | | *Night* | |
|---|---|---|---|---|
| | *Inside* | *Outside* | *Inside* | *Outside* |
| Residential | 50 | 55 | | |
| Bedroom | | | 30<br>45 | 45 |
| Schools | 35 | 55 | | |
| Hospitals | 35<br>30 | | 35<br>30 | 45 (for $L_{Amax}$)<br>40 (for $L_{Amax}$) |
| Concert halls | 100 for a 4-hour period | | 100 for a 4-hour period | |
| Disco | 90 for a 4-hour period | | 90 for a 4-hour period | |

1980s in the northern Member States and somewhat later in the southern Member States. Generally, these immission limits are more detailed and specific about the noise sources, the current noise situation and the kind of living area than the WHO guideline values.

Increasingly, these regulations are being integrated into national abatement laws and are used in land-use plans. Noise immission standards for new developments are normally set by local authorities as part of planning policy and are used as a reference in environmental impact assessments. They serve as a means of ensuring that appropriate measures are taken to minimize the noise impact of a site. Where an acceptable level of noise cannot be achieved, planning permission may be refused or action may be required to improve insulation from the noise sources.

Noise limits have been fixed for aircraft noise to ensure that rules are followed when building new dwellings and other noise sensitive installations close to existing airports and to be taken into consideration for airport capacity expansion. Zones, designed to separate land uses, are established by mapping noise contours and relating permissible land use to ambient noise levels.

In addition to certification limits and land-use planning, it is possible to reduce the effects of noise by acoustical barriers. Acoustical barriers can include wide-ranging measures such as the use of ear protectors for people subjected to high-intensity noise, to soundproofing of buildings and methods for screening the sound source. Specific attention should given to the proper location of ground aircraft engine run-up sites and the orientation of buildings at the airport. A good protection against ground run-up noise might be expected from properly planted trees, as indicated by a study in Japan of the sound-insulating characteristics of wooded areas. The sound attenuation through 100 m of evergreen trees can be 25–30 dB. But the screening efficiency of the trees may suffer from the seasonal changes because their sound-insulating characteristics (at higher frequency) are determined primarily by foliage density (see Chapter 3). Important consideration should be given to selecting species that do not generate a bird hazard for flights.

Noise insulation can lower interior noise levels for residential structures that cannot reasonably be removed from noise-exposed areas. Noise insulation is particularly effective for commercial buildings, including offices and hotels. The degree of insulation required varies from country to country. In some countries there are legal limits for internal noise. For effective noise insulation, it is necessary to have a closed-window condition, which may not be desirable to home-owners in all seasons and which may impose the additional ongoing costs of climate-control systems.

Earth berms or manmade barriers must be both structured and positioned accurately on the ground, to be located between sources of loud ground-level noise on the airport and very close in noise sensitive receptors. They do not mitigate in-flight noise. People appear to hear less noise if they do not see the aircraft on the ground or the maintenance facility that is the source of the noise. A proper positioning of airport buildings can also function as a noise screen for adjacent communities against certain airport activities.

Operational procedures are an important element of the Balanced Approach to noise control around the airports. They include: flight noise abatement procedures at takeoff and landing; preferential runway use; takeoff and/or landing displacement on runways; noise abatement flight tracks/corridors; reverse thrust limitations; flight scheduling and so on. For example, noise abatement flight tracks or corridors effectively concentrate aircraft noise over a small area. This can be very effective if the underlying area is not populated. Reverse thrust limitations may involve the use of the minimum reverse thrust necessary for the safe operation of the aircraft. This is generally interpreted as reverse thrust no greater than reverse idle.

The selection of an appropriate procedure with regard to airport-specific environmental constraints requires the quantification and analysis of the available operational solutions for each runway and departure corridor in terms of noise and/or gaseous emissions. The environmental effects of the procedures depend on the type of aircraft and operating conditions. The assessment of noise effects as part of procedure should therefore be based on actual information regarding the airport fleet mix and geographical position of the airport and its runway(s) with regard to noise sensitive areas.

The ICAO (in PANS-OPS,[34] Part V, Chapter 3) provides recommendations regarding the conditions in which noise abatement procedures can be safely used and the envelope within which main flight parameters defining the procedure can be safely adapted for airport noise mitigation. One procedure called NADP1 is used to mitigate noise at relatively shorter distances and another procedure, called NADP2, to reduce noise at relatively greater distances from the brake release point. Examples of such flight parameters are the height at which engine thrust is reduced and the height at which acceleration and flap/slat retraction are initiated.

Environmental principles may be arranged into a hierarchical set to be applied in developing or analyzing noise abatement operating procedures. The most important is the avoidance of residential areas. In all cases, aviation safety, including system safety through simplified operating arrangements, will be given priority over noise abatement considerations. However, assuming safety conditions have been satisfied, the sole test for moving to a lower level standard is that the higher standard is not operationally practicable.

In some regions of the world, for example, in Europe, there is growing pressure to impose operating restrictions on night flights and hence a curfew, which is an 'easy and ready to use' instrument. A curfew can be either global or partial in nature. It is global when it bans all flights during an identified time period. It is partial when it prohibits the operation of specific types of aircraft, or prevents the use of specific runways, or only affects arrivals (landing) or departures (takeoff). The 'phase-out' of Chapter 2 aircraft from the operation before the 1998 (when 'phase-out' began to be implemented in international flights) has been considered as a curfew, realized in many airports of the world. Even now this curfew exists, for example, in the form of domestic flight restrictions, because 'phase-out' deals with international flights mostly.

In fact the duration and the timing of the curfew depend on parameters such as airport configuration, noise contours/levels, number of people living around the airport, types of aircraft and the nature of night activities at the airport concerned. In addition, cultural considerations and the traditions and lifestyle of people living around the airport could also be taken into account.

There are more than 600 types of curfew implemented worldwide.[35] At present, nine airports in the UK impose a unique type of curfew in the form of Quota Count systems. These are based on a count of aircraft movements against a noise quota according to aircraft noise classification, distinguishing between arrivals and departures. Their effect is to discourage the operation of the noisiest aircraft at these airports, especially for departures, while allowing flexibility in the mix of aircraft.

# 2 The main sources of aircraft noise

There has been a considerable decrease in noise levels from individual aircraft over the past 35 years. The noise levels produced by modern aircraft are about 22 dB lower than those of first generation jet aircraft. This reduction has been achieved as a result of the development of turbofan engines with high bypass ratios, liner technology and turbomachinery source noise reduction. Further aircraft noise reduction will be achieved from improved design of engine systems, decrease in airframe sources and the introduction of noise reduction technologies. Figure 2.1 shows the noise sources in a propulsion system.

## 2.1 Jet noise

The characteristics of noise from a turbofan engine depend upon its construction and the parameters of flow in the duct of the engine. There are many methods and computer programs for predicting aviation noise.[1–9] Here semi-empirical models for predicting the noise generated by the main sources in the turbofan engine are used.

The sound pressure levels $L_j(f)$ of the noise of the turbojet and turbofan resulting from each of $j$ sources (where $j = 1$ represents the coaxial jets, $j = 2$ represents the fan, $j = 3$ represents the turbine, $j = 4$ represents the combustion chamber, and $j = 5$ represents the airframe) in each third-octave frequency band with center frequency $f$ are predicted by the simple relationship (Chapter 4)

$$L_j(f) = L_j(\vec{\lambda}) + \vec{Y}_j, \tag{2.1}$$

where $\vec{\lambda}$ is the vector of parameters used in the $j$th source model (for example, velocity, temperature of the primary and the secondary of coaxial jets), and $\vec{Y}_j$ is an adjustment vector determined from the $j$th source model.[19]

The sound pressure levels due to coaxial shock-free jets without correction for refraction in the far field and effects of atmospheric absorption are

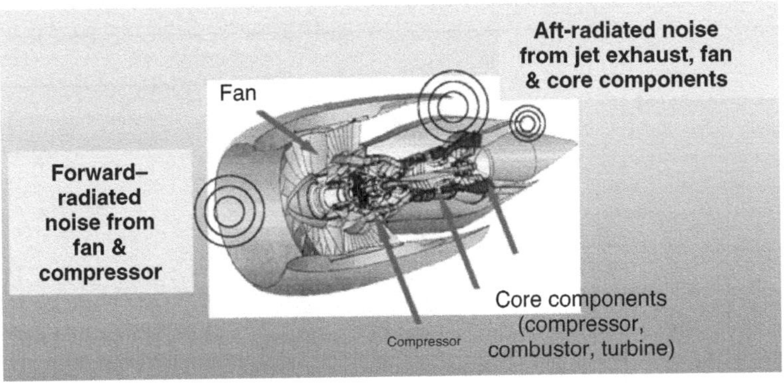


*Figure 2.1* Turbofan engine noise.

calculated from:

$$L_1\left(f,\vec{\lambda}\right) = 10\lg\left(\frac{A_1}{r^2}\right) + 20\lg\left(\frac{p_A}{p_{SA}}\right) + \sum_i \Delta L_{1i}\left(f,\vec{\lambda}\right) + \vec{Y}_1, \qquad (2.2)$$

where $A_1$ is the area of the exit jet, $r$ is the source-to-observer distance, $p_A$, $p_{SA}$ are respectively the ambient and standard sea level atmosphere pressures and $\Delta L_{1i}(f,\vec{\lambda})$ are spectral corrections for coaxial jets. As a rule, a reference distance of 1 m is adopted. $\Delta L_{11}(f,\vec{\lambda})$, the correction of the noise level for reference base turbojet, is a function of the jet Mach number and polar angle from the inlet axis. $\Delta L_{12}(f,\vec{\lambda})$, the additive spectral correction is a function of the jet Mach number, of the polar angle from inlet centerline to exhaust centerline and of the jet temperature and density. $\Delta L_{13}(f,\vec{\lambda})$, the flight velocity correction, is a function of the aircraft flight velocity, jet Mach number and polar angle from the inlet axis. $\Delta L_{14}(f,\vec{\lambda})$, the correction on the coaxial jets, is a function of the primary and secondary velocity, density, temperature of airflow, of the exit jets diameters and of the polar angle from the inlet axis. The functions $\Delta L_{1i}(\vec{\lambda})$ are determined by choosing values of key parameters (for example, Strouhal and Mach number, the enthalpy ratio, the density ratio, geometrical parameters) that provide minimum differences between experimentally measured and calculated values of the jet sound pressure level given by (2.2).[19]

The effects on turbojet and turbofan engine noise of (1) variation in velocity and temperatures of the gases flowing from bypass duct and exhaust nozzle; (2) variation in the fan and compressor pressure ratio; and (3) decreasing turbine pressure ratio are considered next.

Jet noise has been determined for bypass ratio ($m$) values of: 1, 2.5, 4, 6. For illustrative calculations, the thrust is assumed to be 11.5 t and the engine airflow mass is assumed to be 300 kg/s. The engines considered are turbofans with either an air-separation exhaust nozzle or with air mixing. The basic

*Table 2.1* The main parameters for bypass turbojets with air separation exhaust nozzles

| *Parameter* | *Bypass ratio, m* | | | |
|---|---|---|---|---|
| | *1* | *2.5* | *4* | *6* |
| $u_1$, m/s | 525 | 445 | 466 | 461 |
| $u_2$, m/s | 366 | 348 | 297 | 265 |
| $T_1$, K | 910 | 840 | 772 | 760 |
| $T_2$, K | 370 | 362 | 343 | 333 |
| $\pi_F$ | 2.50 | 2.00 | 1.70 | 1.55 |
| $\pi_S$ | 11.20 | 18.00 | 34.00 | 29.70 |
| $\pi_t$ | 5.20 | 12.40 | 18.00 | 15.90 |

*Table 2.2* The main parameters for bypass turbojets with air mixing

| *Parameter* | *Bypass ratio, m* | | | |
|---|---|---|---|---|
| | *1* | *2.5* | *4* | *6* |
| $u_1$, m/s | 440 | 412 | 345 | 332 |
| $u_2$, m/s | 440 | 412 | 345 | 332 |
| $T_1$, K | 543 | 635 | 419 | 395 |
| $T_2$, K | 543 | 635 | 419 | 395 |
| $\pi_F$ | 2.50 | 2.00 | 1.70 | 1.55 |
| $\pi_S$ | 11.20 | 18.00 | 34.00 | 29.70 |
| $\pi_t$ | 5.20 | 12.40 | 18.00 | 15.90 |

parameters of the airflow in these engines are shown in Table 2.1 (for a turbofan with airflow separation) and in Table 2.2 (for a turbofan with mixing of the hot core flow and cold fan flow prior to expansion can).

Figure 2.2 shows typical broadband spectra (plotted against Strouhal number $Sh = fd/u_j$) for the simplest subsonic jet and coaxial subsonic turbofan jets (bypass ratios of 1, 2.2 and 4) with mixing of the hot and cold fan airflow.

The maxima in the third-octave frequency band spectra for turbofan jets are in the range of Strouhal numbers between 0.3 and 0.5 depending on the bypass ratio of the turbofan. Figure 2.3 shows the rate of increase of overall sound pressure level (OASPL) due to a turbofan with respect to the jet speed. The effect of operational jet speeds between 400 and 630 m/s on turbofan noise spectra in the direction of the maximum noise radiation is shown in Fig. 2.4.

For a subsonic jet velocity, the noise spectrum is similar. The variation in third-octave noise level spectra of a turbofan jet with angle between 20° and 150° relative to the inlet centerline is presented in Fig. 2.5. Figure 2.6 shows the variation of the turbofan noise spectra with operating jet exhaust speed

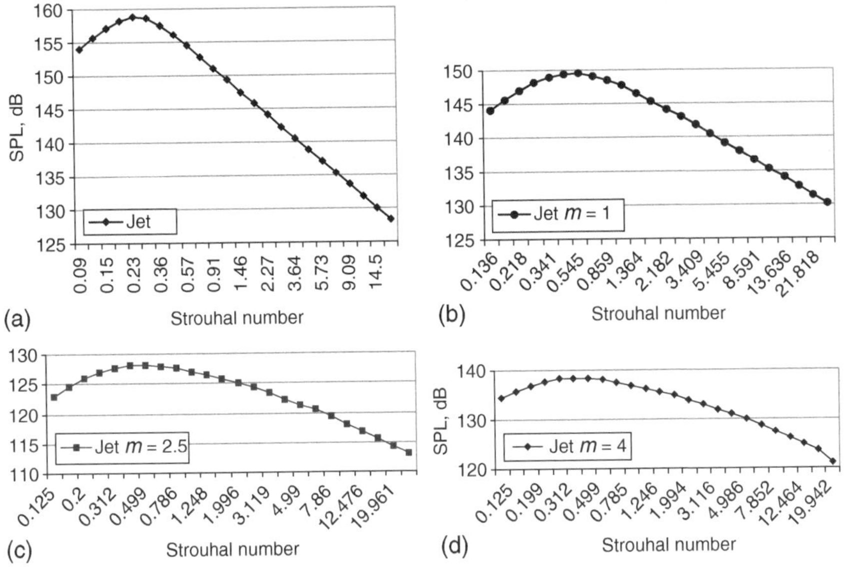


*Figure 2.2* The predicted influence of the engine bypass ratio on jet noise spectrum in the direction of maximum noise radiation: (a) pure jet; (b) $m = 1$; (c) $m = 2.5$; (d) $m = 4$ ($m = 1\ldots4$ for bypass engines with mixed airflow).

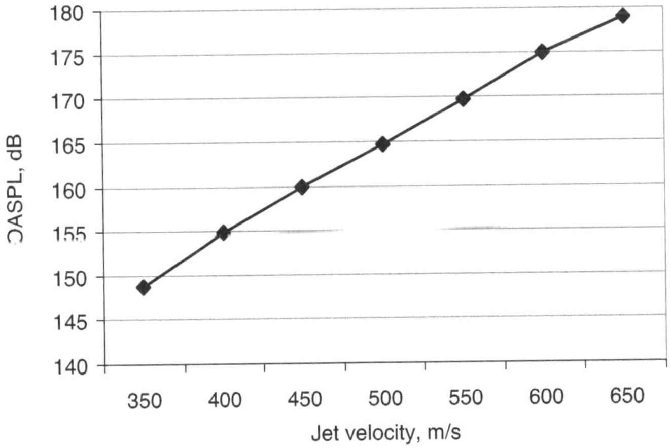


*Figure 2.3* The predicted dependence of jet noise OASPL on the jet velocity ($m = 1$) in the direction of maximum noise radiation.

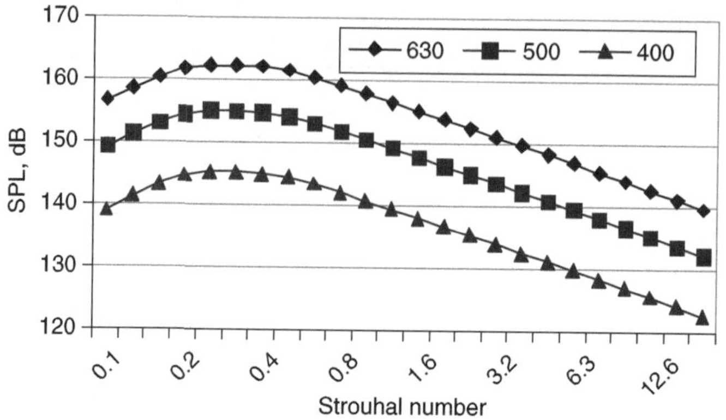


*Figure 2.4* The predicted influence of jet velocity variance on the Strouhal number dependence of the jet noise spectrum ($u_j = 400$–$630$ m/s, $m = 1$) in the direction of maximum noise radiation.

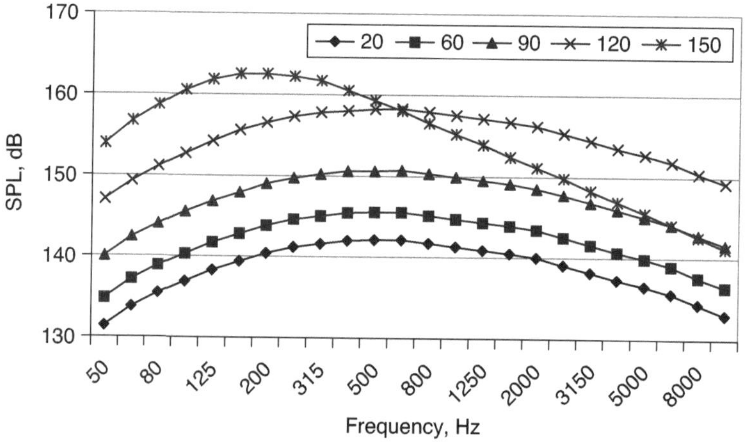


*Figure 2.5* The predicted dependence of the spectrum on the directivity angle of the jet noise radiation ($u_j = 630$ m/s, $m = 1$).

(between 350 and 650 m/s) and angle. These figures show maxima between 140° and 150° from the inlet centerline.

The jet temperature is also an important parameter for jet noise. Figure 2.7 illustrates the variation of the noise spectra with temperature $T$ in the range between 450° and 750° K.

For a turbofan with bypass, there is mixing of the hot core flow and cold fan flow with the atmosphere after jet expansion. The noise levels produced by aircraft with bypass turbofan engines are determined by

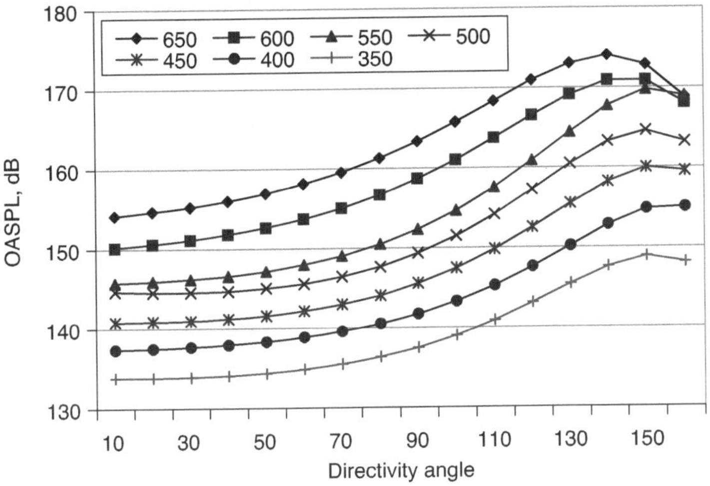


*Figure 2.6* The predicted influence of jet velocity on the noise directivity pattern ($m = 1$).

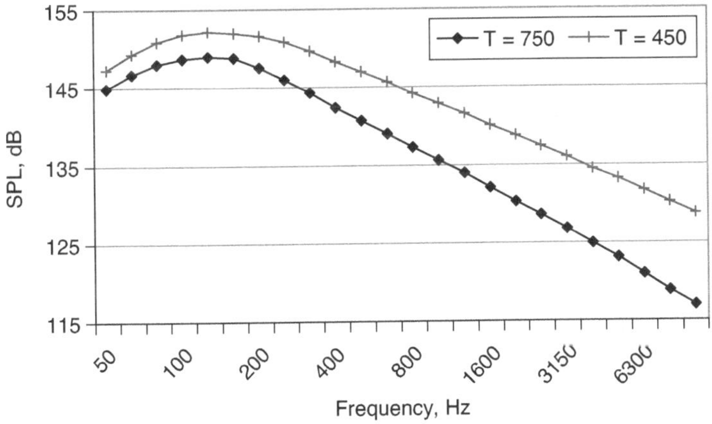


*Figure 2.7* The predicted dependence of the jet spectrum in the direction of maximum noise radiation on the jet temperature.

the nozzle geometry, temperature and velocity of the jets and the flight velocity. In Fig. 2.8, the spectrum amendment for coaxial turbofan jets has been plotted against equivalent Strouhal number $Sh_e = fd_e/u_1$, where $d_e = d_1\sqrt{1+(\overline{d}^2 - 1)\overline{u}^2}$ is the effective jet diameter, $\overline{d} = d_2/d_1$, $\overline{u} = u_2/u_1$; $i = 1$; corresponds to the parameters of the core jet and $i = 2$ corresponds

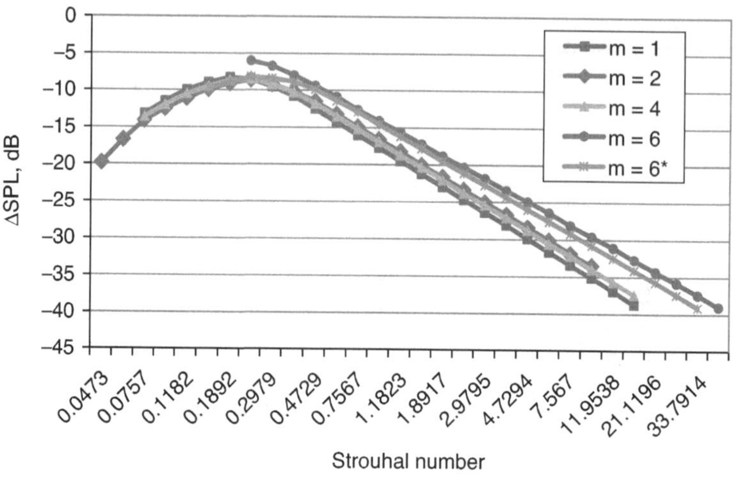


*Figure 2.8* Dependence of the Strouhal number relationship for the bypass jet noise spectrum on bypass ratio ($m = 1\ldots.6$) and co-annular jet velocity ratio (for the curve corresponding to $m = 6$, $u_2/u_1 = 1$, for the curve corresponding to $m = 6^*$, $u_2/u_1 = 0.6$) in the direction of maximum noise radiation.

to the parameters of the bypass jet. For the chosen parameters, the spectra in the direction of maximum noise radiation are similar. The directivity of turbofan bypass jets depends upon the bypass and velocity ratios (see Figs 2.9–2.11). Figure 2.9 shows that, for a given velocity ratio ($u_2/u_1 = 1$), the directivity patterns of engines vary with the bypass ratio. Figure 2.9 also shows that the directivity patterns of engines with the same bypass ratio ($m = 6$) vary with the velocity ratio. The flight velocity of an aircraft is one of the operational parameters influencing noise levels. Figure 2.10 shows the predicted influence of turbofan aircraft flight velocity (between 50 and 90 m/s) on overall SPL as a function of Strouhal number ($m = 6$). Figure 2.11 shows the predicted flight velocity influence on directivity. In each case, Doppler effects influence the noise spectra and directivity characteristics for given bypass jet parameters.

## 2.2 Fan and turbine noise

The fan, compressor and turbine of an aircraft engine generate tonal and broadband noise. Broadband noise results from the interaction of inhomogeneous pressure with turbulent flow. The blade-passage tone and its harmonics for subsonic tip Mach numbers result from the interactions of the pressure fields produced by the flow on the rotor/stator blade rows. There are additional multiple pure tones – 'buzz-saw' noise – which accompany

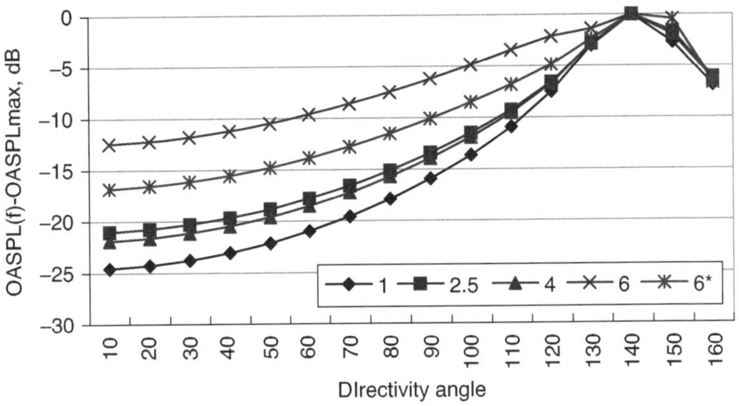


*Figure 2.9* Predicted dependence of the directivity pattern of bypass jet noise on bypass ratio ($m = 1\ldots.6$) and co-annular jet velocity ratios (for the curve corresponding to $m = 6$, $u_2/u_1 = 1$, for the curve corresponding to $m = 6^*$, $u_2/u_1 = 0.6$).

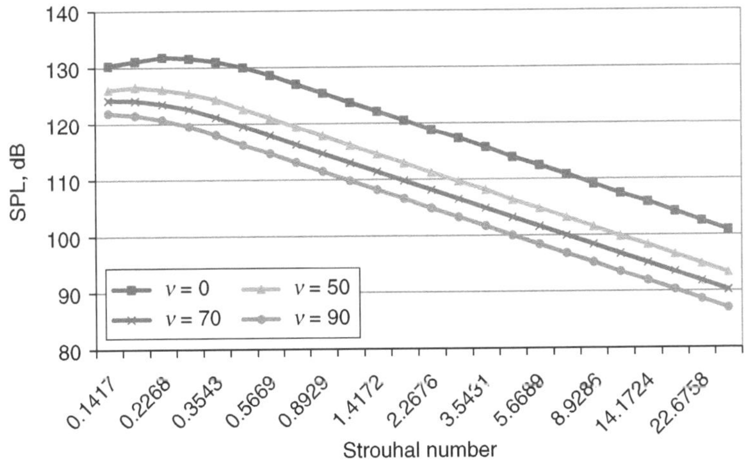


*Figure 2.10* Influence of flight speed on the jet spectrum ($m = 6$).

supersonic tip Mach numbers associated with supersonic flow on the blades and the formation of shock waves. The latter phenomenon is typical during takeoff. To determine the acoustical characteristics of the fan, compressor and turbine, it is necessary to take account of the noise generation, the noise propagation in the duct and the forward acoustic radiation, the rearward radiation from the bypass duct and core of the engine.

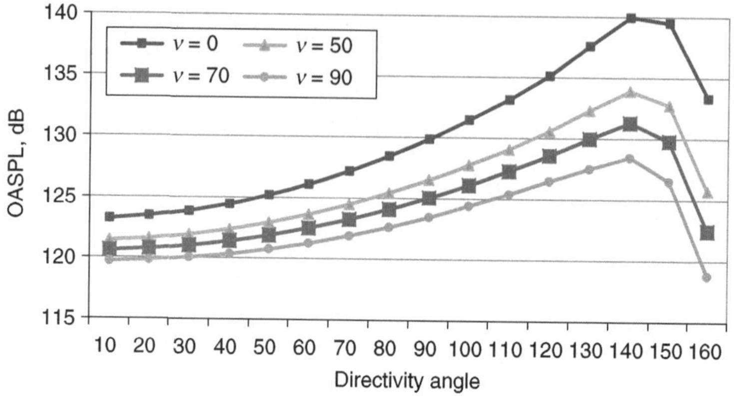


*Figure 2.11* Influence of the flight speed on jet directivity ($m = 6$).

The sound pressure levels of the fan (or turbine) are given by equation (2.3), excluding the effects of atmospheric absorption and assuming that there is no acoustic treatment:

$$L_2(f, \vec{\lambda}) = 10\lg\left(\frac{A_2}{r^2}\right) + \sum_i \Delta L_{2i}(f, \vec{\lambda}) + \vec{Y}_2 \qquad (i = 1 \text{ to } 10) \qquad (2.3)$$

where $A_2$ is the inlet flow area for the fan (which is the exit flow area for the turbine); $\Delta L_{21}(\vec{\lambda})$ and $\Delta L_{22}(\vec{\lambda})$ are, respectively, the noise level corrections determined by the temperature difference $\Delta T$ between the fan and the turbine; $\Delta L_{23}(f, \vec{\lambda})$ is the correction for multiple pure tones; $\Delta L_{24}(f, \vec{\lambda})$ and $\Delta L_{25}(f, \vec{\lambda})$ are, respectively, the spectral corrections for the broadband and tonal noise sources considered as a function of the third-octave frequency band spectrum $f$; $\Delta L_{26}(\vec{\lambda})$ is the correction for the tip-speed Mach number $M_t$; $\Delta L_{27}(f, \vec{\lambda})$, $\Delta L_{28}(f, \vec{\lambda})$ are, respectively, corrections for the directivity of the broadband and tonal noise sources considered as function of the polar angle from inlet centerline to exhaust centerline $\theta$; $\Delta L_{29}(f, \vec{\lambda})$ is the flight velocity correction considered as a function of the aircraft flight velocity; and $\Delta L_{210}(\vec{\lambda})$ is a correction depending on the peculiarity of the turbofan (turbine) construction (rotor-stator axial spacing, type of the mixed-flow exhaust nozzle, vane-blade ratio and so on).[9–11,19] The procedure for predicting the fan (turbine) noise is presented in Fig. 2.12.

In Fig. 2.12, the parameters $a_i$ are empirical constants, obtained from measurements on a specific type of the engine and $f_{bp}$ is the rotor blade passing frequency. The corrections for acoustic treatment of noise inlet radiation, bypass duct and exhaust radiation depend on the characteristics of the inlet (bypass duct) passive liners, fan active noise control and stream liners.

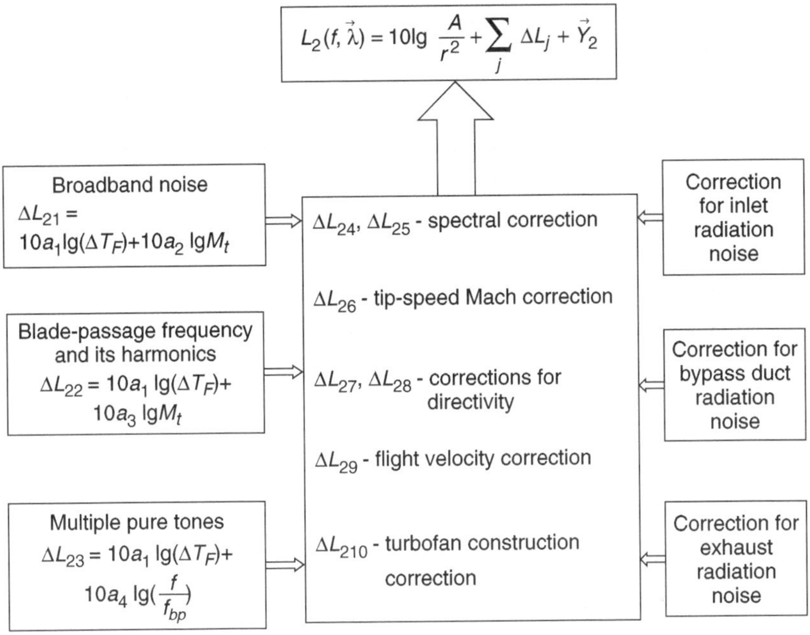


*Figure 2.12* A method for fan (turbine) noise prediction.[19]

The broadband noise of the fan has an aerodynamic origin. Tones are observed at the fundamental frequency and at the second harmonics. Figures 2.13 and 2.14 show typical third-octave fan noise spectra predicted in the forward and backward directions, respectively, for two bypass ratios [(a) $m = 2.5$ and (b) $m = 6$)] and three operational modes of the engines. During aircraft approach before landing, the frequencies of the tones decrease since the rotor rotation decreases.

The directivity patterns of the fan are plotted in Fig. 2.15a and b as predicted for five modes of particular turbofan engines. The maxima of the radiated fan noise are predicted to be between 200 and 400 from the forward direction. There are secondary maxima between 1200 and 1300 in the backward direction.

Figure 2.16 shows predictions of the different directivity patterns of fan and jet for maximum thrust and approach turbofan engine modes, respectively, for bypass ratio $m = 2.5$ (Fig. 2.16a) and $m = 6$ (Fig. 2.16b). Comparison of the fan and bypass jet directivity patterns (Fig. 2.16) enables understanding of the different contributions to the overall directivity properties of turbofan aircraft.

The sources of turbine noise are similar to those in the fan. The generation and transmission of turbine noise occur in regions with high temperature flow from the combustion chambers. A peculiarity of the turbine noise is

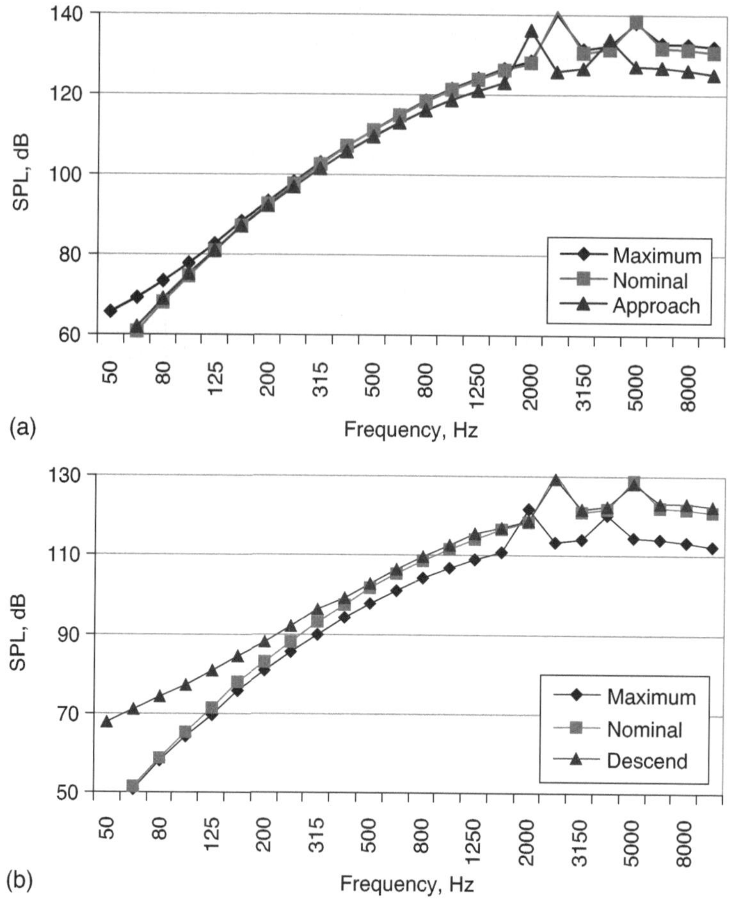


*Figure 2.13* Predicted fan noise spectra in the forward direction of noise radiation for various operational modes: (a) turbofan engine with mixed airflow, $m = 2.5$; (b) bypass engine, $m = 6$.

propagation and radiation from the core exhaust of the turbofan engine. There are the broadband and high-frequency discrete tones in the turbine noise spectra for all operational modes of the turbofan engine, as shown in the predictions in Fig. 2.17. The tonal noise sources (as a rule the tones are at the rotor blade passing frequency) contribute mainly at high frequencies.

The predicted directivity pattern of turbine noise is illustrated in Fig. 2.18. The sound pressure level of the turbine noise radiation in the backward direction is less than that from the fan (compare Figs 2.15b and 2.18) as a result of the fact that a low fraction of the engine airflow will pass through the turbine. For a turbofan engine with high bypass ratio ($m = 6$ in Figs 2.15b

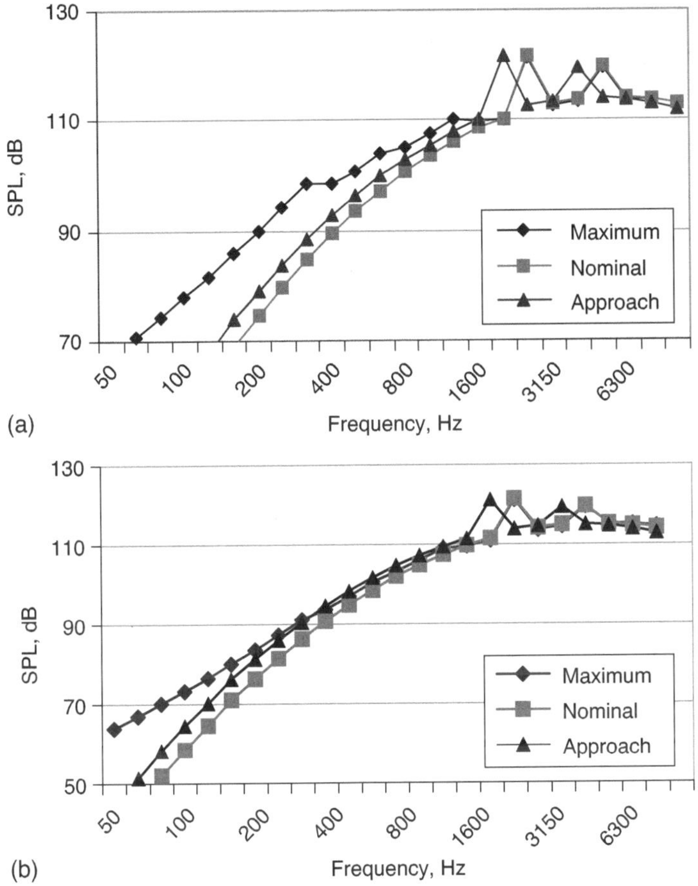


*Figure 2.14* Predicted fan noise spectra in the backward direction of noise radiation for various operational modes: (a) turbofan engine with mixed airflow, $m = 2.5$; (b) bypass engine, $m = 6$.

and 2.18), the part of the engine airflow used in the turbine is the fraction $\frac{1}{(1+m)}$ (14.3 per cent for the data used for Figs 2.15b and 2.18).

## 2.3 Combustion chamber noise

Noise is created during the combustion of the fuel and turbulent airflows in the engine combustion chamber. The low-frequency components of combustion noise propagate through zones with temperature fluctuations, through the pressure jump in the turbine and then radiate through the exhaust nozzle (core noise). Core nose is observed at low engine jet velocity (for example, during aircraft approach). In this case, the noise level of the

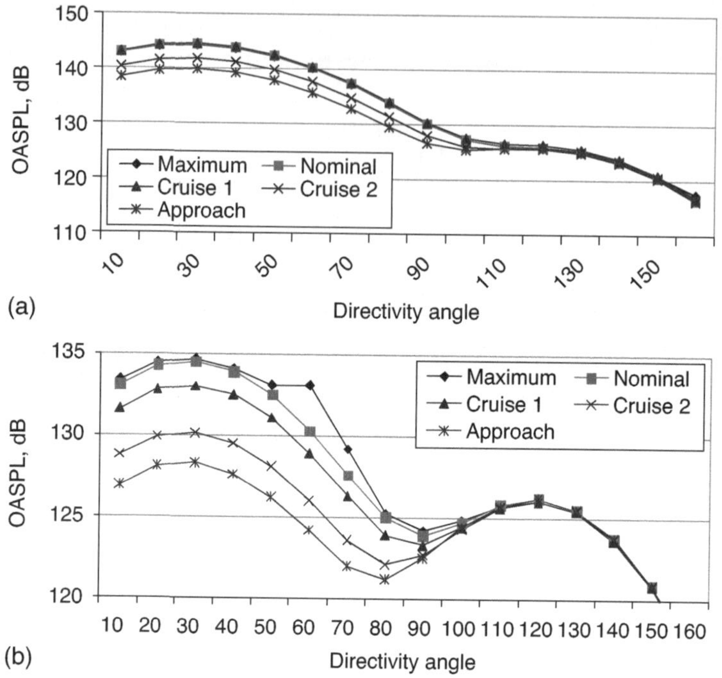


*Figure 2.15* Directivity patterns predicted for five operational engine modes: (a) turbofan engine with mixed airflow, $m = 2.5$; (b) bypass engine, $m = 6$.

engine exceeds that predicted by Lighthill's jet theory, that is, the law of the eighth power of the jet velocity equation (1.17) in Chapter 1.

A semi-empirical model for predicting the noise generated by engine combustion and core noise has the form of equation (2.1) (effects of atmospheric absorption are not included):

$$L_4\left(f,\vec{\lambda}\right) = 10\lg\left(\frac{A_4}{r^2}\right) + \sum_i \Delta L_{4i}\left(f,\vec{\lambda}\right) + \vec{Y}_4, \tag{2.4}$$

where $A_4$ is the combustor cross-sectional area, $\Delta L_{41}(\vec{\lambda}) = 10b_1 \lg G$ is a correction depending on the combustion mass flow rate G, $b_1$ is empirical constant; $\Delta L_{42}(\vec{\lambda})$ is a correction depending on the combustor inlet $T_i$ and outlet $T_o$ temperatures; $\Delta L_{43}(\vec{\lambda})$ are corrections depending on the pressure and temperature drop across the turbine; $\Delta L_{44}(f,\vec{\lambda})$ is a spectral correction for the broadband and tone noise sources considered as a function of third-octave frequency band spectrum f; $\Delta L_{45}(\theta,\vec{\lambda})$, is a directivity correction for the broadband combustion noise sources considered as a function of the polar angle from inlet centerline to exhaust

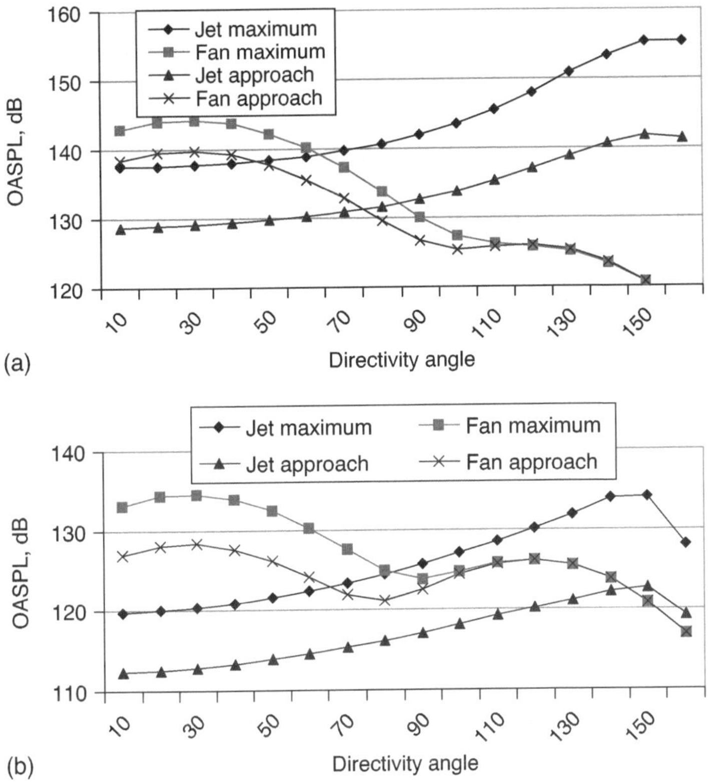


*Figure 2.16* Comparison between jet and fan noise levels in various directions of noise radiation for maximum and approach operational modes: (a) turbofan engine with mixed airflow, $m = 2.5$; (b) bypass engine, $m = 6$.

centerline $\theta$; $\Delta L_{46}(f, \vec{\lambda})$, the flight velocity correction, is considered as a function of the aircraft flight velocity; $\Delta L_{47}(\vec{\lambda})$ is a correction for the particular type of combustor construction (can-type, annular, combustor length and so on).[19]

Figure 2.19 shows the low-frequency broadband combustion noise spectrum for a turbofan engine with $m = 2.5$ at maximum thrust. A typical directivity pattern for combustion chamber noise (as part of the turbofan engine) is shown in Fig. 2.20.

## 2.4 Airframe noise

Airframe noise is the result of complex noise-generating aerodynamic sources on the wing, the horizontal and vertical tail, the flaps, the slat, the gear, the landing gear well dome, the cavities and the airfoil-tip vortex.

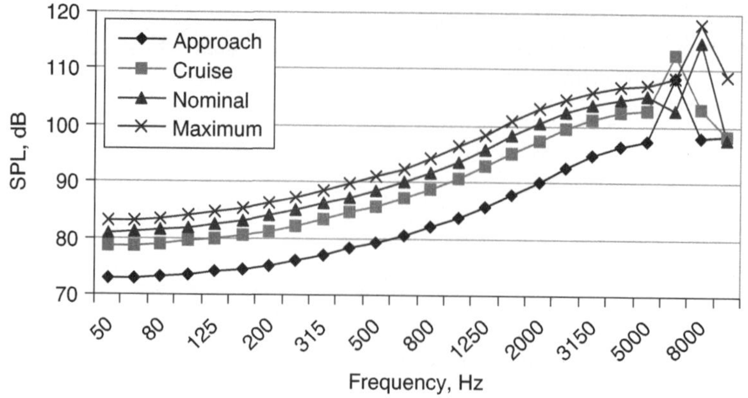


*Figure 2.17* Predicted third-octave turbine noise spectra for four operational engine modes ($m = 6$). Cruise refers to the cruise flight mode.

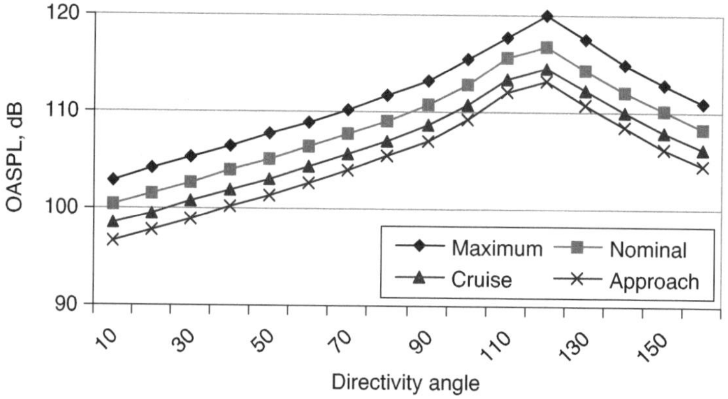


*Figure 2.18* Predicted directivity pattern for turbine noise ($m = 6$).

For short takeoff and landing aircraft, there are additional noise sources, that is, the underwing, upper-surface blowing jets and blown flaps. Airflow interactions with the aircraft fuselage, the gear and the wing devices result in dipole noise sources.

An equation for predicting the sound pressure level arising from airframe noise in third-octave band frequencies may be written in form of the equation (2.1) (effects of atmospheric absorption are not included):

$$L_7\left(f,\vec{\lambda}\right) = -20\lg r + \sum_i \Delta L_{7i}\left(f,\vec{\lambda}\right) + \vec{Y}_7,$$

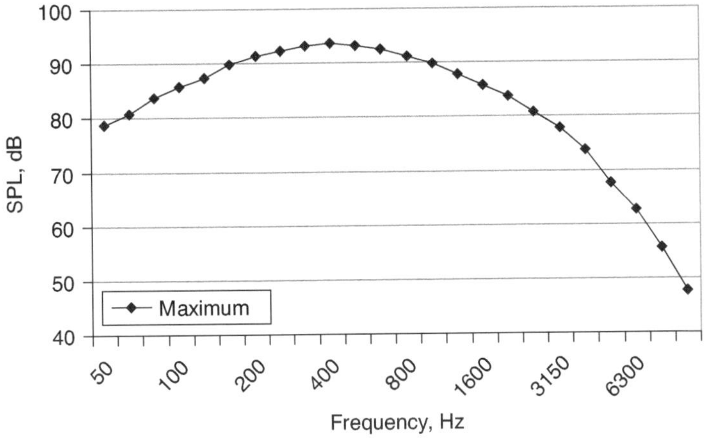


*Figure 2.19* Predicted combustion chamber noise spectra for four operational engine modes ($m = 2.5$).

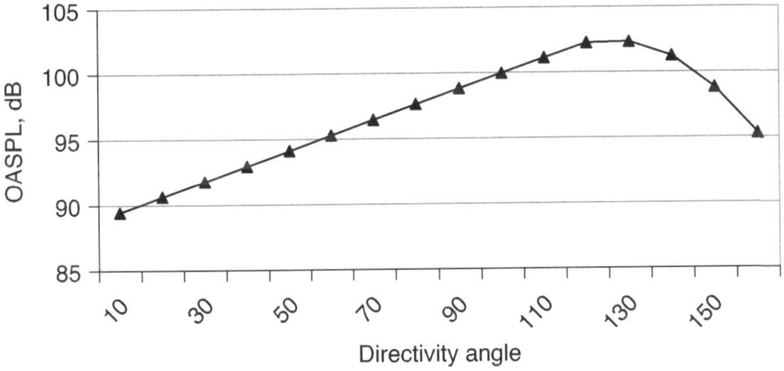


*Figure 2.20* A predicted directivity pattern for combustion chamber noise ($m = 2.5$).

where $\Delta L_{71}(\vec{\lambda}) = 10 d_1 \lg v$ is the flight velocity ($v$) correction; $d_1$ is an experimental constant; $\Delta L_{72}(f, \vec{\lambda})$ is a spectral correction for the broadband noise; $\Delta L_{73}(f, \vec{\lambda})$, a directivity correction for broadband noise sources, is considered to be a function of the observer angle $\theta$; and $\Delta L_{74}(f, \vec{\lambda})$ is a correction for the particular shape and size of the aircraft equipment (for example, the area of the element, the aspect ratio, the flap angle deflection, the landing gear construction and so on).[14,19]

The airframe-induced noise is an important source during aircraft approach, when the aircraft powerplant noise is decreased. Figures 2.21–2.23 show predictions of typical third-octave spectra for wing, flaps and landing gear, respectively. In Fig. 2.21, the wing noise spectrum is predicted as

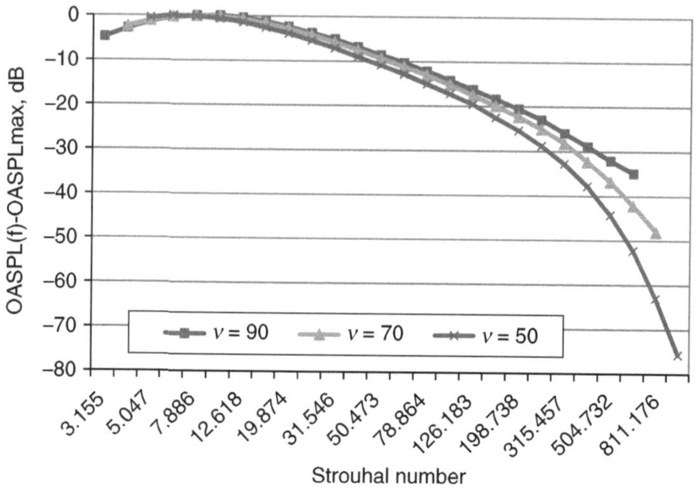


*Figure 2.21* A predicted typical dependence of the wing noise spectrum on Strouhal number.

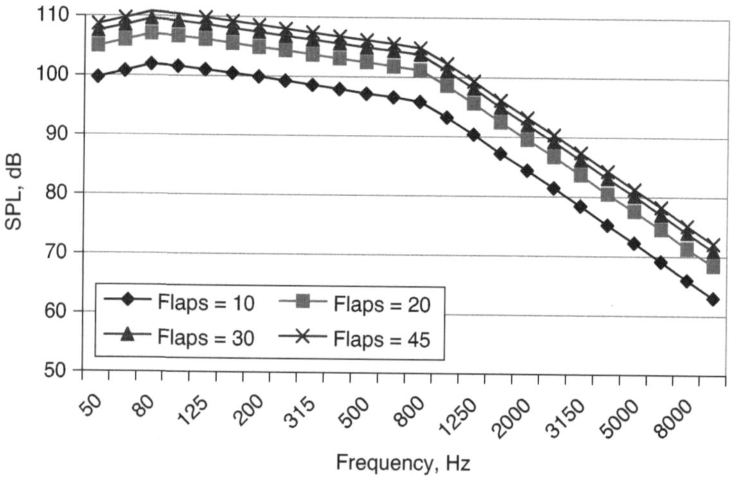


*Figure 2.22* Predicted influence of flap deflection angle on the spectrum of flap noise (for a Yakovlev-42 aircraft, $v = 70$ m/s).

a function of Strouhal number for three values of flight velocity $v$. The Strouhal number is calculated from $Sh = fc_{mac}/v$, where $c_{mac}$ is the length of the wing mean aerodynamic chord. Figure 2.22 shows predicted wing noise spectra for four values of the flap deflection angle. In Fig. 2.23, which shows predictions of noise associated with landing gear for three flight velocities,

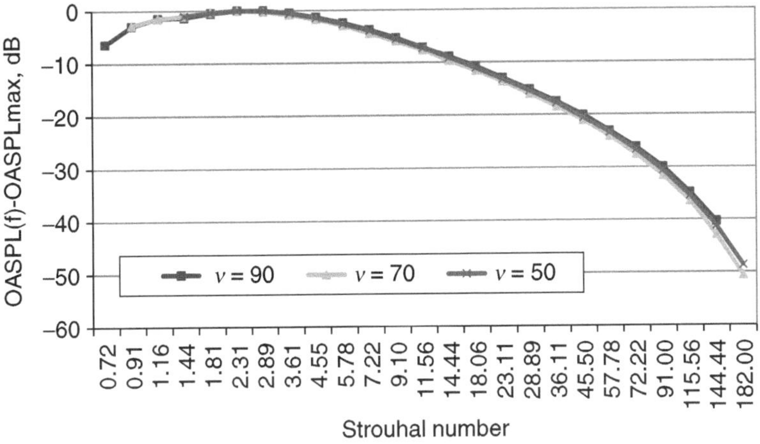


*Figure 2.23* The spectrum of gear noise for various flight speeds (predictions for a Yakovlev-42 aircraft).

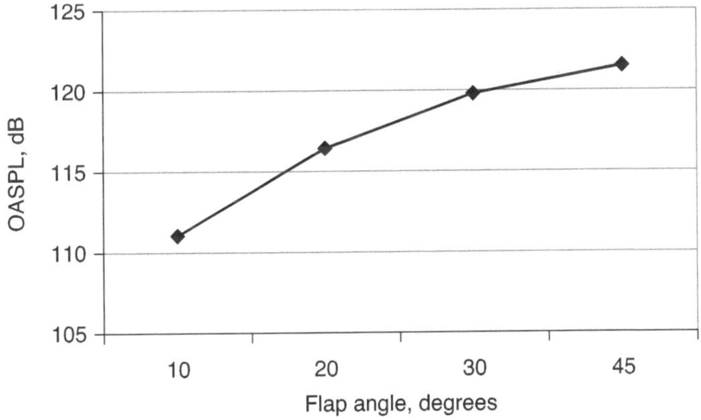


*Figure 2.24* Influence of flap deflection angle on overall flap noise level (prediction for a Yakovlev-42 aircraft, $v = 70$ m/s).

the Strouhal number is calculated from the diameter of the landing gear. It is predicted that the landing gear noise is practically independent of flight velocity. The wing and landing gear noise spectra are broadband and roughly similar. The flaps noise is determined by the flap angle, the flight velocity and the flap design.

Predicted values of OASPL as a function of the flap angle are given in Fig. 2.24.

Figures 2.25 and 2.26 compare the directivity patterns for the airframe noise sources associated with the Tupolev-154M and Yakovlev-42 aircraft

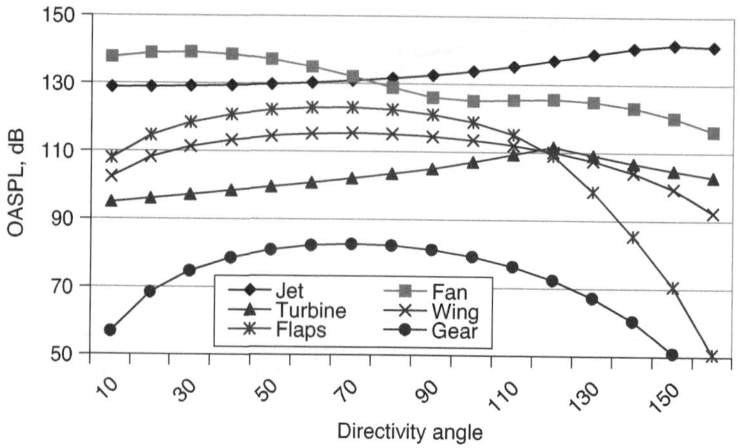


*Figure 2.25* Predicted directivity patterns for the main noise sources on a Tupolev-154M aircraft ($m = 2.5$, $v = 70$ m/s).

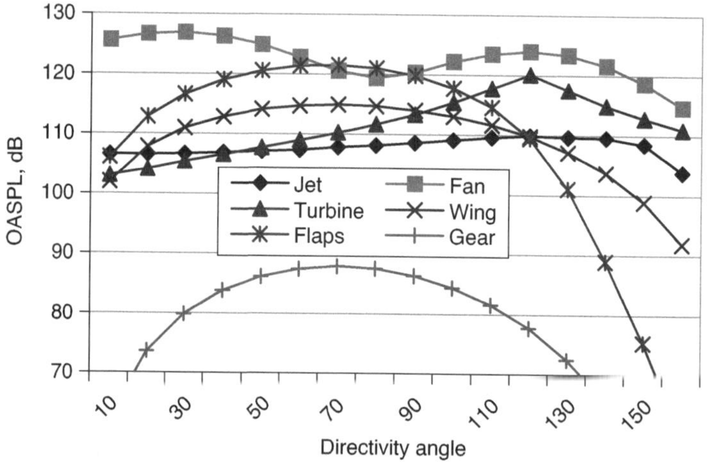


*Figure 2.26* Predicted directivity patterns for main noise sources of Yakovlev-42 aircraft ($m = 6$, $v = 70$ m/s).

with those of the main turbofan sources. Airframe noise sources make significant contributions on aircraft having turbofan engines with a high bypass ratio.

The acoustic model of an aircraft is the sum of particular models for the various noise sources including jet (propeller), fan (compressor), combustion chamber, turbine and airframe. They enable assessment of the aircraft as a complex noise source and investigation of the influence on the overall

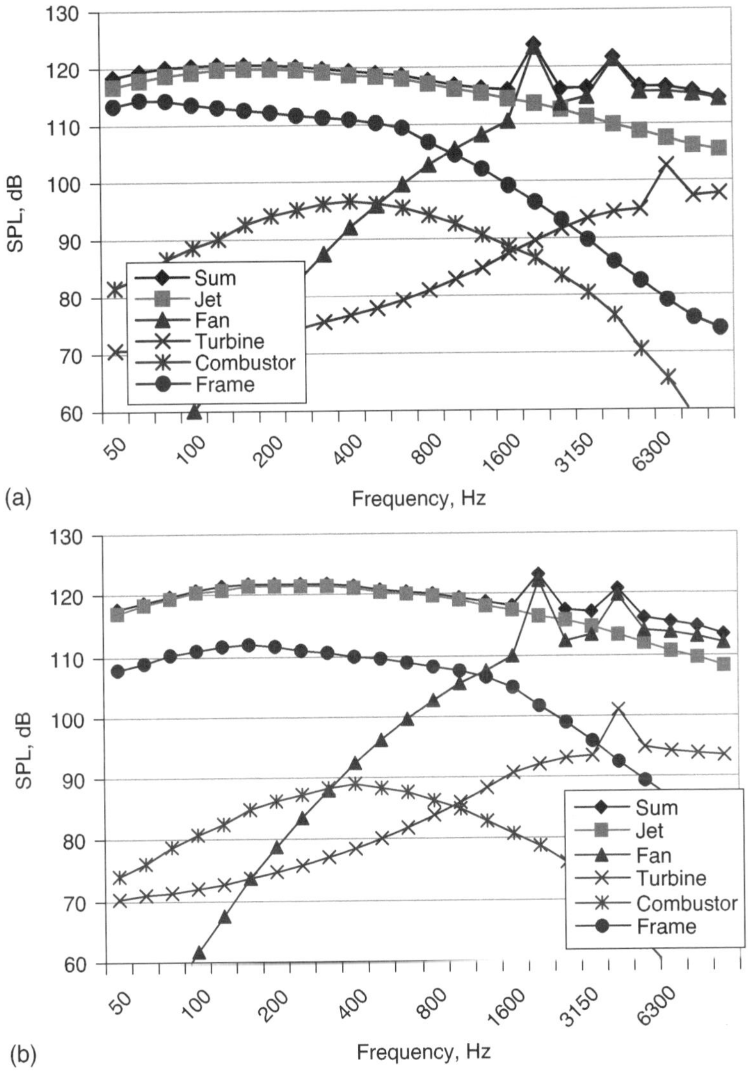


*Figure 2.27* The predicted spectral dependencies of noise sources on various types of aircraft with turbofan engines during landing approach and at a directivity angle of 90°: (a) Il'ushin-86 ($m = 1$, mixed jet, $v = 70$ m/s); (b) Tupolev-154M ($m = 2.5$, mixed jet, $v = 70$ m/s); (c) Il'ushin-96 ($m = 4$, mixed jet, $v = 70$ m/s); (d) Yakovlev-42 ($m = 4$, bypass jet, $v = 70$ m/s).

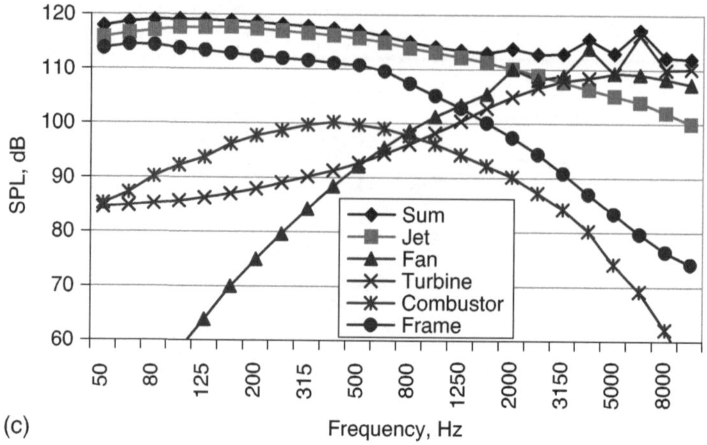


(c)

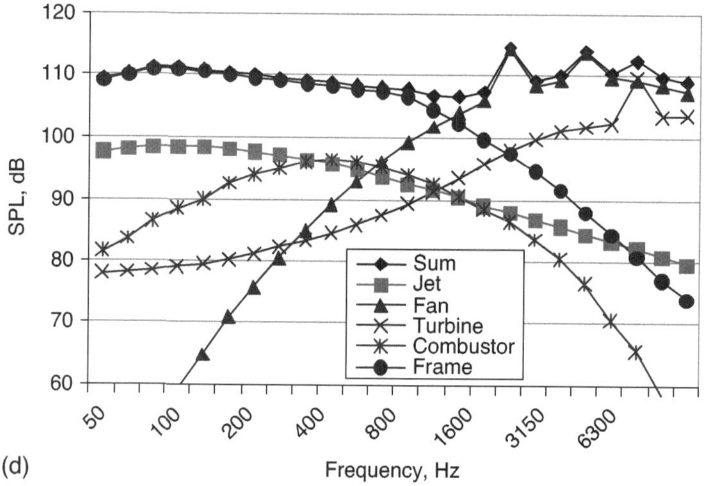


(d)

*Figure 2.27* Cont'd

aircraft acoustic design of powerplant parameters. Fig. 2.27(a–d) show example predictions of the different source contributions to turbofan aircraft noise spectra. In the following chapters, this method is used to evaluate the influence of operational factors on the noise impact of aircraft in the vicinity of airports.

## 2.5 Propeller and helicopter noise

The propeller is the main noise source on a turboprop engine. Thickness noise arises as result of periodic displacement of the air by the volume of the passing blade. The pressure fluctuation due to lift and draft disturbance gives the loading noise of the propeller. When flow over the blade section is transonic, noise is generated.[15,16,18,19] A propeller noise spectrum contains both broadband and harmonic noise. The harmonic noise has components with frequencies given by $f_k = nzk$ (where $k = 1, 2, \ldots, n$ is the rotational speed and $z$ is the number of blades). If there are a small number of blades and subsonic blade section speeds, then the noise is determined mainly by the first two or three harmonics. For such a propeller, the level of broadband noise is less than the first harmonic level by about 10 dB.

On a helicopter, the main contributions to the acoustic field are provided by aerodynamic and mechanical sources. The aerodynamic noise sources include the main rotor, tail rotor and gas turbine engine. The gearbox and transmission are mechanical noise sources, which radiate relatively high frequency sounds. These are attenuated rapidly during propagation in the atmosphere. The main rotor is a source of impulsive noise. The acoustic spectrum of a helicopter (for example, having a single main rotor) includes discrete and broadband noise. The main rotor and tail rotor are responsible for the discrete frequencies. The origin of the broadband noise is the interaction of the blades with atmosphere turbulence and blade-following vortices.

An empirical formula in the form of equation (2.1) is used to predict propeller noise (effects of atmospheric absorption are not included):

$$L_5\left(f,\vec{\lambda}\right) = -20\lg r + \sum_i \Delta L_{5i}\left(f,\vec{\lambda}\right) + \vec{Y}_5,$$

where $\Delta L_{51}(\vec{\lambda})$ is a correction depending on the thrust, the power and tip speed Mach number; $\Delta L_{52}(\vec{\lambda})$ is a correction depending on the number of

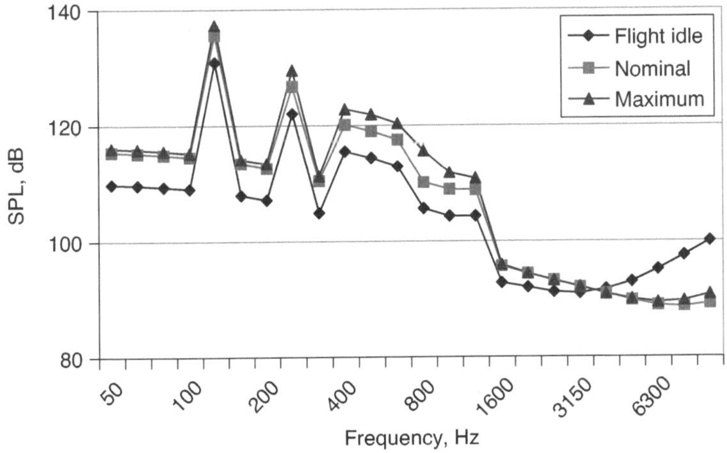


*Figure 2.28* The dependence of the spectrum of the propeller noise of the turboprop engine on the operational mode of the engine.

the blades and the diameter of the propellers; $\Delta L_{53}(f, \vec{\lambda})$ is a directivity correction for the harmonic and broadband noise sources considered as a function of the polar angle from inlet centerline to exhaust centerline $\theta$; $\Delta L_{54}(f, \vec{\lambda})$, is a flight velocity correction and $\Delta L_{55}(f, \vec{\lambda})$ is the spectral correction for the broadband noise. The predicted sound pressure level $L_{56}(f, \vec{\lambda})$ for helicopter includes the main and tail rotor harmonics and broadband noise.[18] The sound pressure levels (SPLs) of the fundamental and harmonic tones depend on the operational mode of the engine.

Typical predicted third-octave spectra from a turboprop propeller for three operational modes are shown in Fig. 2.28.

Although the spectra include the fundamental blade passing frequency and ten harmonics of that frequency, in the cases shown, the broadband levels mask the discrete frequencies resulting from the third and high harmonics. However, the broadband propeller noise is more than 17 dB lower than the blade passing frequency peak level. The flight velocity of the aircraft and the environmental conditions can transform the propeller noise spectrum as a result of Doppler effects, convective amplification of sound and the effects of propagation.

# 3 Aircraft noise propagation

## 3.1 Factors influencing outdoor sound

This chapter details the physical phenomena that determine the extent to which sound propagates outdoors, reviews data that demonstrate their influence and describes methods for predicting their acoustic effects. Propagation of sound outdoors involves geometric spreading, air absorption, refraction associated with wind and temperature gradients and effects of atmospheric turbulence. When the source and receiver are relatively close to the ground, there are interactions with the ground, barriers and buildings, topography and vegetation. These interactions are compounded by various meteorological factors.

### *3.1.1 Spreading losses*

Distance alone will result in wavefront spreading. In the simplest case of a sound source radiating equally in all directions, the intensity $I$ W/m at a distance $r$ m from the source of power $P$ W, is given by

$$I = \frac{P}{4\pi r^2}. \tag{3.1}$$

This represents the power per unit area on a spherical wavefront of radius $r$. In logarithmic form the relationship between sound pressure level $L_p$ and sound power $L_W$, may be written

$$L_p = L_W - 20\log r - 11\,\text{dB}. \tag{3.2}$$

From a point sound source, this means a reduction of 20log2 dB, i.e. 6 dB, per distance doubling in all directions (a point source is omnidirectional). Most sources appear to be point sources when the receiver is at a sufficient distance from them. If the source is directional, then (3.2) is modified by inclusion of the directivity index $DI$.

$$L_p = L_W + DI - 20\log r - 11\,\text{dB} \tag{3.3}$$

The directivity index is 10log*DF* dB, where *DF* is the directivity factor, given by the ratio of the actual intensity in a given direction to the intensity of an omnidirectional source of the same power output. Such directivity may be location induced. A simple case of location-induced directivity arises if an otherwise omnidirectional source, which would usually create spherical wavefronts of sound, is placed on a perfectly reflecting flat plane. Radiation from the source is thereby restricted to a hemisphere. The directivity factor for an omnidirectional source on a perfectly reflecting plane is 2 and the corresponding directivity index is 3 dB. For an omnidirectional source at the junction of a vertical perfectly reflecting wall with a horizontal perfectly reflecting plane, the directivity factor is 4 and the directivity index is 6 dB. It should be noted that these adjustments ignore phase effects and assume incoherent reflections.[1] This means, for example, it is assumed that there is energy doubling rather than pressure doubling above an acoustically hard boundary.

Directivity may also be an inherent characteristic of a source. An example of inherent directivity around an IL-86 aircraft in maximum power mode (defined at a reference distance $R_0 = 1$ m) is shown in Fig. 3.1. Overall A-weighted levels $L_A$, PNL and unweighted levels $L$ are shown as a function of direction normalized to a distance of 1 m.

For an omnidirectional source the points would lie on a circle. The greatest deviations from omnidirectionality and the highest levels for the IL-86 engine occur between 145° and 155° from the forward direction (0°).

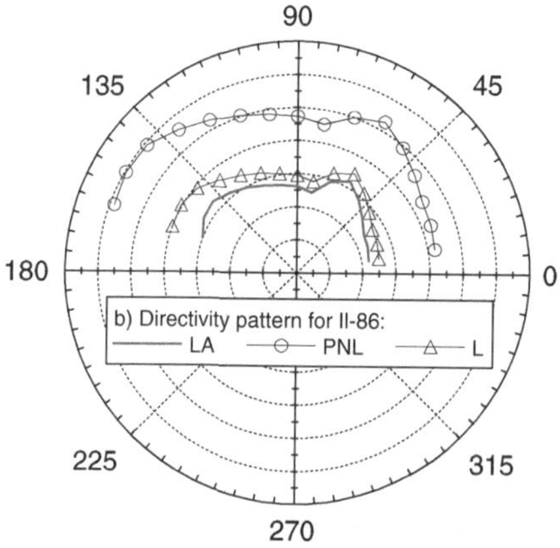


*Figure 3.1* Directivity patterns around an IL-86 aircraft in its maximum power mode (defined at a reference distance $R_0 = 1$ m). The contour grid circles are at intervals of 10dB from 80 dB at the inner circle to 150 dB at the outer circle.[2]

### *3.1.2 Atmospheric sound absorption*

A proportion of sound energy is converted to heat as it travels through the air. There are heat conduction losses, shear viscosity losses and molecular relaxation losses.[3] The resulting air absorption becomes significant at high frequencies and at long range, so air acts as a low pass filter at long range. For a plane wave, pressure $p$ at distance $x$ from a position where the pressure is $p_0$ is given by

$$p = p_0 e^{-\alpha x/2}. \tag{3.4}$$

The attenuation coefficient $\alpha$ for air absorption depends on frequency, humidity, temperature and pressure, and may be calculated using (3.5)–(3.7).[3]

$$\alpha = f^2\left[\left(\frac{1.84\times 10^{-11}}{\left(\frac{T_0}{T}\right)^{\frac{1}{2}}\frac{p_s}{p_0}}\right) + \left(\frac{T_0}{T}\right)^{2.5}\left(\frac{0.10680e^{-3352/T}f_{r,N}}{f^2+f_{r,N}{}^2}+\frac{0.01278e^{-2239.1/T}f_{r,O}}{f^2+f_{r,O}{}^2}\right)\frac{nepers}{m\cdot atm}\right] \tag{3.5}$$

where $f_{r,N}$ and $f_{r,O}$ are relaxation frequencies associated with the vibration of nitrogen and oxygen molecules, respectively, and are given by:

$$f_{r,N} = \frac{p_s}{p_{s0}}\left(\frac{T_0}{T}\right)^{\frac{1}{2}}\left(9+280He^{-4.17\left[(T_0/T)^{1/3}-1\right]}\right), \tag{3.6}$$

$$f_{r,O} = \frac{p_s}{p_{s0}}\left(24.0+4.04\times 10^4 H\frac{0.02+H}{0.391+H}\right), \tag{3.7}$$

where $f$ is the frequency, $T$ is the absolute temperature of the atmosphere in degrees Kelvin, $T_0 = 293.15$ is the reference value of $T$ (corresponding to 20°C), $H = \rho_{sat} r_h p_0/p_s$ being the percentage molar concentration of water vapor in the atmosphere, $r_h$ (per cent) is the relative humidity, $p_s$ is local atmospheric pressure and $p_0$ is the reference atmospheric pressure (1 atm = 1.01325 × 105 Pa) and $\rho_{sat} = 10^{C_{sat}}$, where $C_{sat} = -6.8346^{(T_0/T)^{1.261}} + 4.6151$.

These formulae give estimates of the absorption of pure tones to an accuracy of ±10 per cent for $0.05 < H < 5$, $253 < T < 323$, $p_0 < 200$ kPa.

Outdoor air absorption varies through the day and the year.[4,5] The absolute humidity, $H$, is an important factor in the diurnal variation and generally peaks in the afternoon. Usually, the diurnal variations are greatest in the summer. It should be noted that use of (arithmetic) mean values of atmospheric absorption may lead to overestimates of attenuation when attempting to establish worst-case exposures for the purposes of environmental noise impact assessment. Investigations of local climate statistics, say hourly means over one year, should lead to more accurate estimates of the lowest absorption values.

### *3.1.3 Ground effect*

When source and receiver are elevated but close to the ground, there is interference between sound travelling directly from the source to the receiver and sound reflected from the ground. Sometimes the phenomenon is called ground absorption but, since the interaction of outdoor sound with the ground involves interference, there can be enhancement associated with constructive interference as well as attenuation resulting from destructive interference. Near to non-porous concrete or asphalt, the frequencies of destructive interference may be calculated simply from the source–receiver geometry. If the source and the receiver are near the ground, the first destructive interference occurs at a relatively high frequency, so sound pressure is doubled over a wide range of audible frequencies. Such ground surfaces are described as acoustically hard. Over porous surfaces, such as soil, sand and snow, enhancement tends to occur at low frequencies, since the longer the sound wave, the less able it is to penetrate the pores. However, when the sound penetrates through a porous ground surface, there are changes in amplitude and phase during reflection and the resulting destructive interference can occur over a wide range of frequencies, typically between 200 and 800 Hz. The presence of vegetation tends to make the surface layer of ground, including the root zone, more porous. Snow is significantly more porous than soil and sand. The layer of partly decayed matter on the floor of a forest is highly porous too. Porous ground surfaces are sometimes called acoustically soft. The attenuation due to destructive interference that is observed in excess of that due to wavefront spreading and air absorption is the attenuation due to ground effect. The prediction of ground effect is detailed in Section 3.2.

### *3.1.4 Refraction by wind and temperature gradients*

The atmosphere is constantly in motion as a consequence of wind shear and uneven heating of the Earth's surface (Fig. 3.2). Any turbulent flow of a fluid across a rough solid surface generates a boundary layer. The main focus of interest, from the point of view of community noise prediction, is on the lower part of the meteorological boundary layer called the surface layer. In this surface layer, turbulent fluxes vary by less than 10 per cent of their magnitude but the wind speed and temperature gradients are largest.

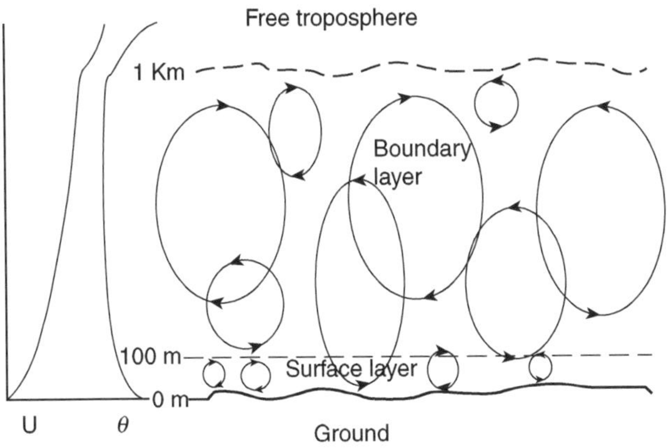


*Figure 3.2* Schematic representation of the daytime atmospheric boundary layer and turbulent eddy structures. The curves on the left show the mean wind speed ($U$) and the potential temperature profiles ($\theta = T + \gamma_d z$, where $\gamma_d = -0.098°C/km$ is the dry adiabatic lapse rate, $T$ is the temperature and $z$ is the height).

In typical daytime conditions, the surface layer extends over 50–100 m and is usually thinner at night. Turbulence may be modelled in terms of a series of moving eddies or 'turbules' with a distribution of sizes.

In most meteorological conditions, the speed of sound changes with height above the ground. Usually, temperature decreases with height (the adiabatic lapse condition). In the absence of wind, this causes sound waves to bend, or refract, upwards. Wind speed adds or subtracts from sound speed. When the source is downwind of the receiver, the sound has to propagate upwind. As height increases, the wind speed increases and the amount being subtracted from the speed of sound increases, leading to a negative sound speed gradient. Downwind, sound is refracted towards the ground. Wind effects tend to dominate over temperature effects when both are present. Temperature inversions, in which air temperature increases up to the inversion height, cause sound waves to refract downwards below that height. Under inversion conditions, or downwind, sound levels decrease less rapidly than would be expected from wavefront spreading alone.

In general, the relationship between sound speed profile $c(z)$, temperature profile $T(z)$ and wind speed profile $u(z)$ in the direction of sound propagation is given by

$$c(z) = c(0)\sqrt{\frac{T(z) + 273.15}{273.15}} + u(z), \tag{3.8}$$

where T is in °C and $u$ is in meters per second.

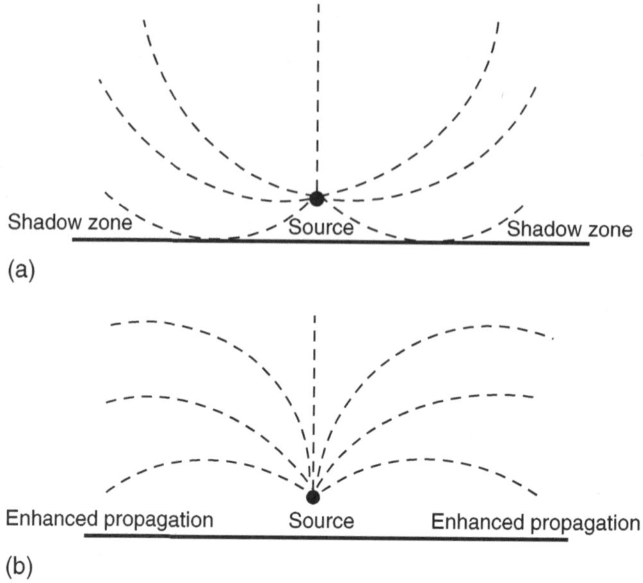


*Figure 3.3* The refraction of sound due to temperature gradients: (a) normal lapse rate condition (temperature decreasing with height); and (b) inversion condition (temperature increasing with height).

Normally, air temperature decreases with height. The corresponding sound rays (which represent high-frequency approximations of the sound propagation) curve away from the ground and there are shadow zones (see Fig. 3.3a) at both sides of the source that are not penetrated by rays. If the air temperature first increases up to some height before resuming its usual decrease with height, then there is an inversion. The height of the inversion is determined by the height at which there is a change of sign in the slope of the temperature gradient. Sound rays from sources beneath the inversion height will tend to be refracted towards the ground (see Fig. 3.3b). This is a favourable condition for sound propagation and may lead to higher levels than would be the case under acoustically neutral conditions. This will also be true for receivers downwind of a source since wind speed increases with height. The opposite may be true upwind of the source if the receiver is within the shadow zone defined by the grazing ray (see Fig. 3.4). If there is a wind gradient, then it tends to dominate over temperature gradient effects.

In many cases where the source or receiver is near to the ground, it is necessary to take into account any ground reflections. However, rather than use plane wave reflection coefficients to describe these ground reflections, a better approximation is to use spherical wave reflection coefficients (see Section 3.2).

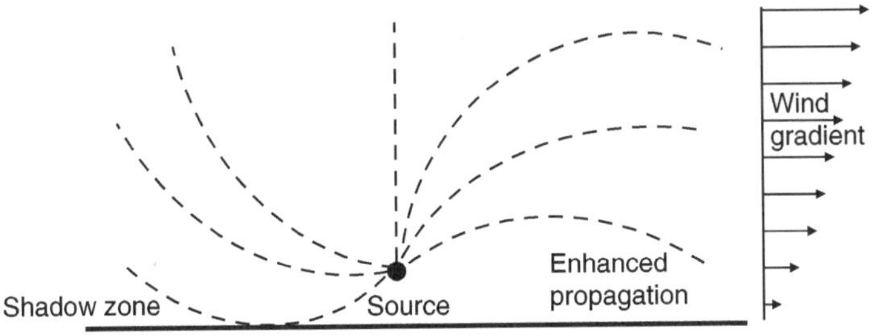


*Figure 3.4* Effects of a wind gradient on sound rays from a point source.

## 3.2 Predicting the ground effect

### *3.2.1 Homogeneous ground*

For most environmental noise predictions, the solid granular framework of porous ground media may be considered to be rigid, so only one wave type (i.e. the wave penetrating the pores) needs to be considered. With this assumption, typically, the speed of sound in the ground ($c_1$) is much smaller than that ($c$) in the air (i.e. $c >> c_1$). The propagation of sound in the air gaps between the solid particles, of which the ground is composed, is impeded by viscous friction. This in turn means that the index of refraction in the ground $n_1 = c/c_1 >> 1$ and any incoming sound ray is refracted towards the normal as it propagates from air into the ground. This type of ground surface is called locally reacting because the air–ground interaction is independent of the angle of incidence of the incoming waves. The acoustical properties of locally reacting ground may be represented simply by its relative normal incidence surface impedance ($Z$), or its inverse (the relative admittance $\beta$) and the ground is said to form a finite impedance boundary. A perfectly hard ground has infinite impedance (zero admittance). A perfectly soft ground has zero impedance (infinite admittance). If the ground is not locally reacting (i.e. if it is externally reacting), the impedance condition is replaced by two separate conditions governing the continuity of pressure and the continuity of the normal component of air particle velocity.

Consider a point source with pressure $P$ and harmonic time dependence ($\exp(-i\omega t)$) near an infinite impedance plane, lying in the plane $z = 0$. The plane is acoustically homogeneous and characterized by the normalized admittance $\beta$ (see Fig. 3.5).

The problem requires a solution of the Helmholtz equation for the sound source over the impedance surface:

$$\Delta P + k^2 P = \Gamma(x, y, z) \tag{3.9a}$$

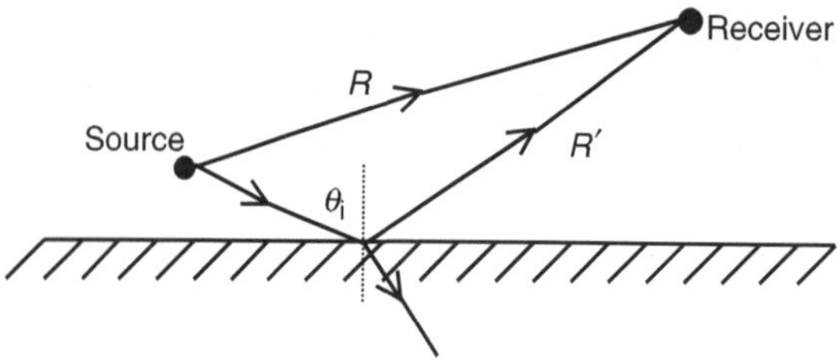


*Figure 3.5* Sound propagation from a point source to a receiver above a ground surface. The direct path length is denoted by $R$ and the reflected path length is denoted by $R'$.

$$\left(\frac{\partial P}{\partial z}+ik\beta P\right)_{z=0}=0 \tag{3.9b}$$

where $k$ is the wave number and $\Gamma(x,y,z)$ represents the source term (in general this is a multipole; see Section 3.2.3) at position $(x,y,z)$. Equation (3.9a) is valid in the half-space $z>0$. It is also necessary to apply the Sommerfeld condition for an outgoing acoustical wave.

The general solution of the boundary problem (3.9) for incident and reflected acoustic pressure can be written using a form corresponding to equation (1.32) in Chapter 1:

$$P=\frac{1}{(2\pi)^{\frac{3}{2}}}\int_{-\infty}^{\infty}\int_{-\infty}^{\infty}d\beta_x d\beta_y\int_{-\infty}^{\infty}\frac{\Gamma(\beta_x,\beta_y,\beta_z)\exp[-i(\beta_x x+\beta_y y+\beta_z z)]}{k^2-\beta_x^2-\beta_y^2-\beta_z^2}d\beta_z$$
$$+\frac{1}{2\pi}\int_{-\infty}^{\infty}\int_{-\infty}^{\infty}A(\alpha,\delta)\exp(-\gamma z-i\alpha x-i\delta y)d\alpha d\delta \tag{3.10}$$

where $\Gamma(\beta_x,\beta_y,\beta_z)$ is the Fourier transformation of the multipole source term:

$$\Gamma(\beta_x,\beta_y,\beta_z)=\frac{1}{(2\pi)^{\frac{3}{2}}}\int_{-\infty}^{\infty}\int_{-\infty}^{\infty}\int_{-\infty}^{\infty}\Gamma(x,y,z)\exp[i(\beta_x x+\beta_y y+\beta_z z)]dxdydz,$$

$\beta_x,\beta_y,\beta_z,\alpha,\delta$ are some complex variables, $\gamma=\sqrt{\alpha^2+\delta^2-k^2}=-i\sqrt{k^2-\alpha^2-\delta^2}$, $A(\alpha,\delta)$ is an unknown function and Re $\gamma>0, z>0$.

Using boundary condition (3.9b), we can solve for the unknown function $A(\alpha, \delta)^6$:

$$A(\alpha,\delta) = A'(\alpha,\delta) + A''(\alpha,\delta),$$

$$A'(\alpha,\delta) = \frac{i\sqrt{2\pi}\left(\sqrt{k^2-\alpha^2-\delta^2}-k\beta\right)\Gamma\left(\alpha,\delta,\sqrt{k^2-\alpha^2-\delta^2}\right)}{\sqrt{k^2-\alpha^2-\delta^2}\left(\sqrt{k^2-\alpha^2-\delta^2}+k\beta\right)} \tag{3.11}$$

$$A''(\alpha,\delta) = -\frac{i\rho\omega^2\sqrt{2\pi}\left(1-\dfrac{k\beta}{\sqrt{k^2-\alpha^2-\delta^2}}\right)}{\sqrt{k^2-\alpha^2-\delta^2}\left(1+\dfrac{k\beta}{\sqrt{k^2-\alpha^2-\delta^2}}\right)} x \tag{3.12}$$

$$x\frac{\Gamma\left(\alpha,\delta,\sqrt{k^2-\alpha^2-\delta^2}\right)}{\left\{2\rho\omega^2+i\sqrt{k^2-\alpha^2-\delta^2}\left(1-\dfrac{k^2\beta^2}{k^2-\alpha^2-\delta^2}\right)\left[D\left(\alpha^2+\delta^2\right)^2-\omega^2 m\right]\right\}} \tag{3.13}$$

If the source at $x_0, y_0, z_0$ is a monopole:

$$\Gamma(x,y,z) = -\delta^{(D)}(x-x_0)\,\delta^{(D)}(y-y_0)\,\delta^{(D)}(z-z_0)$$

$$\Gamma(\beta_x,\beta_y,\beta_z) = -\frac{1}{(2\pi)^{3/2}}\exp\left[i\left(\beta_x x_0+\beta_y y_0+\beta_z z_0\right)\right],$$

where $\delta^D(x-x_0), \delta^D(y-y_0), \delta^D(z-z_0)$ are Dirac delta functions.

An exact analytical solution for the total sound field over a homogeneous plane may be written:[6,7]

$$P = \frac{\exp(ikR)}{4\pi R} + \frac{\exp(ikR')}{4\pi R'} + p_\beta, \tag{3.14a}$$

$$p_\beta = -\frac{k\beta\exp[-ik\beta(z+z_0)]}{4}\left[H_0^{(1)}(kr\sqrt{1-\beta^2}) - H_0^{(1)}(s', kr\sqrt{1-\beta^2})\right] \tag{3.14b}$$

where

$$R = \sqrt{(x-x_0)^2+(y-y_0)^2+(z-z_0)^2}$$

$$r = \sqrt{(x-x_0)^2+(y-y_0)^2},\ s' = \vartheta_1' - i\vartheta_0,$$

$$s' = \vartheta_1' - i\vartheta_0,\ ch\vartheta_1' = \frac{R'}{r},\ \cos\vartheta_0 = \frac{1}{\sqrt{1-\beta^2}},\ \sin\vartheta_0 = \frac{i\beta}{\sqrt{1-\beta^2}}.$$

$H_0^{(1)}(s', kr\sqrt{1-\beta^2})$ is the zeroth order incomplete Hankel function of the first kind.[8] Asymptotic expansions of the Hankel functions for

$$\left|kr\sqrt{1-\beta^2}\sin(\vartheta_1' - i\vartheta_0)\right| > 1, \left|kr\sqrt{1-\beta^2}\right| < \pi$$

and $0 < \vartheta_1' < \pi$ can be used to obtain a simple form of solution for the pressure over a homogeneous impedance plane:

$$p_\beta \approx -\frac{\beta \exp(ikR')}{2\pi R(\beta + \cos\theta_i)}\left[1 - \frac{i(1+\beta\cos\theta_i)}{R'k(\beta+\cos\theta_i)^2}\right] \tag{3.15}$$

where $\theta_i$ is the angle of incidence (see Fig. 3.5).

A widely used but more computationally intensive solution for a point source over an impedance plane can be written[9]

$$p(x,y,z) = \frac{e^{ikR}}{4\pi R} + \frac{e^{ikR'}}{4\pi R'} + \Phi_p + \phi_s \tag{3.16}$$

where

$$\Phi_p \approx 2i\sqrt{\pi}\left(\frac{1}{2}kR'\right)^{1/2} \beta e^{-w^2} \operatorname{erfc}(-iw)\frac{e^{ikR'}}{4\pi R'}. \tag{3.17}$$

and $w$, sometimes called the numerical distance, is given by

$$w \approx \frac{1}{2}(1+i)\sqrt{kR'}(\cos\theta + \beta). \tag{3.18}$$

$\phi_s$ represents a surface wave and is small compared with $\phi_p$ under most circumstances. It is included in careful computations of the complementary error function. erfc($x$).[10,11] On the other hand it is not apparent in (3.15). Conditions under which the surface wave is important are discussed later.

After rearrangement using (3.16) and (3.17), the sound field due to a point monopole source above a locally reacting ground becomes

$$p(x,y,z) = \frac{e^{ikR_1}}{4\pi R} + \left[R_p + (1-R_p)F(w)\right]\frac{e^{ikR_2}}{4\pi R'} \tag{3.19}$$

where $F(w)$, sometimes called the boundary loss factor, is given by

$$F(w) = 1 + i\sqrt{\pi}w\exp\left(-w^2\right)\operatorname{erfc}(-iw) \tag{3.20}$$

and describes the interaction of a spherical wavefront with a ground of finite impedance.[12] The term in the square bracket of (3.12) may be interpreted as the spherical wave reflection coefficient

$$Q = R_p + (1 - R_p)\,F(w), \tag{3.21}$$

which can be seen to involve the plane wave reflection coefficient $R_p$ and a 'correction'. The second term of $Q$ allows for the fact that the wavefronts are spherical rather than plane. Its contribution to the total sound field has been called the ground wave, in analogy with the corresponding term in the theory of AM radio reception.[13] It represents a contribution from the vicinity of the image of the source in the ground plane. If the wavefront is plane ($R_2 \to \infty$) then $|w| \to \infty$ and $F \to 0$. If the surface is acoustically hard, then $|\beta| \to 0$, which implies $|w| \to 0$ and $F \to 1$. If $\beta = 0$, corresponding to a perfect reflector, the sound field consists of two terms: a direct wave contribution and a wave from the image source corresponding to specular reflection and the total sound field may be written

$$p(x, y, z) = \frac{e^{ikR_1}}{4\pi R} + \frac{e^{ikR_2}}{4\pi R'}.$$

This has a first minimum corresponding to destructive interference between the direct and ground-reflected components when $k(R' - R) = \pi$, or $f = c/2(R' - R)$. Normally, for source and receiver close to a perfectly reflecting plane, this destructive interference is at too high a frequency to be of importance in outdoor sound prediction. The higher the frequency of the first minimum in the ground effect, the more likely that it will be destroyed by turbulence.

For $|\beta| << 1$ but at grazing incidence ($\theta = \pi/2$), so that $R_p = -1$ and

$$p(x, y, z) = 2F(w)\,e^{ikr}/r \tag{3.22}$$

the numerical distance, $w$, is given by

$$w = \frac{1}{2}(1 + i)\,\beta\sqrt{kr}. \tag{3.23}$$

If the plane wave reflection coefficient had been used instead of the spherical wave reflection coefficient for grazing incidence, it would have led to the prediction of zero sound pressure when both source and receiver are on the ground. Equation (3.19) is the most widely used analytical approximation for predicting sound fields above a locally reacting ground in a homogeneous atmosphere. There are many other accurate asymptotic and numerical solutions available but no significant numerical differences between various predictions have been revealed for practical geometries and typical outdoor ground surfaces.

### 3.2.2 *The surface wave*

The solution given by (3.14) and (3.15) might be a useful approximation since it avoids the need for computing the complementary error function of the complex argument that appears in (3.19). However, like (3.19), it does not explicitly include the surface wave component; see (3.16). In (3.19), however, it is part of the calculation of the complementary error function. Despite the fact that it is implicit in such calculations of the sound field due to a point source over an impedance boundary, the surface wave represents a separate physical wave entity propagating close to and parallel to the porous ground surface. It produces elliptical motion of air particles as the result of combining motion parallel to the surface with that normal to the surface in and out of the pores. The surface wave decays with the inverse square root of range rather than inversely with range as is true for other components. At grazing incidence on an impedance plane, with normalized admittance $\beta = \beta_r + \mathrm{j}\beta_x$, the condition for the existence of the surface wave is

$$\frac{1}{\beta_x^2} > \frac{1}{\beta_r^2} + 1. \qquad (3.24)$$

For a surface with large impedance (i.e. where $|\beta| \to 0$), the condition is simply that the imaginary part of the ground impedance (the reactance) is greater than the real part (the resistance). This type of surface impedance is possessed by cellular or lattice layers placed on smooth hard surfaces. Surface waves due to a point source have been generated and studied extensively over such surfaces in laboratory experiments.[14–16] The outdoor ground type most likely to produce a measurable surface wave is a thin layer of snow over a frozen ground. By using blank pistol shots in experiments over snow, Albert has confirmed the existence of the predicted type of surface wave outdoors.[17]

There are some cases where it is not possible to model the ground surface as an impedance plane (i.e. $n_1$ is not sufficiently high to warrant the assumption $n_1 >> 1$). In this case, the refraction of sound wave depends on the angle of incidence as sound enters into the porous medium. This means that the apparent impedance depends not only on the physical properties of the ground surface but also, critically, on the angle of incidence. It is possible to define an effective admittance, $\beta_e$ defined by:

$$\beta_e = \varsigma_1 \sqrt{n_1^2 - \sin^2\theta}. \qquad (3.25)$$

where $\varsigma_1 = \rho/\rho_1$ is the ratio of air density to the (complex) density of the rigid-porous ground.

This allows use of the same results as before but with admittance replaced by the effective admittance for a semi-infinite non-locally reacting ground.[18] There are some situations where there is a highly porous surface layer above a relatively non-porous substrate. This is the case with forest floors consisting of partly decomposed litter layers above relatively high flow resistivity soil, with freshly fallen snow on a hard ground or with porous asphalt laid on a non-porous substrate. The minimum depth, $d_m$ for such a multiple layer ground to be treated as a semi-infinite externally reacting ground to satisfy the above condition depends on the acoustical properties of the ground and the angle of incidence. We can consider two limiting cases. If the propagation constant within the surface layer is denoted by $k_1 = k_r - jk_x$, and for normal incidence, where $\theta = 0$, the required condition is simply

$$d_m > 6/k_x. \tag{3.26}$$

For grazing incidence, where $\theta = \pi/2$, the required condition is

$$d_m > 6\left[\sqrt{\frac{(k_r^2 - k_x^2 - 1)^2}{4} + k_r^2 k_x^2} - \frac{k_r^2 - k_x^2 - 1}{2}\right]^{\frac{1}{2}}. \tag{3.27}$$

It is possible to derive an expression for the effective admittance of a ground with an arbitrary number of layers. However, sound waves can seldom penetrate more than a few centimeters in most outdoor ground surfaces. Normally, lower layers contribute little to the total sound field above the ground and consideration of ground structures consisting of more than two layers is not required for predicting outdoor sound. Nevertheless, the assumption of a double-layer structure[18] has been found to enable improved agreement with data obtained over snow.[19] It has been shown that, in cases where the surface impedance depends on angle, replacing the normal surface impedance by the grazing incidence value is sufficiently accurate for predicting outdoor sound.[20]

### 3.2.3 *Multipole sources near the ground*

At near and intermediate distances, an aircraft engine must be considered as a multipole source. Although multipole sources radiate a much weaker sound field than the corresponding monopole source, there are many situations when the multipole source strength is very large and there is no significant contribution from monopole radiation. For example, sound radiation by a vibrating sphere or an unenclosed unbaffled loudspeaker can be represented by a dipole field. An approximate expression for the sound field due to a vertical dipole near an impedance plane (see Fig. 3.6) has been derived and

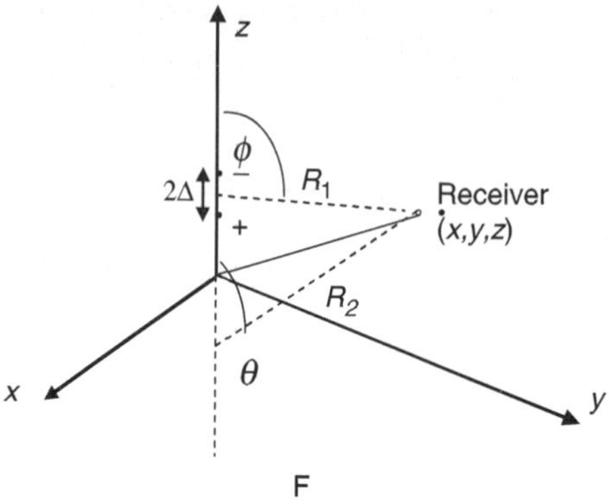


*Figure 3.6* A vertical dipole represented by a pair of out-of-phase monopoles separated by 2Δ.

has a similar form to (3.19)[9] (see Chapter 5).

$$p_v = \frac{S_1 \cos\gamma_l}{4\pi}\left\{\cos\phi\left[\frac{1-ikR_1}{R_1^2}\right]e^{ikR_1} + R_p\cos\theta\left[\frac{1-ikR_2}{R_2^2}\right]e^{ikR_2} + \cos\mu_p(1-R_p)F(w)\left[\frac{1-ikR_2}{R_2^2}\right]e^{ikR_2}\right\} \tag{3.28}$$

where $S_1$ is the source strength of the dipole, $\varphi$ and $\theta$ are the angles between the dipole axis and the direct and image rays and $\cos\mu_p = -\beta$.

Figures 3.7a and b show predictions (solid lines) of the sound field due to a vertical dipole, above a fairly high impedance surface, as a function of horizontal range at 1000 and 100 Hz. The source and receiver heights are located at 2.0 and 1.0 m above the ground, respectively. The dotted lines represent the predicted sound levels due to a monopole source at the same source–receiver geometry and frequencies as the case of vertical dipole. The source strength for the dipole is normalized to give the same sound level as the vertical dipole at 0 m range. In general, the interference pattern of the dipole is essentially the same as that of the monopole source but the level is somewhat lower. Compared with the corresponding spectra for a monopole or a horizontal dipole, they exhibit greater interference effects.

Noise from a jet engine source can be idealized by either a randomly orientated longitudinal or a randomly orientated lateral quadrupole. Equivalent expressions to (3.28) are available for arbitrarily inclined dipoles and quadrupoles[9] (see Chapter 5). Taherzadeh and Li have investigated the directivity pattern of a simple quadrupole and have predicted a significant influence of the impedance of ground surface on the directivity of jet noise.[21]

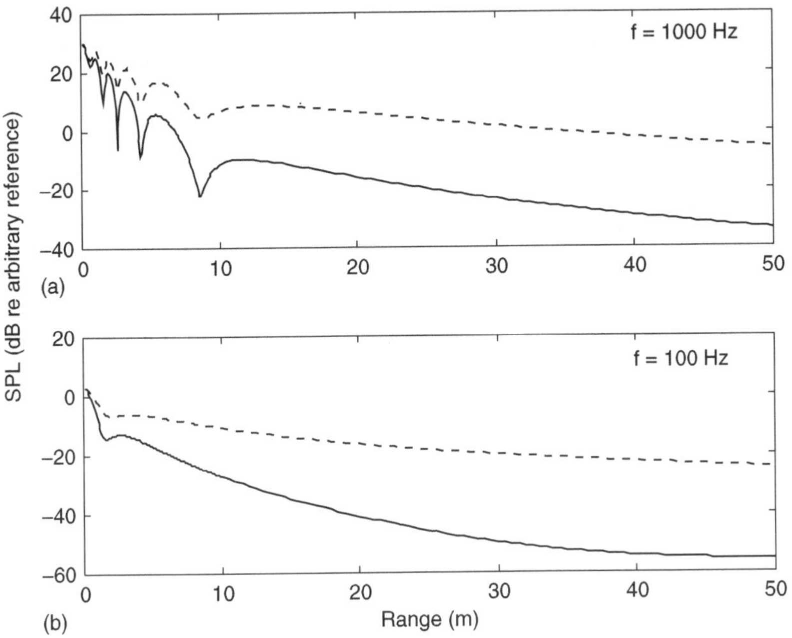


*Figure 3.7* The predicted sound pressure level (SPL) from a vertical dipole (solid line) above an impedance ground. The predicted sound level due to a monopole (dotted line) is also shown. The source and receiver heights are 2.0 m and 1.0 m, respectively. (a) $f = 1000$ Hz; (b) $f = 100$ Hz.

This is a consequence of near cancellation of the intensity field at the near-grazing angles for both longitudinal and lateral quadrupoles.

### *3.2.4 Ground impedance models*

The most important characteristic of a porous ground surface that affects its acoustical character is its flow resistivity. Soil scientists tend to refer to air permeability, which is proportional to the inverse of flow resistivity. Flow resistivity is a measure of the ease with which air can move in and out of the ground. It represents the ratio of the applied pressure gradient to the induced volume flow rate per unit thickness of material and has units of Pa s/m$^2$. If the ground surface has a high flow resistivity, it means that it is difficult for air to flow through the surface. Typically, the flow resistivity increases as porosity decreases. For example, conventional hot-rolled asphalt has a very high flow resistivity (10 million Pa s/m$^2$) and negligible porosity, whereas drainage asphalt has a volume porosity of up to 0.25 and relatively low flow resistivity (< 30 kPa s/m$^2$). Soils have volume porosities of between 10 and 40 per cent. Wet compacted silt may have a porosity as low as 0.1 and rather high flow resistivity (4 million Pa s/m$^2$). Newly fallen snow has a porosity of around 60 per cent and fairly low flow resistivity ($< 10$ kPa s/m$^2$). The thickness of the surface porous layer may be important whether or not it has an acoustically hard substrate.

A widely used model[22] for the acoustical properties of outdoor surfaces involves a single parameter, the 'effective' flow resistivity, $\sigma_e$, to characterize the ground. According to this single-parameter model, the propagation constant, $k$ and normalized characteristic impedance, $Z$ are given by

$$\frac{k}{k_1} = \left[1 + 0.0978\left(f/\sigma_e\right)^{-0.700} + i0.189\left(f/\sigma_e\right)^{-0.595}\right] \tag{3.29a}$$

$$Z = \frac{\rho_1 c_1}{\rho c} = 1 + 0.0571\left(f/\sigma_e\right)^{-0.754} + i0.087\left(f/\sigma_e\right)^{-0.732} \tag{3.29b}$$

This model may be used for a locally reacting ground as well as for an extended reaction surface. Better agreement with some grassland data has been obtained by assuming that the ground surface is that of a hard-backed porous layer of thickness $d$[23] such that the surface impedance $Z_S$ is given by

$$Z_S = Z\coth\left(-ikd\right). \tag{3.29c}$$

On the other hand, there is considerable evidence that (3.29a) tends to overestimate the attenuation within a porous material with high flow resistivity such as soil.

A model based on an exponential change in porosity with depth has been suggested.[24,25] It has enabled better agreement with measured data than that given by equation (3.29b) for the acoustical properties of some outdoor ground surfaces. The two adjustable parameters are another [i.e. different from that in (3.29)] effective flow resistivity ($\sigma_{eff}$) and the effective rate of change of porosity with depth ($\alpha_e$). The impedance of the ground surface is predicted by

$$Z = 0.436(1+i)\sqrt{\frac{\sigma_{eff}}{f}} + 19.74i\frac{\alpha_e}{f}. \tag{3.30}$$

More sophisticated theoretical models for the acoustical properties of rigid-porous materials introduce porosity, the tortuosity (or 'twistiness') of the pores, factors related to pore shape[26] and multiple layering. Models that introduce viscous and thermal characteristic dimensions of pores[27] are based on a formulation by Johnson *et al.*[28] Recently, it has been shown that it is possible to obtain explicit relationships between the characteristic dimensions and grain size by assuming a specific microstructure of identical stacked spheres.[29] Another development allowing for a log-normal pore size distribution, while assuming pores of identical shape,[30] is based on the work of Yamamoto and Turgut[31] on a model for the acoustical properties of underwater sediments.

A standard method for obtaining ground impedance is the template method based on short-range measurements of excess attenuation.[32] Values of parameters deduced from data according to (3.29) and (3.30) show that there can be a wide range of parameter values for 'grassland'[9] (see Chapter 4). It is possible to deduce surface impedance directly by fitting measurements of complex sound pressure.[33]

### 3.2.5 *Effects of surface roughness*

Some surface impedance spectra derived directly from measurements of complex excess attenuation over uncultivated grassland show that the surface impedance tends to zero above 3 kHz.[34] The effects of incoherent scatter from a randomly rough porous surface may be used to explain these measurements. As long as the characteristic roughness dimensions are small compared with the incident sound wavelengths, there is coherent forward scatter. As the incident frequency increases, eventually incoherent scatter dominates. Using a 'boss' approach, an approximate 'effective' admittance for grazing incidence on a hard surface containing randomly distributed two-dimensional (2-D) roughness normal to the roughness axes, may be written[6] (see Chapter 3) as

$$\beta \approx \left(\frac{3V^2k^3b}{2}\right)\left(1+\frac{\delta^2}{2}\right)+iVk(\delta-1) \qquad (3.31)$$

where $V$ is the roughness volume per unit area of surface (equivalent to mean roughness height, which is assumed small compared with the sound wavelength), $b$ is the mean center-to-center spacing, $\delta$ is an interaction factor depending on the roughness shape and packing density and $k$ is the wave number. An interesting consequence of (3.31) is that a surface that would be acoustically hard (i.e. zero admittance, if smooth has 'effective' admittance greater than zero at grazing incidence when rough). The real part of the admittance allows for incoherent scatter from the surface and varies with the cube of frequency and the square of the roughness volume per unit area. A 'boss' approach has been used to predict the effective normalized surface admittance of a porous surface containing 2-D roughness. For a randomly rough porous surface near grazing incidence[9] (see Chapter 3) it is possible to obtain the following approximation:

$$Z_r \approx Z_s - \left(\frac{\langle H\rangle R_s}{\gamma\rho_0 c_0}\right)\left(\frac{2}{\nu}-1\right) \qquad (\mathrm{Re}(Z_r) \geq 0) \qquad (3.32)$$

where $\nu = 1+\frac{2}{3}\pi\langle H\rangle$, $< H >$ is the root mean square roughness height and $Z_s$ is the impedance of the porous surface if it were smooth. This can be used with an impedance model or measured smooth surface impedance to predict the effect of surface roughness for long wavelengths.

Potentially, cultivation practices have important influences on ground effect since they change the surface properties. Aylor[35] noted a significant change in the excess attenuation at a range of 50 m over a soil after disking without any noticeable change in the meteorological conditions. Another cultivation practice is sub-soiling. It is intended to break up soil compaction 300 mm or more beneath the ground surface caused, for example, by repeated passage of heavy vehicles. It is achieved by creating cracks in the compacted layer by means of a single- or double-bladed tine with sharpened leading edges. Sub-soiling has only a small effect on the surface profile of the ground. Plowing turns the soil surface over to a depth of about 0.15 m. Measurements taken over cultivated surfaces before and after sub-soiling and plowing have been shown to be consistent with predicted effects of the resulting changes in surface roughness and flow resistivity.[25,36]

### 3.2.6 *Effects of impedance discontinuities*

De Jong[37] considered Pierce's formulation of sound diffraction from a wedge[38] with different surface acoustic impedances on either side. He then allowed the wedge to fold outwards and derived his expressions for the propagation of sound from a point source over an impedance discontinuity. De Jong's expression can be re-written as

$$P_t = \frac{e^{ikR_1}}{R_1}\left\{1 + (Q_b - Q_a)\frac{e^{-i\pi/4}}{\sqrt{\pi}} \cdot \frac{R_1}{R_d} F\left(\sqrt{k(R_d - R_1)}\right)\right\}$$

$$+ \frac{e^{ikR_2}}{R_2}\left\{Q_{a,b} \pm (Q_b - Q_a)\frac{e^{-i\pi/4}}{\sqrt{\pi}} \cdot \frac{R_2}{R_d} F\left(\sqrt{k(R_d - R_2)}\right)\right\} \tag{3.33}$$

where $R_1$ and $R_2$ are the direct and image ray paths from the source to the receiver, respectively, $R_d$ is the source-discontinuity-receiver path. Subscripts $a$ and $b$ refer to the two impedance surfaces and $Q_{a,b}$ is the appropriate spherical reflection coefficient with $Q_{a,b}$ being equal to $Q_a$ together with the positive sign in the expression if the point of specular reflection falls in region $b$, and equal to $Q_b$ together with the negative sign, if the point of specular reflection falls in region $a$. The wave number is denoted by $k$ and $F(x)$ is the Fresnel integral function. It is defined by

$$F(x) = \int_x^\infty e^{it^2}\,dt. \tag{3.34}$$

Figure 3.8 shows predictions of A-weighted levels based on (3.33) and (3.34) for propagation from a monopole source (with a spectrum corresponding to that of the Il'ushin-86 engine) over an impedance discontinuity between grass and concrete at varying distances from the source. The overall

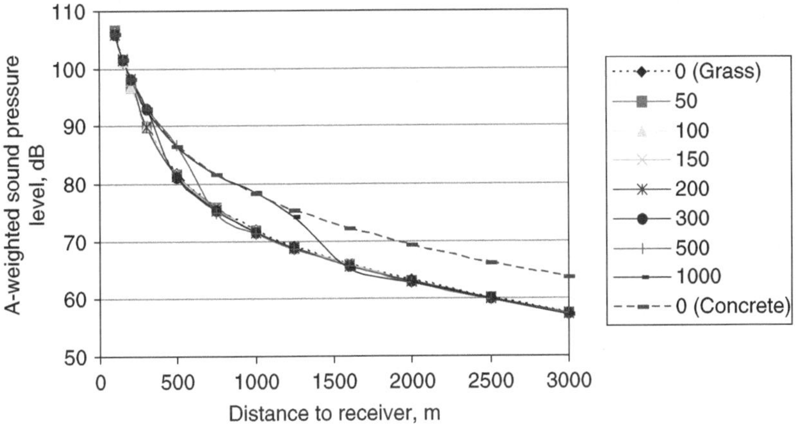


*Figure 3.8* Predicted variation of A-weighted sound pressure level with range from an Il'ushin-86 engine mounted over concrete for various distances to a point of impedance discontinuity with grass. The upper line represents the levels predicted for 100 per cent concrete covering and the lower line those predicted for 100 per cent grass.

tendency is for a transition between the levels predicted over concrete to the levels predicted over grass. These predictions are relevant to later discussions of noise from ground operations (See Section 3.3.3).

### 3.2.7 *Computation of lateral attenuation*

Computation of the influence of the ground effect on noise parameters at a given receiving point for separate frequency bands can be performed in terms of transmission loss:

$$TL = 10\lg\left(\frac{p_t p_t^*}{p_d p_d^*}\right) \tag{3.35}$$

or through the formula for lateral attenuation:[39]

$$\Delta L_{LAT} = 10lg\Big\{1 + S^2|Q_i|^2 + 2S|Q_i|\{\sin[\alpha(\Delta R/\lambda_i)]/[\alpha(\Delta R/\lambda_i)\} \cos[\beta\Delta R/\lambda_i + \delta]/(\Delta R/\lambda_i)\Big\} \tag{3.36}$$

where $Q_i = |Q_i|e^{i\lambda i}, S = R_1/R_2, \Delta R = (R_2 - R_1), \alpha = \pi(\Delta f/f_i), \Delta f$ is the width of frequency band, $f_i$ is the central frequency of the band, $\lambda_i$ is wavelength, $\beta = 2\pi[1+(df/f_i)^2/4]^{1/2}$ and $i$ subscripts indicate values at the band central frequency.

For third-octave bands, $\alpha = 0.725$, $\beta = 6.325$.

## 3.3 Comparisons of measured and predicted ground effects

### *3.3.1 Short range*

Figures 3.9 and 3.10 show a measured excess attenuation spectrum and the real and imaginary parts of the deduced normalized surface impedance spectra together with corresponding predictions based on equations (3.15) and (3.29b) with an effective flow resistivity of 190 kPa s/m$^2$ and 150 kPa s/m$^2$, respectively. The measurements were made using a point source ($1\frac{1}{2}$ inch compression driver) over a lawn. The temperature was around 200°C, and the wind speed was less than 1 m/s with dense cloud cover.

Note that the predicted values of the real part of impedance are larger than those predicted for the imaginary part above 630 Hz but, frequently, the values of the imaginary part of impedance deduced from data are larger than the corresponding real part.

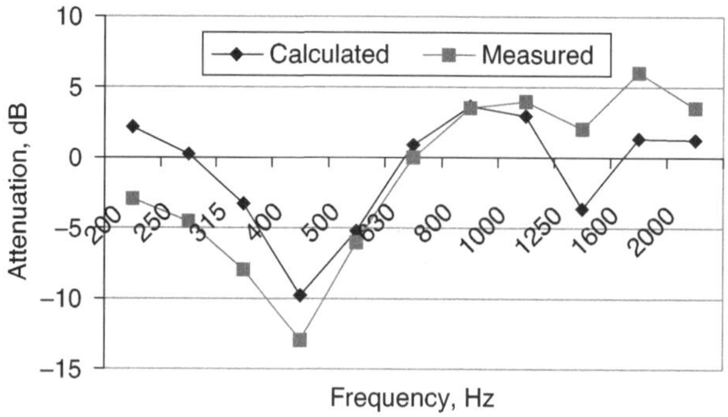


*Figure 3.9* Excess attenuation spectra measured over a lawn with source height 1.5 m, receiver height 0.5 m and horizontal separation 4 m and predicted using (3.15) and (3.29b) with an effective flow resistivity of 190 kPa s/m$^2$.

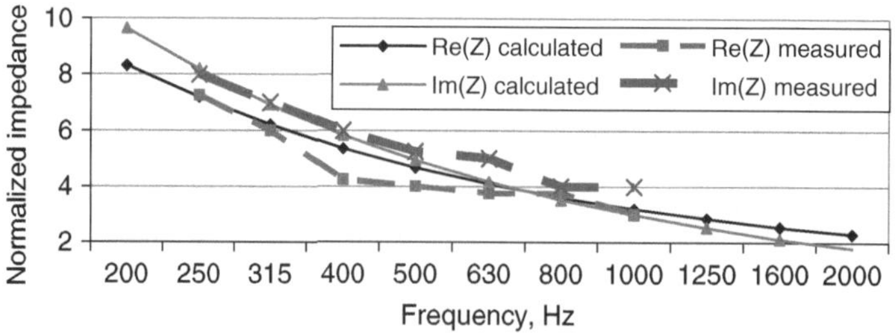


*Figure 3.10* Impedance spectra deduced from complex short-range data (by numerical solution of (3.19)) and calculated from equation (3.29b) assuming an effective flow resistivity of 150 kPa s/m$^2$.

### 3.3.2 *Parkin and Scholes' data*

Pioneering studies of the combined influences of the ground surface and meteorological conditions[40,41] were carried out using a fixed Rolls-Royce Avon jet engine as a source at two airfields (Hatfield and Radlett) in the UK. At Hatfield there was grass up to 5.1 cm high near the runway and surrounding land on which grass was 20.3 cm high. The measurement site was 304.8 m wide and 1127.8 m long. The wind speeds and temperatures were monitored at two heights and therefore it was possible to deduce something about the wind and temperature gradients during the measurements. Perhaps because the role of turbulence was not appreciated, the magnitude of turbulence was not monitored. Nevertheless, this was the first research to note and quantify the change in ground effect with type of surface.

Examples of the resulting data, quoted as the difference between sound pressure levels at 19 m (the reference location) and more distant locations corrected for the decrease expected from spherical spreading and air absorption, are shown in Fig. 3.11. During slightly downwind conditions with low wind speed (< 2 m/s) and small temperature gradients (< 0.01°/m), the ground attenuation over grass-covered ground at Hatfield, although

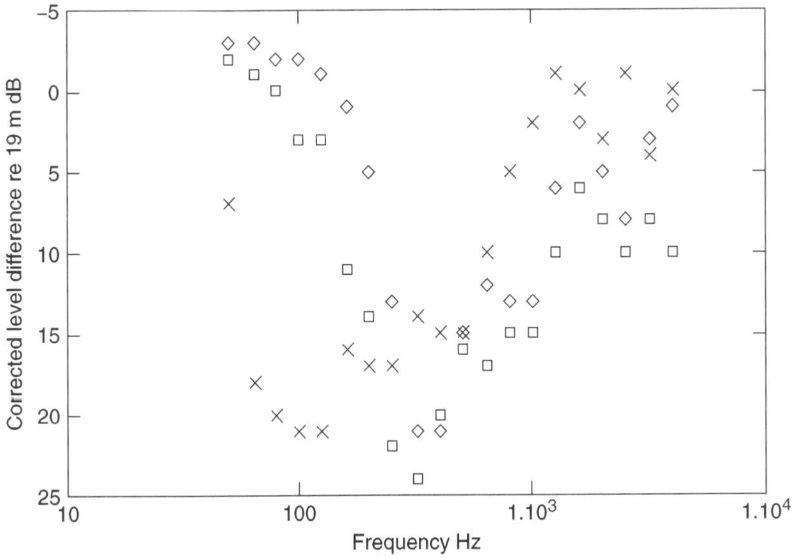


*Figure 3.11* Parkin and Scholes' data for the level difference between microphones at a height of 1.5 m and at distances of 19 m and 347 m from a fixed jet engine source (nozzle–center height 1.82 m) corrected for wavefront spreading and air absorption. □ and ◇ represent data over airfields (grass-covered) at Radlett and Hatfield, respectively, with a positive vector wind between source and receiver of 1.27 m/s (5 ft/s); × represent data obtained over approximately 0.15-m thick (6–9 in.) snow at Hatfield and a positive vector wind of 1.52 m/s (6 ft/s).

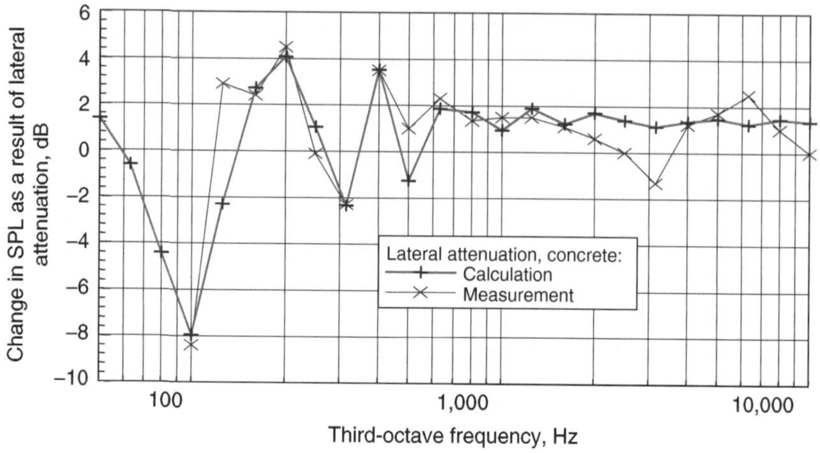


*Figure 3.12* Comparison between predictions and measurements[42] at an engine testing facility with a concrete ground surface. The receiver height is 1.5 m, the assumed point source height is 4.5 m and the horizontal distance between the engine and microphone is 50 m. The prediction has assumed an impedance for concrete given by a semi empirical one-parameter model (3.38b) with effective flow resistivity $\sigma_e = 20,000$ kPa s/m$^2$.

still a major propagation factor (i.e. over 15 dB near 400 Hz), was less than that over the other grass-covered ground at Radlett and its maximum value occurred at a higher frequency. Snowfall during the period of the measurements also enabled observation of the large change resulting from the presence of snow at low frequencies (i.e. over 20 dB attenuation in the 63 Hz and 125 Hz third-octave bands.

### *3.3.3 Noise from aircraft engine testing*

Figure 3.12 shows measured third-octave data obtained at 50 m from an aircraft engine being tested over a concrete stand and predictions of lateral attenuation (3.36) for a point source over a hard surface.

Noise measurements have been made to distances of 3 km during aircraft engine run-ups with the aim of defining noise contours in the vicinity of airports.[2] Measurements were made for a range of power settings during several summer days under weakly refracting weather conditions (wind speed < 5 m/s, temperature between 20 and 25°C). Between seven and ten measurements were made at every measurement station [in accordance with International Civil Aviation Organization (ICAO) Annex 16 requirements] and the results were averaged. Example results are shown in Fig. 3.13.

It has been shown that these data are consistent with nearly acoustically neutral (i.e. zero sound speed gradient) conditions.[2] Note that at 3 km, the

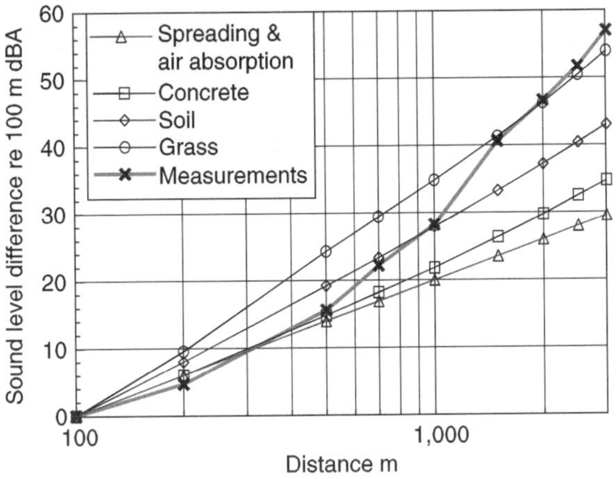


*Figure 3.13* Measured differences between the A-weighted sound level at 100 m and those measured at ranges up to 3 km in the direction of maximum jet noise generation (~40° from exhaust axis) during the testing of an Il-86 aircraft's engine and predictions for levels due to a point source at the engine center height assuming spherical spreading plus air absorption and various types of ground.

measured levels are more than 30 dB less than would be expected from wavefront spreading and air absorption only. Up to distances of between 500 and 700 m from the engine, the data suggest attenuation rates near to the predictions for 'concrete' or 'spherical spreading plus air absorption'. Beyond 700 m the measured attenuation rate is nearer to the predictions for 'soil' or between the 'soil' and 'grass' predictions. These results are consistent with the fact that the run-ups took place over the concrete surface of an apron and further away (i.e. between 500 and 700 m in various directions) the ground surface was 'soil' and/or 'grass' with consequent differences in ground effect.

Noise from ground operations is discussed further in Chapter 4.

## 3.4 Shadow zones

There are distinct advantages in assuming that wind and temperature gradients give rise to a linear effective sound speed profile and ignoring the vector wind since this assumption leads to circular ray paths and relatively tractable analytical solutions. With this assumption, the effective sound speed, $c$, can be written

$$c(z) = c_0(1 + \zeta z) \qquad (3.37)$$

where $\zeta$ is the normalized sound velocity gradient, $(dc/dz)/c_0$, and $z$ is the height above ground. If it is also assumed that the source–receiver distance and the effective sound speed gradient are sufficiently small so that there is only a single 'ray bounce' (i.e. a single ground reflection between source and receiver), it is possible to use a simple adaptation of the formula (3.19) replacing the geometrical ray paths defining the direct and reflected path lengths by curved ones. Consequently, the sound field is approximated by

$$p = \left\{\exp(-jk_0\xi_1) + Q\exp(-jk_0\xi_2)\right\}/4\pi d \tag{3.38a}$$

where $Q$ is the appropriate spherical wave reflection coefficient, $d$ is the horizontal separation between the source and receiver, $\xi_1$ and $\xi_2$ are, respectively, the acoustical path lengths of the direct and reflected waves. These acoustical path lengths can be determined by[43,44]

$$\xi_1 = \int_{\phi_<}^{\phi_>} \frac{d\phi}{\varsigma \sin\phi} = \varsigma^{-1}\log_e\left[\tan(\phi_>/2)/\tan(\phi_</2)\right] \tag{3.38b}$$

and

$$\xi_2 = \int_{\theta_<}^{\theta_>} \frac{d\theta}{\varsigma \sin\theta} = \varsigma^{-1}\log_e\left[\tan(\theta_>/2)\tan^2(\theta_0/2)\Big/\tan(\theta_</2)\right] \tag{3.38c}$$

where $\phi(z)$ and $\theta(z)$ are the polar angles (measured from the positive $z$-axis) of the direct and reflected waves. The subscripts $>$ and $<$ denote the corresponding parameters evaluated at $z_>$ and $z_<$ respectively, $z_> \equiv \max(z_s, z_r)$ and $z_< \equiv \min(z_s, z_r)$.

The computation of $\phi(z)$ and $\theta(z)$, requires the corresponding polar angles ($\phi_0$ and $\theta_0$) at $z = 0$.[45] Once the polar angles are determined at $z = 0$, $\phi(z)$ and $\theta(z)$ at other heights can be found by using Snell's law:

$$\sin\vartheta = (1 + \varsigma z)\sin\vartheta_0 \tag{3.39}$$

where $\vartheta = \phi$ or $\theta$. Substitution of these angles into (3.38b) and (3.38c) and, in turn, into (3.38a), makes it possible to calculate the sound field in the presence of a linear sound velocity gradient.

For downward refraction, additional rays will cause a discontinuity in the predicted sound level because of the inherent approximation used in ray-tracing. It is possible to determine the critical range, $r_c$, at which there are two additional ray arrivals.

For $\varsigma > 0$, this critical range is given by

$$r_c = \frac{\left\{\left[\sqrt{(\varsigma z_>)^2 + 2\varsigma z_>} + \sqrt{(\varsigma z_<)^2 + 2\varsigma z_<}\right]^{2/3} + \left[\sqrt{(\varsigma z_>)^2 + 2\varsigma z_>} - \sqrt{(\varsigma z_<)^2 + 2\varsigma z_<}\right]^{2/3}\right\}^{3/2}}{\varsigma}. \tag{3.40}$$

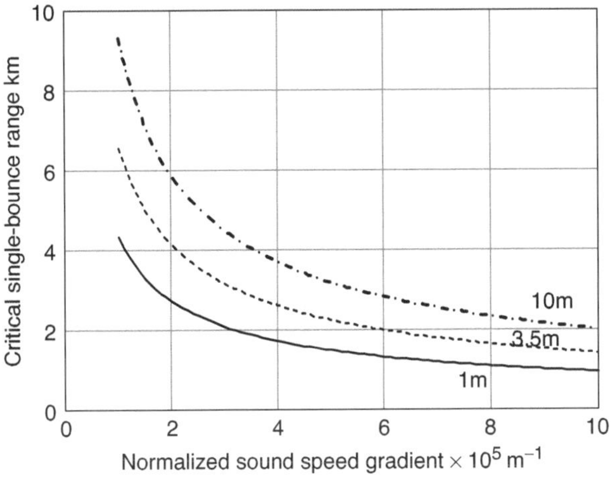


*Figure 3.14* Maximum ranges for which single-bounce assumption is valid for linear sound speed gradient based on (3.37) and assuming equal source and receiver heights of 1 m (solid line); 3.5 m (broken line) and 10 m (dot-dash line).

Figure 3.14 shows that, for the source and the receiver at 1 m height, horizontal separations of less than 1 km and a normalized sound speed gradient of less than 0.0001/m (corresponding, for example, to a wind speed gradient of less than 0.1/s), then it is reasonable to assume a single ground bounce in ray tracing. The range of validity of the single-bounce assumption increases as the source and receiver heights increase.

A negative sound gradient means upward refraction and the creation of a sound shadow at a distance from the source that depends on the gradient. The presence of a shadow zone means that the sound level decreases faster than would be expected from distance alone. A combination of slightly negative temperature gradient, strong upwind propagation and air absorption has been observed, in carefully monitored experiments, to reduce sound levels, 640 m from a 6-m high source over relatively hard ground, by up to 20 dB more than expected from spherical spreading.[46] Since shadow zones can be areas in which there is significant excess attenuation, it is important to be able to locate their boundaries.

For upward refracting conditions, ray-based predictions are incorrect when the receiver is in the shadow and penumbra zones. Nevertheless, the shadow boundary can be determined from geometrical considerations. For a given source and receiver heights, the critical range, $r_c'$, is determined from

$$r_c' = \frac{\sqrt{(\varsigma' z_>)^2 + 2\varsigma' z_>} + \sqrt{\varsigma'^2 z_<(2z_> - z_<) + 2\varsigma' z_<}}{\varsigma'}, \tag{3.41}$$

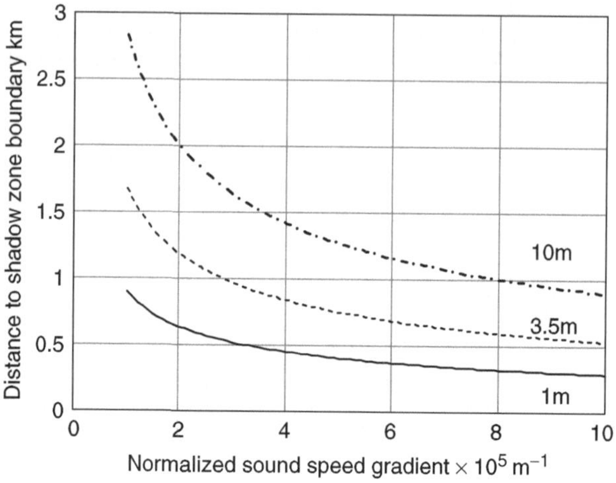


*Figure 3.15* Distances to shadow zone boundaries according to (3.41) for a linear sound speed gradient based on (3.37) assuming equal source and receiver heights: 1 m (solid line); 3.5 m (broken line) and 10 m (dot-dash line).

where $\varsigma' = \frac{|\varsigma|}{1-|\varsigma|z_>}$.

Figure 3.15 shows that, for source and receiver heights of 1 m and a normalized sound speed gradient of 0.0001/m, the distance to the shadow zone boundary is about 300 m. Not surprisingly, the distance to the shadow zone boundary is predicted to increase as the source and receiver heights are increased.

A good approximation of (3.41), for the distance to the shadow zone, when the source is close to the ground and $\zeta$ (the normalized sound speed gradient) is small, is

$$r_c = \left[2c_0 \Big/ \left(-\frac{dc}{dz}\right)\right]^{1/2} \left(\sqrt{h_s} + \sqrt{h_r}\right) \tag{3.42}$$

where $h_s$ and $h_r$ are heights of source and receiver, respectively, and $dc/dz$ must be negative for a temperature-induced shadow zone.

Conditions of weak refraction may be said to exist where, under downward refracting conditions, the ground reflected ray undergoes only a single-bounce and, under upward refracting conditions, the receiver is within the illuminated zone.

When a wind is present, the combined effects of temperature lapse and wind will tend to enhance the shadow zone upwind of the source, since wind speed tends to increase with height. Downwind of the source, however, the wind will counteract the effect of temperature lapse, and the shadow zone will be destroyed. In any case, an acoustic shadow zone is never as complete as an optical one would be, as a result of diffraction and turbulence. In the presence of wind with a wind speed gradient of *du/dz*, the formula for the distance to the shadow zone boundary is given by

$$r_c = \left[2c_0 \bigg/ \frac{du}{dz}\cos\beta - \frac{dc}{dz}\right]^{1/2} \left(\sqrt{h_s} + \sqrt{h_r}\right) \tag{3.43}$$

where $\beta$ is the angle between the direction of the wind and the line between source and receiver.

Note that there will be a value of the angle $\beta$, (say $\beta c$), given by $\frac{du}{dz}\cos\beta_c = \frac{dc}{dz}$ or

$$\beta_c = \cos^{-1}\left(\frac{dc}{dz} \bigg/ \frac{du}{dz}\right) \tag{3.44}$$

at and beyond which there will not be a shadow zone. This represents the critical angle at which the effect of wind counteracts that of the temperature gradient.

## 3.5 Classification of meteorological effects

There is a considerable body of knowledge about meteorological influences on air quality in general and the dispersion of plumes from stacks in particular. Plume behavior depends on vertical temperature gradients and hence on the degree of mixing in the atmosphere. Vertical temperature gradients decrease with increasing wind. The stability of the atmosphere in respect of plume dispersion is described in terms of Pasquill classes. This classification is based on incoming solar radiation, the time of day and wind speed. There are six Pasquill classes (A–F), which are defined in Table 3.1.

Data are recorded in this form by meteorological stations and so, at first sight, it is a convenient classification system for noise prediction. Class A represents a very unstable atmosphere with strong vertical air transport (i.e. mixing). Class F represents a very stable atmosphere with weak vertical transport. Class D represents a meteorologically neutral atmosphere. Such an atmosphere has a logarithmic wind speed profile and a temperature gradient corresponding to the normal decrease with height (adiabatic lapse rate). A meteorologically neutral atmosphere occurs for high wind speeds and large values of cloud cover. This means that a meteorologically neutral atmosphere may be far from acoustically neutral. Typically, the atmosphere is unstable by day and stable by night. This means that classes A–D might be appropriate classes by day and D–F by night. With practice, it is possible to estimate Pasquill Stability Categories in the field, for a particular time and season, from a visual estimate of the degree of

*Table 3.1* Pasquill (meteorological) stability categories

| *Wind speed*[a] *(m/s)* | *Daytime incoming solar radiation (mW/cm$^2$)* | | | | *1 hour before sunset or after sunrise* | *Night-time cloud cover (octas)* | | |
|---|---|---|---|---|---|---|---|---|
| | > 60 | *30–60* | < 30 | *Overcast* | | *0–3* | *4–7* | *8* |
| ≤ 1.5 | A | A–B | B | C | D | F or G[b] | F | D |
| 2.0–2.5 | A–B | B | C | C | D | F | E | D |
| 3.0–4.5 | B | B–C | C | C | D | E | D | D |
| 5.0–6.0 | C | C–D | D | D | D | D | D | D |
| > 6.0 | D | D | D | D | D | D | D | D |

[a]Measured to the nearest 0.5 m/s at 11 m height.
[b]Category G is an additional category restricted to night-time with less than 1 octa of cloud and a wind speed of less than 0.5 m/s.

*Table 3.2* CONCAWE meteorological classes for noise prediction

| *Meteorological category* | *Pasquill stability category and wind speed (m/s) – positive is towards receiver* | | |
|---|---|---|---|
| | *A, B* | *C, D, E* | *F, G* |
| 1 | $v < -3.0$ | – | – |
| 2 | $-3.0 < v < -0.5$ | $v < -3.0$ | – |
| 3 | $-0.5 < v < +0.5$ | $-3.0 < v < -0.5$ | $v < -3.0$ |
| 4[a] | $+0.5 < v < +3.0$ | $-0.5 < v < +0.5$ | $-3.0 < v < -0.5$ |
| 5 | $v > +3.0$ | $+0.5 < v < +3.0$ | $-0.5 < v < +0.5$ |
| 6 | – | $v > +3.0$ | $+0.5 < v < +3.0$ |

[a]Category with assumed zero meteorological influence.

cloud cover. A modified form of the Pasquill classification of meteorological conditions has been adopted as a classification system for noise prediction schemes.[47,48] It is clear from Table 3.1 that the meteorologically neutral category (C), while being fairly common in a temperate climate, includes a wide range of wind speeds and is therefore not very suitable as a category for noise prediction. In the CONCAWE scheme,[47] this problem is addressed by defining six noise prediction categories based on Pasquill categories (representing the temperature gradient) and wind speed. There are 18 sub-categories depending on wind speed. These are defined in Table 3.2.

CONCAWE category 4 is specified as one in which there is zero meteorological influence (i.e. CONCAWE category 4 is equivalent to acoustically neutral conditions). The CONCAWE scheme requires octave band analysis. Meteorological corrections in this scheme are based primarily on analysis of the Parkin and Scholes' data[40,41] together with measurements made at several industrial sites. The excess attenuation in each octave band for each category tends to approach asymptotic limits with increasing distance. Values at 2 km for CONCAWE categories 1 (strong wind from

*Table 3.3* Values of the meteorological corrections for CONCAWE categories 1 and 6

| *Octave band center frequency (Hz)* | *63* | *125* | *250* | *500* | *1000* | *2000* | *4000* |
|---|---|---|---|---|---|---|---|
| Category 1 | 8.9 | 6.7 | 4.9 | 10.0 | 12.2 | 7.3 | 8.8 |
| Category 6 | −2.3 | −4.2 | −6.5 | −7.2 | −4.9 | −4.3 | −7.4 |

*Table 3.4* Estimated probability of occurrence of various combinations of wind and temperature gradient

| *Temperature gradient* | *Zero wind* | *Strong wind* | *Very strong wind* |
|---|---|---|---|
| Very large negative | Frequent | Occasional | Rare or never |
| Large negative | Frequent | Occasional | Occasional |
| Zero | Occasional | Frequent | Frequent |
| Large positive | Frequent | Occasional | Occasional |
| Very large positive | Frequent | Occasional | Rare or never |

receiver to source, hence upward refraction) and 6 (strong downward refraction) are listed in Table 3.3.

Wind speed and temperature gradients are not independent. For example, very large temperature and wind speed gradients cannot coexist. Strong turbulence associated with high wind speeds does not allow the development of marked thermal stratification. Table 3.4 shows a rough estimate of the probability of occurrence of various combinations of wind and temperature gradients.[46]

With regard to sound propagation, the component of the wind vector in the direction between source and receiver is most important, so wind categories must take this into account. Moreover, it is possible to give more detailed but qualitative descriptions of each of the meteorological categories (wind, *W*, and temperature gradient, TG; see Table 3.5).

In Table 3.6, the revised categories are identified with qualitative predictions of their effects on noise levels. The classes are not symmetrical around zero meteorological influence. Typically, there are more meteorological condition combinations that lead to attenuation than lead to enhancement. Moreover, the increases in noise level (say 1–5 dB) are smaller than the decreases (say 5–20 dB). Using the values at 500 Hz as a rough guide to the likely corrections on overall A-weighted broadband levels, it is noticeable that the CONCAWE meteorological corrections are not symmetrical around zero. The CONCAWE scheme suggests meteorological variations of between 10 dB less than the acoustically neutral level for strong upward refraction between source and receiver and 7 dB more than the acoustically neutral level for strong downward refraction between source and receiver.

*Table 3.5* Meteorological classes for noise prediction based on qualitative descriptions

| | |
|---|---|
| W1 | Strong wind (> 3–5 m/s) from receiver to source |
| W2 | Moderate wind (≈ 1–3 m/s) from receiver to source, or strong wind at 45° |
| W3 | No wind, or any cross wind |
| W4 | Moderate wind (≈ 1–3 m/s) from source to receiver, or strong wind at 45° |
| W5 | Strong wind (> 3–5 m/s) from source to receiver |
| TG1 | Strong negative: daytime with strong radiation (high sun, little cloud cover), dry surface and little wind |
| TG2 | Moderate negative: as TG1 but one condition missing |
| TG3 | Near isothermal: early morning or late afternoon (e.g. one hour after sunrise or before sunset) |
| TG4 | Isothermal – zero temperature gradient, occurs very rarely |
| TG5 | Moderate positive: night-time with overcast sky or substantial wind |
| TG6 | Strong positive: night-time with clear sky and little or no wind |

*Table 3.6* Qualitative estimates of impact of meteorological condition on noise levels

| | *W1* | *W2* | *W3* | *W4* | *W5* |
|---|---|---|---|---|---|
| TG1 | – | large attenuation | Small attenuation | Small attenuation | – |
| TG2 | large attenuation | Small attenuation | Small attenuation | Zero meteorological influence | Small enhancement |
| TG3 | Small attenuation | Small attenuation | Zero meteorological influence | Small enhancement | Small enhancement |
| TG4 | Small attenuation | Zero meteorological influence | Small enhancement | Small enhancement | Large enhancement |
| TG5 | – | Small enhancement | Small enhancement | Large enhancement | – |

## 3.6 Typical sound speed profiles

Outdoor sound prediction requires information on wind speed, direction and temperature as a function of height near to the propagation path. These determine the sound speed profile. Ideally, the heights at which the meteorological data are collected should reflect the application. If this information is not available, then there are alternative procedures. It is possible, for example, to generate an approximate sound speed profile from temperature and wind speed at a given height using Monin–Obukhov similarity theory[49] and to input this directly into a prediction scheme.

According to this theory, the wind speed component (m/s) in the source–receiver direction and temperature (°C) at height $z$ are calculated from the

*Table 3.7* Definitions of parameters used to describe Monin–Obukhov profiles

| | | |
|---|---|---|
| $u^*$ | Friction velocity (m/s) | Depends on surface roughness |
| $z_M$ | Momentum roughness length | Depends on surface roughness |
| $z_H$ | Heat roughness length | Depends on surface roughness |
| $T^*$ | Scaling temperature (K) | The precise value of this is not important for sound propagation. A convenient value is 283 K |
| $k$ | Von Karman constant | $= 0.41$ |
| $T_0$ | Temperature at zero height (°C) | Again it is convenient to use 283 K |
| $\Gamma$ | Adiabatic correction factor | $= -0.01$°C/m for dry air Moisture affects this value but the difference is small |
| $L$ | Obukhov length (m) $> 0 \rightarrow$ stable, $< 0 \rightarrow$ unstable | $= \pm \frac{u_*^2}{kgT^*}(T_{av} + 273.15)$, the thickness of the surface or boundary layer is given by $2L$ m |
| $T_{av}$ | Average temperature °C | It is convenient to use $T_{av} = 10$ so that $(T_{av} + 273.15) = \theta_0$ |
| $\psi_{\mathrm{M}}$ | Diabatic momentum profile correction (mixing) function | $= -2\ln\left(\frac{(1+\chi_M)}{2}\right) - \ln\left(\frac{(1+\chi_M^2)}{2}\right) +$ $2\arctan(\chi_M) - \pi/2$, if $L < 0 = 5(z/L)$, if $L > 0$ |
| $\psi_{\mathrm{H}}$ | Diabatic heat profile correction (mixing) function | $= -2\ln\left(\frac{(1+\chi_H)}{2}\right)$, if $L < 0 = 5(z/L)$, if $L > 0$ or for $z \leq 0.5$L |
| $\chi_M$ | Inverse diabatic influence or function for momentum | $= \left[1 - \frac{16z}{L}\right]^{0.25}$ |
| $\chi_H$ | Inverse diabatic influence function for momentum | $= \left[1 - \frac{16z}{L}\right]^{0.5}$ |

values at ground level and other parameters as follows:

$$u(z) = \frac{u_*}{k}\left[\ln\left\{\frac{z + z_M}{z_M}\right\} + \psi_M\left(\frac{z}{L}\right)\right] \tag{3.45a}$$

$$T(z) = T_0 + \frac{T_*}{k}\left[\ln\left\{\frac{z + z_H}{z_H}\right\} + \psi_H\left(\frac{z}{L}\right)\right] + \Gamma z \tag{3.45b}$$

where the various parameters are defined in Table 3.7.

The associated sound speed profile, $c(z)$, is calculated from

$$c(z) = c(0)\sqrt{\frac{T(z) + 273.15}{273.15}} + u(z). \tag{3.45c}$$

For a neutral atmosphere, $1/L = 0$ and $\psi_{\mathrm{M}} = \psi_{\mathrm{H}} = 0$. The associated sound speed profile, $c(z)$, is calculated from (3.37).

Note that the resulting profiles are valid in the surface or boundary layer only but not at zero height. In fact, the profiles given by the above equations, sometimes called Businger–Dyer profiles,[50] have been found to be in good agreement with measured profiles up to 100 m. This height range is relevant to sound propagation over distances up to 10 km.[51] However, improved profiles are available that are valid for greater heights. For example,[52]

$$\psi_M = \psi_H - 7\ln(z/L) - 4.25/(z/L) + 0.5/(z/L)^2 - 0.852 \text{ for } z > 0.5L. \tag{3.46}$$

Often, $z_M$ and $z_H$ are taken to be equal. The roughness length varies, for example, between 0.0002 (still water) and 0.1 (grass). More generally, the roughness length can be estimated from the Davenport classification.[53]

Figure 3.16 gives examples of sound speed (difference) profiles, $c(z) - c(0)$, generated from equations (3.45) through (3.46) using

(a) $z_M = z_H = 0.02, u^* = 0.34, T^* = 0.0212, T_{av} = 10, T_0 = 6,$

(giving $L = -390.64$)

(b) $z_M = z_H = 0.02, u^* = 0.15, T^* = 0.1371, T_{av} = 10, T_0 = 6,$

(giving $L = -11.76$)

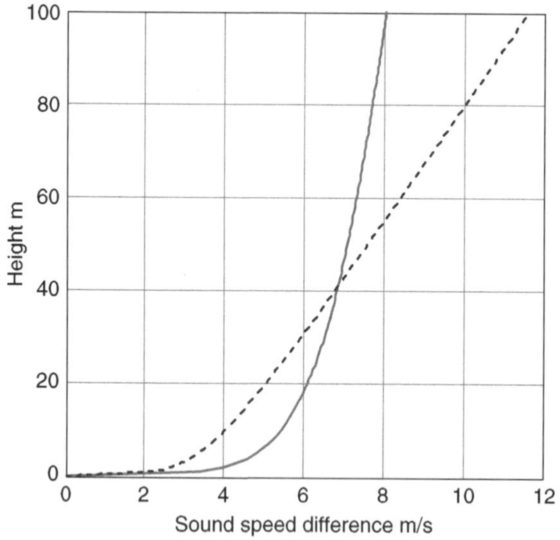


*Figure 3.16* Two downward refracting sound speed profiles relative to the sound speed at the ground obtained from similarity theory. The continuous curve is approximately logarithmic corresponding to a large Obukhov length and to a cloudy, windy night. The broken curve corresponds to a small Obukhov length as on a calm clear night and is predominantly linear away from the ground.

and $\Gamma = -0.01$. These parameters are intended to correspond to a cloudy windy night and a calm clear night, respectively.[54]

Salomons *et al.*[55] have suggested a method for obtaining the remaining unknown parameters, $u^*$, $T^*$ and $L$ from the relationship

$$L = \frac{u_*^2}{kgT_*} \tag{3.47}$$

and the Pasquill Category ($P$).

From empirical meteorological tables, approximate relationships between the Pasquill class $P$, the wind speed $u_{10}$ at a reference height of 10 m and the fractional cloud cover $N_c$ have been obtained. The latter determines the incoming solar radiation and therefore the heating of the ground. The former is a guide to the degree of mixing. The approximate relationship is

$$\begin{aligned} P(u_{10}, N_c) = {} & 1 + 3[1 + \exp(3.5 - 0.5u_{10} - 0.5N_c)]^{-1} && \text{during the day} \\ & 6 - 2[1 + \exp(12 - 2u_{10} - 2N_c)]^{-1} && \text{during the night} \end{aligned} \tag{3.48}$$

A proposed relationship between the Obukhov length $L$ m as a function of $P$ and roughness length $z_0 < 0.5$ m is

$$\frac{1}{L(P, z_0)} = B_1(P)\log(z_0) + B_2(P) \tag{3.49a}$$

where

$$B_1(P) = 0.0436 - 0.0017P - 0.0023P^2 \tag{3.49b}$$

and

$$\begin{aligned} B_2(P) = {} & \min(0, 0.045P - 0.125) && \text{for } 1 \le P \le 4 \\ & \max(0, 0.025P - 0.125) && \text{for } 4 \le P \le 6. \end{aligned} \tag{3.49c}$$

Alternatively, values of $B_1$ and $B_2$ may be obtained from Table 3.8.

*Table 3.8* Values of the constants $B_1$ and $B_2$ for the six Pasquill classes

| | *Pasquill class* | | | | | |
|---|---|---|---|---|---|---|
| | A | B | C | D | E | F |
| $B_1$ | 0.04 | 0.03 | 0.02 | 0 | −0.02 | −0.05 |
| $B_2$ | −0.08 | −0.035 | 0 | 0 | 0 | 0.025 |

Equations (3.44) give

$$L = L(u_{10}, N_c, z_0). \tag{3.50}$$

Also, $u_{10}$ is given by equation (3.32) with $z = 10$ m, i.e.

$$u(z) = \frac{u_*}{k}\left[\ln\left\{\frac{10 + z_M}{z_M}\right\} + \psi_M\left(\frac{10}{L}\right)\right] \tag{3.51}$$

Equations (3.37), (3.50) and (3.51) may be solved for $u^*$, $T^*$ and $L$. Hence, it is possible to calculate $\psi_M$, $\psi_H$, $u(z)$ and $T(z)$.

Figure 3.17 shows the results of this procedure for a ground with a roughness length of 0.1 m and two upwind and downwind daytime classes defined by the parameters listed in the caption.

As a consequence of atmospheric turbulence, instantaneous profiles of temperature and wind speed show considerable variations with both time and position. These variations are eliminated considerably by averaging over a period of the order of 10 minutes. The Monin–Obukhov or Businger–Dyer models give good descriptions of the averaged profiles over longer periods.

The Pasquill Category C profiles shown in Fig. 3.17 are approximated closely by logarithmic curves of the form

$$c(z) = c(0) + b\ln\left[\frac{z}{z_0} + 1\right] \tag{3.52}$$

where the parameter $b$ ($> 0$ for downward refraction and $< 0$ for upward refraction) is a measure of the strength of the atmospheric refraction. Such logarithmic sound speed profiles are realistic for open ground areas without obstacles particularly in the daytime.

A better fit for night-time profiles is obtained with power laws of the form

$$c(z) = c(0) + b(z/z_0)^\alpha \tag{3.53}$$

where

$$\alpha = 0.4(P - 4)^{\frac{1}{4}}.$$

The temperature term in the effective sound speed profile given by (3.45c) can be approximated by truncating a Taylor expansion after the first term to give

$$c(z) = c(T_0) + \frac{1}{2}\sqrt{\frac{\kappa R}{T_0}}[T(z) - T_0] + u(z). \tag{3.54}$$

When combined with (3.37) this leads to a linear dependence on temperature and a logarithmic dependence on wind speed with height. By comparing

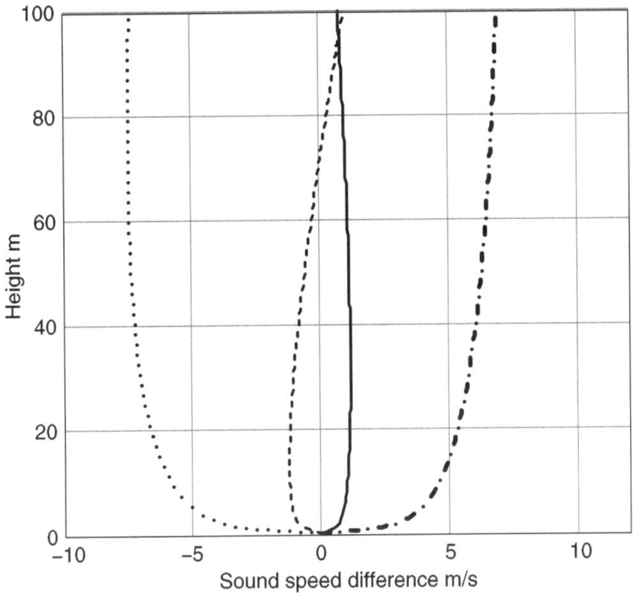


**Key**

| Line type | Wind speed at 10 m in m/s | Cloud cover in octels | Pasquill class | Direction |
|---|---|---|---|---|
| ———————— | 1 | 0 | A | downwind |
| – – – – – – – – | 1 | 0 | A | upwind |
| — — — — — | 5 | 4 | C | downwind |
| ························ | 5 | 4 | C | upwind |

*Figure 3.17* Two daytime sound speed profiles (upwind – dashed and dotted; downwind – solid and dash-dot) determined from the parameters listed in Table 3.8.

with 12 months of meteorological data obtained at a 50-m high meteorological tower in Germany, Heimann and Salomons[56] have found that (3.54) is a reasonably accurate approximation to vertical profiles of effective sound speed even in unstable conditions and in situations where Monin–Obukhov theory is not valid. By making a series of sound level predictions (using the Parabolic Equation method) for different meteorological conditions, it was found that a minimum of 25 meteorological classes are necessary to ensure 2 dB or less deviation in the estimated annual average sound level from the reference case with 121 categories.

There are simpler, linear-segment profiles deduced from a wide range of meteorological data that may be used to represent worst-case noise conditions (i.e. best conditions for propagation). The first of these profiles may be calculated from a temperature gradient of +15°C/km from the surface to 300 m and 8°C/km above that assuming a surface temperature of 20°C. This type of profile can occur during the daytime or at night downwind due to wind shear in the atmosphere or a very high temperature inversion. If this is considered too extreme, or too rare a condition, then a second possibility is a shallow inversion. A shallow inversion occurs regularly at night. A typical depth is 200 m. The profile may be calculated from a temperature gradient of +20°C/km from the surface to 200 m and −8°C/km above that assuming a surface temperature of 20°C.

The prediction of outdoor sound propagation also requires information about turbulence.

## 3.7 Sound propagation in a turbulent atmosphere

Sound propagating through a turbulent atmosphere will fluctuate both in amplitude and phase as a result of fluctuations in the refractive index caused by fluctuations in temperature and wind velocity. When predicting outdoor sound, it is usual to refer to these fluctuations in wind velocity and temperature rather than the cause of the turbulence. The amplitude of fluctuations in sound level caused by turbulence initially increases with increasing distance of propagation, sound frequency and strength of turbulence, but reaches a limiting value fairly quickly. This means that the fluctuation in overall sound levels from distant sources (e.g. line-of-sight from an aircraft at a few kilometers) may have a standard deviation of no more than about 6 dB.[57]

There are two types of atmospheric instability responsible for the generation of turbulent kinetic energy: shear and buoyancy. Shear instabilities are associated with mechanical turbulence. High wind conditions and a small temperature difference between the air and ground are the primary causes of mechanical turbulence. Buoyancy or convective turbulence is associated with thermal instabilities. Such turbulence prevails when the ground is much warmer than the overlying air, as, for example, on a sunny day. The irregularities in the temperature and wind fields are directly related to the scattering of sound waves in the atmosphere.

Fluid particles in turbulent flow often move in ‘loops’ (see Fig. 3.2) corresponding to swirls or eddies. Turbulence can be visualized as a continuous distribution of eddies in time and space. The largest eddies can extend to the height of the boundary layer (i.e. up to 1–2 km on a sunny afternoon. However, the outer scale of usual interest in community noise prediction is of the order of meters. In the size range of interest, sometimes called the inertial sub-range, the kinetic energy in the larger eddies is transferred continuously to smaller ones. As the eddy size becomes smaller, virtually all of the energy is dissipated into heat. The length scale at which viscous dissipation processes begin to dominate for atmospheric turbulence is about 1.4 mm.

Shadow zones due to atmospheric refraction are penetrated by sound scattered by turbulence and this sets a limit of the order of 20–25 dB to the reduction of sound levels within the sound shadow.[58]

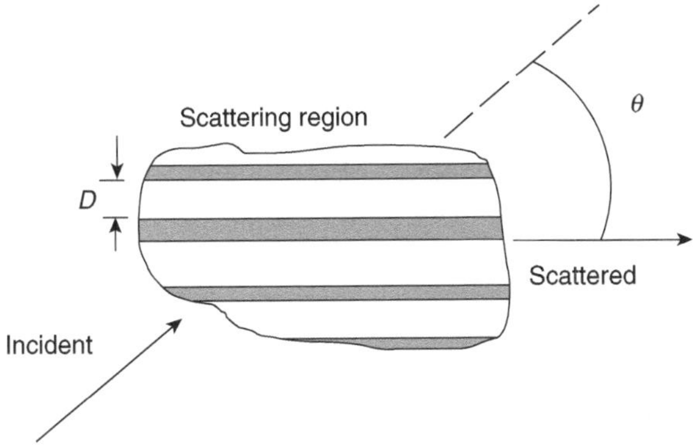


*Figure 3.18* Bragg reflection condition for acoustical Scattering by turbulence.

The size of eddies of most importance to scattering into the shadow zone, may be estimated by considering Bragg diffraction.[59] To scatter sound with wavelength $\lambda$ through angle $\theta$ (see Fig. 3.18), the important scattering structures have a spatial periodicity $D$ satisfying

$$\lambda = 2D\sin\left(\theta/2\right). \tag{3.55}$$

For a frequency of 500 Hz and a scattering angle of 10°, this predicts that the important eddy size for scattering into the shadow zone is 4 m.

When source and receiver are such that acoustic waves are propagating nearly horizontally, the (overall) variance in the effective index of refraction $\langle\mu^2\rangle$ is related approximately to those in velocity and temperature by[60]

$$\left\langle\mu^2\right\rangle = \frac{\langle u'^2\rangle}{c_0^2}\cos^2\phi + \frac{\langle v'^2\rangle}{c_0^2}\sin^2\phi + \frac{\langle u'T\rangle}{c_0}\cos\phi + \frac{\langle T^2\rangle}{4T_0^2} \tag{3.56}$$

where $T$, $u'$ and $v'$ are the fluctuations in temperature, horizontal wind speed parallel to the mean wind and horizontal wind speed perpendicular to the mean wind, respectively. $\phi$ is the angle between the wind and the wavefront normal. Use of similarity theory gives[61]

$$\left\langle\mu^2\right\rangle = \frac{5u_*^2}{c_0^2} + \frac{2.5u_*T_*}{c_0T_0}\cos\phi + \frac{T_*^2}{T_0^2}, \tag{3.57}$$

where $u^*$ and $T^*$ are the friction velocity and scaling temperature ($= -Q/u^*$, $Q$ being the surface temperature flux), respectively.

Typically, during the daytime, the velocity term in the effective index of refraction variance always dominates over the temperature term. This is true, even on sunny days, when turbulence is generated by buoyancy rather than shear. Strong buoyant instabilities produce vigorous motion of the air.

Situations where temperature fluctuations have a more significant effect on acoustic scattering than velocity fluctuations occur most often during clear, still nights.

Although the second term (the covariance term) in (3.56) may be almost as important as the temperature term,[61] it is often ignored for the purpose of predicting acoustic propagation. Estimations of the fluctuations in terms of $u^*$ and $T^*$ and the Monin–Obukhov length $L$ are given by,[62] for $L > 0$ (stable conditions, e.g. at night)

$$\sqrt{\langle u'^2\rangle} = \sigma_u = 2.4u_* \,, \; \sqrt{\langle v'^2\rangle} = \sigma_v = 1.9u_* \,, \; \sqrt{\langle T^2\rangle} = \sigma_T = 1.5T_*$$

for $L < 0$ (unstable conditions, e.g. daytime)

$$\sigma_u = \left(12 - 0.5\frac{z}{L}\right)^{1/3} u_*, \; \sigma_\mathrm{v} = 0.8\left(12 - 0.5\frac{z}{L}\right)^{1/3} u_*,$$

$$\sigma_\mathrm{T} = 2\left(1 - 18\frac{z}{L}\right)^{-1/2} T_*.$$

For line-of-sight propagation, the mean squared fluctuation in the phase of plane sound waves (sometimes called the strength parameter) is given by[63]

$$\Phi^2 = 2\left\langle \mu^2 \right\rangle k_0^2 XL$$

where $X$ is the range and $L$ is the inertial length scale of the turbulence. Alternatively, the variance in the log-amplitude fluctuations in a plane sound wave propagating through turbulence is given by[64]

$$\left\langle \chi^2 \right\rangle = \frac{k_0^2 X}{4}\left(L_T \frac{\sigma_T^2}{T_0^2} + L_v \frac{\sigma_v^2}{c_0^2}\right)$$

where $L_T$, $\sigma_T^2$ and $L_v$, $\sigma_v^2$ are integral length scales and variances of temperature and velocity fluctuations, respectively.

There are several models for the size distribution of turbulent eddies. In the Gaussian model of turbulence statistics, the energy spectrum $\phi_n(K)$ of the index of refraction is given by

$$\phi_n(K) = \left\langle \mu^2 \right\rangle \frac{L^2}{4\pi} \exp\left(-\frac{K^2 L^2}{4}\right) \tag{3.58}$$

where $L$ is a single scale length (integral or outer length scale) proportional to the correlation length (inner length scale) $\ell_G$, that is,

$$L = \ell_G \sqrt{\pi}/2,$$

Although, as shown below, it provides a poor overall description of the spectrum of atmospheric turbulence,[63] the Gaussian model has some utility in theoretical models of sound propagation through turbulence, since it allows many results to be obtained in simple analytical form.

In the von Karman spectrum, known to work reasonably well for high Reynolds number turbulence, the spectrum of the variance in index of refraction is given by

$$\phi_n(K) = \left\langle \mu^2 \right\rangle \frac{L}{\pi \left(1 + K^2 \ell_K^2\right)} \tag{3.59}$$

where $L = \ell_K \sqrt{\pi}\,\Gamma(5/6)/\Gamma(1/3)$.

Figure 3.19 compares the spectral function $\left(K\phi(K)/\left\langle \mu^2 \right\rangle\right)$ given by the von Karman spectrum for $\left\langle \mu^2 \right\rangle = 10^{-2}$ and $\ell_K = 1\,\mathrm{m}$ with two spectral functions calculated assuming a Gaussian turbulence spectrum, respectively, for $\left\langle \mu^2 \right\rangle = 0.818 \times 10^{-2}$ and $\ell_G = 0.93\,\mathrm{m}$ and $\left\langle \mu^2 \right\rangle = 0.2 \times 10^{-2}$ and $\ell_G = 0.1\,\mathrm{m}$. The variance and inner length scale for the first Gaussian spectrum have been chosen to match the von Karman spectrum exactly for the low wave numbers (larger eddy sizes). It also offers a reasonable representation near the spectral peak. Past the spectral peak and at high wave numbers, the first Gaussian spectrum decays far too rapidly. The second Gaussian spectrum clearly matches the von Karman spectrum over a narrow range of

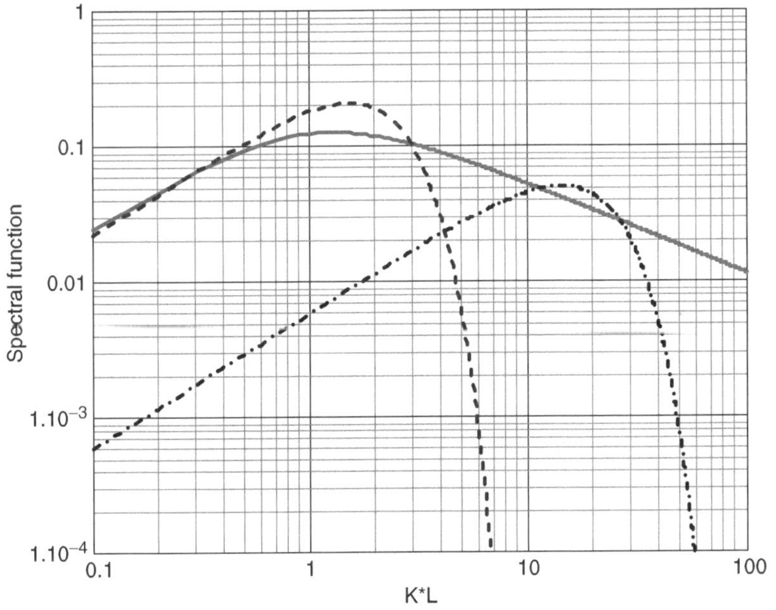


*Figure 3.19* A von Karman spectrum of turbulence and two Gaussian spectra chosen to match it at low wave numbers and over a narrow range of high wave numbers respectively.

smaller eddy sizes. If this happens to be the wave number range of interest in scattering from turbulence, then the Gaussian spectrum may be satisfactory.

Most recent calculations of turbulence effects on outdoor sound have relied on estimated or best-fit values rather than measured values of turbulence parameters. Under these circumstances, there is no reason to assume spectral models other than the Gaussian one. Typically, the high wave number part of the spectrum is the main contributor to turbulence effects on sound propagation. This explains why the assumption of a Gaussian spectrum results in best-fit parameter values that are rather less than those that are measured.

Turbulence destroys the coherence between direct and ground-reflected sound and consequently reduces the destructive interference in ground effect. Equation (3.19) may be modified[65] to obtain the mean-squared pressure at a receiver in a turbulent but acoustically neutral (no refraction) atmosphere

$$\left\langle p^2\right\rangle = \frac{1}{R_1^2} + \frac{|Q|^2}{R_2^2} + \frac{2|Q|}{R_1R_2}\cos\left[k\left(R_2 - R_1\right) + \theta\right]T \tag{3.60}$$

where $\theta$ is the phase of the reflection coefficient, ($Q = |Q|e^{-j\theta}$ ), and $T$ is the coherence factor determined by the turbulence effect.

Hence the sound pressure level, $P$, is given by:

$$P = 10\log_{10}(\langle p^2\rangle). \tag{3.61}$$

For a Gaussian turbulence spectrum, the coherence factor, $T$, is given by[65]

$$T = e^{-\sigma^2(1-\rho)} \tag{3.62}$$

where $\sigma^2$ is the variance of the phase fluctuation along a path and $\rho$ is the phase covariance between adjacent paths (e.g. direct and reflected).

$$\sigma^2 = A\sqrt{\pi}\left\langle\mu^2\right\rangle k^2 RL_0, \tag{3.63}$$

of the index of refraction, and $L_0$ is the outer (inertial) scale of turbulence.

The coefficient $A$ is given by:

$$A = 0.5\ R > kL_0^2 \tag{3.64a}$$

$$A = 1.0\ R < kL_0^2 \tag{3.64b}$$

The parameters $\langle\mu 2\rangle$ and $L_0$ may be determined from field measurements or estimated. The phase covariance is given by

$$\rho = \frac{\sqrt{\pi}}{2}\frac{L_0}{h}erf\left(\frac{h}{L_0}\right) \tag{3.65}$$

where $h$ is the maximum transverse path separation and $erf(x)$ is the Error function defined by:

$$erf(x) = \frac{2}{\sqrt{\pi}} \int_0^x e^{-t^2} dt. \tag{3.66}$$

For a sound field consisting only of direct and reflected paths (which will be true at short ranges) in the absence of refraction, the parameter $h$ is given by:

$$\frac{1}{h} = \frac{1}{2}\left(\frac{1}{h_s} + \frac{1}{h_r}\right) \tag{3.67}$$

where $h_s$ and $h_r$ are the source and receiver heights, respectively. Daigle[66] uses half this value to obtain better agreement with data.

When $h \to 0$, then $\rho \to 1$and $T \to 1$. This is the case near grazing incidence. For $h \to$ large, then $T \to$ maximum. This will be the case for a greatly elevated source and/or receiver. The mean squared refractive index may be calculated from the measured instantaneous variation of wind speed and temperature with time at the receiver. Specifically $\langle \mu^2 \rangle = \frac{\sigma_w^2 \cos^2 \alpha}{C_0^2} + \frac{\sigma_T^2}{4T_0^2}$, where $\sigma_w^2$ is the variance of the wind velocity, $\sigma_T^2$ is the variance of the temperature fluctuations, $\alpha$ is the wind vector direction and $C_0$ and $T_0$ are the ambient sound speed and temperature, respectively. Typical values of best-fit mean squared refractive index are between $10^{-6}$ for calm conditions and $10^{-4}$ for strong turbulence. A typical value of $L_0$ is 1m but in general a value equal to the source height should be used.

Figure 3.20 shows example results of computations of excess attenuation spectra using (3.60) through (3.67). Note that increasing turbulence reduces the depth of the main ground effect dip.

Figure 3.21 shows comparisons between Parkin and Scholes' data[40] and predictions based on (3.60) to (3.67). The inclusion of turbulence results in improved predictions, particularly near the ground effect dip in the spectrum.

Figures 3.22a and b show comparisons between measured data and theoretical predictions using a full wave numerical solution Fast Field Program (FFP) with and without turbulence[9]. The data in Fig. 3.22a were obtained with a loudspeaker source[67] in strong upwind conditions modeled by the logarithmic sound speed gradient[68]

$$c(z) = 340.0 - 2.0 \ln\left(z/6 \times 10^{-3}\right) \tag{3.68}$$

Those in Fig. 3.22b were obtained with a fixed jet engine source during conditions with a negative temperature gradient (modeled by two linear

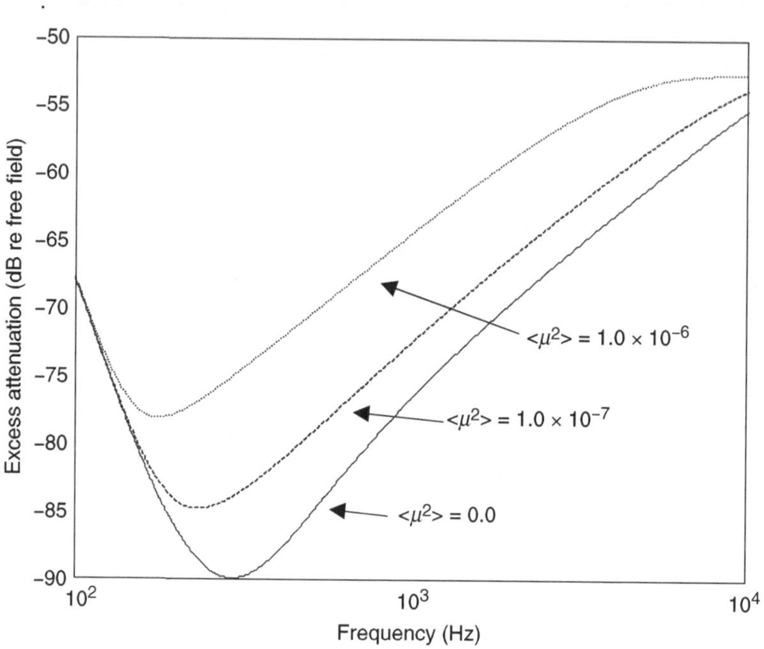


*Figure 3.20* Excess attenuation vs. frequency for a source and receiver above an impedance ground in acoustically neutral atmosphere predicted by (3.55)–(3.62) for three values of $< \mu^2 >$ between 0 and $10^{-6}$. The assumed source and receiver heights are 1.8 and 1.5 m respectively, and the assumed separation is 600 m. A two-parameter impedance model (3.30) has been used with values of 30,000.0 Ns/m$^4$ and 0.0/m.

segments) and a wind speed that was more or less constant with height at 4 m/s in the direction between receivers and source.

The agreement obtained between the predictions from the FFP including turbulence and the data, while not perfect, represents a large improvement over the results of calculations without turbulence and, in the case of Fig. 3.22a, is similar to that obtained with predictions given by the parabolic equation method.[68]

## 3.8 Sound propagation over noise barriers

### *3.8.1 Deployment of noise barriers*

Purpose-built noise barriers have become a very common feature of the urban landscape of Europe, the Far East and America. In the USA, over 1200 miles of highway noise barriers were constructed in the year 2001 alone. Most noise barriers are installed in the vicinity of transportation and industrial noise sources to shield nearby residential properties. At airports

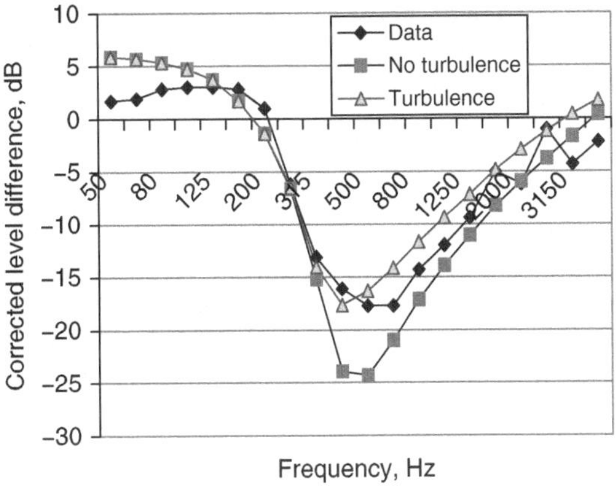


*Figure 3.21* Comparison of Parkin and Scholes' experimental data for the corrected horizontal level difference between 19 m and 347.47 m with predictions of (3.60)–(3.67) without and with turbulence ($\sigma^2 = 2 \cdot 10^{-6}, L_0 = 1$ m) and ground impedance given by (3.29b) with an effective flow resistivity of 300 kPa s/m$^2$.

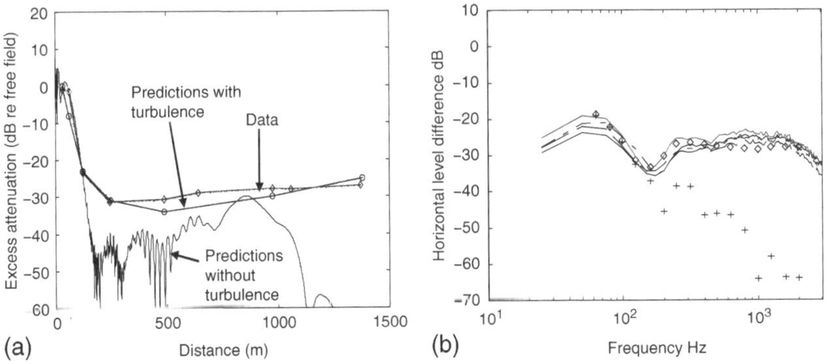


*Figure 3.22* (a) FFP predictions with and without turbulence[6] (see Chapter 10) compared with data for the propagation of a 424 Hz tone as a function of range out to 1.5 km[67] and (b) broadband data to 2500 Hz for four 26-second averages (lines) of the spectra of the horizontal level differences between sound level measurements at 6.4 m high receivers, 152.4 m and 762 m from a fixed jet engine source (nozzle exit centered at 2.16 m height) compared to FFP predictions with (diamonds) and without (crosses) turbulence[9] (see Chapter 10).

they are used to reduce the impact of aircraft engine testing and ground run-up noise. Noise barriers are cost-effective only for the protection of large areas including several buildings and are rarely used for the protection of individual properties. Noise barriers of usual height (up to 3 m) are generally ineffective in protecting the upper levels of multi-storey dwellings.

In the past two decades, environmental noise barriers have become the subject of extensive studies, the results of which have been consolidated in the form of national and international standards and prediction models.[69–71] Extensive guides to the acoustic and visual design of noise barriers are available.[72,73] Some issues remain to be resolved relating to the degradation of the noise barrier performance in the presence of wind and temperature gradients, the influence of localized atmospheric turbulence, temporal effects from moving sources, the influence of local vegetation, the aesthetic quality of barriers and their environmental impact.

### *3.8.2 Single-edge diffraction*

A noise barrier works by blocking the direct path from the noise source to the receiver. The noise then reaches the receiver only via diffraction around the barrier edges. The calculation of barrier attenuation is therefore mainly dependent on the solution of the diffraction problem. Exact integral solutions of the diffraction problem were available as early as the late nineteenth century[74] and early twentieth century.[75] For practical calculations, however, it is necessary to use approximations to the exact solutions. Usually, this involves assuming that the source and receiver are more than a wavelength from the barrier and the receiver is in the shadow zone, which is valid in almost all applications of noise barriers. As long as the transmission loss through the barrier material is sufficiently high, the performance of a barrier is dictated by the geometry (see Fig. 3.23).

The Kirchhoff–Fresnel approximation[76] and the geometrical theory of diffraction[77] for wedges and thick barriers have been used for deriving

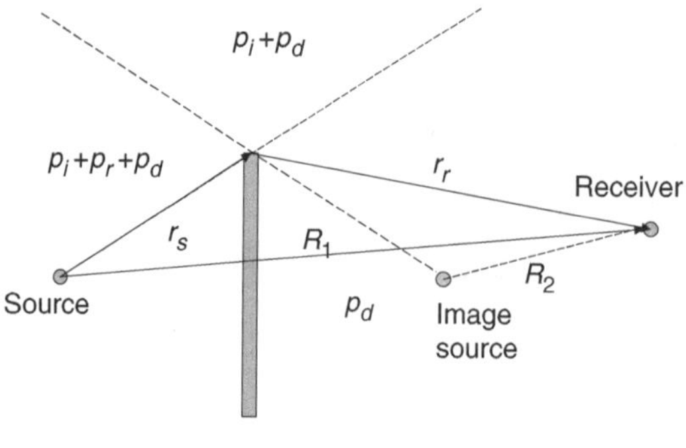


*Figure 3.23* Diffraction of sound by a thin barrier.

practical formulae for barrier calculations. For a rigid wedge barrier, the solution provided by Hadden and Pierce[78] is relatively easy to calculate and highly accurate. A line integral solution, based on the Kirchhoff–Fresnel approximation[79] describes the diffracted pressure from a distribution of sources along the barrier edge and has been extended to deal with barriers with jagged edges.[80] There is also a time domain model.[81]

The total sound field in the vicinity of a semi-infinite half plane depends on the relative position of the source, the receiver and the thin plane. The total sound field $p_T$ in each of three regions shown in Fig. 3.23 is given as follows.

In front of the barrier:

$$p_T = p_i + p_r + p_d \tag{3.69a}$$

Above the barrier:

$$p_T = p_i + p_d \tag{3.69b}$$

In the shadow zone:

$$p_T = p_d. \tag{3.69c}$$

Fresnel numbers of the source and image source are denoted, respectively, by $N_1$ and $N_2$, are defined as follows:

$$N_1 = \frac{R' - R_1}{\lambda/2} = \frac{k}{\pi}(R' - R_1) \tag{3.70a}$$

and

$$N_2 = \frac{R' - R_2}{\lambda/2} = \frac{k}{\pi}(R' - R_2) \tag{3.70b}$$

where $R' = r_s + r_r$ is the shortest path source-edge-receiver.

The attenuation ($Att$) of the screen, sometimes known as the insertion loss, $IL$, is often used to assess the acoustics performance of the barrier. It is defined as follows:

$$Att = IL = 20\lg\left(\left|\frac{p_w}{p_{w/o}}\right|\right) dB, \tag{3.71}$$

where $p_w$ and $p_{w/o}$ is the total sound field with or without the presence of the barrier. Note that the barrier attenuation is equal to insertion loss only in the absence of ground effect.

Maekawa[82] has provided a chart that expresses the attenuation of a thin rigid barrier based on the Fresnel number ($N_1$) associated with the source.

The chart was derived empirically from extensive laboratory experimental data but use of the Fresnel number was suggested by the Kirchhoff–Fresnel diffraction theory. Maekawa's chart extends into the illuminated zone where $N_1$ is taken to be negative. Maekawa's chart has been proved to be very successful and has become the *de facto* standard empirical method for barrier calculations. Many of the barrier calculation methods embodied in national and international standards[70,71] stem from this chart. There have been many attempts to fit the chart with simple formulae. One of the simplest formulae[83] is

$$Att = 10\lg(3 + 20N_1)\,\mathrm{dB} \tag{3.72}$$

The Maekawa curve can be represented mathematically by[84]

$$Att = 5 + 20\lg\frac{\sqrt{2\pi N_1}}{\tanh\sqrt{2\pi N_1}}. \tag{3.73}$$

An improved version of this result, using both of the Fresnel numbers defined in (3.70), is[85]

$$Att = Att_s + Att_b + Att_{sb} + Att_{sp} \tag{3.74a}$$

where

$$Att_s = 20\lg\frac{\sqrt{2\pi N_1}}{\tanh\sqrt{2\pi N_1}} - 1, \tag{3.74b}$$

$$Att_b = 20\lg\left[1 + \tanh\left(0.6\lg\frac{N_2}{N_1}\right)\right], \tag{3.74c}$$

$$Att_{sb} = \left(6\tanh\sqrt{N_2} - 2 - Att_b\right)\left(1 - \tanh\sqrt{10N_1}\right), \tag{3.74d}$$

$$Att_{sp} = -10\lg\frac{1}{(R'/R_1)^2 + (R'/R_1)}. \tag{3.74e}$$

The term $Att_s$ is a function of $N_1$, which is a measure of the relative position of the receiver from the source. The second term depends on the ratio of $N_2/N_1$, which depends on the proximity of either the source or the receiver to the half plane. The third term is only significant when $N_1$ is small and depends on the proximity of the receiver to the (geometrical) shadow boundary. The last term, a function of the ratio $R'/R_1$, accounts for the diffraction effect due to spherical incident waves.

### 3.8.3 *Effects of the ground on barrier performance*

Equations (3.72) to (3.74) predict only the amplitude of sound and do not include wave interference effects. Such interference effects result from the

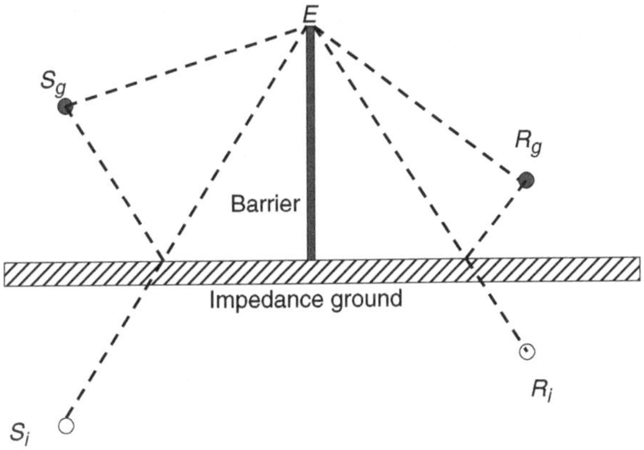


*Figure 3.24* Diffraction by a barrier on impedance ground.

contributions from different diffracted wave paths in the presence of ground (see also Section 3.2) or from the vertical edges of a finite length barrier (see Section 3.8.4). Now we consider the interaction between barrier and ground effects.

Consider a source $S_g$ located at the left side of the barrier, a receiver $R_g$ at the right side of the barrier and $E$ is the diffraction point on the barrier edge (see Fig. 3.24).

The sound reflected from the ground surface can be described by an image of the source $S_i$. On the receiver side, sound waves will also be reflected from the ground. This effect can be considered in terms of an image of the receiver $R_i$. The pressure at the receiver is the sum of four terms which correspond to the sound paths $S_gER_g, S_iER_g, S_gER_i$ and $S_iER_i$. If the surface is a perfectly reflecting ground, the total sound field is the sum of the diffracted fields of these four paths,

$$P_T = P_1 + P_2 + P_3 + P_4$$

where

$$P_1 = P(S_g, R_g, E), P_2 = P(S_i, R_g, E), P_3 = P(S_g, R_i, E), P_4 = P(S_i, R_i, E).$$

$P(S, R, E)$ is the diffracted sound field due to a thin barrier for given positions of source $S$, receiver $R$ and the point of diffraction at the barrier edge $E$. If the ground has finite impedance (such as grass or a porous road surface), then the pressure corresponding to rays reflected from these surfaces should be multiplied by the appropriate spherical wave reflection coefficient(s) to allow for the change in phase and amplitude of the wave on reflection as follows:

$$P_T = P_1 + Q_s P_2 + Q_R P_3 + Q_s Q_R P_4$$

where $Q_s$ and $Q_R$ are the spherical wave reflection coefficients for the source and receiver side, respectively. The spherical wave reflection coefficients can be calculated for different types of ground surfaces and source/receiver geometries (see Section 3.2).

Usually, for a given source and receiver position, the acoustic performance of the barrier on the ground is assessed by use of either the excess attenuation ($EA$) or the insertion loss ($IL$). They are defined as follows:

$$EA = SPL_f - SPL_b \tag{3.75}$$

$$IL = SPL_g - SPL_b \tag{3.76}$$

where $SPL_f$ is the free field noise level, $SPL_g$ is the noise level with the ground present and $SPL_b$ is the noise level with the barrier and ground present. Note that in the absence of a reflecting ground, the numerical value of $EA$ (which was called $Att$ previously) is the same as $IL$. If the calculation is carried out in terms of amplitude only, then the attenuation $Att_n$ for each sound path can be directly determined from the appropriate Fresnel number $F_n$ for that path. The excess attenuation of the barrier on a rigid ground is then given by

$$A_T = 10\lg\left(10^{-\left|\frac{Att_1}{10}\right|} + 10^{-\left|\frac{Att_2}{10}\right|} + 10^{-\left|\frac{Att_3}{10}\right|} + 10^{-\left|\frac{Att_4}{10}\right|}\right). \tag{3.77}$$

The attenuation for each path can either be calculated by empirical or analytical formulae depending on the complexity of the model and the required accuracy.

A modified form of the empirical formula for the calculation of barrier attenuation is[70]

$$IL = 10\log_{10}\left(3 + \left(C_2\frac{\delta_1}{\lambda}\right)C_3 K_{met}\right). \tag{3.78}$$

where $C_2 = 20$ and includes the effect of ground reflections; $C_2 = 40$ if ground reflections are taken into account elsewhere. $C_3$ is a factor to take into account of a double diffraction or finite barrier effect, $C_3 = 1$ for a single diffraction and $\delta_1 = (r_s + r_r) - R_1$ (Fig. 3.24). The $C_3$ expression for double diffraction is given later.

The term $K_{met}$ in equation (3.78) is a correction factor for average downwind meteorological effects, and is given by

$$K_{met} = e^{-\frac{1}{2000}\sqrt{\frac{r_s r_r r_o}{2\delta_1}}} \quad \text{for } \delta_1 > 0 \text{ and } K_{met} = 1 \quad \text{for } \delta_1 \le 0.$$

The formula reduces to the simple formula (3.72) when the barrier is thin, there is no ground and if meteorological effects are ignored.

There is a simple approach capable of modeling wave effects in which the phase of the wave at the receiver is calculated from the path length via the top of the screen, assuming a phase change in the diffracted wave of $\pi/4$.[86] This phase change is assumed to be constant for all source–barrier–receiver geometries. The diffracted wave, for example, for the path $S_gER_g$ would thus be given by

$$P_1 = Att_1 e^{-j[k(r_0+r_r)+\pi/4]}. \tag{3.79}$$

This approach provides a reasonable approximation for the many situations of interest where source and receiver are many wavelengths from the barrier and the receiver is in the shadow zone.

For a thick barrier of width $w$, ISO 9613-2[69] provides the following form of correction factor $C_3$ for use in (3.79):

$$C_3 = \frac{\left[1+\left(\frac{5\lambda}{w}\right)^2\right]}{\left[\frac{1}{3}+\left(\frac{5\lambda}{w}\right)^2\right]}$$

where for double diffraction, $\delta_1 = (r_s + r_r + w) - R_1$.

Note that this empirical method is for rigid barriers of finite thickness and does not take absorptive surfaces into account.

### 3.8.4 *Diffraction by finite length barriers and buildings*

All noise barriers have finite length and for certain conditions sound diffracting around the vertical ends of the barrier may be significant. This will be the case for sound diffracting around buildings also. Figure 3.25 shows eight diffracted ray paths contributing to the total field behind a finite-length barrier situated on finite impedance ground. In addition to the four 'normal' ray paths diffracted at the top edge of the barrier (see Fig. 3.24), four more

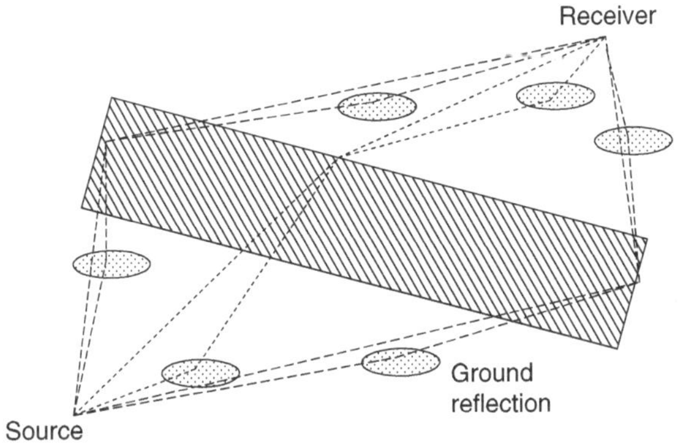


*Figure 3.25* Ray paths around a finite length barrier or building on the ground.

diffracted ray paths result from the vertical edges – two ray paths from each one. The two rays at either side are, respectively, the direct diffracted ray and the diffracted-reflected ray. Strictly, there are two further ray paths at each side which involve two reflections at the ground as well as diffraction at the vertical edge but, usually, these are neglected.

The reflection angles of the two diffracted–reflected rays are independent of the barrier position. They will either reflect at the source side or on the receiver side of the barrier, which is dependent on the relative positions of the source, receiver and barrier. The total field is given by

$$P_T = P_1 + Q_s P_2 + Q_R P_3 + Q_s Q_R P_4 + P_5 + Q_R P_6 + P_7 + Q_R P_8 \quad (3.80)$$

where $P_1 - P_4$ are those given earlier for the diffraction at the top edge of the barrier.

Although accurate diffraction formulas may be used to compute $P_i$ $(i = 1 \ldots 8)$, a simpler approach is to assume that each diffracted ray has a constant phase shift of $\pi/4$ regardless of the position of source, receiver and diffraction point.

To predict the attenuation due to a single building, the double diffraction calculations mentioned earlier could be used. For source or receiver situated in a built-up area, ISO 9613-2[69] proposes an empirical method for calculating the combined effects of screening and multiple reflections. The net attenuation $A_{build}$ dB (< 10 dB) is given by

$$\begin{aligned} A_{build} &= A_{build,1} + A_{build,2} \\ A_{build,1} &= 0.1 B d_0, \\ A_{build,2} &= -10 \log[1 - (p/100)], \end{aligned} \quad (3.81)$$

where $B$ is the area density ratio of buildings (total plan area/total ground area) and $d_0$ is the length of the refracted path from source to receiver that passes through buildings. $A_{build,2}$ is intended to be used only where there are well-defined but discontinuous rows of buildings near to a road or railway, and $p$ is the percentage of the length of facades relative to the total length of the road or railway. As with barrier attenuation, the attenuation due to buildings is to be included only when it is predicted to be greater than that due to ground effect. The ISO scheme offers also a frequency dependent attenuation coefficient (dB/m) for propagation of industrial noise through an array of buildings on an industrial site.

## 3.9 Sound propagation through trees

A mature forest or woodland may have three types of influence on sound. The first is the ground effect. This is particularly significant if there is a thick litter layer of partially decomposing vegetation on the forest floor.

In such a situation the ground surface consists of a thick highly porous layer with rather low-flow resistivity, thus giving a primary excess attenuation maximum at lower frequencies than observed over typical grassland. Secondly, the trunks and branches scatter the sound out of the path between source and receiver. Thirdly, the foliage attenuates the sound by viscous friction. To predict the total attenuation through woodland, Price *et al.*[87] simply added the predicted contributions to attenuation for large cylinders (representing trunks), small cylinders (representing foliage) and the ground. The predictions are in qualitative agreement with their measurements, but it is necessary to adjust several parameters to obtain quantitative agreement. Price *et al.* found that the main acoustical effects of foliage are above 1 kHz and that the attenuation due to foliage increased approximately in a linear fashion with frequency. Figure 3.26 shows a typical variation of attenuation with frequency and linear fits to the foliage attenuation.

Often, the insertion loss of tree belts alongside highways is considered relative to that over open grassland. An unfortunate consequence of the lower-frequency ground effect observed in mature tree stands is that the low-frequency destructive interference resulting from the relatively soft ground between the trees is associated with a constructive interference maximum at important frequencies (near 1 kHz) for traffic noise. Consequently, many narrow tree belts alongside roads do not offer much additional attenuation of traffic noise compared with the same distances over open grassland. A Danish study found relative attenuation of 3 dB in the A-weighted $L_{eq}$ due to traffic noise for tree belts between 15 and

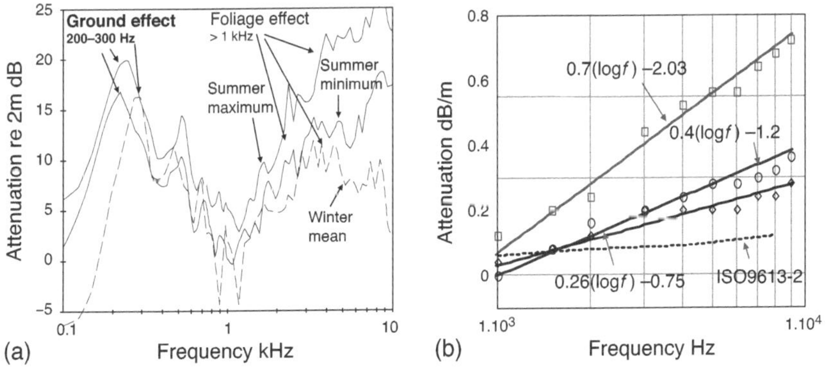


*Figure 3.26* (a) Measured attenuation through alternate bands of Norway spruce and oak (planted in 1946) with hawthorn, roses and honeysuckle undergrowth; visibility less than 24 m. (b) Linear fits and corresponding equations for attenuation above 1 kHz in mixed conifers (squares), mixed deciduous summer (circles) and spruce monoculture (diamonds). Also shown is the foliage attenuation predicted according to ISO 9613-2.[69]

41 m wide.[88] On the other hand, data obtained in the UK[89] indicates a maximum reduction of the A-weighted $L_{10}$ level (the level exceeded for 10 per cent of the time; usually greater than $L_{eq}$) due to traffic noise of 6 dB through 30 m of dense spruce compared with the same depth of grassland. This study also found that the effectiveness of the vegetation was greatest closest to the road. A relative reduction of 5 dB in the A-weighted $L_{10}$ level was found after 10 m of vegetation. For a narrow tree belt to be effective against traffic noise, it is important that: (a) the ground effect is similar to that for grassland; (b) there is substantial reduction of coherence between ground-reflected and direct sound at frequencies of 1 kHz and above; and (c) the vegetation is sufficiently dense so that the attenuation through reverberant and viscous scattering is significant.

If the belt is sufficiently wide, then the resulting greater extent of the ground effect dip can compensate for its low frequency. Through 100 m of red pine forest, Heisler *et al.*[90] found 8 dB reduction in the A-weighted $L_{eq}$ due to road traffic compared with open grassland. The edge of the forest was 10 m from the edge of the highway and the trees occupied a gradual downward slope from the roadway extending about 325 m in each direction along the highway from the study site. Compared with open grassland, Huisman[54] predicted an extra 10 dBA attenuation of road traffic noise through 100 m of pine forest. He also remarked that, whereas downward-refracting conditions lead to higher sound levels over grassland, the levels in woodland are comparatively unaffected. This suggests that extra attenuation obtained through use of trees should be relatively robust to changing meteorology.

Defrance *et al.*[91] compared results from both numerical calculations and outdoor measurements at a pine forest site near a highway for different meteorological situations. A numerical parabolic equation code was developed[92] and adapted to road traffic noise situations[93] where road line sources are modeled as a series of equivalent point sources of height 0.5 m. The data showed a reduction in A-weighted $L_{eq}$ due to the trees of 3 dB during downward-refracting conditions, 2 dB during homogeneous conditions and 1 dB during upward-refracting conditions. The numerical predictions suggest that in downward-refracting conditions, the extra attenuation due to the forest is between 2 and 6 dBA with the receiver at least 100 m away from the road. In upward-refracting conditions, the numerical model predicts that the forest may increase the received sound levels somewhat at large distances but this is of less importance since levels that are at larger distances tend to be relatively low anyway. In homogeneous conditions, it is predicted that sound propagation through the forest is affected only by the scattering by trunks and foliage. Defrance *et al.*[91] concluded that a forest strip of at least 100-m wide appears to be a useful natural acoustical barrier. Nevertheless, both the data and numerical simulations were compared to sound levels without the trees present (i.e. over ground from which the trees had simply been removed). This means that the ground effect both with and without trees would have been similar. This is rarely likely to be the case.

A similar numerical model has been developed recently[94] including allowance for ground effect, wind speed gradient through foliage and assuming effective wave numbers deduced from multiple scattering theory for scattering effects of trunks, branches and foliage. Again, the model predicts that the large wind speed gradient in the foliage tends to refract sound towards the ground and has an important effect, particularly during upwind conditions. However, neither of the PE models[91,94] include back-scatter or turbulence effects. The neglect of the back-scatter is inherent to the PE which is a 'one-way' prediction method. This is not a serious problem for propagation over flat ground because the back-scatter is small, nor over an impermeable barrier because the back-scattered field, though strong, does not propagate through the barrier. However, back-scatter is likely to be significant for a forest. Indeed, acoustic reflections from the edges of forests are readily detectable.

# 4 Methods for aircraft noise prediction

## 4.1 Introduction

Acoustic modeling around airports is intended to satisfy the needs of many users. Its nature ranges between sophisticated noise spectrum modeling and pragmatic noise environment assessment in terms of cumulative noise exposure or, by means of dose–response relationships, predictions of the size of the population annoyed by the noise in the area of concern.[1–3] It must be noted that the form and structure of the noise indices, which must be assessed and investigated around the airport or under the particular flight path, have a dominant influence on the method used for their assessment. Methods for modeling noise radiation, propagation and attenuation include both analytical and semi-empirical techniques. The current tendency is toward less empirical and more analytical techniques. It should be noted that the International Civil Aviation Organization (ICAO) is carrying out analyses of existing models and methods for assessing aircraft noise events and is making proposals for their use.[4,5]

Two approaches to analysis of aircraft noise phenomena have been defined and implemented in computer programs. The first approach is based on the third-octave band spectra noise analysis of any type of aircraft in any mode of flight or during maintenance activities in the vicinity of or inside the airport. The corresponding prediction of the aircraft noise under the flight path or around the aircraft on the ground (run-ups, taxiing, queuing along the runway before the takeoff) is based on the sound wavefront spreading along the shortest distance between the aircraft and the point of noise control. This approach is implemented in well-known computer programs, like ANOPP (USA) and Flula (Switzerland). The BELTRA program (Ukraine) combines two large modules: BELTAS – for noise assessment at points of interest around the source and hence derivation of the directivity pattern of a noise event, and TRANOI – which indicates the need for noise control under the flight paths.

The second approach is based on the concept of 'noise radius' or 'noise–power–distance' and provides calculations of aircraft noise exposure units around the airports or at any noise monitoring point. The values of both basic relationships – 'noise radius' and 'noise–power–distance' – may be obtained from experimental data as well as by calculation (e.g. by using the BELTRA program). The main difference in the second approach is that, in addition to noise-generation effects, it includes the influence of geometrical spreading (through wavefront divergence) and sound absorption along the sound path. The resulting predictions are very specific for particular environmental conditions and do not depend only on flight parameters. Usually they are presented for standard atmosphere (SA) or for SA+10°C conditions, as required for airport noise calculations. The task of deriving an acoustic model for each type of aircraft under consideration has been proposed and solved in a manner that reconciles experimental data with calculation. So the aircraft noise models used in BELTRA are reliable and accurate. The second modeling approach has been utilized in many computer programs including ISOBELL'a (Ukraine), Fanomous (Netherlands), INM and NOISEMAP (USA).

The advantages of both approaches can be combined in one sophisticated tool. For example, both Ukrainian programs, BELTRA and ISOBELL'a, have been combined in the software ISOBELL'a Plus used for decision-making procedures concerning aircraft noise problems. Details of these two approaches are discussed here together with examples of their use for aircraft noise calculations are given.

Models and methods used for assessing environmental noise problems must be based on the noise exposure indices used by relevant national and international noise control regulations and standards. These differ greatly, both in their structure, and in the basic approaches used for their definitions (see Table 1.6 in Chapter 1). On the other hand there is a simple mutual relationship.[5] Values of $EPNL = 100$ and 90 EPNdB were used for noise contour analysis in the current research and they correlate well with values of $L_{Aeq} = 75$ and 65 dB, respectively. The correlation between the values of noise contour areas $S$ and the values of $EPNL$ at the noise monitoring points are also good (see Fig. 4.1a and b) for both takeoff and landing flight stages.

In general, prediction schemes are based on three basic components:

- a noise radiation model corresponding to an aircraft noise emission model;
- a model of sound transmission from source to control point in the form of an aircraft noise propagation model;
- a noise impact model at the control point in the form of an aircraft noise imission model.

It is important to note that ICAO continuously analyzes the available models and methods for estimating aircraft noise impact sources and issues recommendations on their use.[4]

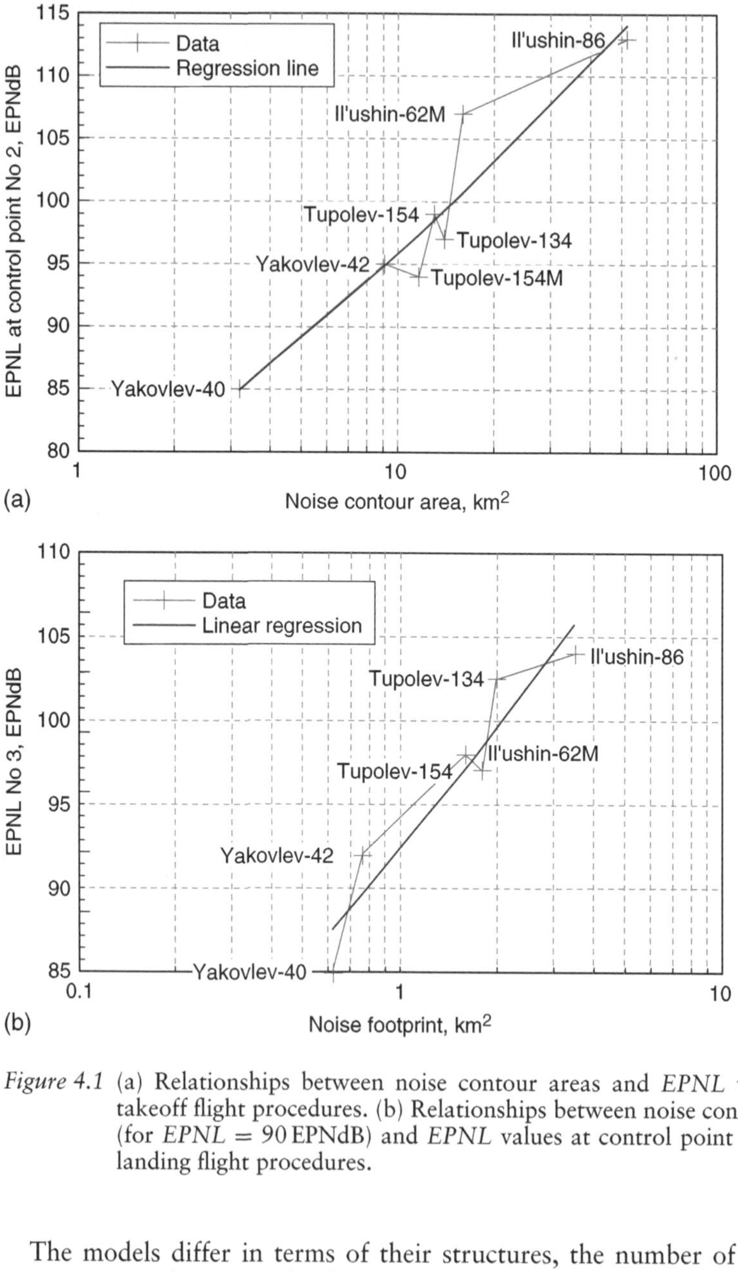


*Figure 4.1* (a) Relationships between noise contour areas and *EPNL* values for takeoff flight procedures. (b) Relationships between noise contour areas (for $EPNL = 90$ EPNdB) and *EPNL* values at control point No. 3 for landing flight procedures.

The models differ in terms of their structures, the number of required parameters and the initial information necessary to implement them. The models can be linked with stages of the life cycle of the aircraft in a balanced approach to the aviation noise (AN) problem (Fig. 4.2).

The three stages of AN modeling are shown in Table 4.1.

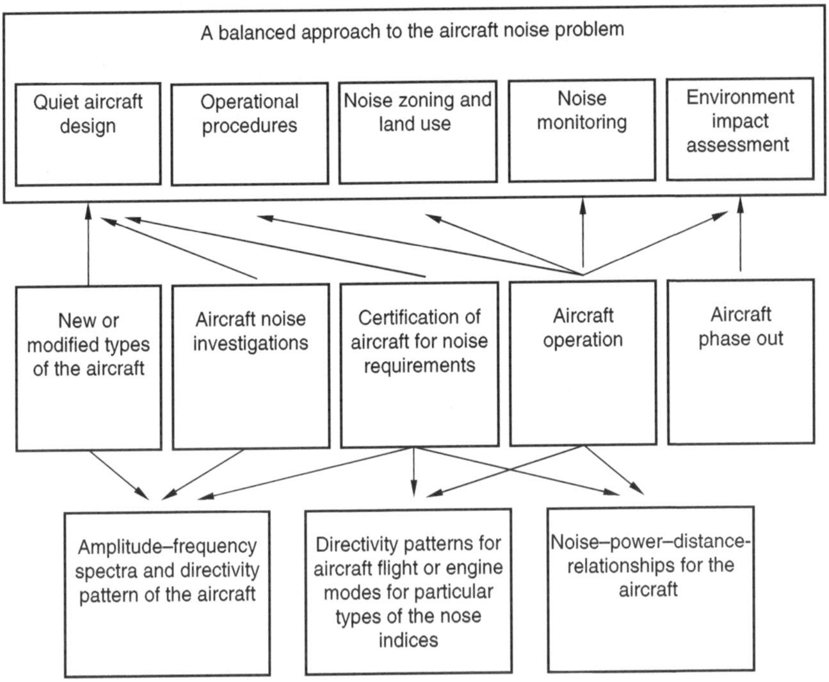


*Figure 4.2* A balanced approach for aviation noise control including stages in the life cycle of the aircraft and in aircraft noise modeling.

*Table 4.1* Relationships between various stages of aviation noise modeling

| *Stage 1* | *Stage 2* | *Stage 3* |
|---|---|---|
| Amplitude–frequency spectrum in any direction of sound propagation | Frequency correction for selected aviation noise criterion + directivity pattern for that criterion | Noise–power–distance relationship (for the specific frequency correction and time averaging associated with the selected aircraft noise index) |
| $L_f = F(R, r, f, \Theta)$ → $\Sigma_f$ → | $L_\Sigma = F(R, r, \Theta)$ → $\int d\Theta$ → | $L_\Sigma = F(R, r)$ |

*Stage 1* is the most complicated. It involves the spectral analysis of the general acoustic field around the aircraft and of the contributions of the component noise sources to the general acoustic field. Accordingly, it provides the basis for technical recommendations on reduction of noise generated by the various sources. It is the foundation for constructing a basic (preliminary) acoustic model of the aircraft of any type. It requires knowledge of the spectral radiation patterns of acoustic sources, defined for diverse directions, environmental conditions and flight mode parameters.

*Stage* 2 includes methods based on measured or calculated noise directivity diagrams, defined for $L_A$ or *PNL*. These change during the flight.

*Stage 3* is the simplest. It is intended to enable estimations of local or regional influences in respect of aircraft noise, including aircraft noise zoning and land-use planning around the airports. It uses the data on flight noise in relationships such as 'noise–power–distance' (NPD-relationships) as detailed later in Section 4.2.

The simplest type of model structure is confined to Stage 3 and includes the definitions of noise footprints (contours for specified values of noise indices and areas, bounded by these contours) for any type of the aircraft and any particular flight mode.

For example, if the noise certification data for the aircraft under consideration are known, a first approximation of the noise footprint area $S$ for any type of the criteria may be found from the linear relationships shown in Fig. 4.1. For the takeoff flight cycle of the aircraft, it may be given by:

$$S = 10^{[(L_1+L_2)/C]+D}, \tag{4.1}$$

where $L_2$ is the certified noise level at monitoring point No. 2 for takeoff and $L_1$ is the sideline noise certified level for monitoring point No. 1, and the coefficients $C$ and $D$ depend on the type and value of the noise footprint (contour). The sum of the certified noise levels $L_1 + L_2$ provide a better correlation than the separate usage of these levels for such a purpose.[6]

According to the noise radius concept, explained further in Section 4.2, a simplified form of the noise footprint is an ellipse (Fig. 4.3); thus the approximation (4.1) has a simple geometric explanation.

The difference between the certified noise level at takeoff ($L_2$) and the level corresponding to the final point on the contour $L$ along the flight (landing or takeoff) axis may be written:

$$L_2 - L = C\lg(a/a_2), \tag{4.2}$$

where the constant C defines the attenuation rate (for spherical spreading its value is near to 20), $a$ is the minimum distance from the flight path to the final point on the contour (Fig. 4.3), $a_2$ is the minimum distance from the takeoff path to the certification point No. 2 (for takeoff). Using the properties of

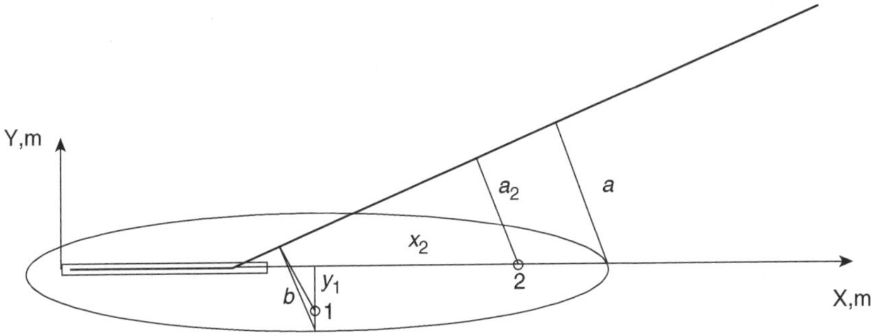


*Figure 4.3* A simplified form of noise footprint having the shape of an ellipse under the takeoff flight path.

similar triangles (see Fig. 4.3), $a/a_2 = x/x_2$ so

$$L_2 - L = C\lg(x/x_2), \tag{4.3}$$

In the same way, for the certification sideline point No. 1, the definition of $b_1$ enables connection of the certificated level $L_1$ with final point of the (maximum half-width) contour $b$:

$$L_1 - L = C\lg(b/b_1), \tag{4.4}$$

and accordingly

$$L_1 - L = C\lg(y/y_1). \tag{4.5}$$

Since area $S$ is proportional to the product $xy$, then, from (4.3) and (4.5), it is proportional to $L_2$ and $L_1$:

$$\lg(S) = (L_2 + L_1)/C + D, \tag{4.6}$$

which is the same as equation (4.1).

As an example of this approach, INM has been used to calculate noise footprints employing the noise certification levels for aircraft listed in Table 4.2 with different types of bypass engines (bypass ratios between 1 and 6), different number of engines and a range of takeoff masses (all the aircraft from the former Soviet Union production). The resulting noise footprints are shown in Table 4.3.

From the calculated footprints for sound exposure levels $SEL = 80$ dBA in Table 4.3, it is possible to show direct relationships between certification noise data and footprint areas. The associated values $D = -4.74$ and $(1/C) = 0.03391$ are obtained with the correlation coefficient 0.9359 and standard deviations 0.3545 and 0.001902, respectively.

*Table 4.2* Measured noise levels, *EPNL*, at certification points for aircraft from the former Soviet Union

| *Type of the aircraft* | *EPNL (EPNdB)* | |
|---|---|---|
| | *Takeoff* | *Landing* |
| Il'ushin -76T | 107.3 | 102.7 |
| Il'ushin-86 | 107.4 | 104.2 |
| Tupolev-154 | 100.1 | 97.8 |
| Il'ushin-62 | 106.9 | 95.2 |
| Tupolev-134 | 95.3 | 101.9 |
| Yakovlev-42 | 93.8 | 93.6 |
| Yakovlev-40 | 88.7 | 85.5 |

If the certification data are not available, *aircraft type grouping* may be used. This defines the appropriate group for the aircraft under consideration depending on the most important parameters, including takeoff mass, number and type of engines in the power plant. For example, in the former Soviet Union method, five such groups were defined (Table 4.4), with group 2 being the reference group.

The averaged noise levels under the flight path were defined for group 2 (Fig. 4.4) and the averaged level differences for the other groups are shown in Table 4.4. The method of aircraft type grouping predicts noise levels with an accuracy of ±5 dBA for takeoff and climbing and ±3 dBA for landing. A similar but more accurate method enables assessment of noise levels under the flight path for every specific group of the aircraft. Figure 4.5 shows an example for four specific groups of the aircraft at takeoff: long range aircraft with four engines (LRA4), long range aircraft with two engines (LRA2), middle range aircraft (MRA) and business class (BC). Its accuracy is about ±3 dBA for takeoff and climbing and ±2 dBA for landing.

## 4.2 An acoustic model of an aircraft

As discussed in Chapter 2, the acoustical characteristics of the aircraft at a control point are represented as a set of third-octave band sound pressure level (*SPL*) spectra defined at 0.5-s intervals. This information is used for calculating aircraft noise indices; usually *SEL* or *EPNL*. For aircraft operating on the ground, the *SPL* spectra (including directivity) are defined at particular distances and directivity angles from the aircraft (or aircraft engine or testing facility). In any case, the set of *SPL* spectra, defined for a specific flight mode (or engine operation mode) and distance to the aircraft, is represented in terms of a noise matrix, which is specific to any type of the aircraft. Each row of the matrix is defined for a specific frequency band and each column is defined for a specific angle of noise radiation. An example is given in Table 4.5.

*Table 4.3* Calculated footprints for $L_{Amax}$ and *SEL*

| *Level (dB)* | *Footprint area for a given $L_{Amax}$ (km$^2$)* | | | | | *Footprint area for given a SEL (km$^2$)* | | | | |
|---|---|---|---|---|---|---|---|---|---|---|
| | *65* | *70* | *75* | *80* | *85* | *80* | *85* | *90* | *95* | *100* |
| Il'ushin-76T | 246.679 | 172.762 | 102.743 | 53.545 | 27.251 | 210.882 | 132.237 | 63.454 | 28.980 | 13.384 |
| Il'ushin-78 | 268.596 | 190.537 | 120.396 | 57.612 | 25.223 | 229.589 | 146.866 | 68.204 | 26.064 | 9.209 |
| Tupolev-154 | 268.596 | 190.537 | 136.487 | 60.799 | 23.733 | 270.810 | 170.748 | 76.322 | 24.441 | 7.458 |
| Tupolev-134 | 306.134 | 199.374 | 105.585 | 48.805 | 22.232 | 254.218 | 139.672 | 56.903 | 22.046 | 9.383 |
| Il'ushin-62 | 194.105 | 128.185 | 69.313 | 35.995 | 17.822 | 156.303 | 84.481 | 39.174 | 16.922 | 7.367 |
| Yakovlev-42 | 86.840 | 48.667 | 26.523 | 13.884 | 7.526 | 59.097 | 28.942 | 13.378 | 6.387 | 3.333 |
| Yakovlev-40 | 51.253 | 25.431 | 12.115 | 5.768 | 2.774 | 28.216 | 10.941 | 4.381 | 1.908 | 0.840 |

*Table 4.4* Aircraft type grouping used for aircraft of the former Soviet Union

| *Group number* | *Type of aircraft* | $\Delta_1$ *[dBA]* |
|---|---|---|
| 1 | Jet or bypass: Il'ushin-86<br>Turboprop: Antonov 22 | +5 |
| 2 | Jet or bypass: Il'ushin-62, Il'ushin-62Ì, Il'ushin-76T, Tupolev-154B, Tupolev-134, B-732<br>Turboprop | 0 |
| 3 | Jet or bypass: Yakovlev-42<br>Turboprop: Antonov-12, Il'ushin-18 | −5 |
| 4 | Jet or bypass: Yakovlev-40<br>Turboprop: Antonov-24, Antonov-26, | −10 |
| 5 | Jet or bypass<br>Turboprop: Antonov-28, L-410 | −15 |

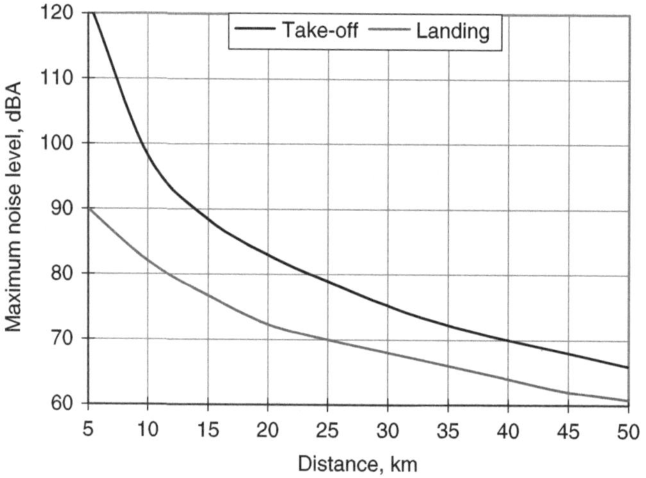


*Figure 4.4* Averaged noise levels for group 2 aircraft under the flight path for takeoff and landing.

Such a matrix method, which defines the directivity patterns of sound radiation for various flight (engine) modes, is based on the results of acoustical measurements made during trials of the aircraft as a whole or of its separate engines. These matrices may be used for noise level calculations under the flight path (e.g. in the TsAGI method or in Flula), and the *SPL* are defined by interpolation between the grid values for measured directions and modes after correcting for distance (usually engine trial measurements are performed along circles with 50, 75 or 100 m radius) and other sound propagation influences.

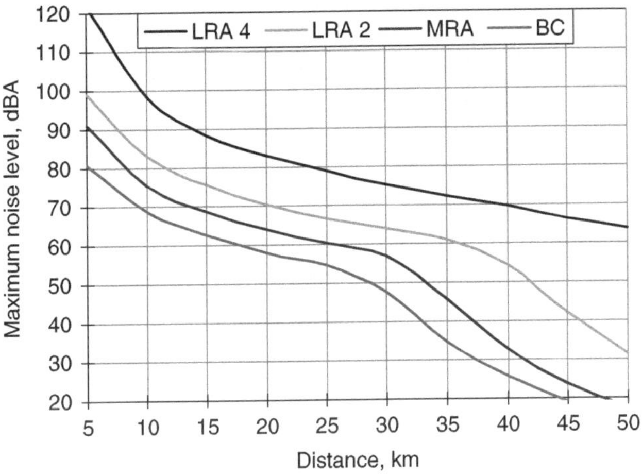


*Figure 4.5* Averaged noise levels for four aircraft groups under the flight path for takeoff: Long Range Aircraft with four engines – (DRA4), long range aircraft with two engines – (LRA2), middle range aircraft – (MRA), and business class – (BC).

In some cases, the noise matrices are obtained experimentally. In others they are obtained by means of calculations based on the models for the particular acoustic sources of interest for the aircraft under consideration. It is impossible to define the characteristics of all phenomena only by means of analytical and semi-empirical models. The most common factors determining or influencing the accuracy of noise matrices are the engine installation effects and noise abatement treatments. Insufficient accuracy of any model for a particular noise source may be considered an additional factor but, even so, very accurate models have been used [perhaps within 1–2 dBA for a total spectrum assessment]. Both measurements and calculations have some disadvantages and the derivation has been formulated to overcome them.

In the general case, the sound pressure level spectrum ($SPL_{jk}$) of aircraft noise of any type in spectral bands $j = 1, \ldots, N_j$, and in some $k$-th direction of sound propagation, where $k = 1, \ldots, N_k$, with reference to previous considerations, can be defined by:

$$SPL_{jk} = SPL_{jkp} + \Delta SPL_{jk}, \tag{4.7}$$

where $SPL_{jpk}$ is the predicted value of $SPL_{jpk}$ resulting from a sum of particular models $SPL_{jki}$ for characteristic noise sources (see Chapter 2), $i = 1, \ldots, N_s$; and $\Delta SPL_{jk}$ are spectral corrections for differences between the predicted $SPL_{jpk}$ and measured values $SPL_{jk}$. For each aircraft of interest

*Table 4.5* Example noise matrix for a specific engine operating in a specific mode at a specific radius

| *Angle (degrees)* | *Frequency band (Hz)* | | | | | | | | | | | | | | | | | | | | | | |
|---|---|---|---|---|---|---|---|---|---|---|---|---|---|---|---|---|---|---|---|---|---|---|---|
| | *50* | *63* | *80* | *100* | *125* | *160* | *200* | *250* | *315* | *400* | *500* | *630* | *800* | *1000* | *1250* | *1600* | *2000* | *2500* | *3150* | *4000* | *5000* | *6300* | *8000* | *10000* |
| 10 | 85.0 | 88.0 | 91.0 | 93.0 | 93.0 | 90.0 | 84.0 | 83.0 | 82.0 | 84.0 | 93.0 | 91.0 | 91.0 | 88.0 | 90.0 | 89.0 | 92.0 | 96.0 | 92.0 | 91.0 | 90.0 | 88.0 | 86.0 | 83.0 |
| 20 | 86.0 | 88.0 | 91.0 | 92.0 | 93.0 | 91.0 | 86.0 | 84.0 | 84.0 | 89.0 | 97.0 | 92.0 | 92.0 | 90.0 | 92.0 | 90.0 | 93.0 | 96.0 | 94.0 | 92.0 | 90.0 | 89.0 | 86.0 | 84.0 |
| 30 | 85.0 | 88.0 | 91.0 | 92.0 | 92.0 | 92.0 | 88.0 | 86.0 | 86.0 | 86.0 | 94.0 | 92.0 | 93.0 | 94.0 | 93.0 | 94.0 | 93.0 | 98.0 | 95.0 | 92.0 | 91.0 | 89.0 | 87.0 | 85.0 |
| 40 | 87.0 | 89.0 | 92.0 | 92.0 | 92.0 | 91.0 | 90.0 | 88.0 | 86.0 | 87.0 | 97.0 | 95.0 | 93.0 | 92.0 | 93.0 | 95.0 | 94.0 | 99.0 | 96.0 | 94.0 | 92.0 | 90.0 | 88.0 | 85.0 |
| 50 | 87.0 | 90.0 | 92.0 | 92.0 | 90.0 | 91.0 | 90.0 | 89.0 | 86.0 | 86.0 | 99.0 | 96.0 | 97.0 | 93.0 | 92.0 | 95.0 | 94.0 | 97.0 | 93.0 | 91.0 | 90.0 | 87.0 | 85.0 | 81.0 |
| 60 | 88.0 | 89.0 | 92.0 | 93.0 | 92.0 | 92.0 | 90.0 | 89.0 | 87.0 | 88.0 | 95.0 | 98.0 | 100.0 | 97.0 | 94.0 | 99.0 | 95.0 | 97.0 | 94.0 | 92.0 | 91.0 | 88.0 | 86.0 | 83.0 |
| 70 | 90.0 | 91.0 | 95.0 | 95.0 | 94.0 | 94.0 | 92.0 | 90.0 | 88.0 | 88.0 | 97.0 | 98.0 | 100.0 | 98.0 | 95.0 | 98.0 | 97.0 | 99.0 | 95.0 | 92.0 | 92.0 | 89.0 | 87.0 | 84.0 |
| 80 | 89.0 | 92.0 | 94.0 | 94.0 | 93.0 | 93.0 | 90.0 | 89.0 | 87.0 | 87.0 | 96.0 | 98.0 | 102.0 | 97.0 | 94.0 | 96.0 | 95.0 | 100.0 | 95.0 | 93.0 | 92.0 | 89.0 | 87.0 | 84.0 |
| 90 | 89.0 | 92.0 | 93.0 | 93.0 | 93.0 | 92.0 | 91.0 | 89.0 | 87.0 | 87.0 | 95.0 | 96.0 | 98.0 | 96.0 | 91.0 | 94.0 | 95.0 | 97.0 | 92.0 | 89.0 | 88.0 | 87.0 | 81.0 | 77.0 |
| 100 | 90.0 | 91.0 | 93.0 | 93.0 | 93.0 | 93.0 | 92.0 | 90.0 | 89.0 | 88.0 | 93.0 | 96.0 | 97.0 | 95.0 | 92.0 | 95.0 | 95.0 | 98.0 | 95.0 | 91.0 | 90.0 | 87.0 | 84.0 | 80.0 |
| 110 | 91.0 | 93.0 | 94.0 | 96.0 | 95.0 | 95.0 | 94.0 | 93.0 | 90.0 | 89.0 | 95.0 | 96.0 | 99.0 | 95.0 | 93.0 | 95.0 | 96.0 | 102.0 | 98.0 | 94.0 | 93.0 | 90.0 | 88.0 | 84.0 |
| 120 | 92.0 | 94.0 | 96.0 | 97.0 | 97.0 | 97.0 | 96.0 | 94.0 | 94.0 | 91.0 | 92.0 | 94.0 | 96.0 | 95.0 | 94.0 | 95.0 | 96.0 | 105.0 | 99.0 | 95.0 | 95.0 | 92.0 | 90.0 | 87.0 |
| 130 | 95.0 | 96.0 | 98.0 | 98.0 | 99.0 | 98.0 | 98.0 | 96.0 | 94.0 | 93.0 | 93.0 | 94.0 | 97.0 | 94.0 | 93.0 | 93.0 | 94.0 | 101.0 | 97.0 | 94.0 | 93.0 | 89.0 | 88.0 | 84.0 |
| 140 | 102.0 | 102.0 | 104.0 | 103.0 | 103.0 | 102.0 | 101.0 | 99.0 | 96.0 | 94.0 | 95.0 | 95.0 | 94.0 | 92.0 | 92.0 | 92.0 | 92.0 | 99.0 | 94.0 | 92.0 | 91.0 | 88.0 | 86.0 | 83.0 |
| 150 | 104.0 | 105.0 | 104.0 | 103.0 | 101.0 | 99.0 | 97.0 | 95.0 | 93.0 | 91.0 | 92.0 | 89.0 | 89.0 | 88.0 | 88.0 | 88.0 | 89.0 | 98.0 | 90.0 | 86.0 | 87.0 | 82.0 | 80.0 | 77.0 |
| 160 | 104.0 | 105.0 | 104.0 | 103.0 | 101.0 | 99.0 | 97.0 | 95.0 | 93.0 | 91.0 | 92.0 | 89.0 | 89.0 | 88.0 | 88.0 | 88.0 | 89.0 | 98.0 | 90.0 | 86.0 | 87.0 | 82.0 | 80.0 | 77.0 |

$SPL_{jpk}$ is defined by

$$SPL_{jkp} = \sum_{i=1}^{N_S} SPL_{jki}. \tag{4.8}$$

The number of acoustic sources that are sufficient for the particular type of the aircraft under consideration depend on the type of the aircraft, the type of the engine in its power plant (see Table 4.6) and the flight mode. The models of acoustic sources are based on semi-empirical models (Chapter 2) that are recommended by ICAO[4] and that have high accuracy for spectral and overall $SPL$ assessment ($\sigma_\Sigma = \pm 1.2$ dBA).

The contribution of every elementary acoustic source to overall $SPL_\Sigma$ in spectrum domain is defined by the formula for the power summation of $SPL_j$ at the point of noise control:

$$SPL_\Sigma(f) = 10\lg \sum_j 10^{0.1SPL_j(f)}. \tag{4.9}$$

For any particular acoustic source:

$$SPL_\Sigma(f) = L_w - \Delta L_\theta - \Delta L_f - \Delta L_v - \Delta L_{int} - \Delta L_{scr} - 20\lg\frac{R}{R_0} - \alpha(R - R_0),$$

where $L_w$ is the $SPL$ for particular frequency band $f$, normalized to the reference distance $R_0$ (usually $R_0 = 1$ m); $\Delta L_\theta$ is the correction for directivity

*Table 4.6* Main acoustic sources on aircraft according to type

| *Engine type* | *Acoustic sources on aircraft* | | | | | | | |
|---|---|---|---|---|---|---|---|---|
| | *Jet* | *By-pass jet* | *Fan, forward sector* | *Fan, backward sector* | *Turbine* | *Combustion chamber* | *Airframe* | *Propeller* |
| Jet engine | ⊕ | | ⊕ | | | | | |
| Bypass engine with mixing chamber and $m \leq 2.5$ | ⊕ | | ⊕ | | | | ⊕ | |
| Bypass engine with mixing chamber and $m > 2.5$ | ⊕ | | ⊕ | | ⊕ | ⊕ | ⊕ | |
| Bypass engine with separate flows and $m \leq 2.5$ | | ⊕ | ⊕ | ⊕ | | | ⊕ | |
| Bypass engine with separate flows and $m > 2.5$ | | ⊕ | ⊕ | ⊕ | ⊕ | ⊕ | ⊕ | |
| Turboprop engine | ⊕ | | | | | | | ⊕ |
| Prop-fan engine | | ⊕ | ⊕ | ⊕ | ⊕ | ⊕ | ⊕ | ⊕ |

of sound radiation; $\Delta L_f$ is the spectral correction; $\Delta L_v$ is the correction for aircraft speed; $\Delta L_{int}$ is the correction for sound interference corresponding to the 'lateral' noise attenuation; $\Delta L_{scr}$ is the correction for sound diffraction corresponding to 'screen' noise attenuation; and a is the sound absorption coefficient in air. The main contributions in the model are from $L_w$, $\Delta L_\theta$, and $\Delta L_f$. All of the contributions are defined by models of the individual acoustic sources listed in Table 4.6.

Spectral corrections are defined as the spectral transfer functions $\Delta SPL$ for the total acoustic model of the aircraft as follows:

$$\Delta SPL_{jk} = SPL_{jko} - SPL_{jkp}, \tag{4.10}$$

where $SPL_{jko}$ are the measured values of $SPL_{jk}$. The observations must be carried out either during flight testing in accordance with noise certification requirements or during engine testing at the outdoor testing facility. Of course, in the latter case, various in-flight effects and airframe acoustic sources are excluded.

In general, $SPL_{jko}$ and $SPL_{jkp}$ are functions of many parameters, so the transfer functions $\Delta SPL_{jk}$ are functions of these parameters too. The main parameters are the flight mode (engine type and thrust) and the direction of noise propagation from source to receiver. If the results of engine noise trials are presented in the form of noise matrices, then it is possible to define the directional relationships for the transfer functions $\Delta SPL_{jk}$. *SPL* spectra may be available from flight noise testing in the directions of maximum magnitude of instantaneous sound levels $L_A(t)$. $PNL(t)$ [or $PNLT(t)$] are the more accessible data in practice. The flight mode relationships can be defined for these directions and then generalized for any direction of noise propagation. To exclude the influence of the sound propagation effects on the transfer functions, the calculations are normalized to a reference distance $R_0$ ($R_0 = 1$ m is usual).

A preliminary or basic acoustic model of the aircraft is obtained as a sum of particular models for characteristic noise sources [see equation (4.7)] for every case (or direction) $k$ of the observed data $SPL_{jko}$ and for each acoustic source considered. The computer programs ANOPP or the program designed in TsAGI[7] can be used for this step. The TsAGI program uses the concept of a noise matrix as the basis for noise assessment at the points of noise control. The sum is corrected by the spectral transfer functions for the total acoustic model of the aircraft of particular type [see equation (4.10)]. The computer program BELTASS combines the previous and last steps.

*NPD* relationships are the most common type of acoustic model for aircraft and they are used in many current methods for calculating noise levels around airports. A graphical portrayal of this approach is shown in Fig. 4.6. The moving aircraft is represented as an axially symmetric noise source, around which cylindrical surfaces with constant noise levels are formed. The central axis of these cylinders is coincident with the flight path

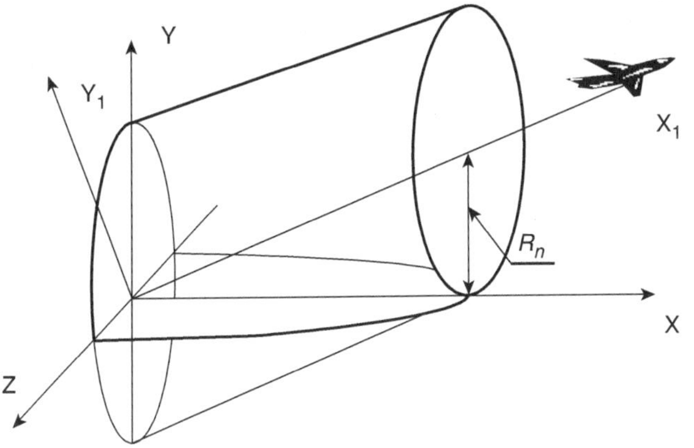


*Figure 4.6* The noise radius, $R_n$, concept for aircraft noise assessment.

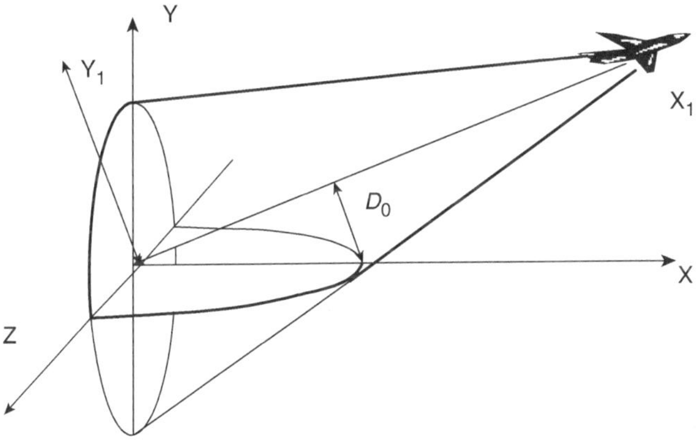


*Figure 4.7* The noise–power–distance and shortest distance, $D_0$, concepts for aircraft noise assessment.

axis and the radius of a particular cylinder surface is defined as the *noise radius* for a given noise level.

The concept of noise radius, $R_n$, is used in the Ukrainian calculation method. It is an extension of the concept of the shortest distance $D_0$ that is used in *NPD* relationships (see Fig. 4.7).

If the necessary data for calculating noise levels at a particular noise radius are not available, their values are obtained by means of the preliminary (or basic) aircraft acoustic model. Examples of $R_n$ relationships are shown in Figs 4.8–4.10.

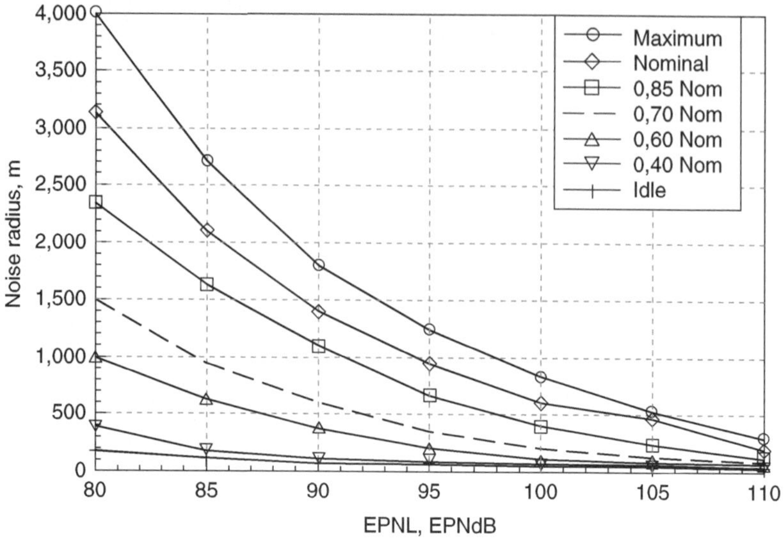


*Figure 4.8* Basic noise radius relationships for a particular aircraft type at a flight velocity of 90 m/s.

Figure 4.8 shows the relationships between noise radius and *EPNL* for a flight velocity of 90 m/s and various flight modes. Figure 4.9 shows the relationship expressed in terms of engine rotor frequencies and the aerodynamic configuration of the aircraft (angle of flap deflection). Figure 4.10 shows the relationship between noise radius and flight velocities for a clean aerodynamic configuration.

The concept of noise radius has some practical advantages. First of all, for investigations of noise footprints corresponding to a single flight path, the coordinates of the boundaries and areas of noise contours are defined simply by the intersection of cylinder surfaces with the Earth's surface. The whole flight path may be divided into separate sections, $k = 1, \ldots, N$, in each of which the flight parameters remain approximately constant, calculations are performed in each section $k$ and the results are summed. The contour of equal noise levels for steady-state climb path is defined in matrix form by the following equations:[8]

$$U^T SU = R_n^2,\ m^T X = K,\ X_0 = \Omega U,\ X_0 = \lambda X. \tag{4.11}$$

The first of these equations represents the cylindrical surface, the second represents the plane ground surface, the third is a transformation to the $X$-coordinate system and the fourth equation represents the rotation of a particular system from the $k$-th segment of the flight path to the initial one.

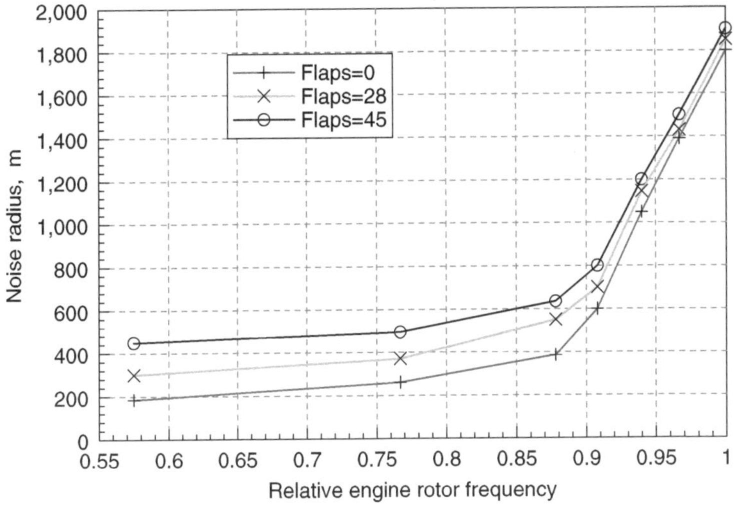


*Figure 4.9* The relationships between noise radius and flight modes expressed in terms of relative engine rotor frequency and flap deflection at flight velocity 90 m/s and a constant value of $EPNL = 90$ EPNdB.

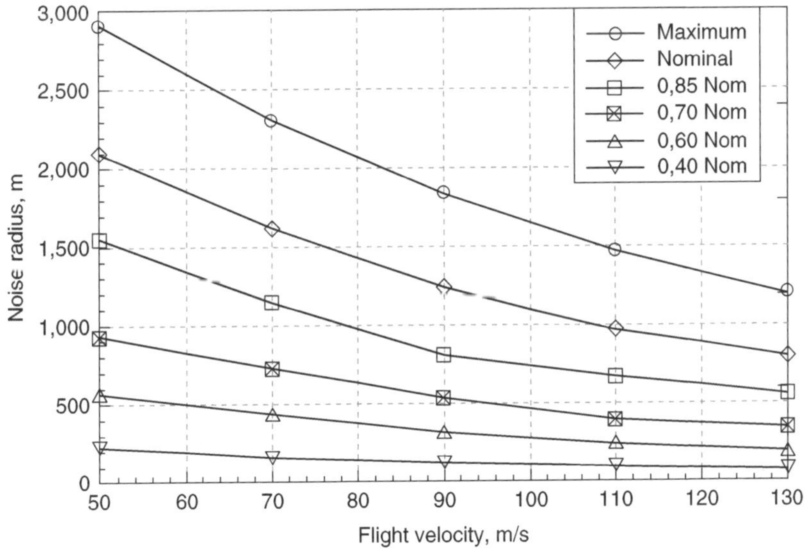


*Figure 4.10* Relationships between noise radius and flight velocity for a particular aircraft type ($EPNL = 90$ EPNdB, flap angle $\delta = 0$).

The matrices $S$, $\Omega$, $\lambda$ are given by:

$$S=\begin{bmatrix}0&0&0\\0&1&1\\0&0&1\end{bmatrix},\Omega=\begin{bmatrix}\cos\Theta&0&-\sin\Theta\\0&1&0\\\sin\Theta&0&\cos\Theta\end{bmatrix},\lambda=\begin{bmatrix}\cos\varphi&-\sin\varphi&0\\\cos\varphi&\sin\varphi&0\\0&0&1\end{bmatrix}, \quad (4.12)$$

where $\Theta$ and $\varphi$ are the angles of slope and of rotation in the horizontal plane for a particular stage of flight path, such that the general equation of the contour is given by:

$$\sin^2\Theta\,(x\cos\varphi - z\sin\varphi)^2 + (x\sin\varphi + z\cos\varphi)^2 = R_n^2. \quad (4.13)$$

For the particular case of zero rotations of each segment of the path, it transforms to the equation for an ellipse:

$$z^2\sin^2\Theta + x^2 = R_n^2. \quad (4.14)$$

The area of each segment $S_k$ of the noise contour corresponding to the flight path section $k$ (with initial and final coordinates $X_i$ and $X_f$, respectively) is defined as follows:

$$S_k = \left[\frac{R_n^2}{\sin\Theta}\arcsin\left(\frac{X\sin\Theta}{R_n}\right) + X\sqrt{R_n^2 - X^2\sin^2\Theta}\right]_{X_i}^{X_f}. \quad (4.15)$$

In ICAO manuals, the aircraft noise index $EPNL$, is defined as:

$$EPNL = \frac{1}{T}\int_{t_1}^{t_2} 10^{0,1PNLT(t)}\,dt. \quad (4.16)$$

The aircraft is represented as a moving point noise with constant velocity $v$. At the point of interest under the flight path:

$$PNL(t) = PNL_{\max} + 10\lg\left[R_n^2 \Big/ \left(R_n^2 + v^2t^2\right)\right], \quad (4.17)$$

where $PNL_{\max}$ is the maximum value of $PNL(t)$ for a given noise event. For the simplified case under consideration, it occurs when the aircraft is directly above the point of interest

$$PNL_{\max} = \underline{PNL} - 20\lg R_n, \quad (4.18)$$

where $\underline{PNL}$ is defined by the flight mode (engine mode, flight velocity and aerodynamic configuration). If $b$ is a constant accounting for sound wave divergence and for atmospheric sound attenuation, and $t$ is the flight time, then, after substituting equations (4.17) and (4.18) into (4.16) and integrating

$$EPNL = \underline{PNL} + 10\lg\left[I(b)/T_0\right] - 10\lg(R_n v), \tag{4.19}$$

where

$$I(b) = \int_{-\varphi_0}^{\varphi_0} \cos^{2(b-1)} \varphi d\varphi, \qquad F_0 = \mathrm{arctg}\left[\mathrm{v\,t_2/R_n}\right]$$

and $t_2$, the time at which $PNL(t) = PNL_{max} - 10$, is given by:

$$t_2 = \sqrt{\left[10^{1/b} - 1\right]} R_n / v.$$

If the further simplification of zero sound absorption is made, then $b = 1$ and formula (4.19) can be written in the following form:

$$EPNL = \underline{PNL} + C_R - 10\lg(R_n v), \tag{4.20}$$

where $C_R$ is a function of sound absorption along the distance between the trajectory and the point of interest and of the reference time duration $T_0$.

This means that the *EPNL* or any other time-integrated sound level is defined as a function of noise radius $R_n$ and velocity $v$ only. Analysis of the latter expression leads to the following conclusions.

(a) The power of the dependence of the time-integrated sound level on the distance tends to 1 for a moving source, not to 2, as it would for instantaneous sound levels like *OASPL*, $L_A$ and *PNL*($t$). It should be noted, however, that atmospheric absorption has been neglected and in real conditions the exponent would be somewhat higher.

(b) The correction to *EPNL* for variations of flight velocity $v$ is equal to the ICAO recommendation:

$$\Delta L_v = 10\lg\left(\frac{v}{v_0}\right), \tag{4.21}$$

where $v_0$ is the reference velocity.

(c) For a constant flight operational mode and magnitude of noise level *EPNL*, the product $R_n v =$ constant. This result has been verified by different numerical investigations for a variety of aircraft types, even after accounting for the influence of atmospheric absorption. At high flight velocities, airframe noise sources contribute to the magnitude of aircraft noise level sufficiently, compared to engine noise sources, so that the relationship $R_n v =$ constant is violated. However, flight velocities do not exceed the critical values at takeoff and landing flight stages in most cases of practical interest, so this relation can be used in calculations and modeling of noise around airports.

## 4.3 Evaluation of an acoustic model of an aircraft

The spectral transfer function $\Delta SPL_{jk}$ is one of the core parameters of the basic acoustic model of an aircraft. It is based on an empirical error minimization approach such that equation (4.10) is revised to have the following form:

$$\Delta SPL_{jk} = SPL_{jko} - SPL_{jkp} - E_j, \tag{4.22}$$

where $E_j$ are the spectral errors, between measured and predicted noise levels, which cannot be included in the transfer function. If the errors $E_j$ have a normal or Gaussian distribution, the likelihood principle can be applied in the form of the minimum value of the sum of squares:

$$\sum_k E_{jk}{}^2 = \sum_k (SPL_{jko} - SPL_{jkp} - \Delta SPL_{jk})^2 = \min. \tag{4.23}$$

Therefore, in the general case, the sums of $(SPL_{jkp} + \Delta SPL_{jk})$ may be performed from the results of linear regression $SPL_{jkr}$, defined by means of the least sum of squares method and $\Delta SPL_{jk}$ would be the systematic portions of differences between $SPL_{jko}$ and $SPL_{jkp}$. Errors $E_j$ would be the non-systematic portions of the differences $(SPL_{jko} - SPL_{jkr})$ so that they can be interpreted as the measures of the precision of the defined solutions. According to the likelihood principle, the following sum

$$M = \sum_k E_{jk}{}^2/\sigma^2 \tag{4.24}$$

must have a $\chi^2$-distribution for $(N_k - 1)$ degrees of freedom, where $\sigma$ is the dispersion of the error distribution, and $N_k$ is the number of directions of sound propagation under consideration ($N_k$ is between 16 and 19 for complete noise matrices, in which all directions are separated uniformly by 10°). This property can be used to assess the assumption about $E_j$.

Thus the acoustic model for an aircraft involves the following steps:

(1) A preliminary acoustic model of the aircraft is obtained as a sum of particular models for characteristic noise sources [see equation (4.9)] for every case (or direction) $k$ of the measurements $SPL_{jko}$ and for each acoustic source considered. The computer program BELTASS is used for this step. For excluding the effects of sound propagation from the

transfer function, it is normalized to a reference distance $R_0$ ($R_o = 1$ m is usual).

(2) Linear regressions are performed to define least sum of squares estimates of $SPL_{jkr}$.

(3) The transfer function and error function are defined by:

$$\Delta SPL_{jkp} = \sum_{jk} W_j[SPL_{jkr} - SPL_{jkp}] \Big/ \sum_j W_j, \tag{4.25}$$

$$E_{jk} = \sum_{jk} W_j[SPL_{jkr} - SPL_{jko}] \Big/ \sum_j W_j, \tag{4.26}$$

where $W$ is a spectral weighting function, every component of which in any band of the spectrum is either 1 or, for bands containing tonal components, greater than 1. At this stage the following equations are useful. The vector of total spectral differences $E$ is defined by

$$E = \left\{ \sum_{jk} W_j \left[SPL_{jkr} - SPL_{jko}\right] \Big/ \sum_j W_j \right\}^{1/2} \tag{4.27}$$

so that

$$E^2 = \Delta SPL_{jk} + E_{jk}^2$$

and the relative error index of agreement $d$ is defined by

$$d_j = 1 - \frac{\sum_j \{W_j[SPL_{jko} - SPL_{jkp}]\}^2}{\sum_j \{W_j \{[SPL_{jkp} - SPL_{jkor}] + [SPL_{jko} - SPL_{jkor}]\}\}^2}, \tag{4.28}$$

where $SPL_{jkor} = \Sigma(W_j\ SPL_{jko})/\Sigma W_j$ is an average estimate of the observed data.[1] The index of agreement $d_j$ is non-dimensional and varies between 0 and 1. If $d_j = 1$, the resulting prediction model is reliable and consistent with the observed data in the $j$th band under consideration. Steps (2) and (3) are realized in a computer program named TRANSFER.

(4) The possible solutions are compared through the sum of squares of residual errors and $\chi^2$-statistics which are calculated in the computer program named TRANSCHI.

Thus the basic acoustic model of an aircraft of any type is derived from the noise matrices and the value of each component of the matrices is defined

*Table 4.7* Spectrum-averaged indices of agreement $d_j$ [see equation (4.28)] for various aircraft and flight modes

| *Flight stage or engine operation mode* | *Turbojets and low bypass turbofans* $(1 \leq m \leq 2.3)$ | *High bypass turbofans* $(2.4 \leq m \leq 5.6)$ |
|---|---|---|
| Takeoff | 0.88 | 0.93 |
| Climbing | 0.86–0.93 | 0.84–0.92 |
| Climbing with throttle-back of engines | 0.83–0.97 | 0.85–0.95 |
| Landing | 0.89–0.97 | 0.84–0.94 |

by formula (4.7). The models are represented in terms of the parameters of aircraft flight (engine) modes and of the state of the ambient environment, so they can be used for any aspect of the aircraft noise problem. Accurate and reliable acoustic models have been obtained for all current types of aircraft and engines (Table 4.7).

This method for deriving an acoustic model of an aircraft has been validated against flight testing data obtained from preliminary noise certification results for the Yakovlev-40 aircraft. Measurements are available for both takeoff and approach stages, so the transfer functions are defined for two flight modes. Figure 4.11a and b shows the upper and lower limits of the spectral differences $E$ between measurements and initial predictions.

The 'No. 11' flights were excluded from the model improvement process because the corresponding data were anomalous. The spectra resulting from the model improvement process $SPL_{jko}$ (1), $SPL_{jkh}$ (2) and $SPL_{jk}$ (3) are shown in Fig. 4.11c. In all cases (of course, without 'No. 11' flights) the index of agreement $d_j$ varies between 0.88 and 0.96 over the spectral bands. The averaged value of index $d_j$ for a 'No. 11' flight = 0.62. For most of the spectrum bands, the probability $P$ that the assessed $\chi^2$-statistic is higher than the $\chi^2$-distribution law is between 0.92 and 1.00, so the reliability of the resulting acoustic model is quite high. Small deviations from these good results are observed in a few low-frequency bands (for which $d_j = 0.77$). Although the ground effects here are substantial, higher accuracy in the overall spectrum has not been achieved despite application of a ground interference model in the prediction procedures, since accurate data about the type of reflecting surfaces and their characteristics were not available.

In Table 4.8, the results of the basic acoustic model are compared with those from summing all contributing acoustic sources for the same Yakovlev-40 and for the approach flight mode data obtained during the noise certification. The differences are quite large and they show the inaccuracy of the modeling tools used for specific acoustic sources, primarily for the fans, which must be dominant for the flight mode under consideration.

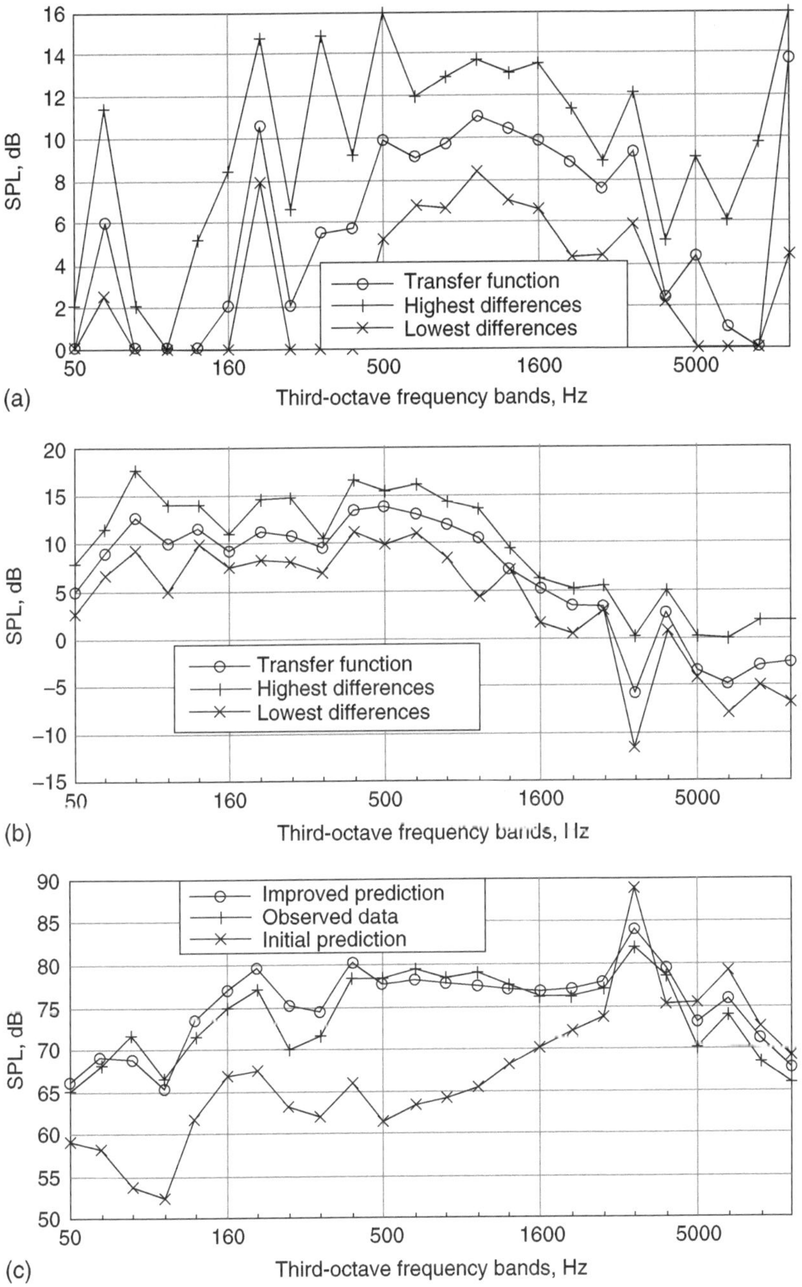


*Figure 4.11* Validation of the acoustic model for a Yakovlev-40 aircraft: (a) takeoff trials; (b) descending trials; (c) spectra resulting from the model improvement process.

*Table 4.8* Comparison of indices of agreement $d_2$[see equation (4.28)] for results of *SPL* modeling of a Yakovlev-40 aircraft during approach flight

| *Acoustic model of the aircraft* | *Number of flight* | | | | |
|---|---|---|---|---|---|
| | *6* | *7* | *8* | *9* | *12* |
| Summation of all contributing acoustic sources (4.9) | 0.63 | 0.71 | 0.52 | 0.86 | 0.67 |
| Basic acoustic model (4.7) | 0.96 | 0.87 | 0.96 | 0.92 | 0.94 |

*Table 4.9* The results of noise radius assessment for $L_{Amax}$

| *Noise radius (m)* | $L_{Amax}$ *(dB) for approach flight mode* | | | $L_{Amax}$ *(dB) for takeoff flight mode* | | |
|---|---|---|---|---|---|---|
| | *1* | *2* | *3* | *1* | *2* | *3* |
| 100 | 91.3 | 93.9 | 92.3 | 101.3 | 97.6 | 99.8 |
| 300 | 79.0 | 80.6 | 80.0 | 87.2 | 82.9 | 88.1 |
| 500 | 72.4 | 72.3 | 73.8 | 79.0 | 73.8 | 81.9 |
| 700 | 67.9 | 66.0 | 69.5 | 73.1 | 66.6 | 77.4 |
| 1000 | 62.9 | 58.1 | 64.8 | 66.7 | 58.7 | 72.5 |
| 1300 | 59.2 | 51.9 | 61.6 | 62.1 | 53.4 | 69.1 |
| 1500 | 57.2 | 48.5 | 59.2 | 59.6 | 50.7 | 66.5 |
| 2000 | 53.1 | 42.2 | 55.2 | 54.5 | 45.7 | 62.1 |
| 2500 | 49.8 | 38.1 | 51.6 | 50.3 | 42.0 | 58.5 |

Various acoustic models of the Yakovlev-40 aircraft type have been used for noise radius assessments. The results shown in Table 4.9 are (1) for the improved model with transfer function; (2) for the preliminary model $SPL_{jkp}$; and (3) for the assessment method currently used in national practice.

The deviations are least between the improved model and the applied national assessment method. The deviations are greater at larger values of noise radius (implying larger distances between the noise source and receiver point), especially for the takeoff mode. However, the error in the observed data approximations used for the definition of relationships for the current national assessment method may be higher at the larger distances. Most of the data are obtained at ranges between 500 and 1000 m at takeoff, and between 100 and 200 m at landing. Usually these results are extrapolated to larger distances.

Approximations of $L_{Amax}$ for the approach flight mode have been obtained as second-order polynomials in $lgR$ and are shown, together with measures of their accuracy ($T$ is the rms deviation), in Table 4.10.

The transmission function deduced from flight test data for an Il'ushin-86 aircraft ($\Delta SPL_{TF}$) is shown in Fig. 4.12. Calculated and measured NPD-relationships for the Il'ushin-86 during flight are shown in Table 4.11.

*Table 4.10* Quadratic approximations of $L_{Amax}$ in terms of $lgR$ for the approach flight

| *Model* | $L_{Amax}$ *formula* | *T* |
|---|---|---|
| Preliminary | $101.158 + 18.442\ lgR - 10.904\,(lgR)^2$ | 0.273206 |
| Improved | $126.02 - 9.95\ lgR - 3.683\,(lgR)^2$ | 0.039562 |
| Previous national noise assessment method | $127.02 - 10.58\ lgR - 3.39\,(lgR)^2$ | |

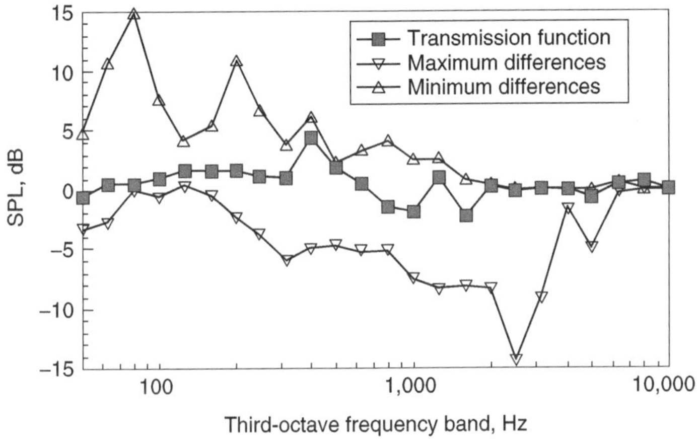


*Figure 4.12* The transmission function derived empirically from aircraft flight testing data for the acoustic model of an Il'ushin-86.

The range of flight heights between 300 and 900 m, used during the flights, is insufficient for construction of NPD-relationships. The recommended range is from 80 to 8000 m.[4] In accordance with the ICAO Technical Manual,[4] noise levels for heights greater than 800 m are determined by extrapolating fits to the data for $H \leq 800$ m taking account of divergence $L_R$ and atmospheric attenuation $L_{ATM}$ . Hence, for $L_{AE}$:

$$L_{AE} = L_{AMAX} + L_{AEH} - L_{AMAXH} + 7.5\lg(d/800), \tag{4.29}$$

where $d$ is the minimum distance from the control point to the flight path (approximately equal to the flight height $H$), sound levels with index '$H$' correspond to measured values and $L_{AMAX}$ is the extrapolated value at distance $d$:

$$L_{AMAX} = 10\lg\left[\sum_j 10^{0.1(L_j - \Delta L_A)}\right],$$

$$L_j = L_{jH} - \Delta L_R - \Delta L_{ATM} \tag{4.30}$$

*Table 4.11* Measured and predicted noise–power–distance (NPD) relationships for flight testing of an Il'ushin-86 aircraft

| *Flight height (m)* | *EPNL (EPNdB) Maximum engine mode* | | | *EPNL (EPNdB) Nominal engine mode* | | |
|---|---|---|---|---|---|---|
| | *Calculated from acoustic model* | *Calculated with zero $\Delta SPL_{TF}$* | *Measured* | *Calculated from acoustic model* | *Calculated with zero $\Delta SPL_{TF}$* | *Measured* |
| 300.0 | 115.4 | 111.9 | 115.0 | 107.7 | 108.0 | 107.6 |
| 400.0 | 113.2 | 110.2 | 113.0 | 105.7 | 106.2 | 105.7 |
| 500.0 | 111.6 | 108.8 | 111.0 | 104.2 | 104.5 | 104.2 |
| 600.0 | 110.4 | 107.9 | 109.8 | 103.0 | 103.4 | 102.9 |
| 700.0 | 109.2 | 106.7 | 109.0 | 101.9 | 102.3 | 102.0 |
| 800.0 | 108.3 | 105.9 | 108.4 | 100.9 | 101.3 | 101.4 |
| 900.0 | 107.6 | 105.2 | 107.8 | 100.1 | 100.6 | 101.0 |

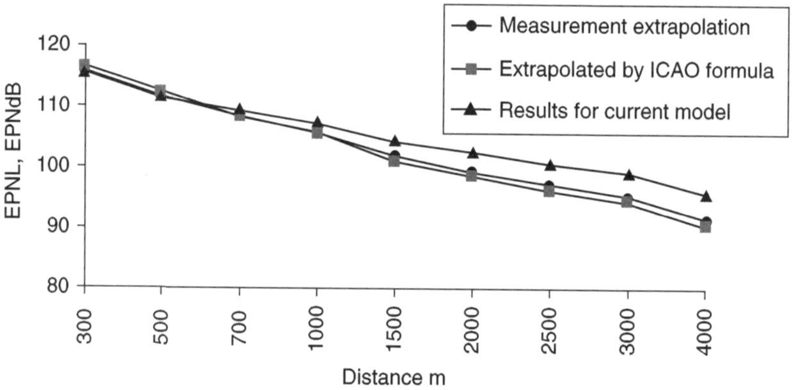


*Figure 4.13* Comparison of noise–power–distance (NPD) relationships of an Il'ushin-86 aircraft predicted by various models.

Figure 4.13 compares NPD-relationships, *EPNL=f(d)*, extrapolated by formula (4.29), predicted using program RADIUS_N and formula (4.30) and predicted by program RADIUS using an acoustic aircraft model for an Il'ushin-86 including the ground effect computed for grass cover.

An algorithm for estimating the noise level at a control point used in the certification procedure is based upon calculated NPD-relationships. For these purposes it is necessary to define the engine mode, conforming to the requirements of the minimum permissible climbing gradient, and distance *d* from control point to flight path. In accordance with data from aircraft flight investigations for an Il'ushin-86 with maximum weight 210 tons, the permissible engine cut-off and permissible climbing gradient 4 per cent are defined by the relative frequency of high-pressure compressor being 87.5 per cent of the maximum value. The distance $d$ defined by flight data is equal to 300 m. Consequently, the noise level at the control point is equal to 107.6 EPNdB. The result of flight noise certification of the Il'ushin-86 for the given control point is 107.4 ± 0,6 EPNdB. This confirms the possibility of using NPD-relationships for estimating noise levels in accordance with certification requirements.

The results of static aircraft engine noise testing have also been used for constructing acoustic aircraft models, estimating NPD-relationships and determining noise levels, corresponding to certification conditions. Compared with in-flight investigations, the static test data do not include the influence of flight speed on generated noise levels. Moreover, there is no standard that regulates the noise-measuring conditions at a testing facility. Two microphone heights ($h_r = 4.5$ and 0.5 m) were used to investigate the interference between direct and reflected sound waves. The maximum interference (first minimum) effect according to the simplified formula $f_0 = Ra_0/(4h_S h_R)$ is located at 420 Hz for the microphone at 4.5 m and at 3780 Hz for $h_R = 0.5$ m. Averaging of engine noise spectra was performed for frequencies less than 800 Hz for data received at height 0.5 m for frequencies greater than 3500 Hz on data for height 4.5 m, and between 800 and 3500 Hz for both heights bearing in mind these interference effects. Equations (4.22–4.28) were used to determine the transmission function $\Delta\ SPL_{TF}$ using engine noise testing data. The resulting $\Delta SPL_{TF}$ for maximum and nominal engine modes are shown in Fig. 4.14. The resulting acoustic models have been used to calculate NPD-relationships for maximum and nominal engine modes (Table 4.12).

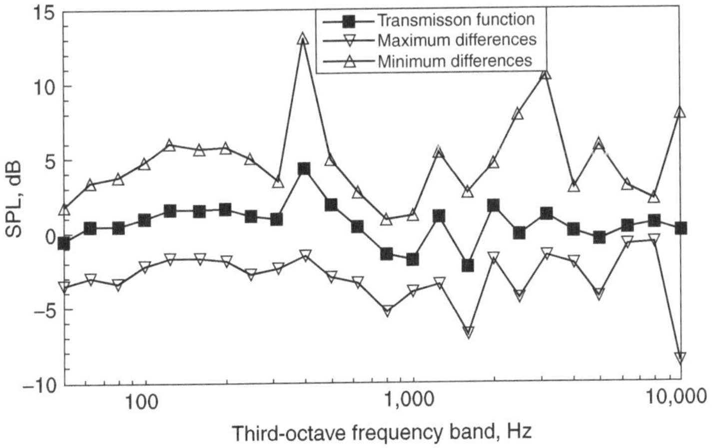


*Figure 4.14* The transmission function derived empirically for the acoustic model of an Il'ushin-86 from aircraft engine noise testing data.

*Table 4.12* Noise–power–distance (NPD) relationships computed for an Il'ushin-86 aircraft from engine test data

| *Flight height (m)* | *EPNL (EPNdB) Maximum engine mode* | | *EPNL (EPNdB) Nominal engine mode* | |
|---|---|---|---|---|
| | *$\Delta SPL_{TF}$ for maximum mode* | *$\Delta SPL_{TF}$ for nominal mode* | *$\Delta SPL_{TF}$ for nominal mode* | *$\Delta SPL_{TF}$ for maximum mode* |
| 300.0 | 115.4 | 114.2 | 108.0 | 109.4 |
| 400.0 | 113.4 | 112.1 | 106.0 | 107.5 |
| 500.0 | 112.0 | 110.7 | 104.6 | 106.1 |
| 600.0 | 110.8 | 109.5 | 103.4 | 105.0 |
| 700.0 | 109.6 | 108.3 | 102.4 | 104.0 |
| 800.0 | 108.8 | 107.2 | 101.4 | 103.2 |
| 900.0 | 107.8 | 106.3 | 100.6 | 102.4 |

In real operational conditions, in particular during takeoff and/or climbing, the aircraft flight procedures differ from certification flight procedures. Adaptation of aircraft noise certification results to operational data is performed using two groups of parameters: flight trajectory parameters and NPD-relationships. For example, in accordance with flight operation guide for Il'ushin-86 aircraft with a takeoff weight of 210 tons, the distance from control point No. 2 to the climbing trajectory is 330 m (with the engine mode at maximum) and the flight speed is about 90 m/s. Using the NPD relationships from Table 4.12, these parameters correspond to noise level of 114.5 EPNdB. Bearing in mind that the result from flight noise certification is $107.4 \pm 0.6$ EPNdB, the engine cut-off effectiveness (i.e. the noise reduction due to the engine thrust reduction realized during the takeoff/climbing procedure) is about 7 EPNdB.

## 4.4 Prediction of noise under the flight path: trajectory models

Trajectory models are intended for use in research and in the assessment of noise levels under the flight path, at one or more reception points of interest. Noise levels are calculated either as time-varying sound spectra or in terms of noise indices, such as *PNL* and *EPNL* (4.16), $L_{Amax}$ or *SEL*:

$$SEL = \frac{1}{T}\int_{t_1}^{t_2} 10^{0,1L_A(t)}dt \qquad (4.31)$$

The indices represent single-event levels, but they may include weightings according to the time of day or night at which the event occurs and/or a weighting of the number of events occurring within the time period. The time limits and weightings are often chosen to correlate with public opinion (as determined from surveys) and noise-control rules.

A noise footprint (a line of constant noise level around a runway, due to the noise of one takeoff and one landing of an aircraft, operating under prescribed conditions, including weather, atmospheric conditions, flight profile, etc.) is useful for noise-impact assessment under the flight path. The flight path is a path of an aircraft through the air, defined in three dimensions, usually with reference to an origin at the start of takeoff roll or at the landing threshold. The elevation of the flight path, showing the variation of aircraft height along the ground track, is a *flight profile*. The vertical projection of the flight path onto the ground plane is the *flight track*.

The noise at points on the ground from aircraft flying into and out of a nearby airport depends on a number of factors. Principal among these are: (1) the types of airplane and engines in their power plant including the thrust, flap and airspeed management procedures used on the airplanes themselves; and (2) the factors affecting sound propagation, including the distances from the points concerned to the various flight paths, the local topography and the weather conditions.

Generally, the noise levels for individual movements are calculated, for given atmospheric conditions, assuming flat terrain, an appropriate acoustic model, or more frequently, appropriate NPD relationships, and airplane performance data. The performance data are for defined values of atmospheric temperature and humidity, airport altitude and wind speed. However, given that the calculated noise footprints depict average conditions over a long period of time, the same basic data are assumed to apply over specified ranges of conditions.

The trajectory model is useful for assessing the efficiency of low noise treatments, low noise flight procedures and takes actual flight rules and circumstances into account in the vicinity of a specified airport. The model is sufficiently sophisticated for specialist investigations of the aircraft noise problem. In principle, a trajectory model consists of three main parts:

- an aerodynamic model for flight path parameters assessment;
- an acoustic model including sound propagation and attenuation effects defined in overall *SPL* form, so that the appropriate models for aircraft noise assessment can be used;
- an acoustic model based on the noise radius approach, in which sound attenuation and propagation effects are included in the form of relationships between the relevant noise criteria and basic parameters of such effects.

Mathematically, a trajectory model is represented by a set of ordinary differential equations derived for the center of mass of the aircraft in a flight velocity coordinate system. It is expressed in matrix form in Chapter 6 of the book.

The following equations describe the three-dimensional motion of a commercial point mass aircraft over a few minutes time span. All vector parameters of the equations are shown in Figs 4.15–4.17.

$$
\begin{aligned}
m\dot{V}_t &= [T\cos(\alpha_t) - D] - mg\sin(\gamma_a) - \dot{w}_x\cos(\gamma_a)\sin(\psi_a)\\
&\quad - \dot{w}_y\cos(\gamma_a)\cos(\psi_a) - \dot{w}_h\sin(\gamma_a)\\
mV_t\dot{\gamma}_a &= [L + T\sin(\alpha_t)]\cos(\varphi_a) - mg\cos(\gamma_a) + \dot{w}_x\sin(\gamma_a)\sin(\psi_a)\\
&\quad - \dot{w}_y\sin(\gamma_a)\cos(\psi_a) - \dot{w}_h\cos(\gamma_a)\\
\dot{h} &= V_t\sin(\gamma_a) + w_h\\
mV_g\dot{\psi}_i &= L[\sin(\varphi_a)\cos(\theta_t) + \cos(\varphi_a)\sin(\gamma_a)\sin(\theta_t)] + D\cos(\gamma_a)\sin(\theta_t)\\
&\quad + T[\sin(\alpha_t)\sin(\varphi_a)\cos(\theta_t) + \sin(\alpha_t)\cos(\varphi_a)\sin(\theta_t)\\
&\quad - \cos(\alpha_t)\cos(\gamma_a)\sin(\theta_t)]\\
\dot{x} &= V_g\sin(\psi_i)\\
\dot{y} &= V_g\cos(\psi_i)\\
\theta_c &= \psi_i - \psi_a
\end{aligned}
\tag{4.32}
$$

where $g$ is the acceleration due to gravity, $m$ is the aircraft mass, $V_t$ is the true airspeed, $\dot{V}_t$ is the rate of change of the true airspeed, $V_g$ is the ground

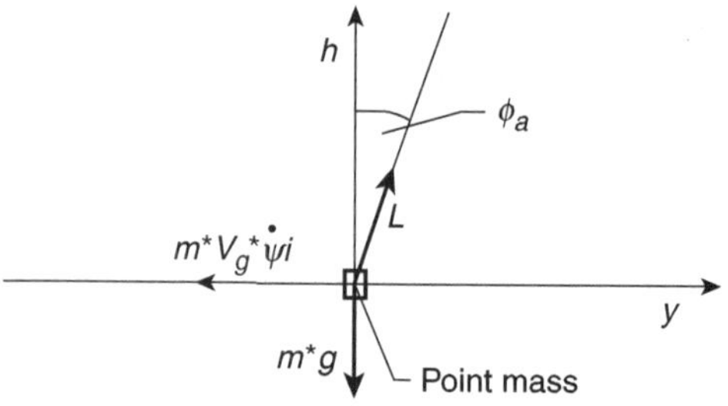


*Figure 4.15* Forces on point mass in $y$, $h$ direction.

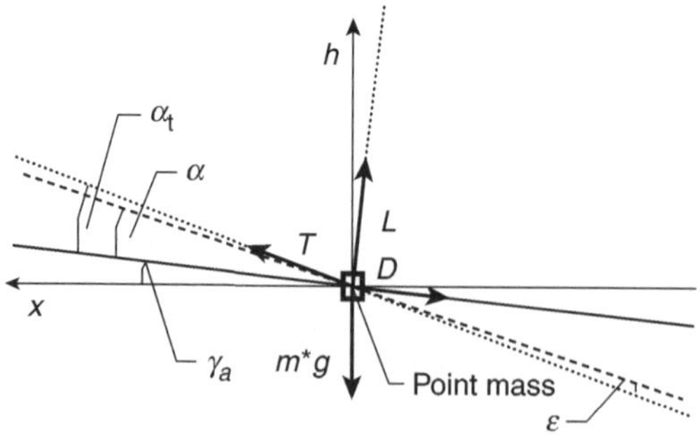


*Figure 4.16* Forces on point mass in $x$, $h$ direction.

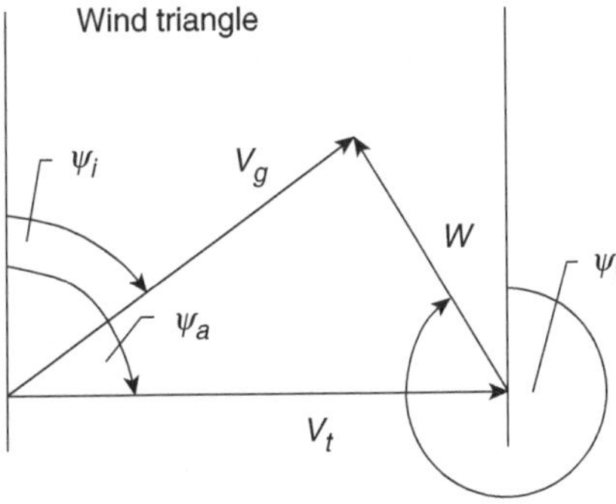


*Figure 4.17* Wind triangle.

speed, $T$ is the thrust, $D$ is the drag, $L$ is the lift, $\alpha$ is the angle of attack (see Fig. 4.16), $\alpha_t = \alpha + \varepsilon$ is the angle between the thrust and the relative wind, $\varepsilon$ is the engine thrust inclination, $\gamma_a$ is the air relative flight path angle, $\dot{\gamma}_a$ is the rate of change of the air relative flight path angle, $\psi_a$ is the aerodynamic heading, $\psi_i$ is the inertial heading, $\dot{\psi}_i$ is rate of change of the inertial heading, $\varphi_a$ is the aerodynamic bank heading, $\theta_c = \psi_i - \psi_a$ is the crab angle, $\dot{w}_x$ is the rate of change of wind component in the $x$-direction, $\dot{w}_y$ is the rate of change of the wind component in the $y$-direction, $w_h$ is the wind component in the $h$-direction, $\dot{w}_h$ is the rate of change of the wind component in the $h$-direction, $\dot{x}$ is the rate of change of position $x$, $\dot{y}$ is the rate of change of position $y$ and $\dot{h}$ is the rate of change of position $h$.

The first assumption is that an aircraft may be represented as a point mass. This means that the movement of the center of gravity is modeled but not the movement around the center of gravity. The drag parameters in the aerodynamic model are 'balanced'. This means that the drag and lift coefficients, from flight test analysis, take into account corrections to balance the aircraft on the longitudinal axis. The equations presented hereafter are valid for all flight phases, including ground phases and aerial phases.

For the takeoff and climbing flight paths, the rotational motion around the center of mass of the aircraft may be excluded. This means that the trajectory model can be considered in the vertical plane only and can be represented by a set of four equations with the following phase parameters: flight velocity $v$, flight path angle $\Theta$, longitudinal coordinate $x$ and vertical coordinate $y$. The flight control parameters in a trajectory model of the aircraft are the total engine thrust of the aircraft $T$, the flap angle $\delta$ and the pitch angle $\nu = (\alpha + \Theta - \varphi)$ , where $\alpha$ is the angle of attack, and $\varphi$ is the engine installation angle.

The engine shaft speed of a higher pressure compressor defines the thrust $T$ of the engine directly (using height–velocity diagrams and husk plots for the engine performances under consideration) for the predefined height, velocity of the flight, operation mode of the engine and atmospheric conditions. The pitch angle $\nu$ must be chosen as the control parameter (e.g. instead of angle of attack) because it gives rise to more stable solutions of the set of equations (4.32).[9]

An important assumption (made before[1]) is that vertical and horizontal path specifications are not coupled, so they can be defined separately. There are four simplifying assumptions:

(a) Most of the time (in cruise, cruise–climb, climb–descend) the aircraft are assumed to move in straight lines. Turns are rare and brief compared to the straight segments.
(b) All aircraft are in stable flight and an en-route flight phase: the angle of attack ($\alpha$) and the flight path angle ($\gamma$ or $\Theta$ in some cases) are assumed to be small. Angular rates are also assumed to be small. In this case, the rate of the change of the flight path angle is set to zero, assuming quasi-steady flight.
(c) The change of the aircraft mass due to fuel burned during a simulated situation may be ignored.
(d) The influence of the wind is neglected. A 'real-world' trajectory on which wind has an influence, is not the object of the current simulation.

The flight path angle $\gamma$ is small; therefore, the true airspeed $V_t$ and the ground speed $V_g$ are set equal. For example, at takeoff/climbing or at descending before the landing (Fig. 4.18), the equations of motion may be written as follows:

$$\begin{aligned} &\frac{dv}{dt} = \frac{P}{m}\cos(\nu - \Theta - \varphi) - C_{Xa}\frac{\rho v^2 S}{2m} - g\sin\Theta; \\ &\frac{d\Theta}{dt} = \frac{P}{mv}\sin(\nu - \Theta - \varphi) + C_{Ya}\frac{\rho v^2 S}{2m} - \frac{g}{v}\cos\Theta; \\ &\frac{dx}{dt} = v\cos\Theta; \quad \frac{dy}{dt} = v\sin\Theta. \end{aligned} \tag{4.33}$$

Initial control parameters for takeoff procedures are the engine power setting (for simplicity defined as the relative value of rotor frequency, $\overline{n}$), pitch angle $\nu$ (the solutions for pitch angle are more stable than for angle of attack) and angle of flap deflection $\delta$.

Aerodynamic drag and lift information is derived from data for each aircraft in the aircraft database. This consists of basic lift and drag data for reference conditions of angles of attack, Mach numbers and configurations. These values can be modified for all kinds of operational conditions and types of flight.

For noise assessment during the approach–landing flight stages the same approach is used. The flight path is observed without rotation around the vertical axis, so it is observed in vertical plane only and the equation system

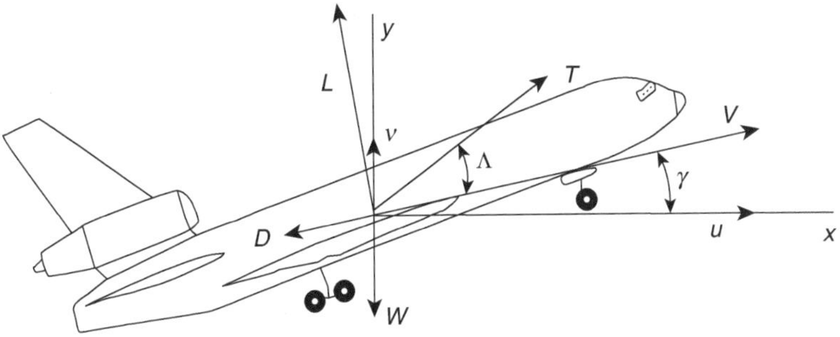


*Figure 4.18* Forces on aircraft during a climb.

is the same as for takeoff/climbing. If one considers the flight along the glide-slope only in this case, the slide-slope angle is not assumed to be constant and second equation transforms from a differential to an algebraic one.

During ground phases of the motion, the aerodynamic gradient in equation (4.32) is replaced by a constant runway slope and the bank angle $\varphi_a$ is assumed to be zero. During the rolling parts of takeoff and landing, the angle of attack ($\alpha$) is assumed to be constant and during rotation it is assumed to be a function of time. More detailed peculiarities of aircraft aerodynamic models are not considered here.

The *simplified Powers' methodology* uses only one equation of motion throughout the aircraft takeoff.[10,11] It is found through the balance of forces on the aircraft in the horizontal direction. The forces on the aircraft are shown in Fig. 4.19 where, for the simplified Powers' methodology, the thrust deflection angle, Λ, is zero. Summing the forces and introducing non-dimensional coefficients for lift and drag yields:

$$dv/dt = g/W[(P - \mu W) + 0.5\rho S(\mu C_L - C_D)u^2].$$

With Powers' assumption of constant thrust, this equation of motion may be integrated analytically for the time and the distance needed to accelerate between two velocities $V_a$ (the initial velocity) and $V_b$ (the final velocity). The resulting time and distance equations are:

$$t_b - t_a = \frac{G}{\overline{V}} \ln\left[\frac{\left(\overline{V} + V_b\right)\left(\overline{V} - V_a\right)}{\left(\overline{V} - V_b\right)\left(\overline{V} + V_a\right)}\right] \quad (4.34)$$

$$X_b - X_a = -G \ln\left|\frac{\overline{V}^2 - V_b^2}{\overline{V}^2 - V_a^2}\right| \quad (4.35)$$

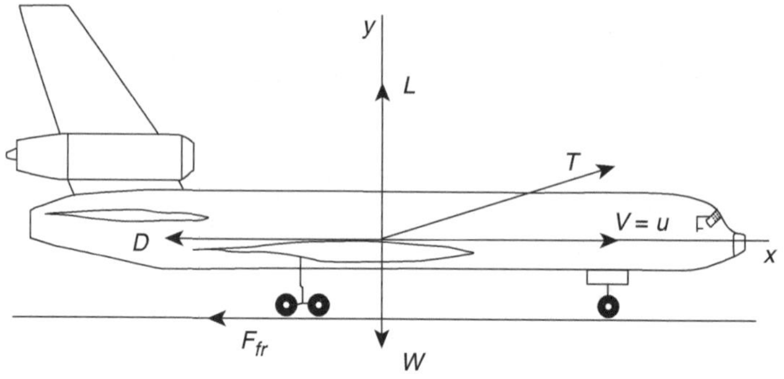


*Figure 4.19* Forces on aircraft force during a ground roll.

where

$$G = \frac{W/S}{\rho g(C_D - \mu C_L)} \quad \text{and} \quad \overline{V} = \sqrt{\frac{T - \mu W}{\rho S(C_D - \mu C_L)}}.$$

It is critical to note that, although the above equations are specifically for ground roll, Powers uses them with aerodynamic climb input to approximate the times and distances of the climb segment as well. Thus, the two-dimensional qualities of climb-out are not represented in Powers' solution. As will be seen, this assumption works fairly well for 'conventional' commercial aircraft, but seriously misrepresents the actual takeoff for aircraft like High Speed Civil Transport (HSCT). This is one of the largest potential problems with the simplified Powers' methodology.

The time-and-distance equations, (4.34) and (4.35), of the simplified Powers' method, may be used in the following algorithm to provided all engines operating (AEO) takeoff solutions:

(1) Given the basic aircraft geometry and aerodynamics, calculate $V_s$.
(2) Calculate $V_{lo}$ by applying user input, usually the FAR 25 limitations.
(3) Calculate $V_2$ by applying user input, usually the FAR 25 limitations.
(4) Calculate the ground distance between $V = 0$ and $V = V_{lo}$ as well as the time taken, using equations (4.34) and (4.35).
(5) Calculate the climb distance between $V = V_{lo}$ and $V = V_2$ as well as the time taken, using equations (4.34) and (4.35).

The same approach was proposed by Miele[11] to define the length and time variables of the ground roll. The incremental length of takeoff roll, for example, is

$$\Delta L = V_{OTP}^2/g/(\tau - \mu) \times (A_1 - A_0). \tag{4.36}$$

The incremental time of the roll is

$$\Delta T = V_{OTP}/g/(\tau - \mu) \times (B_1 - B_0). \tag{4.37}$$

where $V_{OTP}$ is unstick velocity, $\tau$ is thrust-to-weight ratio and $\mu$ is the friction coefficient.

Coefficients $A, B$ for takeoff roll are defined by:

$$A_1 = ln(1 + y \times U^2)/(2 \times y);$$

$$y = (\mu \times C_{YP} - C_{XP})/C_{YOTP}/(\tau - \mu);$$

$$U = V/V_{OTP},$$

$$y < 0 \; S_Y = (-y)^{1/2}$$

$$B_1 = ln(1 + S_Y \times U)/[(2 \times y) \times (1 - S_Y \times U)],$$

$$y > 0 \; S_Y = y^{1/2}$$

$$B_1 = arctg\,(S_Y \times U)/S_Y,$$

$$A_0 = A_1; B_0 = B_1.$$

This algorithm of Miele,[11] based on (4.36) and (4.37), is used in flight path calculations. For example, it is used in the NOITRA model.

However, two remarks must be made. Such models have been derived as a system of differential equations, but it has been shown that a simplified system of algebraic equations is sufficient for aircraft noise assessment needs. The flight trajectory may be represented by means of a set of linear and arc segments, in which flight parameters are defined as approximately constant. The ground tracks of the airplane are also represented by straight-line segments and arcs of circles (see Figs 4.20 and 4.21). Airplane flight profiles are required in order to allow the determination of slant distances from the observation points to the flight paths. The variations of engine thrust, or other noise-related thrust parameter, and airplane speed along the flight path are also required. The slant distances and thrusts are then used for entry into and interpolation of the noise–power–distance data.

Investigations into the influence of the non-stationary flight parameters on noise indices under the flight path[12] show that it is unimportant. So, for aircraft noise assessment, the flight path may be presented as a sum of linear stages inside which the control parameters are constant and phase parameters are defined from linear equations.

For example, the flight profile and flight speeds during takeoff are defined for the following three common simple situations:

(1) Equivalent takeoff roll, $S_g$, is the distance along the runway from the start of the takeoff roll to the intersection point of the runway

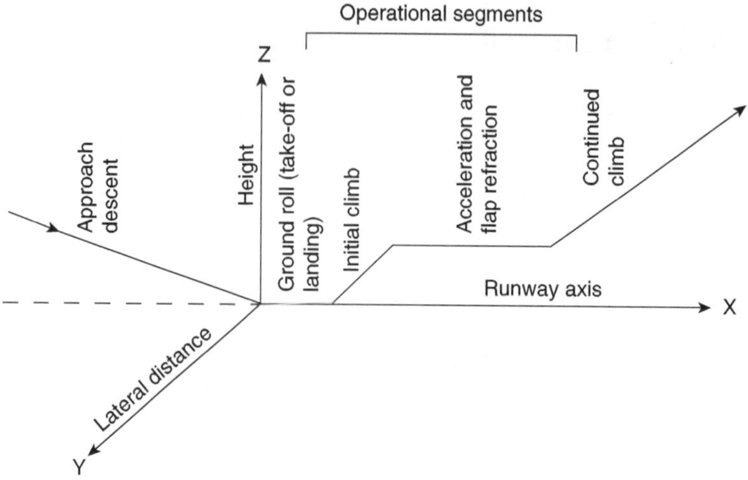


*Figure 4.20* Typical flight path segments for performance calculations.

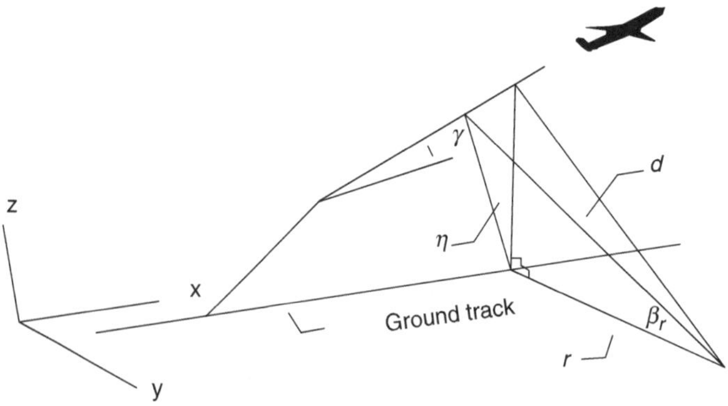


*Figure 4.21* Identification of the different distances and angles used for calculating the sound level or sound exposure level and for calculating ground attenuation.

and the initial climb path projected downwards (see Fig. 4.22) and is given by:

$$S_g = p\theta[f_W(W/\delta)]^2/(X_{NTO/\delta}), \tag{4.38a}$$

$$f_w = (V_{EAS} - V_w)/(V_{EAS} - 4.1). \tag{4.38b}$$

In equation (4.38), $V_w$ is the wind velocity and $V_{EAS}$ is the mean air speed over the initial climb segment.

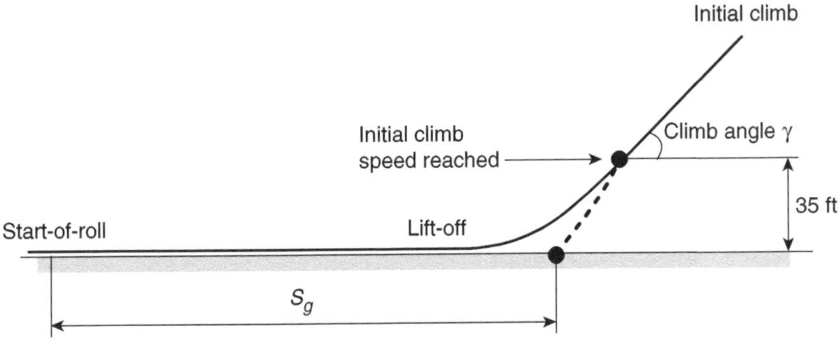


*Figure 4.22* Schematic of aircraft takeoff showing equivalent takeoff roll.

The speed over a particular flight path segment is given as follows:

$$V_{EAS} = Q\sqrt{W}, \quad (4.39)$$

where $Q$ is the flight speed coefficient for which different values are applicable during the climb ($Q_{CL}$), flap retraction ($Q_{FR}$) and approach ($Q_{AP}$).
The relationship between the true and equivalent airspeeds is given by:

$$V_{TAS} = V_{EAS}/\sqrt{\sigma}, \quad (4.40)$$

where $\sigma$ is the ratio of air density to the SA sea-level value.

(2) The climb (or descent) angle of the flight path is determined from:

$$\gamma = \sin^{-1}\{(f/f_w)[X_N/W) - R]\}, \quad (4.41)$$

where $R$ is the climb/descent performance coefficient and $f$ is an acceleration coefficient over the flight path segment.
For an accelerated climb from position 1 to 2:

$$f = \{1 + (V_{TAS2}^2 - V_{TAS1}^2)/[2g(\Delta h)]\}^{-1}. \quad (4.42)$$

For a climb at constant $V_{EAS}$ expressed in meters per second:

$$f = 1 + 5.2 \cdot 10^{-6} V_{EAS}^2. \quad (4.43)$$

The angle $\gamma$ takes a positive value during the climb and a negative value during descent. For a flap retraction segment, the climb angle should be approximated by the average of the values of the coefficient $R$ at the beginning and end of the segment. If the rate of climb $R_C$ is given, the climb angle becomes

$$\gamma = \sin^{-1}[R_C/(V_{TAS}f_w)]. \tag{4.44}$$

If a constant attitude is specified, the climb angle should be assumed constant for the purpose of flight-path schematization. During climb or descent, the horizontal distance covered in a flight segment is determined from:

$$S = \Delta h/tg\gamma. \tag{4.45}$$

While the airplane is accelerating in level flight, the horizontal distance covered is given by:

$$S = f_w(V_{TAS2}^2 - V_{TAS1}^2)/2g[X_N/W - R]. \tag{4.46}$$

The shortest distance (noise radius or slant range) to the flight path is given by:

$$d = \sqrt{l^2 + (h\cos\gamma)^2}. \tag{4.47}$$

If the aim of the calculations is to define the noise spectra $SPL$ at the point of noise control in an algorithm for that is similar to basic acoustic model of the aircraft:

$$\begin{aligned} SPLj &= SPL_{\sum} - \Delta L_V - \Delta L_{int} - \Delta L_{scr} - 20lgR/R_o - \alpha(R - R_o), \\ SPL_{\sum}(f) &= 10lg\sum 10^{0,1SPLj(f)} + \Delta SPLj; \\ SPL_j(f) &= L_W - \Delta L_{\Theta} - \Delta L_f, \end{aligned} \tag{4.48}$$

where $L_W$, $\Delta L_{\Theta}$, $\Delta L_f$, $\Delta SPLj$ defined for the reference distance $R_o = 1$ of sound source normalization, and the values of $\Delta L_V$, $\Delta L_{int}$, $\Delta L_{scr}$, $R$, $R_o$, $\alpha$ are the same as in (4.9).

It is known, even in aircraft flyover noise measurements, that 'lateral attenuation' and shielding effects are the function of sound propagation geometry, thus spectral features (interference dips and/or peaks) of its influence can vary during a flyover episode.[13]

An example of Il'ushin-86 aircraft noise analysis is presented. The NK-86 engines that are installed in Il'ushin-86 power plant are bypass engines with mixing chamber for jets of exhaust gases and with low bypass parameter ($m \cong 1.3$ for the engine mode of interest). Accordingly, the prevalent noise source under the takeoff trajectory is the jet of exhaust gases and the

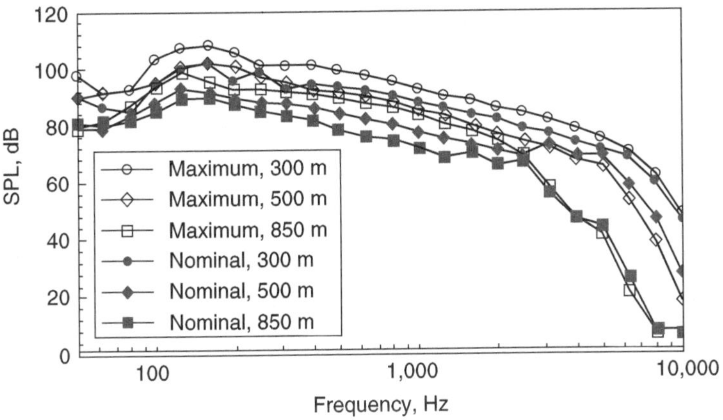


*Figure 4.23* Measured noise spectra for Il'ushin-86 at various flight heights over the control point at the moment of $PNLT_{max}$.

prevalent sources under the landing trajectory are the fan and jet. Measured noise spectra for an Il'ushin-86 aircraft during horizontal flight at diverse heights over the control point and for maximum and nominal engine modes are shown in Fig. 4.23. The corresponding predictions are shown in Fig. 4.24.

The computations were carried out using the program BELTRA based on identified transmission functions for the given flight modes. At the moment of maximum noise level $PNLT_{max}$ at the point of noise control, the results of analysis indicate that the spectral *SPL* of jet noise exceeds the noise from the other acoustic sources on the aircraft by more than 10 dB. The various attenuation phenomena, wavefront divergence $\Delta L_R$, air absorption $\Delta L_{ATM}$ and ground effect $\Delta L_{INT}$ at the measuring microphone position (see Fig. 4.25) for flight height 300 m and noise directivity angle $\Theta_{max} = 136.6°$, as appropriate for calculation of $PNLT_{max}$, have been taken into account in the *SPL* spectra estimation also. The angles $\Theta_{max}$ have been defined by the subprogram TETAMAX (one of the options in the BELTRA program suite) for all of the various noise analyses considered.

Results show that the predictions for transfer functions for every particular engine mode (including all used flight heights) provide minimum value of prediction error (Table 4.13). Thus, this calculation scheme has been proposed for making predictions in any possible case. The results of predictions for maximum and nominal engine modes are shown in Fig. 4.26. Calculated and measured noise event values *EPNL* are compared in Table 4.14.

The trajectory models have been used for the analysis of the influence of various operational factors on aircraft noise levels under the flight paths

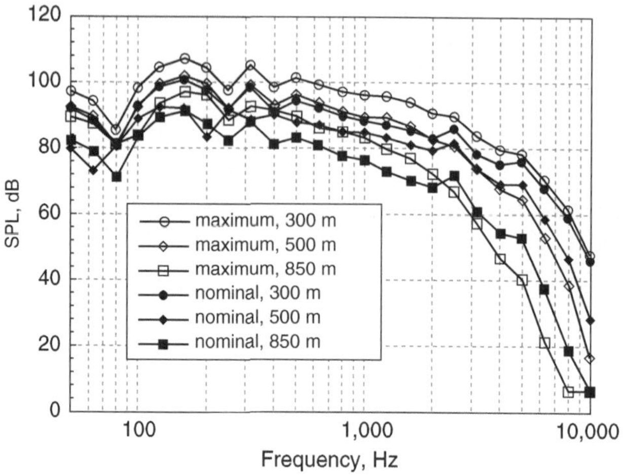


*Figure 4.24* Calculated noise spectra for Il'ushin-86 at various flight heights over the control point at the moment of *PNLTmax*.

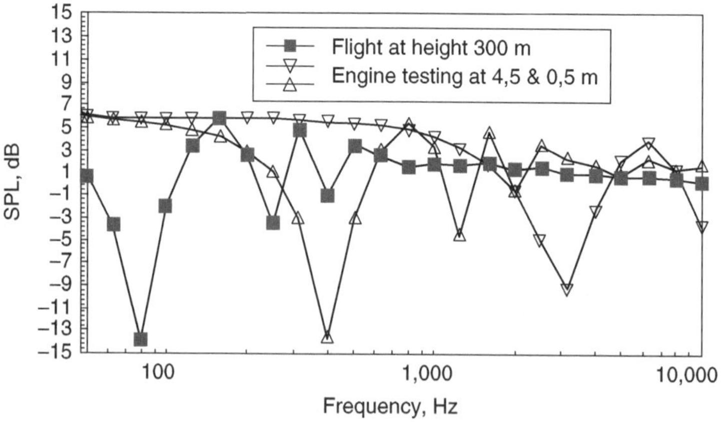


*Figure 4.25* Calculated excess attenuation spectra for sound wave interference ('lateral attenuation') used in the acoustic model of an Il'ushin-86 aircraft.

*Table 4.13* Prediction accuracies for the transfer function of the basic acoustic model of Il'ushin-86 at nominal engine thrust

| *Type of prediction* | *The sum least square errors for every variant of analysis* |
|---|---|
| Overall flight heights and engine modes considered | 299.8 |
| Overall flight heights considered | 226.8 |
| Overall engine modes considered | 191.4 |
| Zero transfer function | 474.7 |

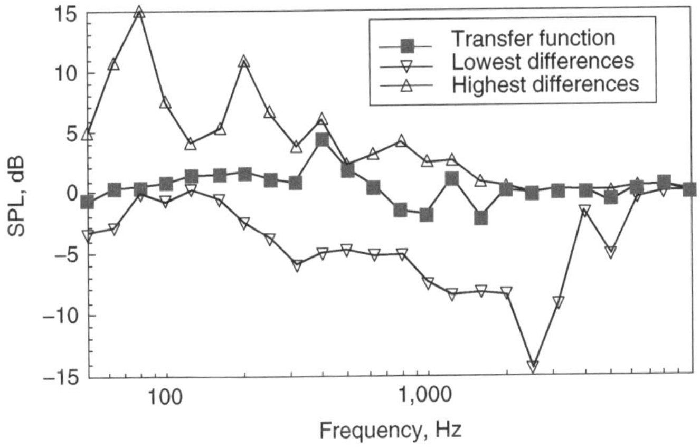


*Figure 4.26* Differences between measured and predicted transfer functions for the Il'ushin-86 acoustic model in accordance with flight tests (for the same conditions as assumed in Fig. 4.23).

*Table 4.14* Predicted and measured noise levels, *EPNL*, for aircraft in operation

| *Type of aircraft* | *Takeoff, monitoring point No. 2* | | *Landing, monitoring point No. 3* | |
|---|---|---|---|---|
| | *Predicted* | *Measured* | *Predicted* | *Measured* |
| Tupolev-154 | 99.2 | 100.1 ± 1.2 | 105.8 | 106.0 ± 0.9 |
| Tupolev-154M | 98.3 | 98.4 ± 0.9 | 100.7 | 102.1 ± 0.5 |
| Tupolev-204 | 97.0 | 96.0 ± 2.6 | 102.2 | 99.9 ± 2.7 |
| Yakovlev-40 | 91.2 | 90.3 ± 3.9 | 98.7 | 97.2 ± 3.8 |
| Yakovlev-42 | 93.8 | 93.4 ± 0.7 | 103.7 | 102.4 ± 1.6 |
| Il'ushin-62M | 100.2 | 102.9 ± 2.5 | 100 | 103.5 ± 3.8 |
| Il'ushin-86 | 107.6 | 107.4 ± 0.6 | 105.7 | 105.1 ± 03 |

during takeoff and landing. First of all, the influence of flight procedures is the subject of great concern. Among them the noise certification flight test procedures are the most important. Corresponding numerical predictions for Soviet aircraft have been validated against experimental data (see Table 4.14). An example is shown in Fig. 4.26.

All predicted data are inside the limits of confidence of the measured values. The analysis of results shows good enough accuracy, especially for the stages of descending and climbing. For the descent before the landing stage, the averaged divergence between results, ±1.4 *EPN*dB, is higher than that for climbing stage. The discrepancies can be attributed to inexact initial information about the parameters for the mode of operation of the engines during descent before landing. This is because the engine modes during descent are highly dependent on flight conditions and the resulting noise levels are difficult to calculate using the available models).

Although the basic acoustic models for the aircraft under consideration have been used, it is possible to use the noise radius approach (NPD-relationships) too. Moreover, the noise radius approach is convenient for footprint assessment for every flight path, takeoff and landing. The relationships between *EPNL* and footprints are presented in Fig. 4.1a and b. The results were achieved assuming normal flight procedures, without the use of noise reducing operations along the flight path.

## 4.5 Effects of ground, atmosphere and shielding by wing and fuselage

### *4.5.1 Ground effects*

Effects due to interference between direct sound and sound reflected at the ground between source and receiver were discussed in Chapter 3. Good examples of these effects have been observed in acoustic measurements at engine testing facilities. Figure 4.27 and Table 4.15 show relevant data.[13] Analyses of the results for engine trials show that at least the first dips and peaks are modeled correctly and these are the most important for noise assessment. The conditions of sound propagation during the testing are fully specified since the heights of the engine and microphones and the distance between them are known and meteorological effects are minimal. Consequently, these data offer a good opportunity for testing models of ground effect. Calculations of excess attenuation have been performed for a monopole noise source. Table 4.15 shows that different values of impedance parameters for reflecting surfaces (e.g. for foam linings and concrete) are predicted to produce different excess attenuation spectra. The acoustical effects of lining the surface with foam materials have been predicted to investigate their potential efficiency for reducing noise at engine testing facilities.

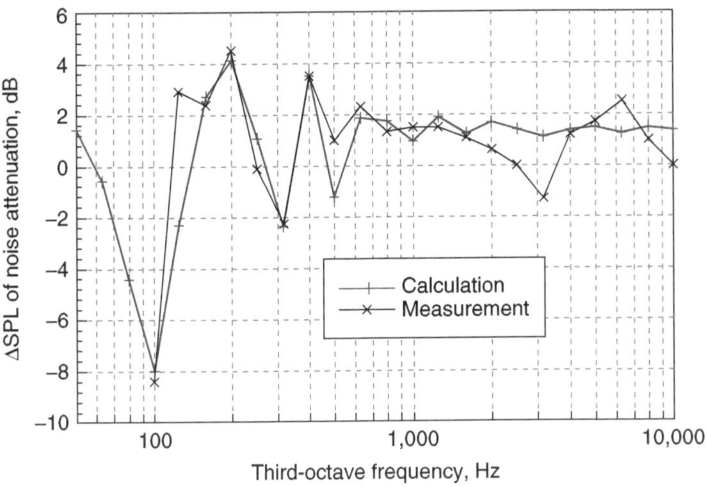


*Figure 4.27* Comparison of noise spectra predictions and measurements at engine testing facility with surface made from concrete.

*Table 4.15* Measured and predicted excess attenuation (dB) for three reflecting surfaces

| *Frequency* | *Foam lining 4 inches* | | *Foam lining 8 inches* | | *Concrete* | |
|---|---|---|---|---|---|---|
| | *Calculation* | *Measurement* | *Calculation* | *Measurement* | *Calculation* | *Measurement* |
| 100 | −6.18 | −5.7 | −2.81 | −3 | −7.96 | −8.4 |
| 125 | −1.06 | 2 | 0.29 | 2 | −2.3 | 2.9 |
| 160 | 2.75 | 2.1 | 2.16 | 1.7 | 2.71 | 2.4 |
| 200 | 3.27 | 3.5 | 1.53 | 1.8 | 3.09 | 3.5 |
| 250 | −0.58 | −1.2 | −1.5 | 10 | 1.06 | −0.1 |
| 315 | −0.91 | −0.5 | 0.15 | −1.2 | −2.34 | −2.3 |
| 400 | 2.22 | 2.2 | 0.54 | 0 | 3.53 | 3.5 |
| 500 | −0.77 | 1.5 | −0.07 | 1.3 | 1.24 | 1 |
| 630 | 0.38 | 1 | −0.21 | 0.8 | 1.87 | 2.3 |
| 800 | 0.57 | 0.7 | 0.05 | 0.1 | 1.72 | 1.35 |
| 1,000 | 0.24 | 0.4 | 0.09 | −1 | 0.95 | 1.5 |
| 1,250 | 0.75 | 0.7 | 0.17 | −1 | 1.9 | 1.5 |
| 1,600 | 0.24 | −0.3 | 0.07 | −0.6 | 1.24 | 1.1 |
| 2,000 | 0.14 | −0.4 | −0.05 | −0.5 | 1.71 | 0.6 |
| 2,500 | 0.17 | −1 | 0.04 | −0.4 | 1.39 | 0 |
| 3,150 | 0.05 | 1 | 0.04 | −0.5 | 1.13 | −1.3 |
| 4,000 | 0.01 | 1 | 0 | −1 | 1.35 | 1.2 |
| 5,000 | 0.08 | 1.5 | 0.02 | −0.9 | 1.47 | 1.7 |
| 6,300 | 0.03 | 1.1 | 0.04 | −0.1 | 1.23 | 2.5 |
| 8,000 | 0.02 | −1 | 0 | −1 | 1.46 | 1 |
| 10,000 | 0.01 | 0.5 | 0.01 | −0.5 | 1.35 | 0 |

### 4.5.2 Refraction effects

Important additional factors influencing aircraft noise levels along the propagation path are sound refraction caused by wind speed or temperature gradients in the atmosphere boundary layer, and diffusion of sound energy by turbulence.

Ray tracing is a high-frequency approximation for wave propagation in refracting media. The conditions for the validity of ray tracing in an unbounded medium are

$$\lambda_0 n' << n^2 \approx 1,\ \lambda_0 \frac{A'}{A} << 1,$$

where $n$ is the index of refraction, $n'$ is the spatial derivative of the index of refraction, $\lambda_0$ is the wavelength, $A$ is the wave amplitude and $A'$ is the spatial derivative of the wave amplitude. These conditions mean that the speed of sound and the amplitude of the wave cannot change significantly in a wavelength.

For the sound velocity profiles of interest, $\lambda_0 n' \approx 10$ at 10 Hz and $10^{-3}$ at 100 Hz. So the criteria are satisfied at 100 Hz but may be marginal at 10 Hz.

For a homogeneous atmosphere, the total acoustic field is formed by the combination of direct and ground-reflected rays with lengths $R$ and $R'$, respectively:

$$\varphi = \frac{\exp(ikR)}{4\pi R} + \{R(\theta) + (1 - R(\theta)F(p_e)\} \frac{\exp(ikR')}{4\pi R'}, \tag{4.49}$$

Interference between the direct and ground-reflected paths results from the phase difference $k(R - R')$. In the presence of downward refraction corresponding to normalized sound speed gradient $a = \frac{1}{c(0)} \frac{dc}{dz}$, the number of rays is given by:

$$N = 2\left(\left[\frac{d - l_1}{2l}\right] + \left[\frac{d - l_2}{2l}\right]\right) + 2, \tag{4.50a}$$

where $[x]$ represents the integer part of $x$ and

$$l = \sqrt{h_r^2 + 2h_r/a},\ l_1 = \sqrt{(h_r + a^{-1})^2 - (h_s + a^{-1})^2},\ l_2 = 2l - l_1.$$

Equation (4.50a) assumes that $h_s \geq h_r$.

The rays in a linear sound speed profile environment can be categorized in different orders, where all rays belonging to the same order experience the same number of ground reflections. For all but the first order there will be four rays in each order. The order ($n$) to which a ray belongs can be calculated from its number using the following formula:

$$n = \left[\frac{N - 1}{4}\right] + 1 \tag{4.50b}$$

In each order containing four rays, there are two rays with as many maxima as the number of ground reflections (equal to the order), one ray with one less, and one ray with one more.

To avoid discontinuities and reduce the necessary computation time, an even simpler model can be obtained using the following equations:

$$N_t = \frac{2d - l_1 - l_2}{l}; \tag{4.50c}$$

$$N = N_t \text{ if } N_t \geq 2$$

$$N = 2 \text{ if } N_t < 2.$$

The parameters of the reflected rays are defined by an algorithm, the main features of which are given below, but which are described in more detail elsewhere.[14–17]

The width of a caustic after $N$ reflections is given by

$$\delta = (4N)^{1/3} \frac{H}{\psi_G}, \tag{4.51}$$

where $H$ is the wave layer thickness and $\psi_G$ is the grazing angle on reflection.

The wave layer thickness is a diffraction parameter for full wave analysis of upward and downward refraction in linear sound velocity gradients and is given by

$$H = (2ak_0^2)^{-1/3},$$

where

$$a = \frac{1}{c(0)} \frac{dc}{dz} = \frac{1}{R_C}. \tag{4.52}$$

$R_C$ corresponds to the radius of curvature of a near horizontal ray.

The distance between reflections and the maximum height for a ray launched at grazing angle $\psi_G$ can be calculated easily since the ray paths for propagation in a linear sound speed gradient are arcs of circles (Fig. 4.28). The separation between reflections $\Delta$ is given by

$$\Delta = \frac{2tg\psi_G}{a}. \tag{4.53}$$

For linear speed gradients, Hidaka *et al.*[15] state that sound rays are arcs of a circle, with radius $R_c$ defined by:

$$R_C = \frac{1}{a\cos\psi_G} \tag{4.54}$$

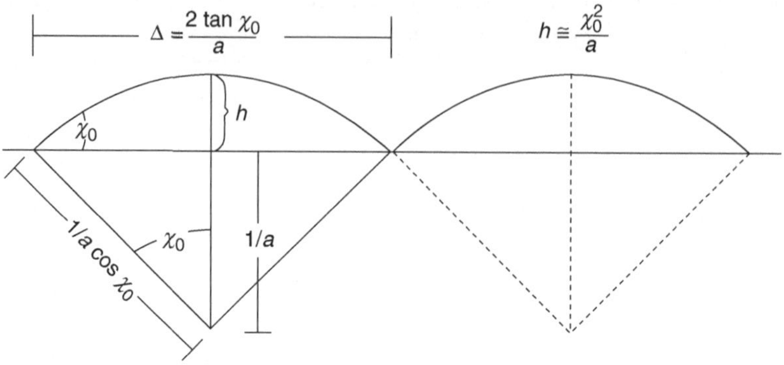


*Figure 4.28* Geometry for ray tracing in an atmosphere with a linear soundspeed gradient[17].

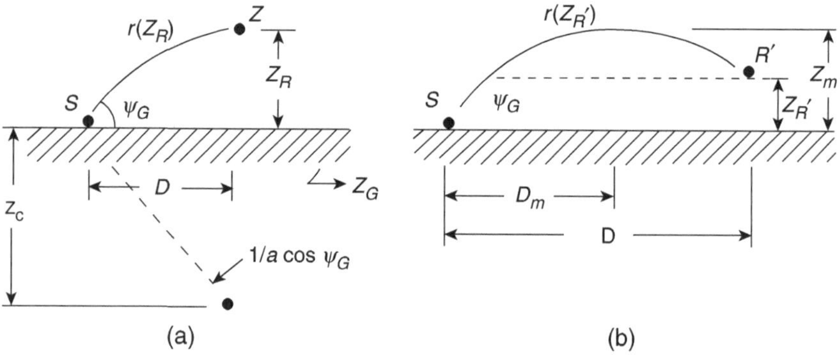


*Figure 4.29* Curvature of the direct path for a positive celerity gradient[17]: (a) receiver before the apex, (b) receiver beyond the apex (angle $\psi$ is the same as angle $\chi_0$ in Fig. 4.29).

where the angle $\psi_G$ (shown in Fig. 4.29) is derived from:

$$\psi_G = arctg\left[\frac{aD}{2} + \frac{z_R(2+az_R)}{2D}\right], \tag{4.55}$$

where $\Delta = 2D_m$.

The center of the arc lies in the horizontal plane, at the height (depth) $z_c$:

$$z_C = -1/a \tag{4.56}$$

Thus, for a source lying on the ground, if the receiver is located before the apex is reached (Fig. 4.30), the length of the curved ray $r(z_R)$ and the

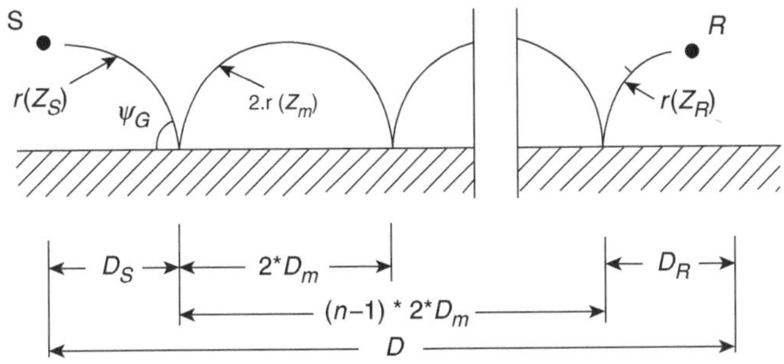


*Figure 4.30* Ray paths for multiple reflections on the ground[17].

propagation time $\tau(z_R)$ can be derived from:

$$r(z_R) = \frac{1}{a\cos\psi_G}\left[\arcsin\{(1+az_R)\cos\psi_G\} - \frac{\pi}{2} + \psi_G\right]; \tag{4.57}$$

$$\tau(z_R) = \frac{1}{2ac_G}\log\left[\frac{f(0)}{f(z_R)}\right],$$

where $c_G$ represents the sound speed at the source (equal to $c_0$ here) and

$$f(z) = \frac{1+\sqrt{1-(1+az)^2\cos^2\psi_G}}{1-\sqrt{1-(1+az)^2\cos^2\psi_G}}. \tag{4.58}$$

When the sound ray reaches its apex, two points ($z_R$ and $z_{R'}$) may lie at the same height; so equations (4.52) to (4.58) cannot be used directly to derive the parameters of the arc of circle related to $z_{R'}$.

Accordingly, the following formulae must be applied:[17]

$$r(z'_R) = 2r(z_m) - r(z_R); \tag{4.59}$$

$$\tau(z'_R) = 2\tau(z_m) - \tau(z_R).$$

Up to now, we have assumed that the source was on the ground and, moreover, that $z_R$ was greater than $z_S$. In the reverse case, when the receiver is on the ground, $z_S$ must be substituted for $z_R$ in the previous equations.

In the more general case, where neither the source nor the receiver are on the ground, the total length and propagation time of the reflected ray can be derived by applying to the reference plane a translation equal to the minimum height of $z_R$ and $z_S$.

When the speed gradient is high, or when the receiver is far away from the source, there is more than one reflection on the ground (see Fig. 4.30).

The location of the first reflection, $x$, (this is shown as $D_s$ in Fig. 4.30) is a solution of the following fourth-order polynomial equation:[18,19]

$$n(n+1)x^4 - (2n+1)Dx^3 + [b_R^2 + (2n^2 - 1)b_S^2 + D^2]x^2 - (2n-1)b_S^2 Dx + n(n-1)b_S^4 = 0 \tag{4.60}$$

where

$$b_i^2 = z_i(2+az_i)/a, i = RorS$$

and

$$D = D_S + D_R + 2(n-1)D_M$$

$n$ being the order of reflection (4.57). For each extra reflection, the geometrical and time parameters are given by:

$$r_i = 2(n-1)r(z_m) + r(z_R) + r(z_S)$$

and

$$\tau_i = 2(n-1)\tau(z_m) + \tau(z_R) + \tau(z_S). \tag{4.61}$$

Each $r_i$ and $\tau_i$ for the rays propagating upward only are calculated using (4.57) to (4.61).

When the receiver moves closer to the ground and further from the source, or when the gradient is increased, the complex conjugate of the roots may become real, which means that additional reflected rays have to be considered. For $n = 2$, two additional reflected rays may appear, and, for $n > 2$, four additional reflected rays may appear for each $n$. Figure 4.31 shows rays including only one ground reflection for the case $h_s = 1.0$ m, $h_r = 1.0$ m, $R_0 = 160.00$ m and based on the application of (4.49) and (4.50). Figures 4.32 and 4.33 and Table 4.16 show rays including one and two ground reflections for $h_s = 1.0$ m, $h_r = 1.0$ m and $R_0 = 190.0$ m.

For an arbitrary number of ground reflections (arbitrary $n$), the possible solutions for $D_R$ are derived by using an algorithm to determine polynomial roots of equation (4.60), for example, by using Laguerre's method or Jenkins–Traub three-stage algorithm. When found, only those that are physically admissible are retained (i.e. those which are real, positive and satisfy $D_R < D$). The number of solutions is a function of the gradient amplitude, of the heights of both source and receiver, and of distance $D$. Thus, for $a > 0$ and $n = 1$, equation (4.60) reduces to a cubic equation, of which at least one root is real. For this value of $n$, when the distance $D$ from the source to the receiver increases, the heights of the source and the receiver decrease, and/or the gradient $a$ increases, the complex solutions become real,

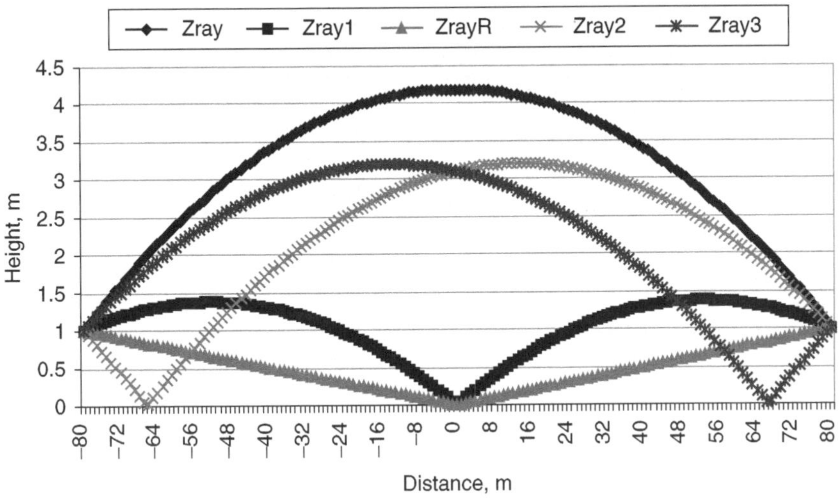


*Figure 4.31* Multiple sound paths with one point of reflection: ZrayR is the reflected ray for the homogeneous case; Zray is the direct ray for downward refraction, Zray1, Zray2 and Zray3 are the reflected rays for the downward refraction case ($a = 0.001$).

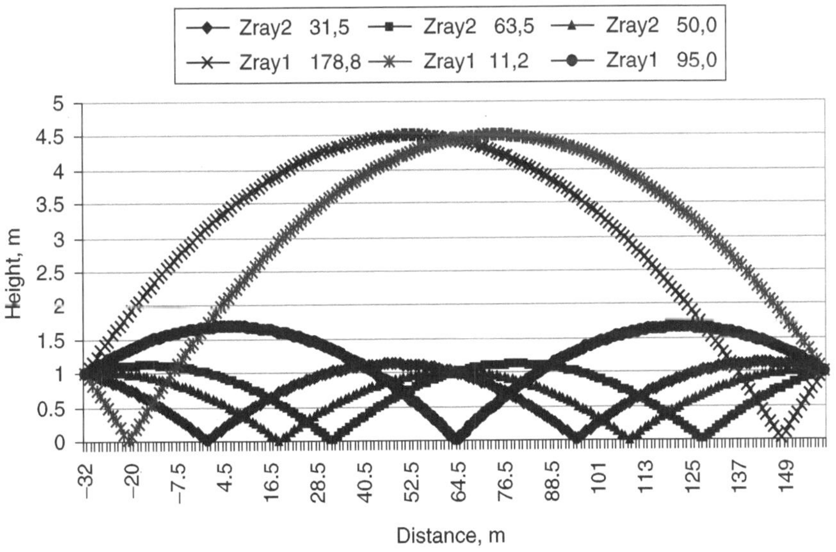


*Figure 4.32* Sound paths with one and two ground reflections: Zray1 are reflected rays with one ground reflection; Zray2 are reflected rays with two ground reflections. The x-co-ordinate of the point corresponds to the distances in meters from the source ($a = 0.001$).

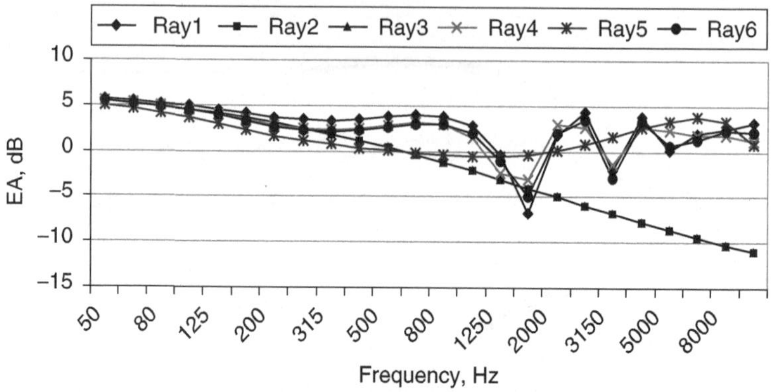


*Figure 4.33* Excess attenuation (EA) along different ray paths during downward refraction. The numbers of the rays correspond to those in Table 4.17.

*Table 4.16* Characteristic distances and incidence angles for refracted rays

| *Ray number* | *Reflection point (m)* | $R_1$ *(m)* | $R_2$ *(m)* | *Incidence angle (degrees)* |
|---|---|---|---|---|
| 1 | 95.0 | 190.28 | 190.08 | 86.7 |
| 2 | 11.2 | 190.28 | 190.28 | 84.6 |
| 3 | 178.8 | 190.28 | 190.28 | 84.6 |
| 4 | 50.0 | 190.28 | 190.06 | 87.4 |
| 5 | 13.5 | 190.28 | 190.26 | 85.3 |
| 6 | 63.5 | 190.28 | 190.07 | 87.3 |

which means that two new rays appear. When $n \geq 2$, the complex roots become real in pairs, meaning that rays appear two at a time.

During aircraft noise assessment under the flight path, refraction due to wind or temperature gradients provides only direct and one reflected ray, modifying, in comparison with homogeneous case, the grazing angle and the phase and transmission loss as shown in Table 4.16 and Fig. 4.33.

Ray trace analysis can be applied, for example, to data obtained during takeoff/climbing trials of an aircraft with two engines under the wing. Some points of noise control were located on opposite sides of the flight path (see Table 4.17).

The horizontal displacement of the side points of noise control from the flight path was 450 m. Points that are placed almost symmetrically either side of the flight path are B5 and B14. Point B12 is on the centerline. The abscissa of all the control points is approximately 942 m. The height of the microphone above the ground, except for B2, B3 and B10, is equal to 1.2 m.

*Table 4.17* Microphone positions along takeoff path

| *No. of microphone position* | *Abscissa of the control point (m)* | *Height of the ground above sea level (m)* | *Ordinate of the control point (m)* | *Height of the microphone above ground (m)* |
|---|---|---|---|---|
| B1 | 925.0 | 592.5 | 6.0 | 1.200 |
| B2 | 960.0 | 492.5 | 6.2 | 0.006 |
| B3 | 953.6 | 492.7 | 6.3 | 0.006 |
| B4 | 947.3 | 492.5 | 6.2 | 1.200 |
| B5 | 941.5 | 492.5 | 6.3 | 1.200 |
| B6 | 935.3 | 492.5 | 6.2 | 1.200 |
| B7 | 925.0 | 392.5 | 6.0 | 1.200 |
| B8 | 925.0 | 292.5 | 5.6 | 1.200 |
| B9 | 960.0 | 42.5 | 6.4 | 0.006 |
| B10 | 954.0 | 42.7 | 6.4 | 0.006 |
| B11 | 948.0 | 42.5 | 6.3 | 1.200 |
| B12 | 942.0 | 42.5 | 6.3 | 1.200 |
| B13 | 936.0 | 42.5 | 6.2 | 1.200 |
| B14 | 942.0 | −407.5 | 5.7 | 1.200 |
| B15 | 936.0 | −407.5 | 5.5 | 1.200 |
| B16 | 925.0 | −479.5 | 5.5 | 1.200 |

*Table 4.18* Meteorological conditions during the flight trials

| *No. of flight* | *Temperature (°C)* | *Pressure (millibars)* | *RH (%)* | *Wind speed (knots)* | *Wind direction (degrees)* | *Cross-wind speed (knots)* |
|---|---|---|---|---|---|---|
| 3 | 16 | 961.4 | 61 | 6.9 | 6 | 2 |
| 11 | 16.1 | 961.6 | 57 | 5.3 | 299 | 5.3 |
| 12 | 16.2 | 961.7 | 57 | 4.3 | 261 | 3.6 |
| 14 | 17 | 961.5 | 55 | 1.7 | 200 | 0.1 |

During the flight trials, the wind speed at height 10 m varied between 1.2 and 7.2 knots (see Table 4.18). Examples of the measured noise level spectra measured at the points listed in Table 4.19 are shown in Figs 4.34–4.37. The maximum cross-wind observed during flight No. 11, was 5.3 knots (2.75 m/s).

For each point of interest, equations (4.49, 4.50) predict a direct ray and a single reflected ray. This is true even for flight No. 12, which was lowest over the point of noise control. For the meteorological conditions of flight No. 12, an appropriate normalized sound speed gradient ($a$) in equation (4.52) was chosen with a maximum possible value of 0.001.

For a non-refracting atmosphere and for slant distance $R_0 = 450.00$ m, the direct ray length $R_1 = 475.56$ m, the reflected ray length $R_2 = 476.34$ m, the horizontal distance to the reflection point $D = 446.54$ m and the grazing

*Table 4.19* Closest approach of the flight path to the centerline between symmetrically placed points of noise control B5 and B12 during the flight trials

| *No. of flight* | *Abscissa (m)* | *Ordinate (m)* | *Height (m)* | *Aircraft pitch angle (degrees)* | *Aircraft speed (knots)* | *Aircraft mach number* | *Corrected engine rating* |
|---|---|---|---|---|---|---|---|
| 3 | 961.192 | 47.706 | 209.378 | 17.880 | 165.166 | 0.259 | 97.935 |
| 11 | 940.554 | 60.183 | 643.973 | 19.264 | 167.014 | 0.269 | 98.170 |
| 12 | 928.064 | 45.143 | 154.632 | 18.149 | 166.750 | 0.261 | 97.835 |
| 14 | 952.015 | 48.871 | 315.824 | 14.963 | 168.389 | 0.266 | 92.358 |

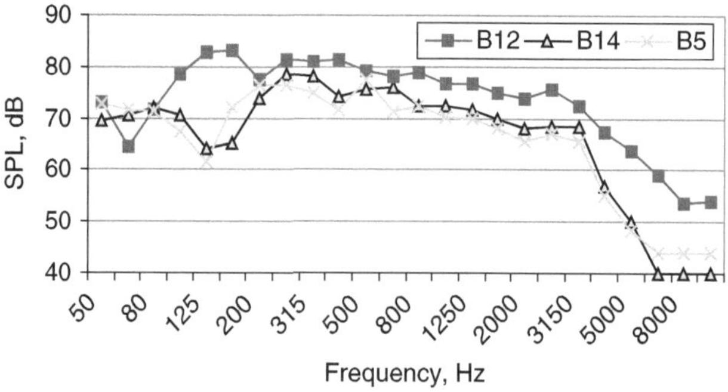


*Figure 4.34* Measured Noise Spectra at control points B12, B14 and B5 during flight No. 14.

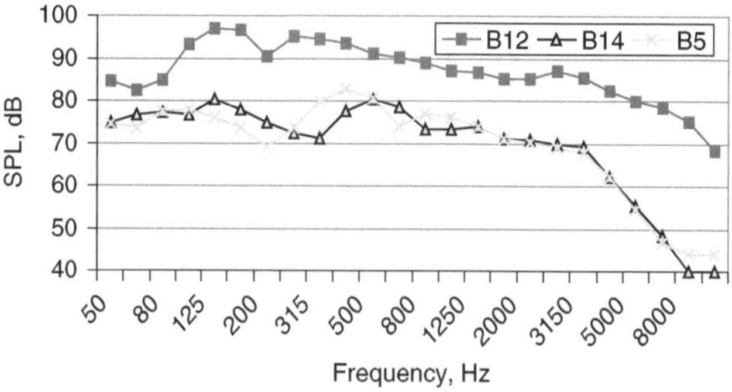


*Figure 4.35* Measured Noise Spectra at control points B12, B14 and B5 during flight No. 12.

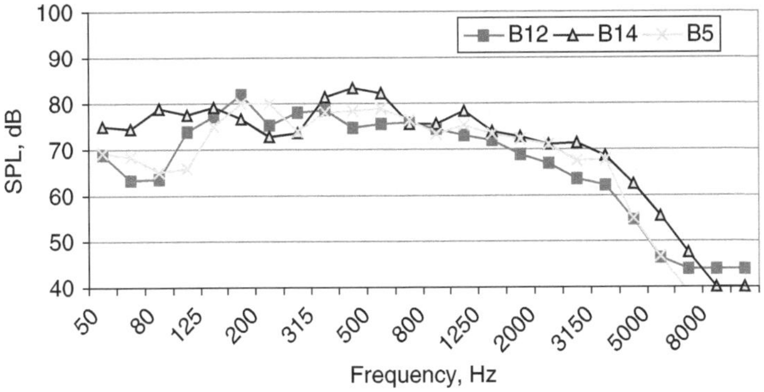


*Figure 4.36* Measured noise spectra at control points B12, B14 and B5 during flight No. 11.

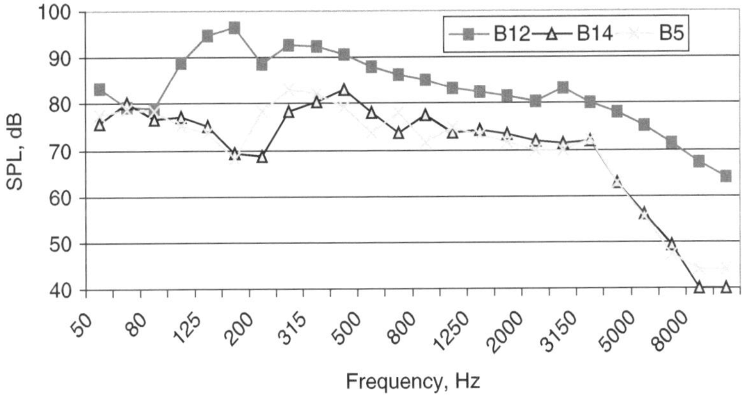


*Figure 4.37* Measured noise spectra at control points B12, B14 and B5 during flight No. 3.

angle $\psi_G = 19.1$ degrees. For the same slant distance during downward refraction, the direct ray length $R_1 = 478.94$ m (see the curved direct ray in Fig. 4.38), the reflected ray length $R_2 = 479.72$ m, the horizontal distance to the reflection point $D = 447.98$ m and the grazing angle $\psi_G = 30.8$ degrees. The implication of this analysis is that, for the trials considered, refraction effects are small. This is demonstrated further by comparing predictions without and with refraction.

The Chien and Soroka model for a monopole above an impedance plane (equation 3.19) for the abovementioned geometry predicts the attenuation shown in Fig. 4.39. The first dip is predicted to be in the frequency band near 200 Hz. The first dip in the measured spectra for point B14 (Fig. 4.37)

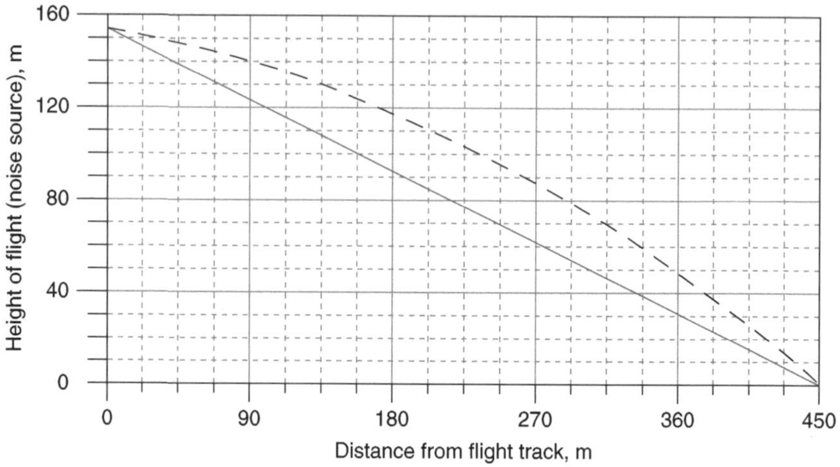


*Figure 4.38* Curvature of the direct ray (broken line) for the case under consideration (flight trials No. 12) in comparison with homogeneous case (continuous line).

is near 200 Hz too. Similar dips at frequencies between 200 and 315 Hz can be observed in Figs 4.35, 4.36 and 4.38. The second predicted ground effect dip near 800 Hz (see Fig. 4.39) is observable in the measured spectra in Figs 4.34–4.37, but the predicted dip is deeper than that measured; this is understandable in view of the fact that total coherence is assumed in the lateral attenuation model. During downward refraction, the grazing angle becomes larger than for a homogeneous atmosphere, which means that the lateral attenuation is less. However, for the aircraft noise trials considered, the refraction effects are predicted to be rather small.

According to (4.49) and (4.50), flight positions more distant from the points of noise control do not produce more than one reflected ray because of their height.

### *4.5.3 Shielding and reflection by wings*

The effects on noise due to the configuration of the aircraft may be of the same magnitude as sound propagation effects (e.g. that due to ground attenuation). Possible configurational effects (mainly resulting in attenuation) are:

- shielding – by wing and tail for noise control points just under the flight path (e.g. at the takeoff/climbing and the approach noise control points) or by fuselage (at the sideline noise control point);
- jet-on-jet attenuation;

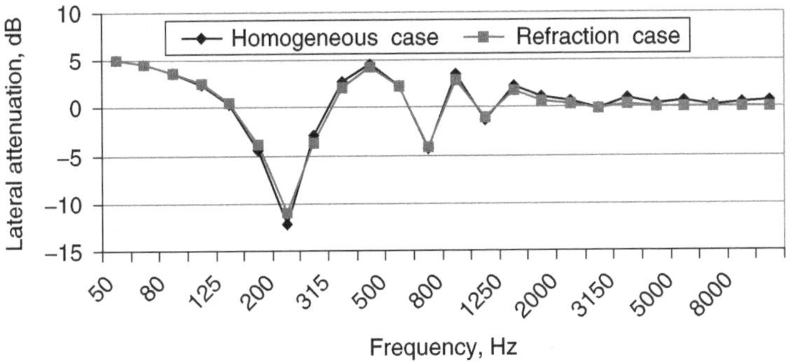


*Figure 4.39* Lateral attenuation predicted for homogeneous and downward refracting atmospheric conditions.

- reflections by wing or/and fuselage. Noise shielding in flight involves diffraction by the solid boundaries near to the source. To simplify matters, we consider diffraction without air flow. In practice, the acoustic sources are not simple to specify.

Geometric shielding can be achieved by means of the aircraft wings, tail-plane and fuselage. Jet noise is particularly difficult to shield efficiently by the wings because it is described often by distributed elementary sources downstream of the wing. Fan noise, which is dominant for modern turbofans, may be shielded by the wing more effectively.

When sound waves spread out, for instance, from the fan, some of them may be scattered by the fuselage. Scattered waves spread out from the fuselage in all directions, distorting and interfering with the incident waves. A model of secondary point sources, set on the cylinder surface is used to calculate the sound pressure level of the engine near the fuselage, and a model of wave propagation in a homogeneous moving layer is used to model the refraction produced by the engine jet.

Experimental investigations[20,21] have shown that wing shielding can have a measurable effect on the attenuation of engine inlet and/or exhaust noise, for certain engine mount configurations. This is an important consideration when predicting noise from business aircraft with aft-mounted engines.

A wing-shielding model employs the Fresnel diffraction theory for a semi-infinite barrier, as described in Maekawa,[22] with modifications, to treat the finite barrier presented by the aircraft wing. The detail of the process for computing the attenuation resulting from wing shielding is described elsewhere.[23] Initially, the wing configuration is described in a local coordinate system with the origin positioned at the engine inlet (point 1; see Fig. 4.40). Then, the local origin and wing coordinates are transformed into a global coordinate system consistent with the observer location in the ground.

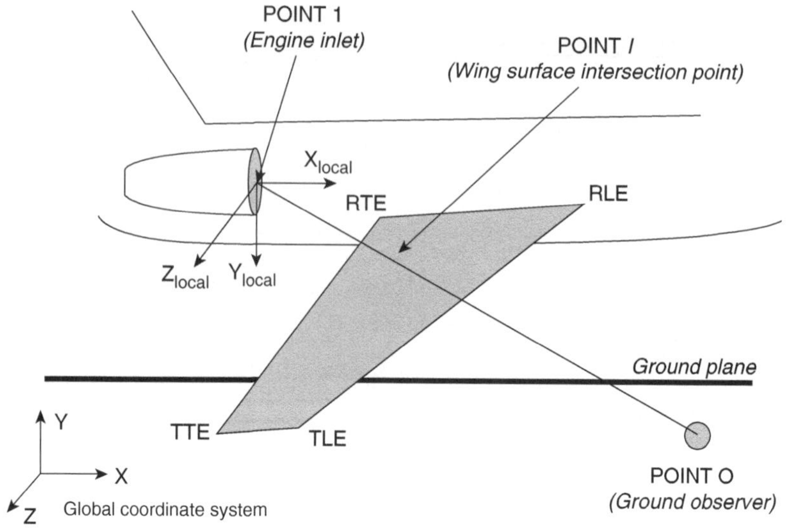


*Figure 4.40* Definition of the wing-engine-observer configuration with local and global co-ordinate systems for the wing-shielding model[23].

This transformation must take into account the aircraft attitude and position at the particular time of the observation.

The intersection point (point I) may or may not be located within the boundaries of the wing surface. But after the intersection point I has been determined, then the point on each wing boundary, which is nearest to point I, must be located, as shown in Fig. 4.41.

The locations of each of these points ($W_{LE}$, $W_{TE}$ and $W_{TIP}$) are computed by solving a set of three equations and three unknowns for each point. For example, the coordinates $x_{WLE}$, $y_{WLE}$ and $z_{WLE}$ may be found for the point $W_{LE}$. The equations are obtained by imposing the following conditions.

(1) The line $IW$ must be perpendicular to the wing boundary. This condition is represented by setting the dot product of the line $IW$ vector and the wing boundary line vector equal to zero, e.g.:

$$(X_I - X_{WLE})(X_{RLE} - X_{TLE}) + (Y_I - Y_{WLE})(Y_{RLE} - Y_{TLE}) + (Z_I - Z_{WLE})(Z_{RLE} - Z_{TLE}) = 0$$

(2) The point W must lie on the wing boundary. This condition is met when the coordinates of the point W satisfy the two-point equation of

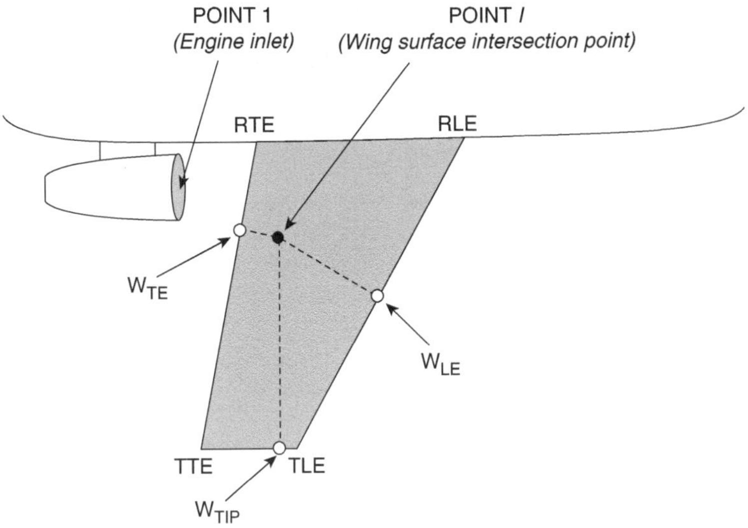


*Figure 4.41* Points W that are closest to point *I* on each diffraction edge wing boundary[23].

the line representing the wing boundary edge, e.g.:

$$(X_{WLE}-X_{RLE})/(X_{TLE}-X_{RLE})-(Y_{WLE}-Y_{RLE})/(Y_{TLE}-Y_{RLE})=0; \tag{4.62}$$

$$(X_{WLE}-X_{RLE})/(X_{TLE}-X_{RLE})-(Z_{WLE}-Z_{RLE})/(Z_{TLE}-Z_{RLE})=0.$$

If point I lies on the wing surface, then Fresnel diffraction theory is used to determine the attenuation. For each diffraction edge, three distances must be computed (see Fig. 4.42):

(1) The direct source–receiver path length, from point 1 to point O, $d_{1O}$.
(2) The distance, $d_{1W}$, from point 1 to the closest point on the diffraction edge, point W.
(3) The distance, $d_{WO}$, from the point W on the diffraction edge to the observer location on the ground at point O.

From these three distances, the difference, $\delta$, in source–receiver path lengths between the direct and diffracted sound fields may be computed, that is,

$$\delta = (d_{1W} + d_{WO}) - d_{1O}, \tag{4.63}$$

Here $\delta > 0$ when point I lies on the wing surface, $\delta = 0$ when point I lies on the wing boundary edge and $\delta < 0$ when point I is beyond the wing surface.

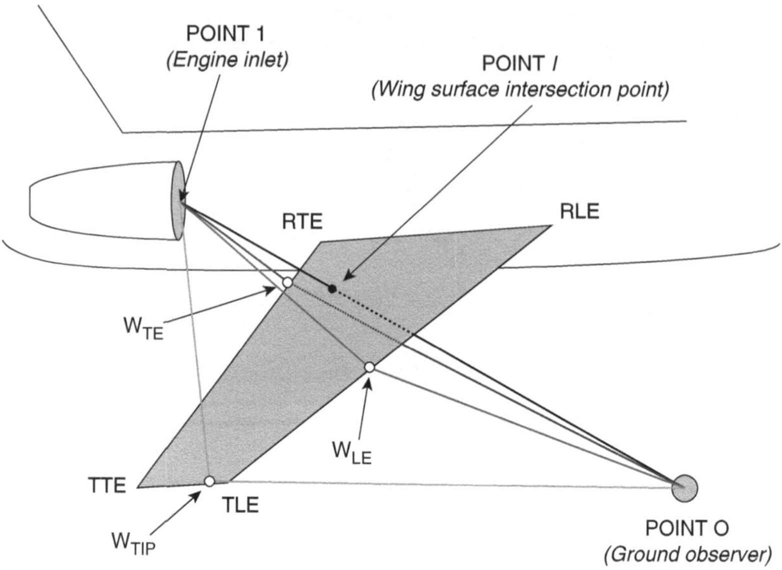


*Figure 4.42* Characteristic distances for wing diffraction assessment[23].

The corresponding Fresnel number is

$$N = 2f_i\delta/c, \tag{4.64}$$

where $f_i$ represents the frequency for each third-octave band, in hertz, and $c$ is the speed of sound in free space. The attenuation is then computed for each third-octave band frequency (see Chapter 3).

Predictions have been made of the diffraction (screening) effect of the wing on a Yakovlev-40 aircraft for sound propagated from the engine installed at the tail of the fuselage (see Figs 4.43 and 4.44). This aircraft has three engines in the power plant: one over the fuselage and one on either side of the fuselage. The side engines are mounted with their flow intakes over the wing such that the length $Delx_{WE} = 0.25$ m along the wing chord at engine half-span $z_{EN}$. The vertical displacement of the intakes over the wing $Dely_{WE} = 1.2$ m (see Fig. 4.44) is sufficiently small to cause a significant diffraction effect.

The output noise predictions are achieved by using the following steps:

(1) Obtain the geometry and received spectra data from the input files.
(2) For each reception time value, calculate the point $I$ and determine if the ray intersects the object of sound shielding or reflection or refraction.

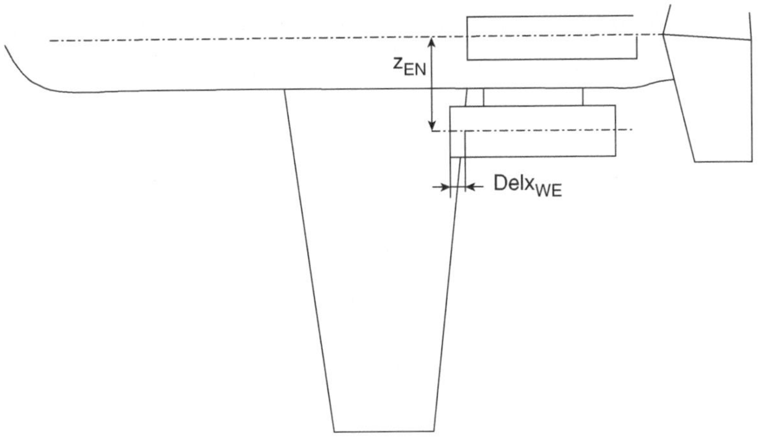


*Figure 4.43* Yakovlev-40 plan view used for calculating wing diffraction.

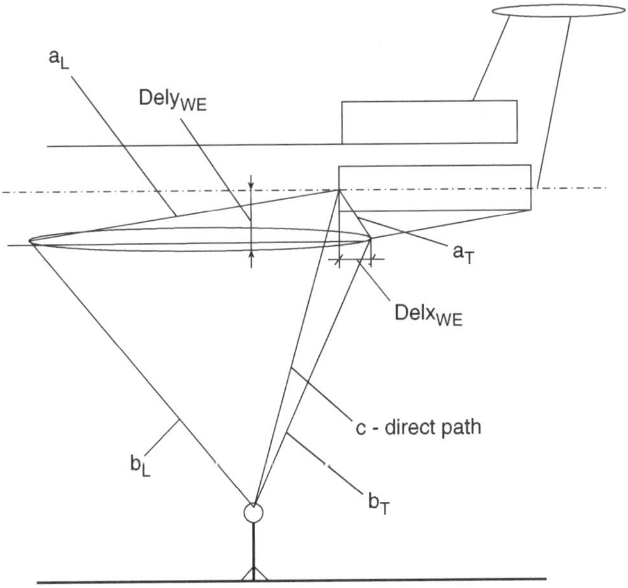


*Figure 4.44* Yakovlev-40 side elevation in flight used for calculating wing diffraction.

(3) Compute the sound effects (mostly attenuation) at the desired values of frequency.
(4) Apply the results to obtain the appropriate value of the received mean-square pressure.

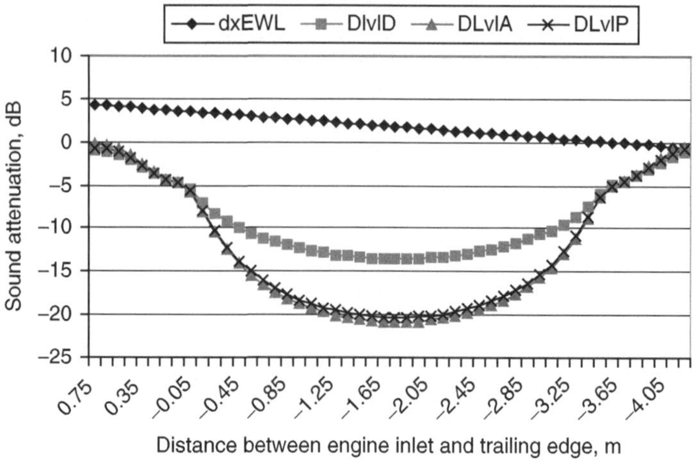


*Figure 4.45* Dependence of predicted attenuation of noise from a Yakovlev-40 aircraft on the distance between engine inlet and wing trailing edge (dxEWL – distance between engine inlet and wing leading edge, m).

The output values are the attenuated mean-square acoustic pressure values as a function of frequency (*SPL*), reception time and observer position. Then the overall *SPL* (*OASPL*), $L_A$ and *PNL* are calculated using well-known algorithms.

Figure 4.44 shows all of the possible sound paths for noise from the engine fan to an observer near to the ground. The length of the direct sound path to noise control point No. 3 is 109.41 m. The path through the trailing edge will be $(a_T + b_T) = 109.44$ m. Thus, for trailing edge diffraction $d_T = 0.06$ m. The path through the leading edge will be $(a_L + b_L) = 111.69$ m. Thus, for leading edge diffraction $d_L = 2.28$ m.

In reality, fan noise is only one of the sources contributing to the overall sound level of the aircraft. Nevertheless, it is interesting to consider the relative predicted contribution of shielded fan noise to the overall sound level for the specific case of a Yakovlev-40 aircraft (Fig. 4.45).

Because the bypass ratio of the engine under consideration is quite small, the contribution of fan noise to the total aircraft noise spectrum is not particularly large and the resulting influence of the wing shielding on the total aircraft noise is much less than that for fan noise alone.

Shielding of fan noise has been calculated for the approach flight of a Yakovlev-40 aircraft and the results of the corresponding noise calculations (*PNL* and $L_A$) at noise control point No. 3 (2000 m before the runway edge) are shown in Fig. 4.46, where the noise levels calculated without (*PNL* and $L_A$) and with ($PNL_{shield}$ and $L_{Ashield}$) sound shielding depend on the distance along the glide-path.

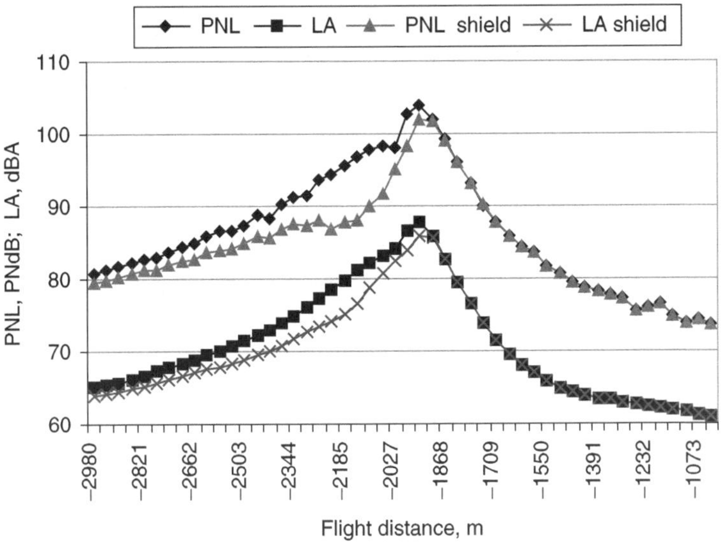


*Figure 4.46* The total leading and trailing edge wing attenuation for a Yakovlev-40 aircraft as a function of distance from runway edge at the moment during the flight when *PNL* = *PNLM*.

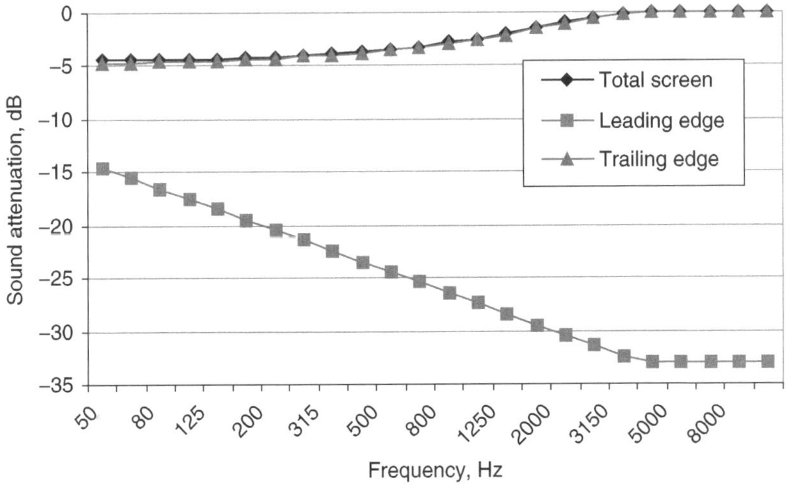


*Figure 4.47* Spectral sound attenuation for a Yakovlev-40 aircraft at the 2000m observation point at the moment of flight, when *PNL* = *PNLM*.

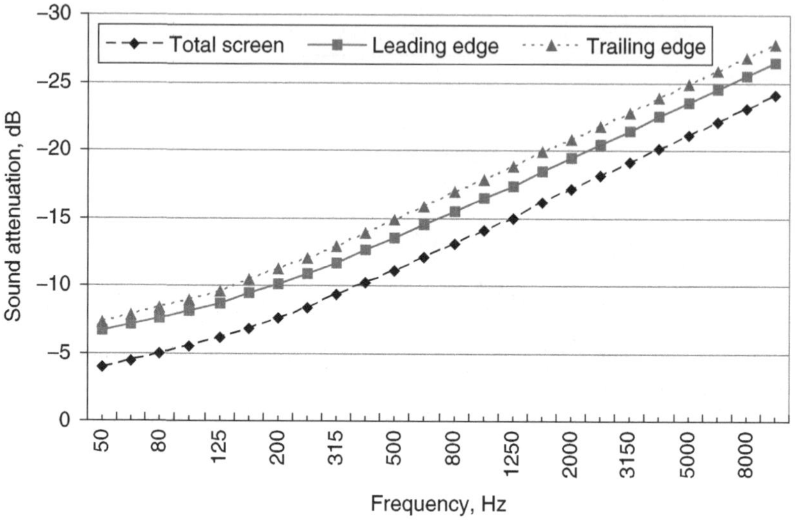


*Figure 4.48* Predicted total, leading and trailing edge wing attenuation spectra at distances between 2200 and 2150 m before the start of the runway.

The maximum screening effect (~5 PNdB and ~3 dBA; see Fig. 4.46) is predicted to occur at distances of between 2200 and 2150 m from the start of the runway. The corresponding spectral sound attenuation is shown in Fig. 4.47. The shielding has the following influences on flight event indices: $\Delta EPNL = 2.4$ EPNdB, $\Delta L_{AMAX} = 2.0$ dB, $\Delta PNLTM = 2.0$ PNdB.

During flights *without* sound shielding by the wings:

$$EPNL = 97.8\,EPNdB, L_{AMAX} = 87.8\,dB,\ \ PNLTM = 103.9\,PNdB,$$

whereas during flights *with* sound shielding by wings:

$$EPNL = 95.4\,EPNdB, L_{AMAX} = 85.8\,dB,\ \ PNLTM = 101.9\,PNdB.$$

The predictions in Fig. 4.46 show that, at the moment, $t$, of the flight when $PNL = PNLM$, the shielding effect $\Delta PNL$ is less than at distances between –2200 and –2150 m (the corresponding spectral sound attenuation is shown in Fig. 4.48). This is because the intersection point I for the latter situation is located at the middle of the chord of the wing thereby providing maximum shielding effect. Qualitatively and quantitatively, the spectral effect (Fig. 4.48) is the same as in ANOPP calculations.[23]

For calculating *wing reflection* effects, the origin of the local coordinate system is chosen to be at the engine exhaust (see Fig. 4.49). The wing (flap or flap tab) configuration is described in a local coordinate system with the origin positioned at the engine inlet (point 1). For the propagation

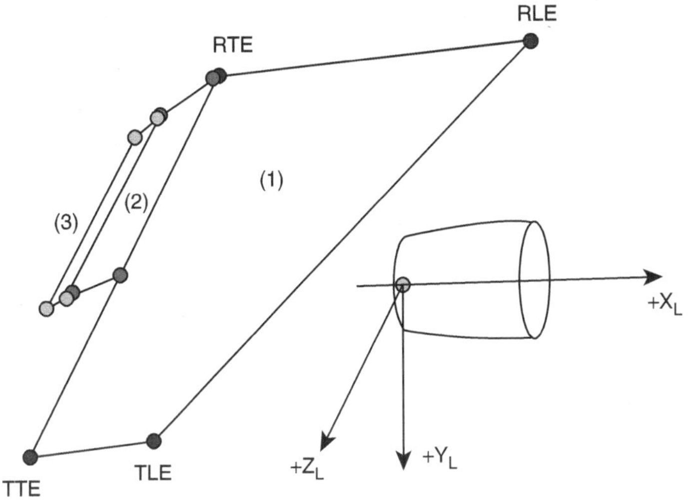


*Figure 4.49* The reflecting panel geometry is defined in terms of panel boundary points similar to those for wing shielding calculations shown in Figs. 4.41–4.43.

calculation, the local coordinate system origin is assumed to be at the aircraft location specified in the body coordinate system. The coordinates at the wing root leading edge, wing root trailing edge, wing tip leading edge and wing tip trailing edge, relative to the location of the local coordinate system origin must be specified.

The local origin and wing coordinates are transformed into a global coordinate system consistent with the observer location on the ground, as in the shielding model. This transformation must take into account the aircraft orientation and position at the particular time of the observation.

The reflecting panels that represent the wing/flap system are modeled as planar surfaces. The Cartesian coordinates are defined for the root leading edge (RLE), root trailing edge (RTE), tip leading edge (TLE) and tip trailing edge (TTE; see Fig. 4.49). Strictly speaking, three points are enough to define the plane of the wing, for example, RLE, RTE and TLE.

The reflection model uses an image source point, which is positioned on the normal to the reflecting plane at a distance equivalent to that between the source and the reflecting plane, but on the opposite side of the reflecting plane. The points *S* (source point) and *IS* (image source point) are equidistant from the reflecting panel. The normal between points *S* and *IS* does not have to intersect the wing panel itself, only the plane extending through the wing panel. Nevertheless, it is necessary that point W, the reflection point, is located inside the edges of the wing (see Fig. 4.50).

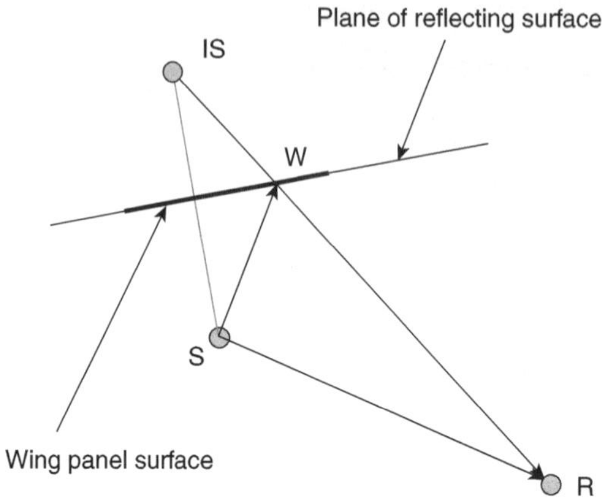


*Figure 4.50* The reflection point *W* must be inside the wing panel boundaries for a reflection contribution[23].

The coordinates of the reflection point W must be tested against the coordinates of the three nearest boundary points, to be sure that it is actually located on the wing (or flap reflecting surface, if it is considered). If the point W lies outside the wing panel boundaries, no contribution to reflected noise occurs, and the algorithm proceeds to consider the next reflection panel.

If a reflection has occurred for a particular wing panel, the path length of the reflected ray must be determined. The path length is the combined distance from the source to the reflection point (*SW*) and that from the reflection point to the receiver (*WR*), as shown in Fig. 4.51. The reflected ray path length is computed from:

$$d = \sqrt{(\Delta x_r)^2 + (\Delta y_r)^2 + (\Delta z_r)^2}, \tag{4.65}$$

where $\Delta x_I = x_{IS} - x_R$, $\Delta y_I = y_{IS} - y_R$ and $\Delta z_I = z_{IS} - z_R$.

To obtain the pressure at the ground receiver, the source strength of the reflected ray must be subjected to the same propagation effects as the directly radiated ray (i.e. atmospheric attenuation, spherical spreading and ground reflection). After the corrected contribution of each of the reflected rays has been computed, they must be combined with that of the direct ray, to obtain the total *SPL* at the receiver.

The combined contribution of rays from each of the reflecting panels is computed using the following equation, for every third-octave frequency band, *k*:

$$SPL_{k,REF,TOT} = 10 \lg \sum_{IPANEL=1}^{NPANEL} 10^{(SPL_{k,IPANEL}/10)}. \tag{4.66}$$

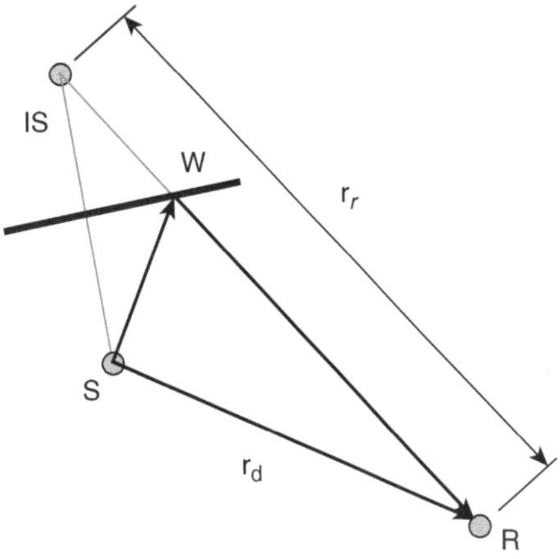


*Figure 4.51* The path length of the reflected ray is equivalent to the distance between the receiver $R$ and the image source $IS$[23].

The total contribution from reflected rays may then be combined with the direct ray contribution, to yield the total $SPL$ at the receiver. For every third-octave band, the total $SPL$ is given by:

$$SPL_{k,TOT} = 10\lg\left[10^{(SPL_{k,DIR}/10)} + 10^{(SPL_{k,REF,TOT}/10)}\right]. \tag{4.67}$$

Sound reflection by the wings is considered in the same manner as sound shielding. Predictions of linear $OASPL$, $L_A$ and $PNL$ have been obtained for a distance of 100 m, which is approximately equal to the distance for the approach noise control monitoring point. The method for the preparation of the predictions is as follows:

(1) obtain the geometry and received spectra data from the input files;
(2) for each reception time value, calculate the point W and determine if the ray intersects the object of sound reflection;
(3) compute the sound effects (mostly attenuation) at the desired values of frequency;
(4) apply the attenuation (or to the appropriate value of the received mean-square pressure.

The output values are the attenuated mean-square acoustic pressure values as a function of frequency ($SPL$), reception time and observer position. The same geometry of the wing as for sound-shielding effects has been analyzed for this case but the engines are considered under the wing.

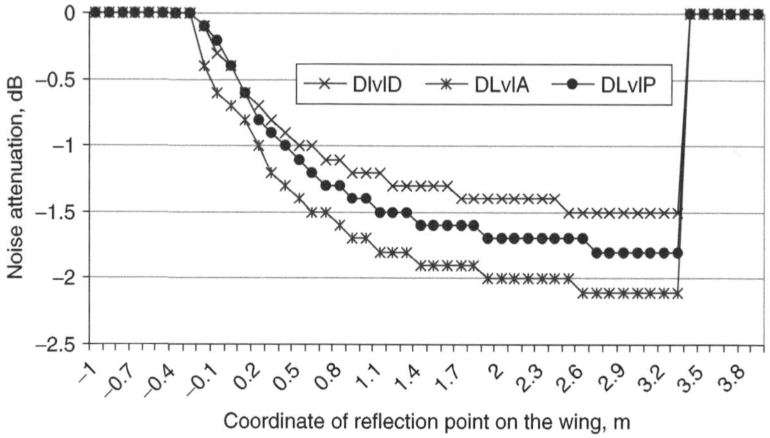


*Figure 4.52* Predicted directivity pattern of the wing reflection effect without refraction by exhaust jet.

The directivity patterns, like those used for the shielding effect for fan noise spectra, have been calculated for the reflection effect, and the results are presented in Fig. 4.52. Small decreases of the *OASPL* (DlvLD in Fig. 4.52), $L_A$ (DlvLA in Fig. 4.52), *PNL* (DlvLP in Fig. 4.53) are observed for wings of moderate size.

### 4.5.4 Refraction through jet exhaust

The influence of exhaust jet refraction on the directivity pattern is shown in Fig. 4.53, where $W_{trans} = 1$ if the point of sound reflection is less than the point of noise control $x_W < x_{Contrl}$ (clear reflection) and $W_{trans}$ is given by expression (4.66) if $x_W > x_{Contrl}$.

The influences of the sound reflection by the wing and the refraction by the jet on flight noise levels have been calculated, and flights along the glide slope have been considered. A comparison of the levels predicted without these effects, for sound reflection alone and for wing reflection and jet refraction, is shown in Fig. 4.54. For the assumed aircraft and engine parameters considered in this example, the main effect (an overall increase in *PNL*) is predicted to be between 2000 m and 1900 m from the noise control point. The qualitative dependence is the same as that obtained elsewhere[23] but there is a quantitative difference, which may be explained by differences in the geometry and engine mode parameters of the aircraft considered.[23]

The predicted changes in flight noise spectra at the *PNLM* point produced by reflections of fan noise are shown in Fig. 4.55. Since the predicted changes

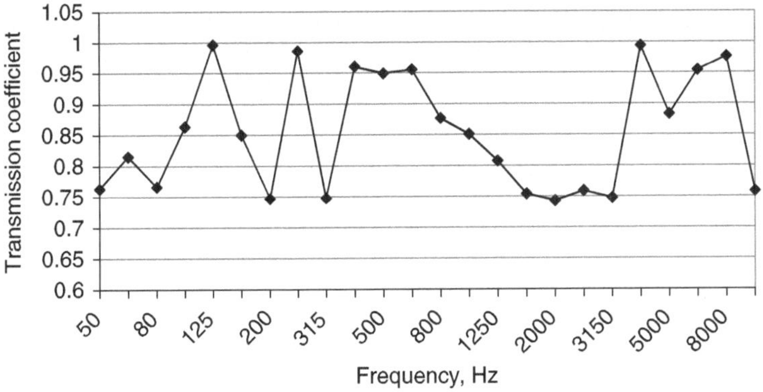


*Figure 4.53* Sound transmission coefficient for the engine jet in descending flight mode.

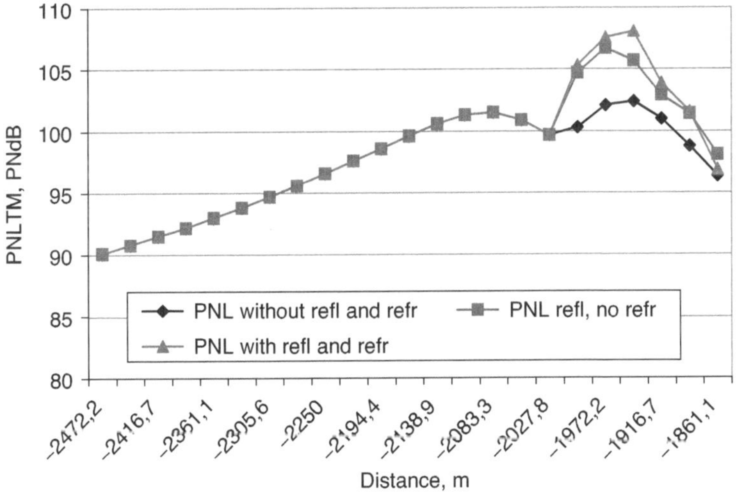


*Figure 4.54* Comparison of *PNL* and $L_A$ indices predicted without reflection and refraction, for sound reflection alone and including both of them.

are observed at high frequencies, the effect on *PNL* is more evident. The assumed engine parameters are those for an engine with moderate bypass ratio (~2.5).

All the predictions obtained using the models and subprogram (written in Fortran) must be validated by comparison with measurements and this is the subject of the next stage of the investigation.

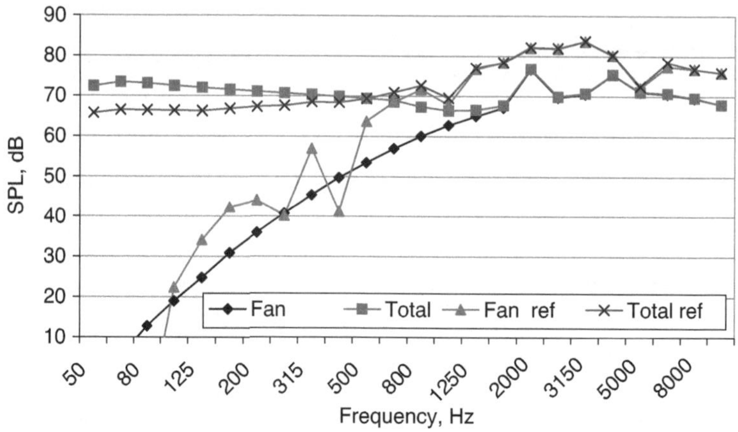


*Figure 4.55* Comparison of noise levels predicted for fan and aircraft (total) without and with (ref) sound reflection and refraction.

The overall noise levels predicted for the analyzed flight event are:

- without wing reflection and jet refraction

$$EPNL = 99.7\text{EPNdB}, L_{AMAX} = 87.0\text{dB}, PNLT_{MAX} = 102.4\text{PNdB};$$

- with wing reflection but no jet refraction

$$EPNL = 101.3\text{EPNdB}, L_{AMAX} = 91.0\text{dB}, PNLT_{MAX} = 106.7\text{PNdB};$$

- with wing reflection and jet refraction

$$EPNL = 102.0\text{EPNdB}, L_{AMAX} = 91.0\text{dB}, PNLT_{MAX} = 108.1\text{PNdB}.$$

These values are comparable with data for noise levels observed during the approach of the tested aircraft.[23] This suggests that the models for sound reflection and refraction are correct and that they have been applied correctly to flight noise prediction.

### *4.5.5 Refraction, interference and comparisons with data*

The effects of sound reflection by aircraft wings along the propagation path have been computed and compared with measurements obtained during flight trials corresponding to noise certification procedures involving takeoff and landing. Sound reflection by the wings has been shown for discrete tones corresponding to the first fan rotor harmonics. Two types of data have been used to test the validity of the models: (1) flight path data for the aircraft under consideration at takeoff and landing stages; and (2) the corresponding noise levels measured at a few noise control points.

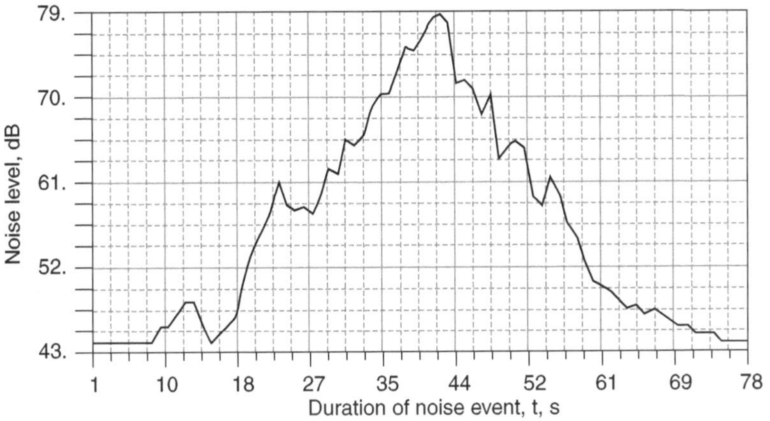


*Figure 4.56* *SPL* (dB) as a function of time (seconds) in the 1600 Hz third-octave band measured at point No. A1 during a noise certification flight.

Analysis of data in the spectrum bands, which includes the discrete tones from fan noise radiation, should reveal the effects of sound reflection by wings. For example, during landing, such a band may be the 1600 Hz third-octave band, which includes the first harmonic of the fan rotor. Figures 4.56 and 4.57 show the measured *SPL* as a function of flight time in the 1600 Hz third-octave band at two noise control points. The corresponding 'C'-weighted predictions without and with wing reflections are shown in Figs 4.58 and 4.59. Dips at times (in seconds) between 16 and 17, 26 and 27, 46 and 47, 53 and 54 are measured and predicted for every event. They are produced by ground (interference) effects and they confirm the validity of the lateral attenuation models used. But, more interestingly, comparisons of predictions in Figs 4.58 and 4.59 with and without wing reflections suggest that the wing reflections may be the cause of the peaks.[23]

For the aircraft under consideration, the design of the engine installation under the wing provides the possibility to influence the reflected ray path (Fig. 4.60), which may reach the microphone just as the aircraft is over the point of noise measurements. The measurements and predictions in Figs 4.56–4.59 suggest that such reflections exist during the time intervals between 3 and 4 s or 7 and 8 s. The reflected ray must pass through the jet exhaust, thus it may be refracted depending on the jet temperature and speed (Fig. 4.61).

To calculate the refraction effect, the engine jet can be treated as a homogeneous moving layer. Consider the reflection of a plane wave by a homogeneous moving layer located between the planes $z = 0$ and $z = d$ (Fig. 4.62). The wave theory of sound propagation in stratified moving media gives models for wave reflection and transmission.[24,25]

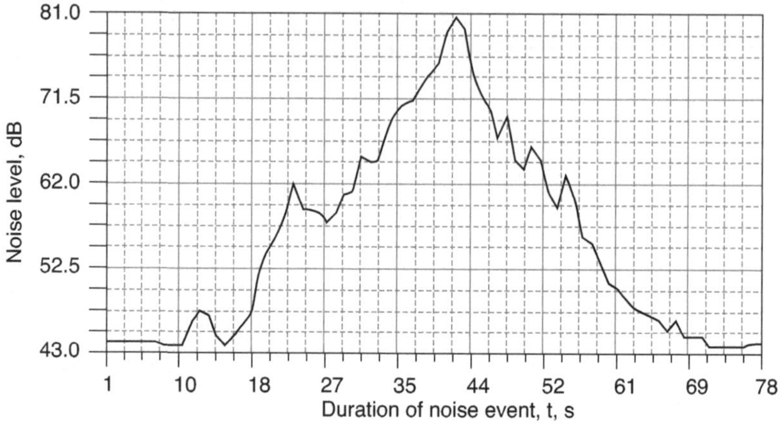


*Figure 4.57* *SPL* (dB) as a function of time (seconds) in the 1600 Hz third-octave band measured at point No. A2 during a noise certification flight.

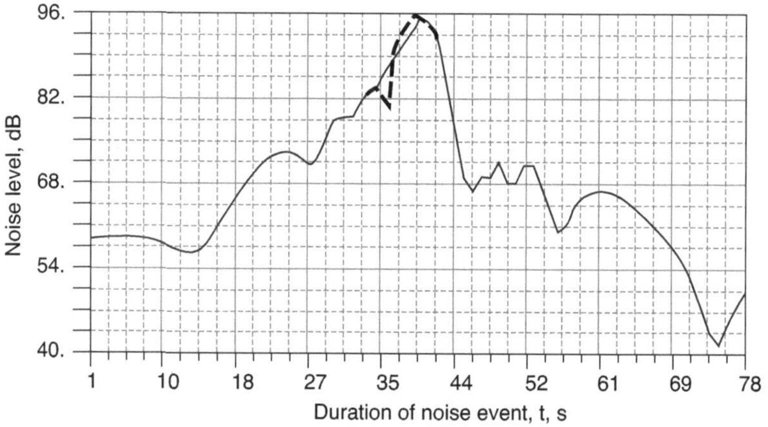


*Figure 4.58* *SPL* (dB) as a function of time (seconds) in the 1600 Hz third-octave band predicted at point No. A1 during a noise certification flight: blue line – *SPL* predictions without the ray reflected by wing; green line – *SPL* predictions including the ray reflected by wing.

The coefficient of sound wave transmission through the layer can be expressed in the form:[25]

$$W = \frac{4Z_2Z_3}{(Z_2 - Z_1)(Z_3 - Z_2)\exp(iq_2d) + (Z_2 + Z_1)(Z_3 + Z_2)\exp(-iq_2d)}, \tag{4.68}$$

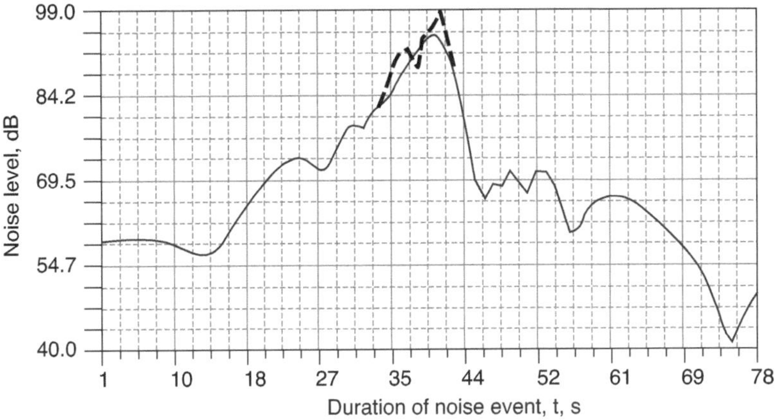


*Figure 4.59* $SPL$ (dB) as a function of time (seconds) in the 1600 Hz third-octave band predicted at point No. A2 during a noise certification flight: continuous line – $SPL$ predictions without the rays reflected by the wings; broken line – $SPL$ predictions including the rays reflected by the wings.

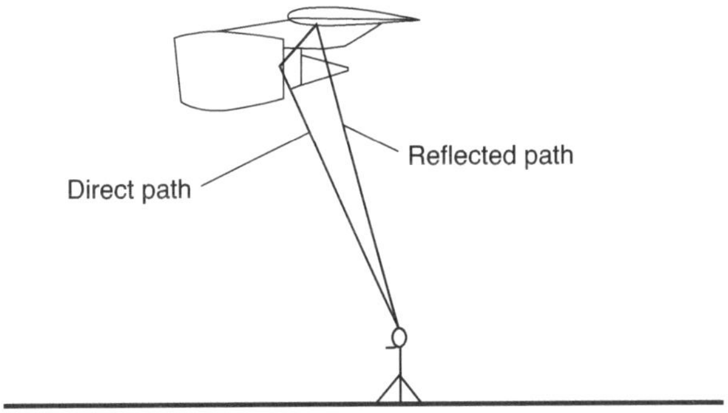


*Figure 4.60* Direct and reflected rays from the wing.

where $Z_i$ is the characteristic impedance of a moving medium, $q_2 = atg\theta_2$. From the equation above it follows that the dependence of the transmission coefficient $W$ on the thickness $d$ of the layer and the vertical wave-number $q_2 = atg\theta_2$ has an oscillatory character.

In the case of $v_1 \neq 0$, $v_3 \neq 0$

$$Z = \frac{\rho c}{\sin\theta(1 + \cos\theta e \cdot v/c)}.$$

For the simple case of constant (frequency-independent) impedance, the problem is reduced to that of plane wave reflection by an interface between

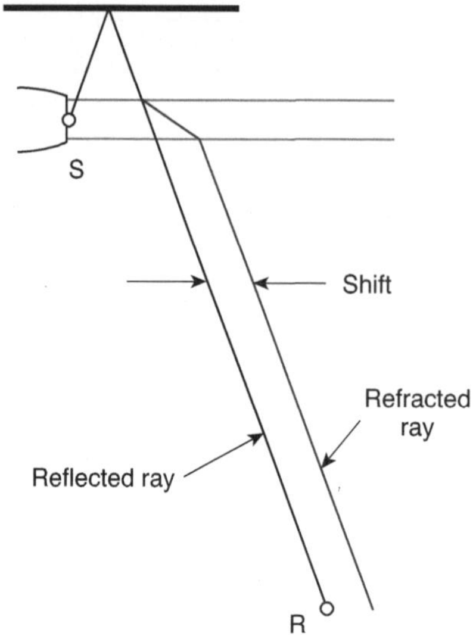


*Figure 4.61* A wing-reflected ray experiences refraction as it passes through the exhaust jet[23].

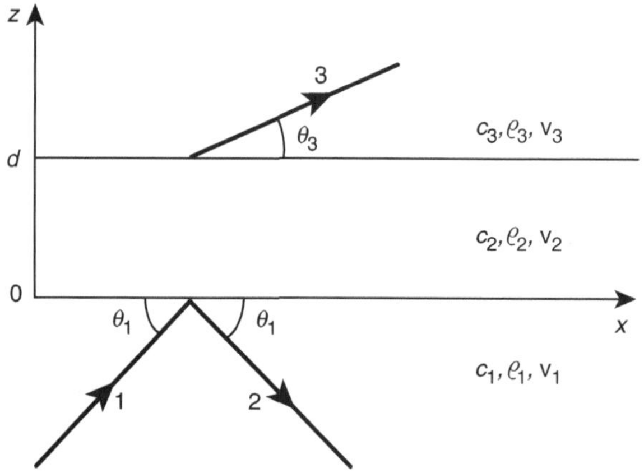


*Figure 4.62* Propagation of a plane wave through a homogeneous moving layer: 1 – incident wave; 2 – reflected wave; 3 – transmitted wave.

moving media, in this case

$$W = 4Z_2Z_3/[(Z_3 + Z_2)(Z_2 + Z_1)], \tag{4.69}$$

where $Z_i = \frac{\rho_i c_i}{\sin\theta_i}$ for $v_1 = v_2 = 0$.

*Table 4.20* Jet exhaust transmission coefficients for wing reflections during landing

| *Frequency (Hz)* | *Transmission coefficient* (W) | | *Absolute value of* W |
|---|---|---|---|
| | *Real part* | *Imaginary part* | |
| 50 | 0.29380 | 0.67664 | 0.73767 |
| 63 | 0.14731 | 0.67888 | 0.69468 |
| 80 | −0.01373 | 0.67694 | 0.67708 |
| 100 | −0.21095 | 0.67925 | 0.71125 |
| 125 | −0.56382 | 0.62070 | 0.83854 |
| 160 | −0.99399 | −0.08807 | 0.99789 |
| 200 | −0.38837 | −0.66680 | 0.77166 |
| 250 | 0.13708 | −0.67872 | 0.69242 |
| 315 | 0.99934 | 0.02917 | 0.99977 |
| 400 | −0.06920 | 0.67751 | 0.68104 |
| 500 | −0.61737 | −0.59739 | 0.85908 |
| 630 | 0.99738 | 0.05825 | 0.99908 |
| 800 | −0.86631 | −0.39520 | 0.95219 |
| 1,000 | 0.20041 | 0.67929 | 0.70824 |
| 1,250 | 0.97585 | −0.17538 | 0.99149 |
| 1,600 | 0.60026 | 0.60539 | 0.85253 |
| 2,000 | −0.41650 | 0.66209 | 0.78219 |
| 2,500 | 0.90904 | −0.33166 | 0.96765 |
| 3,150 | 0.93782 | 0.27734 | 0.97797 |
| 4,000 | −0.03363 | −0.67707 | 0.67790 |
| 5,000 | 0.70449 | −0.54745 | 0.89220 |
| 6,300 | 0.78534 | 0.48389 | 0.92245 |
| 8,000 | −0.96498 | 0.21033 | 0.98764 |
| 10,000 | 0.31424 | −0.67522 | 0.74476 |

Corresponding predictions are shown in Table 4.20.

The interference between the direct and wing-reflected rays from the engine may be calculated using the same model as for the ground effect, but for an acoustically hard surface (i.e. $\sigma_e = 100,000$ kPa s/m$^2$ in the Delany and Bazley impedance model [see equation (3.29)].

The resulting theory of wing-reflected engine noise propagation predicts a change in the propagation effects around the point of maximum *OASPL* as shown in Table 4.21.

Similar effects have been found in predictions and measurements of the spectra in particular bands during other flight tests (Figs 4.63–4.70).

The effect of engine installation geometry has been calculated using the example of the discrete tone radiated by fan rotor during landing. The observed value of the effect is of the same order as predicted. For broadband noise, the same effect is present but it is masked by oscillations produced by variations of engine mode and meteorological conditions.

*Table 4.21* Predicted sound divergence, air absorption and lateral attenuation effects in the 1600 Hz–band of the spectrum around the point of maximum *OASPL* (point corresponding to time = 41 s)

| *Time (s)* | *37* | *38* | *39* | *40* | *41* | *42* | *43* |
|---|---|---|---|---|---|---|---|
| Divergence (dB) | 45.0 | 43.3 | 41.6 | 40.3 | 40.0 | 41.0 | 42.5 |
| Absorption (dB) | 1.12 | 0.92 | 0.76 | 0.65 | 0.63 | 0.70 | 0.84 |
| Lateral attenuation (dB) | 3.21 | 2.57 | 2.57 | 3.30 | 3.31 | 3.30 | 2.84 |
| Reflection by wing effect (dB) | 0.0 | −1.9 | −5.8 | 0.3 | 4.1 | 0.1 | 0 |

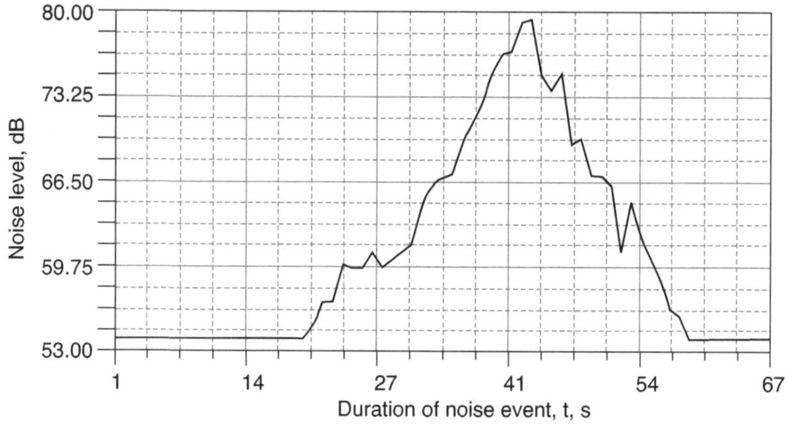


*Figure 4.63* *SPL* (dB) as a function of time (seconds) in the 1600 Hz third-octave band measured at point No. A1 during over-flight No. 4 test 77.

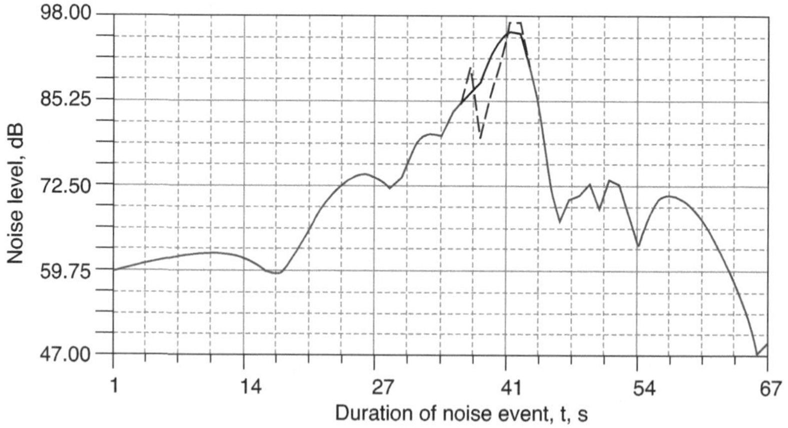


*Figure 4.64* *SPL* (dB) as a function of time (seconds) in the 1600 Hz third-octave band predicted at measurement point No. A1: blue line – *SPL* predicted without the ray reflected by wing; green line – *SPL* predicted including the ray reflected by wing.

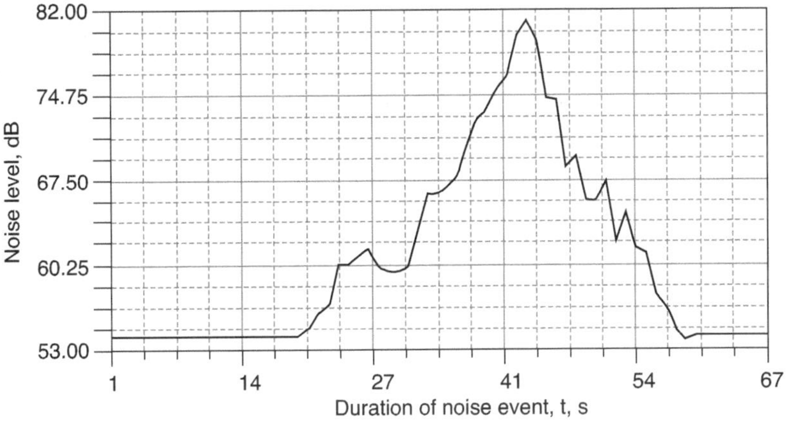


*Figure 4.65* *SPL* (dB) as a function of time (seconds) in the 1600 Hz third-octave band measured at point No. A1 during over-flight No. 4 test 77.

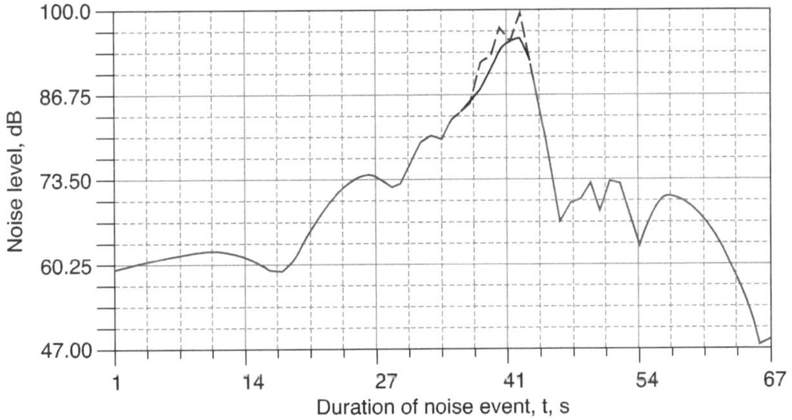


*Figure 4.66* *SPL* (dB) as a function of time (seconds) in the 1600 Hz third-octavem band predicted at measurement point No. A2: blue line – *SPL* predictions without the rays reflected by wings; green line – *SPL* predictions including the rays reflected by wings.

### *4.5.6 Scattering of sound by the fuselage*

A long rigid cylinder of radius $a$ may be considered for modeling the scattering of sound by the fuselage (Fig. 4.71). The point $M(r, \varphi, z)$ is the observation point (the location of the noise receiver). The pressure at

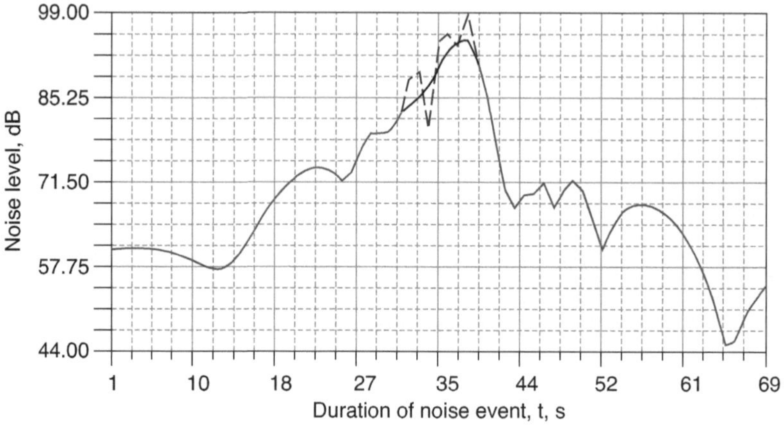


*Figure 4.67* *SPL* (dB) as a function of time (seconds) in the 1600 Hz third-octave band predicted at measurement point No. A1 for over-flight No. 3 test 77: blue line – *SPL* predictions without the rays reflected by wings; green line – *SPL* predictions including the rays reflected by wings.

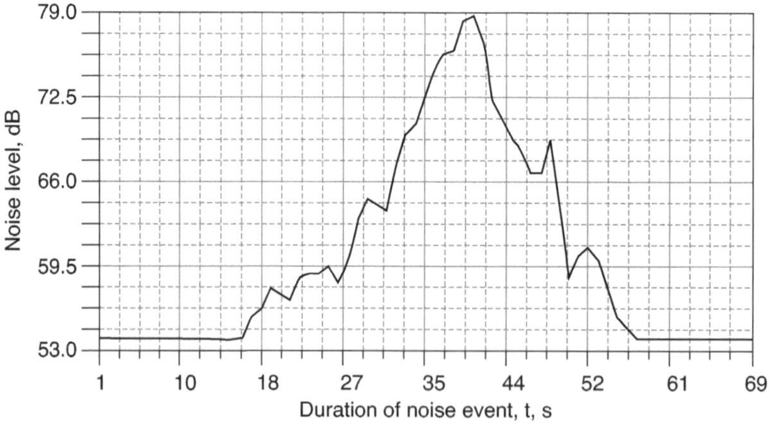


*Figure 4.68* *SPL* (dB) as a function of time (seconds) in the 1600 Hz third-octave band measured at point No. A1 during over-flight No. 3 test 77.

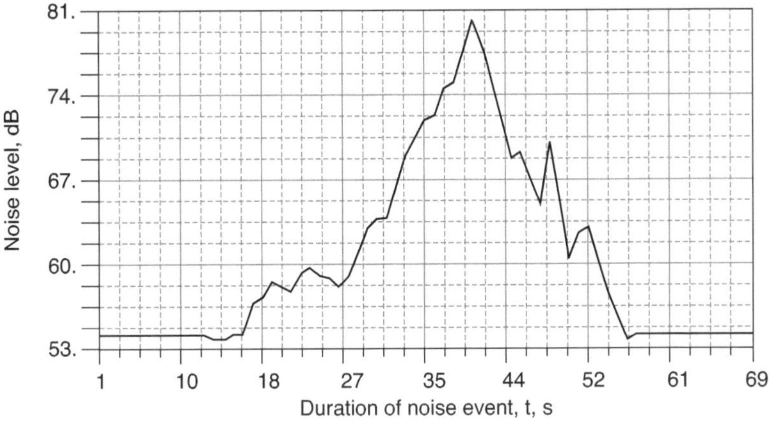


*Figure 4.69* *SPL* (dB) as a function of time (seconds) in the 1600 Hz third-octave band measured at point No. A2 during over-flight No. 3 test 77.

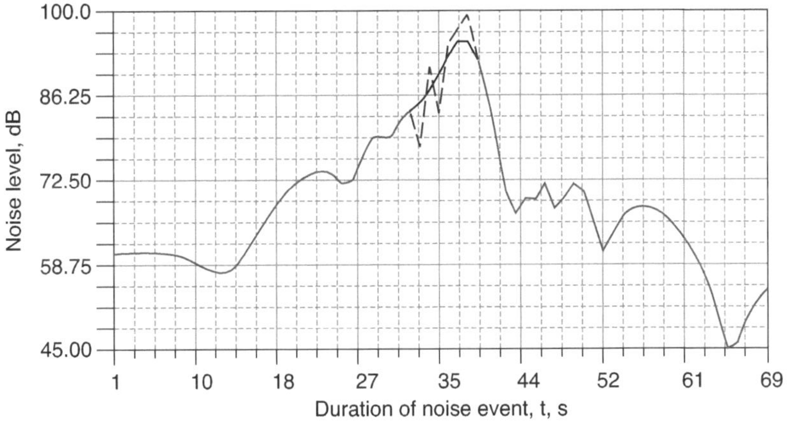


*Figure 4.70* *SPL* (dB) as a function of time (seconds) in the 1600 Hz third-octave band predicted at measurement point No. A2 for over-flight No. 3 test 77: blue line – *SPL* predictions without the rays reflected by wings; green line – *SPL* predictions including the rays reflected by wings.

distance $R$ from the cylinder is given by ($kR >> 1$):

$$p(R,\varphi,\beta) = \frac{Q\rho_0 c}{2\pi^2 a\cos\beta}\frac{\exp[ik(R - z_0\sin\beta)]}{R} \sum_{n-0}^{\infty}\varepsilon_n(-i)^n\frac{[\cos n(\varphi-\varphi_1)+\cos n(\varphi-\varphi_2)]}{H_n^{(1)'}(ka\cos\beta)}, \tag{4.70}$$

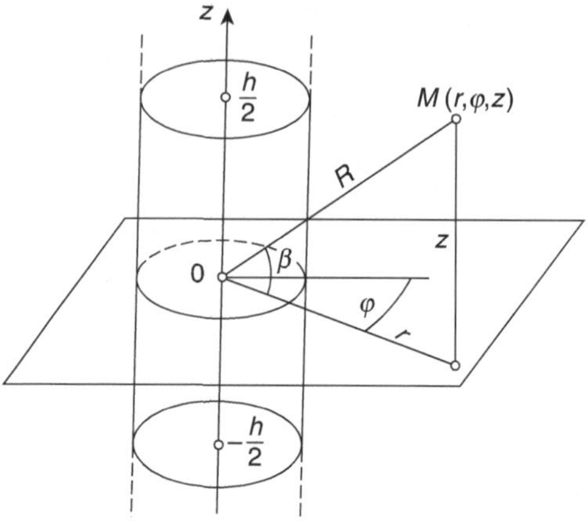


*Figure 4.71* Fuselage scattering scheme in cylindrical co-ordinates.

where angles $\beta$ and $\varphi$ are shown in Fig. 4.71. The angles $\varphi_1$ and $\varphi_2$ define the position of the source and receiver; $Q$ is the sound power of the source; $\rho_0 c$ is the characteristic impedance of the air; and $H_n^{(1)'}$ represents the derivative of the Hankel function of the first order.

The directivity pattern of the effect (the absolute value of which is defined by expression (4.67) is given by:

$$R(\varphi) = \left| \frac{p_0(\varphi)}{p_0(\varphi_1)} \right|, \tag{4.71}$$

where the source is considered to be on the opposite side to the receiver point. Thus $\varphi_1 = 0$ and $\varphi_2 = 180$.

The influences of sound frequency, longitudinal direction to receiver point and cylinder radius $a$ have been investigated numerically. If the angle $\beta$ becomes larger, the effect becomes smaller because the shadow from the cylinder decreases. Predictions for $a = 2$ m and various values of $\beta$ are shown in Fig. 4.72. For higher values of the angle $\varphi$, the shadow effect of the cylinder is higher too.

## 4.6 Prediction of aircraft noise during ground operations

There are two types of noise propagation relevant to aircraft noise assessment procedures. In *air-to-ground propagation*, the height of the noise source is much greater than the assumed height of a noise receiver (usually 1.2 m) because the airplane is in flight (Section 4.4). *Ground-to-ground propagation* is relevant to: (1) assessment of noise from aircraft on the ground since, for this case, the height of the noise source is equal to the height

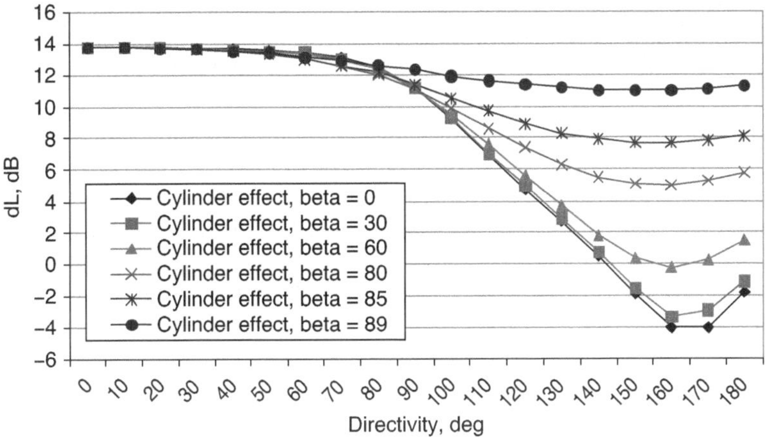


*Figure 4.72* Dependence on $\beta$ of the predicted reduction of *OASPL* (directivity pattern 4.71) due to the cylinder shielding, for $a = 2$ m.

of engine installation; and (2) noise from aircraft engines testing facilities since usually in this testing the height is between 2 and 5 m above the ground. It is obvious that sound wave interference, refraction and diffraction effects are stronger in the ground-to-ground case of noise propagation.

Results of surveys around 16 airports have shown that the most irritating noise inside the territory of the airport, in the opinion of 73 per cent of the interrogated aviation personnel, is that from engine testing operations (engine run-ups). Relatively small fractions of interrogated aviation personnel were disturbed by aircraft takeoff and landing (9 per cent), aircraft taxiing (7.5 per cent), auxiliary power units (4 per cent), noise of ground service equipment that facilitate aircraft operation and maintenance (3.5 per cent) or other noise sources (3 per cent). In contrast, surveys of community response in the vicinity of airports have found that aircraft takeoff and landing operations are the predominant sources of annoyance.

At the same time, it is common for engine run-ups to be performed at locations near to the airport boundaries and not far from the surrounding habitations. Engine modes that are used during the trials include all possible operational modes for the considered engine, from the lowest to the maximum power modes, including that usually used during the takeoff (on a runway and during the first stage of the flight). Thus, the problem of run-up noise is included in AN calculations because it may have an important influence on the results of AN contour assessment for the airport under consideration. The difficulties concerning the separate analyses of directivity of sound radiation by engines, sound reflection (interference), refraction and diffraction effects, are overcome by including all of them in one quantity – the directivity pattern of the engine (or aircraft as a whole) the values of which are used in AN calculations. Examples of the forms of directivity patterns for aircraft with specific types of engines are shown in Fig. 4.73a–c.

Noise levels from aircraft engine trials, engine run-ups, etc., at a particular point $(x, z)$ of the calculation grid are derived from the formula:

$$L(x,z) = L_{DP} + \Delta L_M + \Delta L_t + \Delta L_s, \tag{4.72}$$

where $L_{DP}$ is the noise level derived from the appropriate directivity pattern, $L_t$ includes the noise generated at maximum engine thrust in the particular direction of noise propagation (defined by the angle of directivity), spherical divergence, absorption and extra attenuation along the propagation path with distance $R$ (using the directivity patterns, like those shown in Fig. 4.73). $L_{DP}$ is interpolated between the distances between 1 m and 3000 m, and

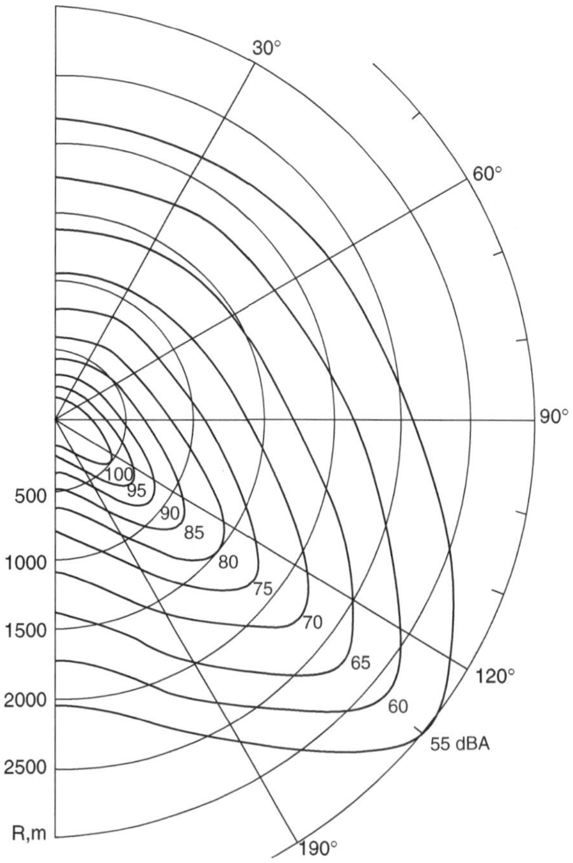


*Figure 4.73* (a) Jet engine noise directivity pattern; (b) bypass engine noise directivity pattern; (c) turboprop engine noise directivity pattern.

extrapolated for distances over 3000 m. $\Delta L_M$ is the correction for the difference between the engine mode and the maximum thrust. $\Delta L_t$ is a correction for the duration of run-up/trials procedures and $\Delta L_s$ is the attenuation produced by any noise screens. Time corrections $\Delta L_t$ are made only when calculating equivalent sound levels like $L_{Aeq}$ or $L_{DN}$. They are zero for $L_{Amax}$ assessments.

Equation (4.72) is used for computing overall noise indices and is not appropriate for spectral *SPL*. Thus the approach realized by equation (4.72) is in accordance with the second level of AN modeling (Section 4.1). The *SPL* spectra vary greatly with directivity angle (in the same manner as the maximum sound level in Fig. 4.73b for a low bypass engine NK-86), as do other contributions to the total level at point of noise control.

The fact that 'lateral attenuation' is a function of sound propagation geometry has been recognized in aircraft flyover noise measurements too. Spectral features (interference dips and/or peaks) can vary during a flyover event. It has also been recognized that spectral features of noise generation by aircraft can influence AN assessments, but limitations in calculation

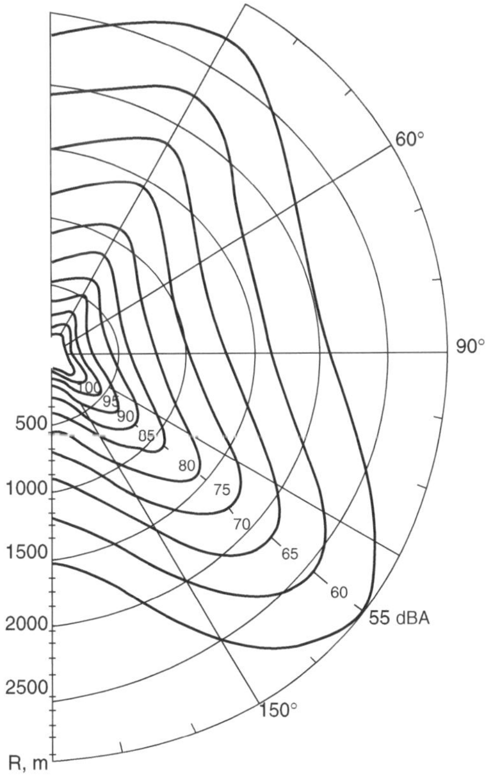


*Figure 4.73 Continued*

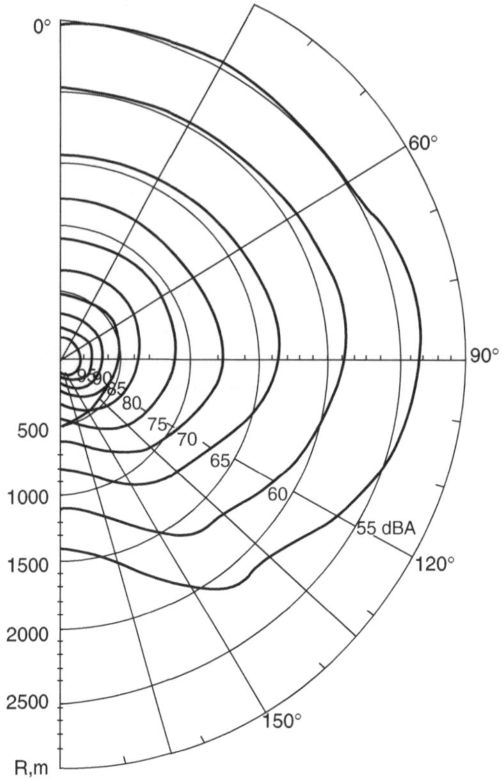


*Figure 4.73 Continued*

possibilities have not enabled incorporation of these variations in traditional prediction methods.

Nevertheless, AN assessment at the point of noise control should account for the impedance characteristics of the reflecting surfaces under consideration. In reality there are no absolutely rigid reflecting surfaces and the amplitude and phase of the reflected wave are affected during sound wave interactions with surfaces. The phenomenon is described by use of the acoustic impedance $Z$, which is one of parameters that characterizes the acoustical properties of the reflecting surface under consideration. Specific (per unit area) acoustic (surface) impedance is defined as the complex ratio of the effective sound pressure at a point of an acoustic medium to the effective particle velocity, normal to the surface at the point and has units of $Ns/m^3$. It is usual to state the surface impedance normalized to the impedance of the air, $Z = Z_s/\rho c$, where $\rho$ is the air density and $c$ is the speed of sound in air.

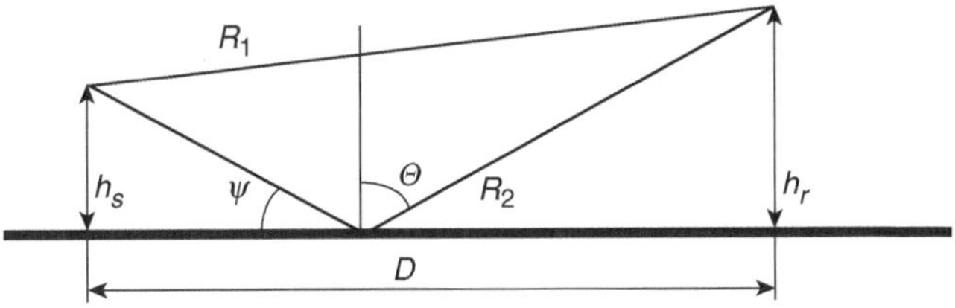


*Figure 4.74* Geometrical parameters used in predicting the sound field from a source and receiver above a reflecting plane.

For monopole type sources, the spectral interference effect $\Delta SPL$ in each frequency band, $\Delta f$, may be defined by:[26]

$$\Delta SPL = 10\lg\{1 + (\frac{R_1}{R_2})Q^2 + \frac{2R_1}{R_2}|Q|\frac{\sin[k(R_2 - R_1)\Delta f]}{k(R_2 - R_1)}$$

$$\cos[k(R_2 - R_1) + \varphi]\}, \quad (4.73a)$$

where $Q$ is the spherical wave reflection coefficient of the surface and the distances $R_2$ and $R_1$ are shown in Fig. 4.74.

The spherical reflection coefficient $Q$ is defined by:

$$Q = Q_p + (1 - Q_p)F(p_e), \quad (4.73b)$$

where $Q_p$ is the plane wave reflection coefficient predicted by

$$Q_p = \frac{Z\sin\psi - 1}{Z\sin\psi + 1} = \frac{\beta - \cos\theta}{\beta + \cos\theta}, \quad (4.73c)$$

The angles $\theta$ and $\psi$ are also shown in Fig. 4.74.

$F(p_e)$ is the boundary loss function. It is a function of the *numerical distance*, $p_e$, and represents the result of interaction between the curved wavefront with the flat reflecting surface:

$$F(p_e) = 1 + ip_e^{1/2}\exp(-p_e)\mathrm{erfc}(-ip_e);$$

$$p_e = (ikR_2/2)^{1/2}[\beta + \cos\theta]. \quad (4.73d)$$

For certain forms of ground impedance, the boundary loss factor includes a surface wave term. This gives rise to the possibility that low-frequency peaks in the excess attenuation spectra may have a magnitude that exceeds +6 dB (i.e. more than the expected pressure doubling over a perfectly reflecting surface).

Validation of the ground effect model has been performed in several ways. First, the results of predictions have been compared with data.

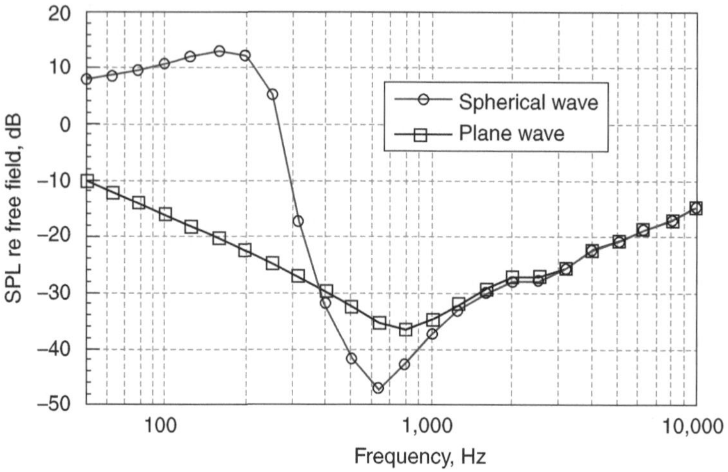


*Figure 4.75* Comparison of excess attenuation spectra predicted by spherical and plane sound wave models (distance between source and receiver $R = 20\,\text{m}$, heights of source and receiver $H_s = H_r = 1\,\text{m}$).

Second, the predictions derived using (4.73) have been compared with predictions of more complicated numerical solutions.[26–28] For example, in Table 4.22 the predictions of equation (4.73) are compared with results of numeral integration of the exact Sommerfeld solution and with results of the asymptotic solution of Kawai.[29] For these predictions, the reflecting surface admittance (admittance is the inverse of impedance) has been described by the Thomasson model,[30] that is,

$$\beta = [1 + i10000/(2\pi f)]^{-1/2}\, tg\left\{[1 + i10000/(2\pi f)]^{1/2}\frac{f}{800}\right\}/i.$$

This admittance is such that a large low frequency surface wave component is evident below 200 Hz (see Fig. 4.75).

The various predictions from equation (4.73) for a spherical wave reflection corresponding to those in Table 4.22 are compared with those for a plane wave in Fig. 4.75. The importance of the boundary loss factor $F(p_e)$ in a given case depends on the source–receiver geometry. For Fig. 4.75, the elevation angle $\beta_0$ is close to zero since the source and receiver heights are relatively small. If the elevation angle is increased beyond 5–10 degrees, the dominance of boundary loss factor is reduced and the excess attenuations for spherical and plane waves are numerically the same.

To examine the influence of excess attenuation on values of noise indices, it is necessary to consider the noise source spectrum. The prediction model for excess attenuation is realized in subroutine LATER of the programs NOITRA, BELTASS and BELTRA. It enables prediction of the spectral

*Table 4.22* Comparison of numerical predictions of various 'lateral attenuation' models for $R = 20$ m, $h_s = h_r = 0.1$ m over ground impedance given by Thomasson[30]

| *Frequency (Hz)* | *Numerical exact solution* | *Kawai solution* | *Equation (4.73)* |
|---|---|---|---|
| 50 | 7.90 | 7.90 | 7.89 |
| 63 | 8.60 | 8.60 | 8.59 |
| 79 | 9.53 | 9.53 | 9.50 |
| 100 | 10.68 | 10.69 | 10.69 |
| 126 | 11.98 | 11.99 | 12.00 |
| 159 | 13.00 | 13.00 | 13.03 |
| 200 | 12.38 | 12.38 | 12.34 |
| 252 | 5.71 | 5.71 | 5.20 |
| 317 | −16.50 | −16.48 | −17.52 |
| 400 | −32.11 | −32.09 | −32.04 |
| 504 | −42.02 | −41.97 | −41.88 |
| 635 | −47.06 | −47.11 | −47.07 |
| 800 | −42.68 | −42.82 | −42.87 |
| 1008 | −37.36 | −37.42 | −37.32 |
| 1270 | −33.20 | −33.23 | −33.16 |
| 1600 | −30.04 | −30.06 | −30.00 |
| 2016 | −28.08 | −28.10 | −28.04 |
| 2540 | −27.91 | −27.93 | −27.85 |
| 3200 | −25.82 | −25.82 | −25.74 |
| 4032 | −22.47 | −22.47 | −22.50 |
| 5080 | −21.06 | −21.06 | −21.00 |

components $\Delta SPL_{INT}(f)$ in any frequency band according to the geometry of interest.

Equation (4.73) predicts the mean-squared pressure at a receiver in a turbulent but acoustically neutral atmosphere (i.e. without refraction).

The coefficient $F$ determines the coherence between the direct and ground-reflected rays, which give rise to the interference effect. It is the product of several coherence coefficients:

$$F = F_f F_{\Delta\tau} F_c F_r F_{sc}, \tag{4.74}$$

where $F_f, F_{\Delta\tau}, F_c, F_r, F_{sc}$ are coherence coefficients corresponding, respectively, to frequency band averaging, averaging due to fluctuating refraction, partial incoherence (e.g. turbulence), surface roughness and scattering zones.

For third-octave bands:[30]

$$F_f = \frac{\sin[k(R_2 - R_1)0.115]}{k(R_2 - R_1)0.115}.$$

If $\tau_1$ is the travel time corresponding to the direct ray and $\tau_2$ to the reflected ray, a solution can be obtained for fluctuating refraction $F_{\Delta\tau}$, for a fixed wave number $k$ if it is assumed that the travel time difference $\tau_2 - \tau_1$ varies

*Table 4.23* Turbulence parameter values for various weather conditions

| *Weather conditions* | *Turbulence parameter* |
|---|---|
| Sunny, light wind (< 2 m/s) | $5 \times 10^{-6}$ |
| Sunny, moderate wind (2–4 m/s) | Between 9 and $10 \times 10^{-6}$ |
| Sunny, strong wind (> 4 m/s) | Between 15 and $25 \times 10^{-6}$ |
| Overcast, light wind (< 2 m/s) | $3 \times 10^{-6}$ |
| Overcast, moderate wind (2–4 m/s) | Between 8 and $9 \times 10^{-6}$ |
| Overcast, strong wind (> 4 m/s) | Between 15 and $25 \times 10^{-6}$ |

within the range $\Delta\tau_{22} - \Delta\tau_1$ and is uniformly distributed within this range. Hence:

$$F_f = \frac{\sin[\pi(\Delta\tau_2 - \Delta\tau_1)f]}{\pi(\Delta\tau_2 - \Delta\tau_1)f}.$$

A somewhat more complex method for estimating the effect of atmospheric turbulence has been proposed.[31] However, the accuracy of this method is not known. A simpler empirical expression[32] is

$$F_c = \exp\{-[\eta k(R_2 - R_1)]^2\}, \tag{4.75}$$

where the turbulence parameter $\eta$ is the mean square fluctuation of the refraction index $\overline{\mu}^2$ (typical values are listed in Table 4.23).

Outside the shadow area, the rapid fluctuations of wind and temperature lead to fluctuations in the speed of each sound ray and, as a consequence, in its phase and amplitude. The turbulence effect is maximal where interference predominates. Daigle[33] developed a model using a factor of partial correlation $\Gamma$ (or $T$ for a Gaussian turbulence spectrum) based on meteorological parameters, such as the (outer) turbulence scale $L_T$, the separation distance between rays $\rho$ and the mean square fluctuation of the refraction index $\overline{\mu}^2$.

For a spherically symmetric source having unit pressure at 1 m the mean squared received pressure is given by[34]

$$\overline{p}^2 = \frac{1}{r_d^2} + \frac{|F|^2}{r_r^2} + \frac{2|F|}{r_r r_d}\cos[k(r_r - r_d) + \theta]T,$$

where $T = \exp[-\sigma^2(1-\rho)]$, $\sigma^2$is the variance of the phase fluctuation along the path, $\rho$ is the phase covariance between path, $F = |F|\exp(i\theta)$ is the complex spherical wave reflection factor, $k$ is the wave number as usual. The factor $\sigma^2$ is given by

$$\sigma^2 = \frac{\sqrt{\pi}}{2}\overline{\mu}^2 k^2 L L_0, \; for\ L > kL_0^2,$$

where $\overline{\mu}^2$ is the variance of index of refraction, $L$ is the horizontal path length, $L_0$ is the Gaussian turbulence scale. The phase covariance between the path is given by

$$\rho = \sqrt{\pi}\frac{L_0}{2h} erf(\frac{h}{L_0}),$$

where $h$ is the maximum path separation and $erf(x)$ is an error function.

In general, for the mean-squared pressure, one can obtain

$$\overline{p}^2 = \frac{\rho c W}{4\pi}\{\frac{1}{r_d^2} + \frac{|F|^2}{r_r^2} + \frac{2|F|}{r_r r_d}\cos[k(r_r - r_d) + ArgF]\Gamma\}, \qquad (4.76)$$

with function $\Gamma$ defined as:

$$\Gamma(R,\rho) = \exp\{\frac{\sqrt{\pi}}{2}\overline{\mu}^2 k^2 L_T[1-\varphi(\frac{\rho}{L_T})\frac{\rho}{L_T}]\},$$

where

$$\varphi(\frac{\rho}{L_T}) = \int_0^{\rho/L_T} \exp(-t^2)dt.$$

In practice, the value of the length scale $L_T$ is close to 1 m, and the mean-squared refractive index $\overline{\mu}^2$ has values ranging from $10^{-7}$ for little turbulence to $10^{-5}$ when the turbulence is strong. Research carried out by Daigle *et al.*[35] has shown that the best agreement of predictions with the data was obtained when the separation distance between rays $\rho$ was chosen as the half of the maximum separation. This solution only applies when the refraction is sufficiently weak such that there is only one direct and one reflected ray. It cannot be used when there is strong negative refraction or when the receiver lies in the shadow zone.[36–39]

In Fig. 4.76, predictions of equation (4.76) for Gaussian turbulence are compared with Parkin and Scholes' data for the difference in level between horizontally separated microphones, one of which was 19 m from a fixed jet engine noise source under acoustically neutral atmospheric conditions.[37] A constant value ($2 \times 10^{-6}$) has been used for the phase fluctuation variance ($\sigma$).

Figure 4.77 compares the results of predictions using different values of $\sigma$ and data for various distances from source to receiver and acoustically neutral weather conditions. Good agreement between Parkin and Scholes' data[40,41] and predictions was obtained for values of $\sigma$ between $2 \times 10^{-7}$ and $7 \times 10^{-7}$.

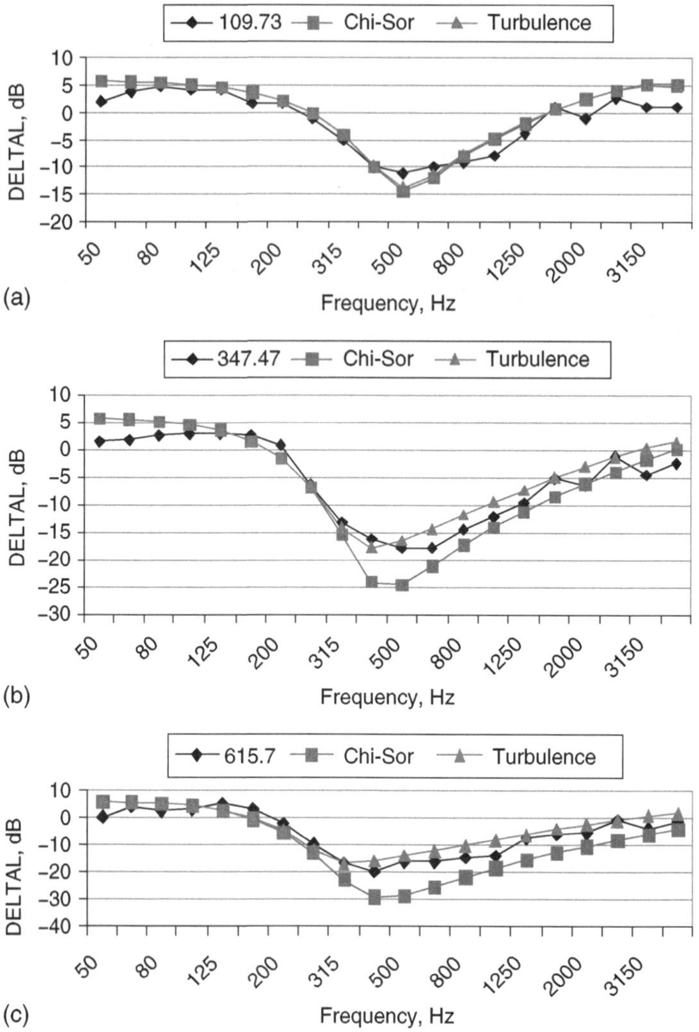


*Figure 4.76* Comparison of Parkin and Scholes' data for the corrected horizontal level difference between a reference level at 19 m and received levels at various other distances from a fixed jet engine at 2 m height over grassland and predictions of eq. (4.6.5) with assumed variance of the phase fluctuation and length scale $\sigma = 2 \cdot 10^{-6}$, $L_0 = 1$ m respectively. The distances from source to receiver are: (a) 109.73 m, (b) 195.07 m, (c) 347.47 m, (d) 615.7 m and (e) 1097.28 m.

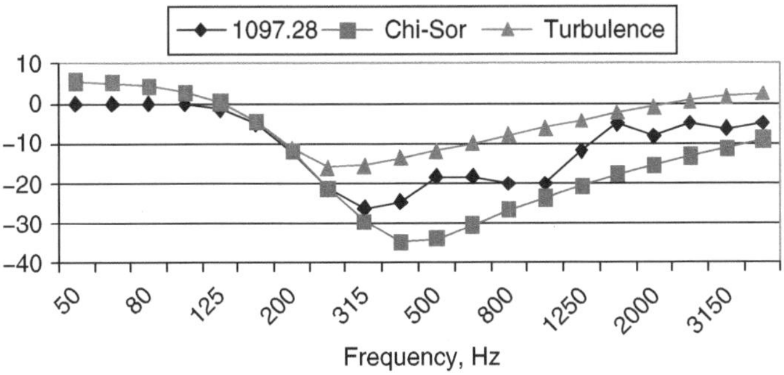


(d)

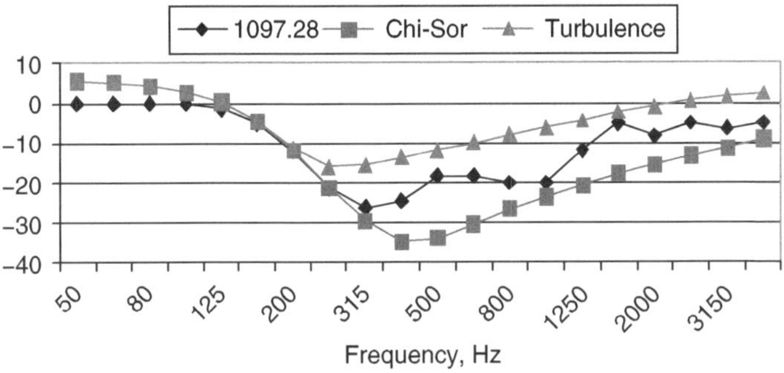


(e)

*Figure 4.76 Continued*

Many methods are available for calculating the ground (interference) effect. The ISO 9613/2 method[38] is based on the expressions listed in Table 4.24, where:

$$a(h) = 1.5 + 3.0 \exp[-0.12(h-5)^2]p$$

$$+ 5.7 \exp(-0.09h^2)[1 - \exp(-2.810^{-6}s)];$$

$$b(h) = 1.5 + 8.6 \exp(-0.09h^2)p;$$

$$c(h) = 1.5 + 14.0 \exp(-0.46 - h^2)p;$$

$$d(h) = 1.5 + 5.0 \exp(-0.9h^2)p.$$

$h$ corresponds to either the source height $h_s$ or the receiver height $h_r$ as the case may be and $s$ is the horizontal distance from source to receiver,

$$p = 1 - \exp(-s/50);$$

$$q = 0, \quad \text{for} \quad s < 30(h_r + h_s) \text{ or}$$

$$q = 1 - \frac{30\left(h_r - h_s\right)}{s} \quad \text{when} \quad s \geq 30(h_r + h_s).$$

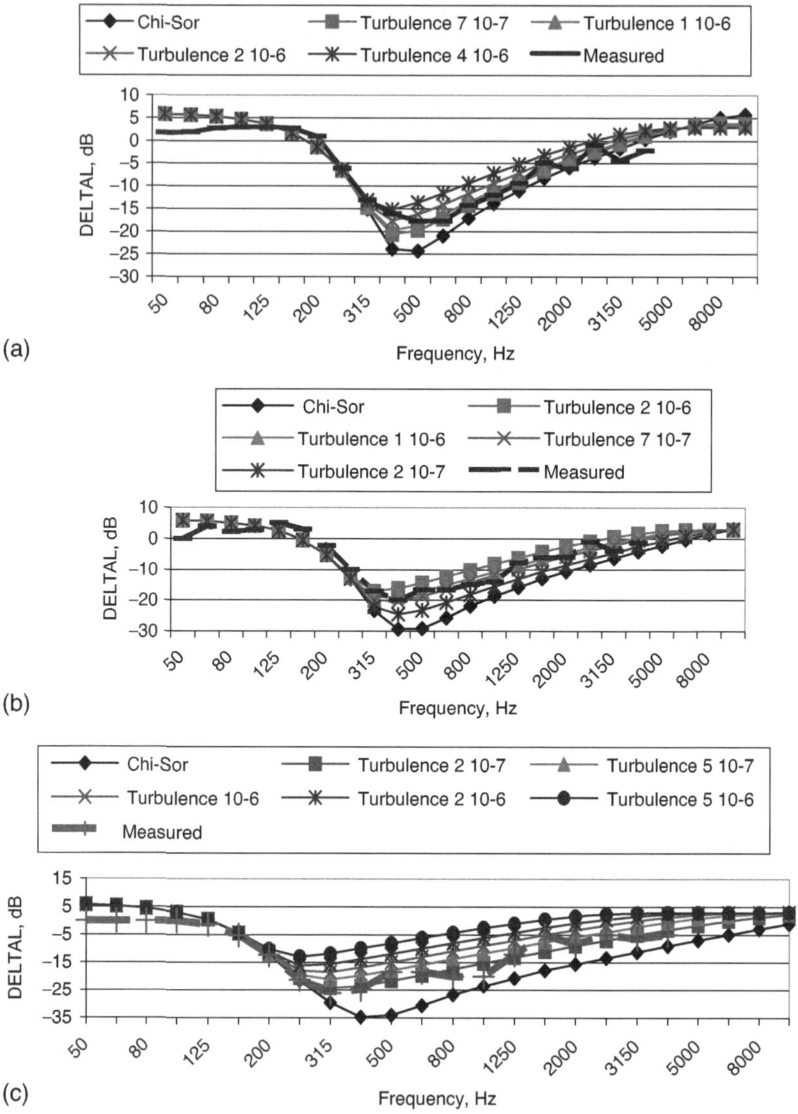


*Figure 4.77* Comparisons of the Parkin and Scholes' corrected sound pressure level difference data, for acoustically neutral weather conditions and receiver separations of (a) 347.47 m (b) 615.7 m and (c) 1097.28 m, with predictions of equation (4.6.5) using various values for the variance of the phase fluctuations.

*Table 4.24* ISO 9613/2 ground effect calculation (symbols are defined in the main text)

| *Octave band center frequency (Hz)* | *Source or receiver zone attenuation ($A_s$ or $A_r$)* | *Mid-ground attenuation ($A_m$)* |
|---|---|---|
| 16 | 1.5 | $3q$ |
| 31.5 | 1.5 | $3q$ |
| 63 | 1.5 | $3q$ |
| 125 | $1.5+\{G_s \text{ or } G_r\}\cdot a'(h)$ | $3q\cdot(1-G_m)$ |
| 250 | $1.5+\{G_s \text{ or } G_r\}\cdot b'(h)$ | $3q\cdot(1-G_m)$ |
| 500 | $1.5+\{G_s \text{ or } G_r\}\cdot c'(h)$ | $3q\cdot(1-G_m)$ |
| 1000 | $1.5+\{G_s \text{ or } G_r\}\cdot d'(h)$ | $3q\cdot(1-G_m)$ |
| 2000 | $1.5(1-\{G_s \text{ or } G_r\})$ | $3q\cdot(1-G_m)$ |
| 4000 | $1.5(1-\{G_s \text{ or } G_r\})$ | $3q\cdot(1-G_m)$ |
| 8000 | $1.5(1-\{G_s \text{ or } G_r\})$ | $3q\cdot(1-G_m)$ |

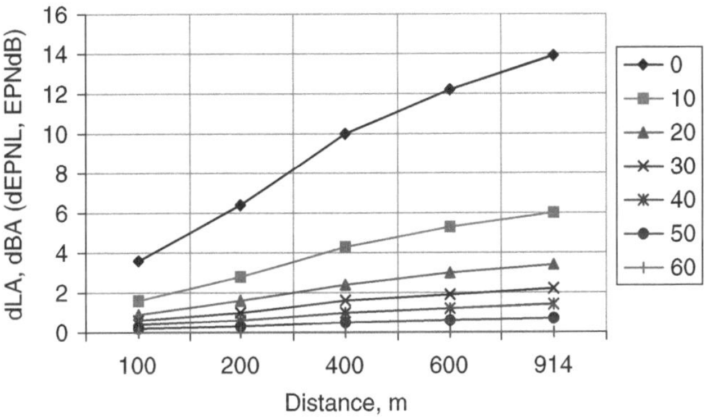


*Figure 4.78* SAE/ICAO predictions of the 'lateral attenuation' of aircraft noise for various values of grazing angle.

The parameter $G$ is defined as the ground hardness indicator, or absorption coefficient, equal to 1 for hard ground and to 0 for very soft ground. For mixed ground, values should be input between 0 and 1.

Frequently, for aircraft noise level calculations the SAE 1751 standard method is used (Fig. 4.78 and Table 4.25).

This method is used in FAA INM (both in version 5.2 and in version 6.0), in the former USSR prediction scheme and in the current Ukrainian one (realized in software ISOBELL'a). The corresponding formula for attenuation due to ground effect is

$$\Delta L_{INT} = [\Delta_D \times \Delta_\beta]/13.86 \tag{4.77}$$

$$\Delta_D = 15.09 \times (1.0 - 1.0/\exp(0.00274 \times D))$$

$$\Delta_\beta = 3.96 - 0.066 \times \beta + 9.9/\exp(0.13\beta)$$

where the distance $D$ (m) and grazing angle $\beta$ are defined in Fig. 4.75.

*Table* 4.25 Values of SAE/ICAO predictions corresponding to Fig. 4.79

| *Distance D (m)* | *Grazing angle β (degrees)* | | | | | | |
|---|---|---|---|---|---|---|---|
| | *0.0* | *10.0* | *20.0* | *30.0* | *40.0* | *50.0* | *60.0* |
| 100.0 | 3.6 | 1.6 | 0.9 | 0.6 | 0.4 | 0.2 | 0.0 |
| 200.0 | 6.4 | 2.8 | 1.6 | 1.0 | 0.6 | 0.3 | 0.0 |
| 400.0 | 10.0 | 4.3 | 2.4 | 1.6 | 1.0 | 0.5 | 0.0 |
| 600.0 | 12.2 | 5.3 | 3.0 | 1.9 | 1.2 | 0.6 | 0.0 |
| 914.0 | 13.9 | 6.0 | 3.4 | 2.2 | 1.4 | 0.7 | 0.0 |

Another well-known method for calculating lateral attenuation is that used in the UK for aircraft noise assessment.[42] According to this method:

$$\Delta L_{INT} = K lg[D/152]; \tag{4.78}$$

$$K = 26.6 + 1861(0.06 - \sin^2\beta) \text{ for } \beta < 4.8;$$

$$K = 26.6 \text{ for } \beta \geq 14.8$$

where distance $D$ is measured in feet.

The difference between the results of calculations using SAE/ICAO and UK methods varies between −3.5 and +1.5 dB.

All of the above methods are empirical and represent averages over a great number of sources and ambient conditions. Some of them are recommended for use with particular noise indices. For example, the SAE/ICAO prediction is used for *SEL* and *EPNL* calculations without any reference to the particular spectrum and directivity pattern of the noise source, in spite of the fact that each index is defined by very different frequency weighting function. Moreover, the physical type of acoustic source and the nature of the covering of the reflecting surface are neglected in this method. Evidence that these factors are important for AN calculations is that the USA FAA INM model includes the NOISEMAP model (developed by USA, Department of Defence (DOD)). If the NPD curves for airplanes are used in a same form as in NOISEMAP, the formula for ground effect interference $\Delta L_{INT}$ must also be used in NOISEMAP.

Given that many aircraft noise calculations are performed for the summer season, which is the busiest time for flight operations at any airport worldwide, soft ground covering around the airports, such as grass, may be used for defining the type and values of the ground effect. Indeed, equations (4.77) and (4.78) have been defined for the summer season and assume soft ground cover.

Noise measurements have been made during aircraft engine run-ups to define the noise directivity patterns necessary for AN calculations in the vicinity of an airport. Measurements have been provided for a large number of aircraft types and the results grouped according to the three characteristic types of engines now in operation (Fig. 4.73). They are used in various

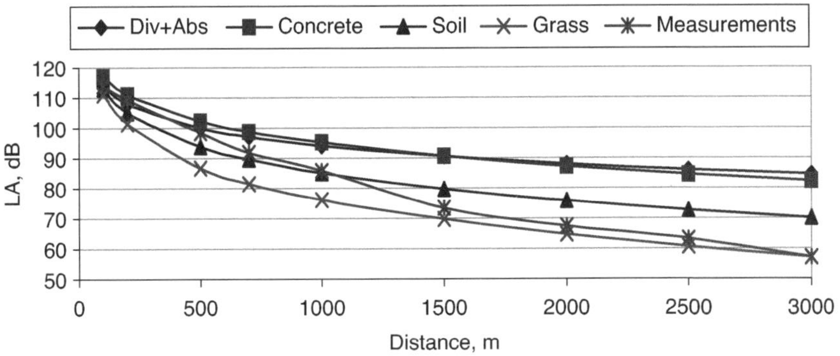


*Figure 4.79* Average sound levels measured as a function of distance from an Il'ushin-86 aircraft engine in the direction of maximum jet noise generation compared with various predictions: spherical spreading plus air absorption and ground effects for concrete, soil and grass.

methods for assessment of AN levels around engine run-ups at airports. Moreover, this approach is used in Flula, versions 2 & 3, Switzerland, even for AN flyover level assessment.

Measurements, previously used for AN contour assessment, have been used to investigate the contributions of the effects defined above: divergence, absorption and excess attenuation. Figure 4.79 and Table 4.26 present data for an NK-86 engine in the direction of maximum jet noise generation ($\theta \cong 140°$). Measurements were carried out on several summer days during near calm weather conditions (wind speed < 5 m/s, temperature between 20 and 25°C). The measurements were not intended for any particular propagation effects analysis and only included overall sound levels ($OASPL$ and $L_A$). In spite of this, the results are useful for providing a preliminary analysis of ground attenuation of noise propagated to the point of noise control from aircraft engine at run-ups. In Fig. 4.79, the line 'Div + abs' represents the sum of reductions due to sound wave spreading (strongly proportional to the value $R^{-2}$) and air absorption (proportional to $\alpha(R - R_0)$. The other lines in this figure represent predictions of noise propagation over three kinds of ground surface.

In the ground effect predictions, the ground surface impedance has been described by the Delany and Bazley model.[39]

$$Z = 1 + 9.08\left(\frac{1000f}{\sigma}\right)^{-0,75} + i11.9\left(\frac{1000f}{\sigma}\right)^{-0,73}.$$

The flow resistivity values (kPa s/m$^2$) assumed for 'hard soil', 'grass' and 'concrete' are 2500, 250 and 20,000, respectively.

The data are in good accordance with the predicted influence of the local reflecting surface characteristics. Near the engine source (at distances

*Table 4.26* Measured and predicted sound levels as a function of distance from an Il'ushin-86 aircraft engine in the direction of maximal jet noise generation

| *Distance (m)* | *A-weighted sound level, $L_A$ (dBA)* | | | | |
|---|---|---|---|---|---|
| | *Measurements* | *Predicted* | | | |
| | | *Divergence + absorption* | *Divergence + absorption + reflection from concrete covering* | *Divergence + absorption + reflection from soil covering* | *Divergence + absorption + reflection from grass covering* |
| 100 | 114 | 114 | 117 | 113 | 111 |
| 200 | 109.3 | 108 | 111 | 105 | 101.4 |
| 500 | 98.4 | 100 | 102.2 | 93.8 | 86.7 |
| 700 | 91.8 | 97.1 | 98.8 | 89.7 | 81.6 |
| 1000 | 85.8 | 94 | 95.2 | 85.1 | 34.82 |
| 1500 | 73.4 | 90.5 | 90.6 | 79.7 | 41.3 |
| 2000 | 67.4 | 88 | 87.3 | 75.8 | 46.26 |
| 2500 | 63.3 | 86 | 84.5 | 72.7 | 50.39 |
| 3000 | 57 | 84.5 | 82.3 | 70 | 53.96 |

*Table 4.27* Measured and predicted A-weighted sound level differences with respect to levels at 100 m from an Il'ushin-86 aircraft engine in the direction of maximum fan noise generation

| *Distance (m)* | *$\Delta L_A$ (dBA)* | | | | | |
|---|---|---|---|---|---|---|
| | *Measurements* | *Divergence + absorption* | *Concrete (dipole)* | *Concrete (monopole)* | *Soil (dipole)* | *Soil (monopole)* |
| 100 | 0 | 0 | 0 | 0 | 0 | 0 |
| 200 | 8.3 | 6 | 5.3 | 8 | 10 | 7.3 |
| 500 | 19.1 | 14 | 15.6 | 20.6 | 27.6 | 14.9 |
| 700 | 23.5 | 16.9 | 19.6 | 21 | 34 | 17.2 |
| 1000 | 28.2 | 20 | 24.2 | 21 | 40 | 19.5 |
| 1500 | 33 | 23.6 | 30 | 22.2 | 46.7 | 22.7 |
| 2000 | 40 | 26.1 | 34.4 | 23.8 | 51 | 24.9 |

of < 500–700 m), the measurements are close to the predictions for 'concrete'. At greater distances the data lie near to the predictions for 'soil' or between predictions for 'soil' and 'grass'. These results are consistent with the fact that the run-ups were arranged on the concrete surface of an apron, whereas at further distances the ground surface was soil and/or grass.

For the purposes of more detailed analysis all of the data have been prepared in the form of level differences with respect to the reference distance of $R = 100$ m. These level differences are presented in Fig. 4.80 and Table 4.27.

The corresponding data and predictions for propagation from a bypass engine fan and a turboprop engine in the direction of maximum fan noise

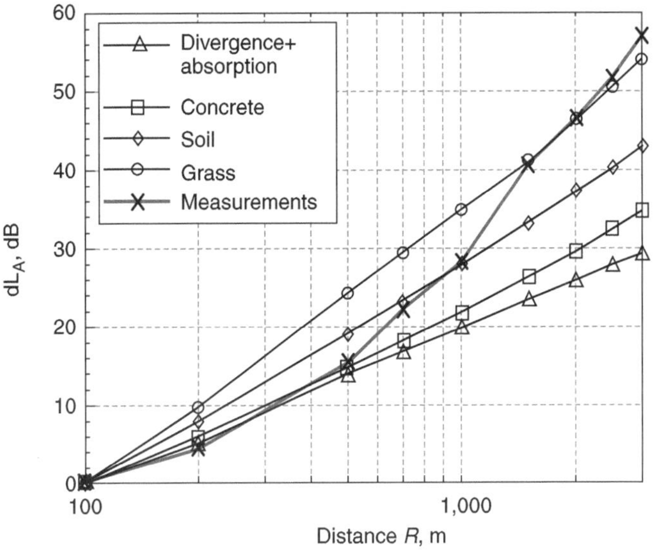


*Figure 4.80* Predicted and measured sound level differences with respect to levels at 100 m for noise propagating from an Il-86 aircraft engine in the direction of maximum jet noise generation.

generation and the direction of maximum propeller noise generation are presented in Figs 4.81 and 4.82 and in Tables 4.28 and 4.29, respectively.

All of the predictions in Figs 4.80–4.83 and the associated data in Tables 4.26–4.28 were calculated as third-octave spectra and then used to obtain the noise indices of interest in AN analysis.

In other shorter range sound propagation predictions,[13] an example of which is shown in Fig. 4.83, the point source is assumed to be located 30 m from the receiver and at 0.2 m above the ground. The receiver is 1.5 m above the ground. Using the Delany and Bazley impedance model,[39] three values of flow resistivity (40, 160 and 630 kPa s/m$^2$) have been considered.

The predicted overall ground effects in various directions of noise generation are shown in Table 4.29, which includes the results of ground effects calculations for two types of reflecting surfaces (grass and concrete). Differences in the A-weighted level corrections to free field values exceed 10 dB. Indeed, these results differ significantly from the results obtained by the recommended ISO 9613/2 or SAE/ICAO methods. The predicted ground effect spectra used to obtain the results listed in Table 4.29 are shown in Fig. 4.84.

Acoustic screens may be used inside airports for mitigating noise from aircraft ground operations and maintenance (around engine run-ups, along the runway or taxi-ways, etc.). The effects of screens are assessed by means

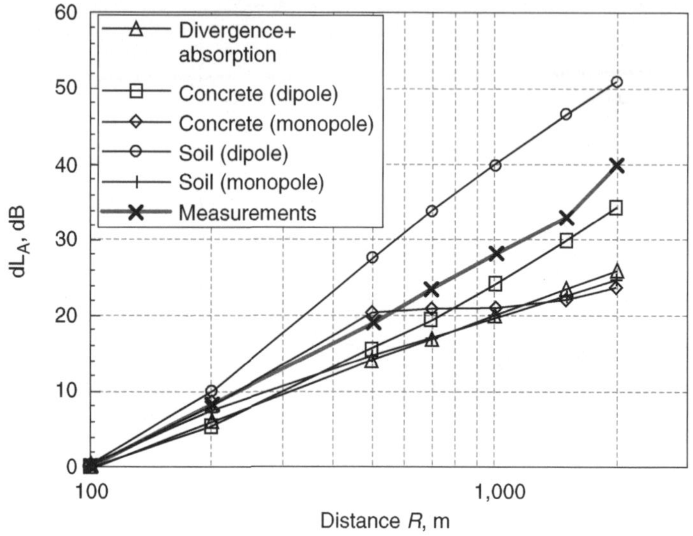


*Figure 4.81* Predicted and measured sound level differences with respect to levels at 100 m for noise propagating from an Il-86 aircraft engine in the direction of maximum fan noise generation.

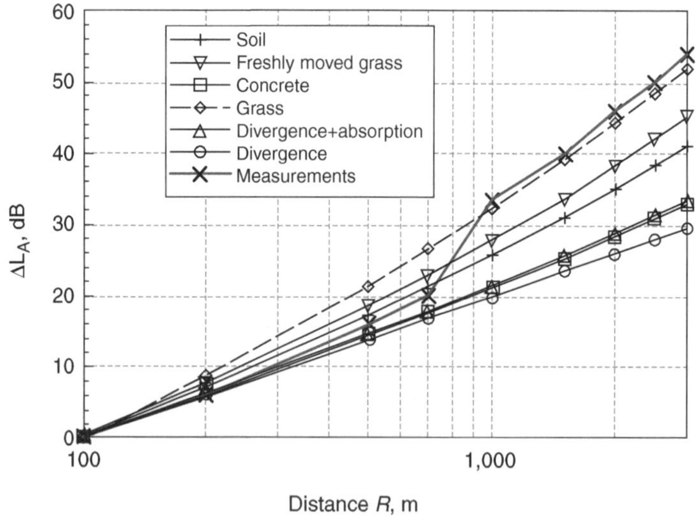


*Figure 4.82* Measured and predicted sound level differences with respect to levels at 100 m for noise propagating from an AI-24 engine (turboprop type, installed as the powerplant on an Antonov-24 aircraft) in the direction of maximum propeller noise generation.

*Table 4.28* Measured and predicted A-weighted sound level differences as a function of distance from 100 m for the propeller noise from a turboprop engine (AI–24) in the direction of maximum noise generation

| *Distance (m)* | *$\Delta L_A$ (dBA)* | | | | | | |
|---|---|---|---|---|---|---|---|
| | *Measurements* | *Divergence* | *Divergence + absorption* | *Soil* | *Freshly mown grass* | *Concrete* | *Grass* |
| 100 | 0 | 0 | 0 | 0 | 0 | 0 | 0 |
| 200 | 6 | 6 | 6.2 | 7.1 | 7.7 | 6.05 | 8.7 |
| 500 | 16 | 14 | 14.7 | 17.2 | 18.6 | 14.45 | 21.5 |
| 700 | 20 | 16.9 | 18 | 21.2 | 22.9 | 17.65 | 26.6 |
| 1000 | 33.5 | 20 | 21.5 | 25.6 | 27.8 | 21.15 | 32.35 |
| 1500 | 40 | 23.5 | 25.7 | 30.9 | 33.7 | 25.31 | 39.3 |
| 2000 | 46 | 26 | 28.8 | 35 | 38.3 | 28.4 | 44.4 |
| 2500 | 50 | 28 | 31.4 | 38.3 | 42 | 30.9 | 48.5 |
| 3000 | 54 | 29.5 | 33.5 | 41.2 | 45.3 | 33.05 | 51.9 |

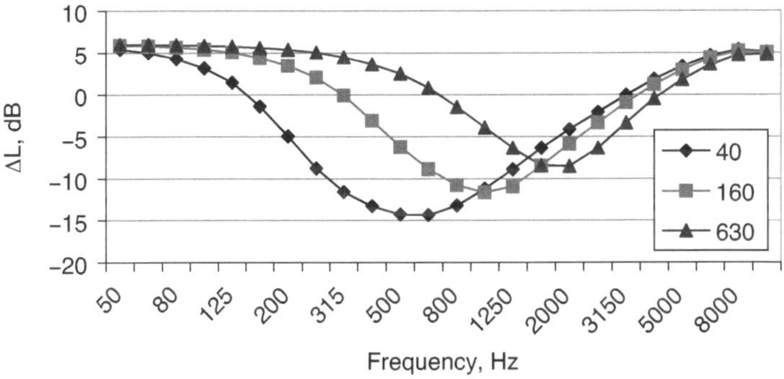


*Figure 4.83* Sound pressure level relative to free field predicted for three flow resistivity classes according to the Delany and Bazley model [4.39]. The units of flow resistivity are kPasm$^{-2}$.

of models that account for: (1) the effects of sound diffraction at screen edges;[22] (2) interference between direct and reflected waves from the various impedance surfaces; and (3) the particular noise spectra generated by the aircraft. Spectral efficiencies of screens must be calculated for different kinds of noise propagation.[40]

Example predictions of *OASPL* ($\Delta L_{in}$) and $L_{Amax}(\Delta L_A)$ for different types of noise sources under identical conditions are shown in Table 4.30 ($\Delta L = L_{withoutsreen} - L_{withscreen}$). The assumed geometry of the screen and its location corresponds to the use of a 12-m high screen for aircraft noise control near a hotel in the vicinity of an airport. The length of the screen is about 500 m; the center of the coordinate system is at the

*Table 4.29* Predicted influence of the ground surface on the noise directivity pattern of Il'ushin–86 engines

| *Direction with respect to forward axis (degrees)* | *Without surface influence* | *With grass-covered surface* | *With concrete-covered surface* | $\Delta_{without-with\ grass}$ | $\Delta_{without-with\ concrete}$ |
|---|---|---|---|---|---|
| 10.0 | 89.405 | 82.161 | 83.462 | 7.2 | 5.95 |
| 20.0 | 90.239 | 83.069 | 84.117 | 7.15 | 6.1 |
| 30.0 | 90.957 | 83.553 | 84.504 | 7.4 | 6.4 |
| 40.0 | 93.927 | 85.161 | 85.691 | 8.75 | 8.2 |
| 50.0 | 97.438 | 87.706 | 87.737 | 9.7 | 9.2 |
| 60.0 | 100.759 | 90.623 | 90.303 | 10.1 | 10.45 |
| 70.0 | 97.909 | 87.924 | 88.101 | 10 | 9.1 |
| 80.0 | 92.474 | 83.442 | 85.350 | 9 | 7.1 |
| 90.0 | 93.893 | 84.838 | 86.629 | 9 | 7.25 |
| 100.0 | 94.019 | 85.408 | 87.685 | 8.6 | 6.4 |
| 110.0 | 94.462 | 86.409 | 89.130 | 8 | 4.3 |
| 120.0 | 95.296 | 87.403 | 90.645 | 7.8 | 4.6 |
| 130.0 | 96.406 | 88.942 | 92.986 | 7.5 | 3.4 |
| 140.0 | 97.717 | 91.890 | 96.334 | 5.8 | 1.4 |
| 150.0 | 97.217 | 93.557 | 97.339 | 3.6 | −0.1 |
| 160.0 | 95.837 | 93.690 | 96.811 | 2.1 | −1.0 |

left side of the screen, axis OZ is directed along the screen, axis OX is perpendicular to the screen; the coordinates of noise source are −175, 325, 2; the coordinate of the receiver are 113, 175, 5, where the format corresponds to $(x, z, y)$-coordinates, respectively, and the screen is installed at $x = 0$. The calculations have been performed in the spectral domain and spectral results have been transformed to noise indices. In this way both the type of noise spectra and the types of reflecting surfaces are considered in calculations. Three types of reflection conditions have been considered: (1) reflection is absent; (2) reflection is from an acoustically hard surface; and (3) reflection is from a surface with continuous impedance corresponding to a grass covering (Delany and Bazley model[39] with an effective flow resistivity of 300 kPa s/m$^2$).

The assumptions behind the software algorithm used to obtain these results include the following.

(1) The acoustical efficiency of the screen depends upon the Fresnel number $N$

$$N = 2\Delta/\lambda; D = a + b - d; -3 \leq N \leq 100,$$

where $\Delta$ is the difference between the length of the propagation path from source to the top of the screen and from the top of the screen to

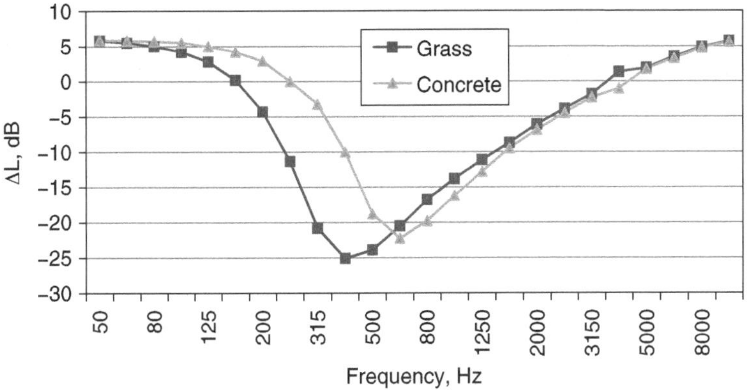


*Figure 4.84* Predicted ground effect spectra for grass and concrete used in obtaining the predictions shown in Table 4.29.

*Table 4.30* Predicted influence of noise source type and ground reflections on screening efficiency for source at height 2 m and 175 m from a 12-m high screen and receiver at height 5 m and 113 m beyond the screen

| *Type of noise source* | *Character of sound reflection* | | |
|---|---|---|---|
| | *Without reflection* $\Delta L_A/\Delta L_{in}$ | *Hard ground* $\Delta L_A/\Delta L_{in}$ | *Impedance ground* $\Delta L_A/\Delta L_{in}$ |
| Turbojet engine | 12.9/10.9 | 8.3/6.1 | 12.0/7.9 |
| Bypass engine | 15.5/8.8 | 10.8/3.8 | 13.7/5.7 |
| Pink noise | 13.8/8.5 | 9.1/3.5 | 12.6/5.9 |

the receiver, $(a+b)$, and that directly through the screen $(d)$; $\lambda$ is the wavelength.

(2) The noise level reduction by a screen at any frequency band is defined by approximations of Maekawa's solutions:

$$\Delta L = 1.1518211 \exp[0.5493061(N+3)] - 1.1468337 \text{ for } -3 \le N < 0;$$

$$\Delta L = 5 + 20\lg\left(\sqrt{2\pi\,|N|}\Big/ th\sqrt{2\pi\,|N|}\right) \text{ for } 0 \le N \le 100.$$

The value of Fresnel number $N$ is negative when the line-of-sight passes over the screen. If $N < -3$, diffraction effects are neglected.

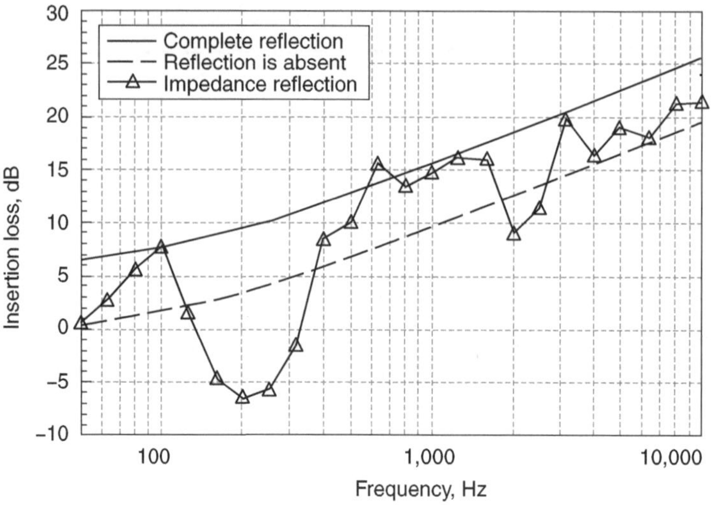


*Figure 4.85* Predicted acoustic screen efficiency as a function of frequency for three types of sound wave reflections at the ground around the screen.

The differences between the predictions in Table 4.30 for various types of aircraft are of the same order as the differences arising from various types of reflecting surfaces around the screens.

The predicted spectral efficiencies of the screen are shown in Fig. 4.85. The presence of an acoustically hard surface is predicted to increase the acoustical efficiency over the whole frequency range. The predictions, including the presence of a continuous grass-like surface, vary between those for no surface and a hard surface except in some frequency bands. For example, in bands with frequencies between 160 and 250 Hz (Fig. 4.85), the efficiency of the screen is predicted to be negative. This is a consequence of the fact that the presence of the screen reduces the ground effect in this frequency band. This is the reason also for the prediction of reduced efficiency over grassland near 2 kHz.

In most cases the ground surfaces differ on either side of the screen. The predicted influences of a combination of surface types on both sides of the screen are presented in Fig. 4.86. The impedance characteristics for these surfaces are given elsewhere.[28,41]

The predicted differences in spectral efficiencies cause predicted differences in $\Delta L_A$ of between 1 and 2 dB. But more interesting is the predicted spectral shape of results, particularly for noise from aircraft with propeller engines, because their main acoustic energy is located at low frequencies in discrete tones (either at the first or first two harmonics), where interference effects may be strong, so the efficiency of the screen may be decreased.

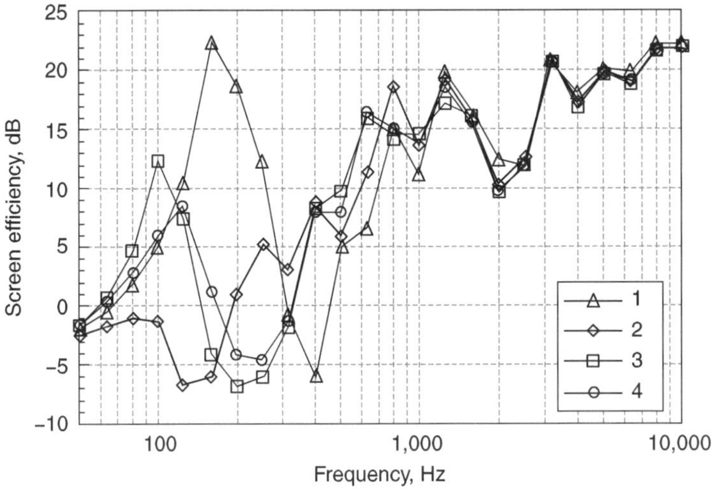


*Figure 4.86* Predicted influence of different ground surfaces either side of a noise barrier on its efficiency: 1 – concrete/concrete; 2 – concrete/grass; 3 – freshly raked soil/grass; 4 – sharp sand with moisture <10%/grass. The assumed source–barrier–receiver configuration is the same as that for Fig. 4.85 and Table 4.30.

The need to predict the influences of screen type and excess attenuation effects has resulted in the screen effect generator (SCREEN-generator) and excess attenuation generator (LATER-generator) in the software tools BELTASS, BELTRA, ISOBELLA'a. These sub-routines enable calculations to be made for any AN assessment.

## 4.7 Prediction of noise in the vicinity of an airport

A complete airport noise model is the main tool for decision-making about noise abatement processes (e.g noise-zoning and appropriate land use) or for the needs of Environmental Impact Statements around airports. It combines the specific features of flight path and of ground aircraft noise models that have been presented in Sections 4.4 and 4.5. Usually such models are implemented in computer programs offering various output possibilities.[4,5] Airport operations generally include different types of aircraft, various flight and on-ground procedures and a range of operational weights. Important input parameters are the atmosphere temperature, pressure and humidity, which may influence both the flight performances of the aircraft and the sound propagation. In addition, a large quantity of aircraft specific data and airport operational information are required to compute the noise of each individual operation.

Typically, the final results of noise computations are presented as noise contours. Noise contours are drawn on a map and illustrate how the specific noise index varies from location to location as the result of a given aircraft traffic pattern at an airport. Normally they are obtained by interpolation between discrete predicted values of the noise index at the intersection points of a regular observation grid centered on the airport.

When it is necessary to analyze the noise climate around the airport or in a particular region, or to compare situations with several noise sources in the region, it is necessary to take into account parameters other than acoustical ones. So some kind of composite indicator must be used for aircraft noise impact assessment and investigations around airports. A widely used indicator is the amount of population exposed by noise. This can be calculated by counting the number of dwellings exposed to a certain noise level (inside the zone between two specified noise contours) and multiplying that by the average number of inhabitants per dwelling. Often these data are grouped in classes of 5 dB (i.e. the difference between the noise indicator values on the outer and inner noise contours of the zone under consideration is equal to 5 dB).

Values of aircraft noise levels or specific indicators around airports are the result of many factors: the acoustical characteristics of the aircraft; the intensity of flight traffic around the airport; the scheme of routes and tracks (both on the airfield and for departure and arrival); the distribution of aircraft between routes; the recommended operational procedures used on various routes for each type of aircraft; the operational factors including the in-flight mass of the aircraft, meteorological characteristics and runway characteristics; the presence of acoustic screens; the topographic conditions in the airport location; and any other features that might cause diffraction and interference between propagated and reflected sound waves.

Calculations are usually repeated at each of the series of points around the airport and then interpolations are made to trace outlines of equal noise index values (noise 'contours') which are then used for study purposes. Noise levels from individual aircraft movement at the points $(x, z)$ of a calculation grid are derived from the following formula:

$$L(x,z) = L_{Rn} + \Lambda(\beta, R) + \Delta L_{\Theta} + \Delta L_v + \Delta L_t + \Delta L_s \tag{4.79}$$

where $L_{Rn}$ is the noise level from the noise radius relationship (*Rn*-relationship), described previously in Section 4.2 and distance $R$ (or $D_0$) between the aircraft and a point on a grid defined on the surface around the airport; $\Lambda$ is the excess attenuation of sound along the propagation path lateral to the direction of aircraft movement with distance $R$ and angle of incidence $\beta$; $\Delta L_{\Theta}$ is the correction for the directivity of the noise source; $\Delta L_v$ is the correction for the flight velocity changes [see equation (4.79)]; $\Delta L_t$ is the correction for changes in the duration of the highest sound levels for the considered noise event; and $\Delta L_s$ is the attenuation by various screens.

The above process is repeated at the same point for all the movements of all the airplane types occurring within the time period over which the noise contours are to be calculated and then again at all the other grid points. In an airport noise study, it may not be practicable to account for each airplane type separately when calculating flight profiles and noise levels. In such cases, different airplane types having similar noise characteristics and also similar performance at a particular airport may be categorized or grouped, effectively, as a single type at that airport (see Section 4.4). For such categorized or grouped types on a given route, the calculations need only be made once and the resultant noise levels at the grid points are then factored according to the number of movements of the type in the noise index summation.

Several factors can affect the accuracy of the modeling. Principal among these are wind and temperature gradients and variability in the operational procedures employed during takeoff. Current models do not include any allowance for wind and temperature effects, even though these can cause significant changes in ground-to-ground attenuation and can even result in shadow zones.

The quality of the noise contours will also depend on the choice of the grid spacing, especially in zones where rapid changes occur in the noise contours. Interpolation errors on the noise contours are minimized by close grid-spacing, but this increases the computation time as the noise index then has to be calculated at a large number of grid points. Studies have shown that a grid spacing of about 300 m offers a good compromise between the computation time and the accuracy of the interpolated noise contours (i.e. leading to standard deviations of less than 0.5 dB for low and medium noise contours).

Before the noise exposure from the total traffic at a particular calculation point can be determined, the sound exposure level or maximum level has to be calculated for each individual aircraft operation. If the purpose is to determine the maximum level from the total traffic, this should be done according to the specified national noise index, for example, $L_{Amax}$ in the Ukraine. However, it is recommended to use the envelope of the noise footprint for the noisiest airplane on each track.

If the purpose is to determine the equivalent sound pressure level $L_{Aeq}$, or its variant including penalties for noise from night flights or evening and night flights, the sound exposure levels *SEL* ($L_{AE}$) for each individual operation on an average day are added on an energy basis. The time periods for determining an average day are defined in the national methods. The sound exposure level for each operation is weighted for the time of day and, in some countries, for the time of week in accordance with the national method. The summation is defined as follows:

$$L_{Aeq,W} = 10\log\left[\frac{1}{T}\sum_{j=1}^{N} w10^{\frac{L_{AEj}}{10}}\right],$$

where $L_{AE,j}$ is the sound exposure level from the $j$th aircraft operation out of a total of $N$ operations, $w$ is the weighting factor depending on the time of day and, in some countries, time of week, $T$ is the reference time for $L_{Aeq}$ in seconds. If the reference time is 1 day (24 hours), $T = 86{,}400$ s.

In view of the large number of variables and of the simplifications usually made in the calculations, it is desirable to standardize procedures for computing airport noise contours. There is a need to provide an outline for such a standard method, to identify the major aspects and to supply specifications in respect of each of these. A complication is that the calculation methods, recommended by the ICAO, have allowed different Member States to use different noise descriptors and scales as the bases for their respective noise indices.

There are a number of noise-generating activities at operational airports that are excluded from the calculation procedures mentioned here. These include taxiing, engine testing and use of auxiliary power units. In practice, the effects of these activities are unlikely to affect the noise contours in regions beyond the airport boundary.

For an airport noise study, the calculation procedures include the following steps:

(1) determination of the noise levels from individual airplane movements at observation points around the airport;
(2) addition or combination of the individual noise levels at the respective points according to the formulation of the chosen noise index; and
(3) interpolation and plotting of contours of selected index values.

At all airports, the patterns of operations vary from day to day, depending on the weather, scheduling and many external factors. Typically, the noise index for which the contours are calculated is defined in terms of long term average daily values over a period of some months. It follows that the contours intended to show noise exposures around an airport defined in terms of such an index should similarly depict average conditions over a long period of time. The traffic and operational patterns used in the study are then selected accordingly.

The noise levels for individual movements are calculated, for given atmospheric conditions and assuming a flat terrain, from NPD and airplane performance data (see Section 4.4). The conditions for the noise data are defined by atmospheric attenuation rates, for which the yearly averages drawn from several major world airports are assumed. The performance data are for defined atmospheric temperature and humidity, airport altitude and wind speed. However, given that the calculated noise contours depict average conditions over a long period of time, the same basic data are assumed to apply over specified ranges of conditions. The forms of presentation, methods of derivation and reference conditions for the airplane data are given in Section 4.4.

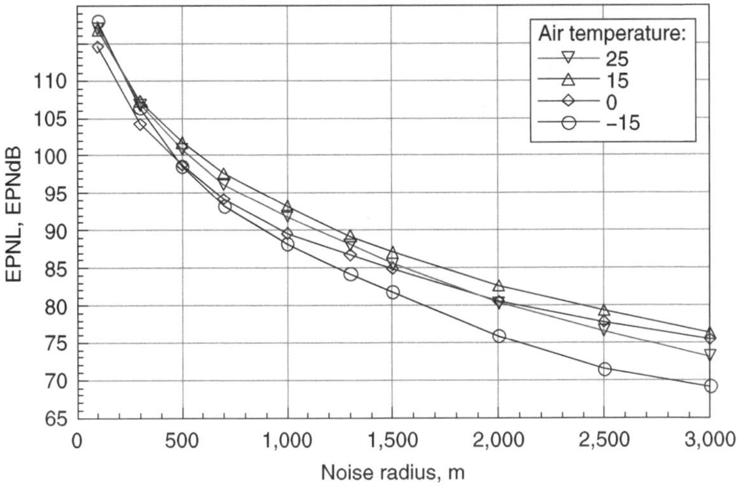


*Figure 4.87* Predicted dependence of the noise-radius relationship on ambient air temperature for the maximum thrust engine mode.

In Section 4.2 it is proved that NPD-relationships (or $R_n$-relationships) depend on many factors: aircraft type, flight mode, meteorological characteristics, etc. For planning purposes, it is necessary to use them in accordance with the most unfavourable conditions. Figure 4.87 compares predictions of $L_{Rn}$ for 25°C (unfavourable propagation conditions) and for other possible temperatures.

The differences for the same engine mode are so large, nearly 10 EPNdB at 2.5 km, that it suggests a need to examine the dependence of NPD-relationships on temperature changes – seasonal or diurnal. The simple adjustment, accounting for ambient air temperature influence through air acoustic impedance changes used in some calculation methods, appears to be insufficient. Moreover, the changes in NPD-relationships shown in Fig. 4.87 are not monotonic. This is because the generation effects (noise source contributions) change as well as the absorption effect. Prevention of such underlying inconsistencies is a subject of great current concern. To overcome them requires a combination of various noise modeling approaches.

The main idea for an improved method is that the basic acoustic model of each aircraft, used in the assessment procedure, must include the combination of sources that contribute to the noise power of the aircraft, including corrections for discrepancies between observed and calculated values. Such a correction is proposed here in the form of transfer function $\Delta SPL_{jk}$ (Section 4.3). Since the models of the main noise sources include their dependencies on flight mode parameters, it is possible to obtain the appropriate values of the NPD-relationships (or other forms of the basic tools, necessary for noise level assessment, in accordance with the modeling approach) taking account of ambient parameter values.

The influence of atmospheric parameters is examined in detail in Chapter 5. For example, the influence of temperature on aircraft noise levels is the result of the change in flight path parameters, the change in the absorption parameters of atmosphere and the change in noise generation by the main acoustical sources. The corresponding changes in *noise radius* are shown in Fig. 4.87. An increase in temperature means that the corresponding air density decreases. This causes a decrease in the air consumption ratios within the engine flow tracts. As a result, the thrust decreases but the engine jet velocity increases. The aircraft takeoff velocities and takeoff lengths increase, and the takeoff gradients and heights under the control surfaces decrease. As the ambient atmospheric temperature varies between −20 and +30°C, the *EPNL* at the monitoring point under the takeoff path is predicted to change by between 8 and 14 EPNdB, and the noise contour area *S* is predicted to change by between 20 and 30 per cent (relative to the area *S* for SA conditions). Figure 4.88 shows examples, for a modern aircraft with high bypass turbofan engines, of the influence of ambient temperature on flight path heights associated with takeoff trajectories and the corresponding noise levels under these trajectories.

These predictions confirm the necessity for accounting for the influence of certain operational factors when calculating noise levels around airports. If these calculations are connected with noise zoning and land-use planning, the worst (from the noise point of view) possible operational conditions must be considered. If models are used for monitoring purposes, NPD-relationships must be derived in accordance with actual values of meteorological parameters in the routine engine mode. It requires the use of specific calculation modules (e.g. RADIUS in the ISOBELL'a software, including application of basic acoustic models of the aircraft of particular types with identified transfer functions). The software module is called NPD-generator (or RADIUS-generator). Flight paths for the aircraft under consideration are built with the so-called 'FLIGHTPATH-generator', which assumes the common flight dynamics models (as e.g. in FAA INM). The flight paths shown in Fig. 4.88 have been calculated accordingly.

Aircraft flight profiles are required to determine the slant distances from the observation points to the flight paths (Fig. 4.16). The variations of engine thrust, or other noise-related thrust parameters, and aircraft speed along the flight path are also required (Section 4.1). The slant distances and thrusts are then used for entry into and interpolation of the NPD data. For noise contour computations, the takeoff and approach flight paths are assumed to be represented by a series of straight-line segments illustrated in Fig. 4.15. The ground tracks of the aircraft are also represented by straight-line segments and arcs of circles. Where possible, flight profiles, associated engine thrust information and aircraft speeds

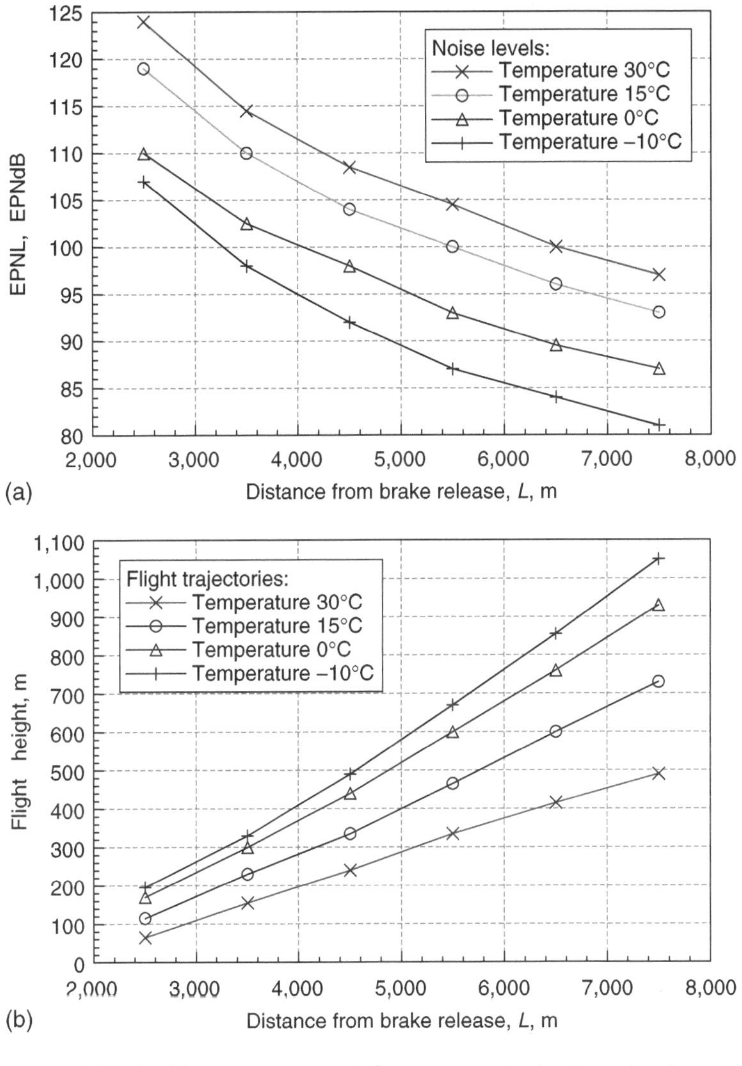


*Figure 4.88* Ambient temperature influence on noise levels under the trajectories (a) and on flight path parameters (b) for aircraft with high bypass turbofans.

should be supplied for each airplane type as it undergoes the following reference flight procedures:

(1) ICAO Noise Abatement Takeoff Procedure A and/or Procedure B at 85 per cent of maximum takeoff mass;
(2) ICAO Annex 16 Noise Compliance Approach at 90 per cent of maximum landing mass, but with the normal flap setting.

All aircraft performance information should be derived for reference conditions as follows:

- International Standard Atmosphere (ISA) atmospheric conditions;
- runway altitude is at sea-level;
- no runway slope;
- 4 m/s headwind, with no wind gradient;
- airplane takeoff mass 85 per cent of maximum takeoff mass;
- airplane landing mass 90 per cent of maximum landing mass;
- all engines operating; and
- normal airplane configurations.

Unless an aircraft manufacturer specifies otherwise, the performance information (i.e. coefficients derived from the reference flight profile or provided by the manufacturer) can be used as given, without correction, over a range of conditions as follows:

- air temperature is $< 30°C$;
- any runway altitude, provided the temperature and altitude are within the engine flat-rating range;
- wind speed is $< 8$ m/s; and
- all practical operational airplane masses.

Modeling of the noise at ground locations near the airport runway during the takeoff roll requires several modifications of the basic NPD data. The modifications result from the fact that the aircraft on the ground accelerates from essentially zero velocity to its initial climb speed, whereas the basic data are representative of overflight operations at constant airspeed [see equation 4.21 and Fig. 4.10]. To accommodate these differences, consideration must be given to changes in generated sound resulting from jet relative-velocity effects, varying directivity patterns from the moving airplane, the modified effective duration with increased speed and extra attenuation of sound during over-ground propagation at near zero elevation angles. Also directivity patterns $\Delta L_\Theta$ are necessary for both taxiing aircraft and engine testing.

Using the co-ordinate system shown in Fig. 4.89, the noise level for negative values of $x$ (i.e. behind the start-of-roll point) is computed as follows:

$$\Delta L_\Theta = 51.44 - 1.553\Theta + 0.015147\Theta^2 - 0.000047173\Theta^3 \text{ for } 90° < \Theta < 148.4° \quad (4.80a)$$

$$\Delta L_\Theta = 339.18 - 2.5802\Theta - 0.0045545\Theta^2 + 0.000044193\Theta^3 \text{ for } 148.4° < \Theta < 180° \quad (4.80b)$$

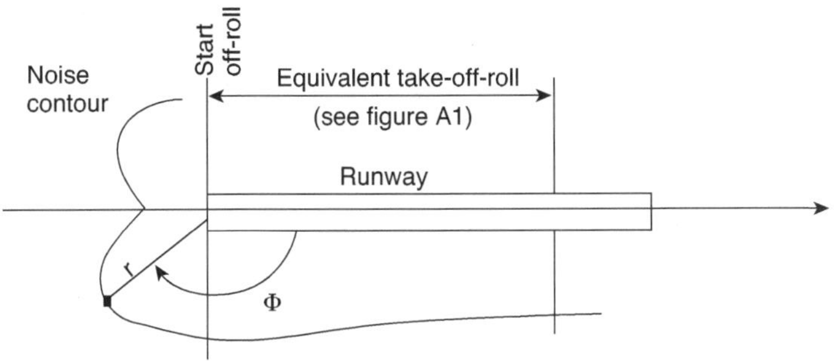


*Figure 4.89* Geometry for on-ground aircraft noise contour assessment.

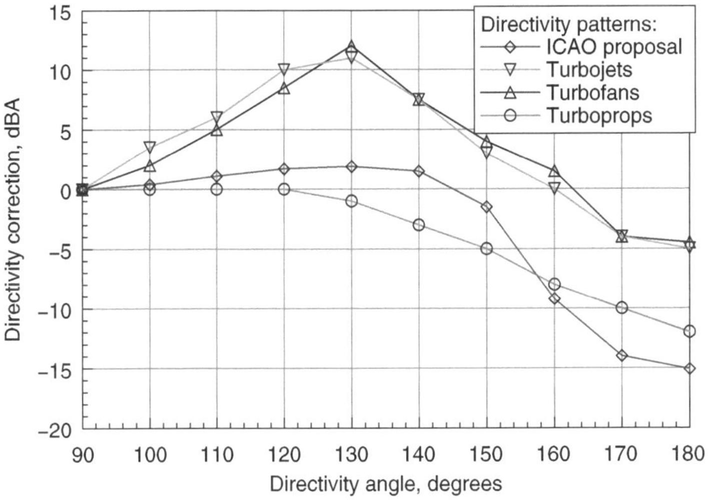


*Figure 4.90* Generalized directivity corrections for characteristic engine types.

Equation (4.80) is general for the whole aviation transport. As one can see from Fig. 4.27, the directivity patterns of noise generation are specific to each engine type. Directivity corrections for characteristic types of engines are presented in Fig. 4.90. Differences between specific directivity patterns and the generalized relationship proposed by ICAO[43] may be as large as 10 dBA in certain directions.

All relationships for the excess attenuation due to the ground effect $\Delta L_{int}$ are based on approximate solutions of the reflection of a spherical sound wave from the locally reacting plane surface. The appropriate formulas for the ground effect calculations from ISO 9613/2 are proposed for two kinds of ground conditions without reference to source parameters. In fact,

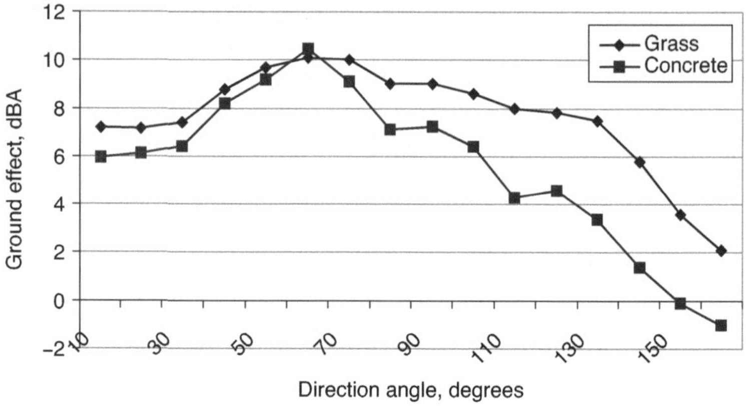


*Figure 4.91* Ground effect as a function of the angle of engine noise radiation.

differences between the predicted attenuation effects on overall A-weighted levels $L_A$ can be as much as 12 dB due to the spectrum variations between aircraft engine types, or even for one type of the engine but in various directions (see Fig. 4.91). The magnitude of the predicted variation is the same or even greater as that for $L_{Rn}$ shown in Fig. 4.87. The spectra generated by aircraft engines are defined for use in basic aircraft acoustic models (Section 4.3).

Differences between the predicted magnitudes of lateral attenuation are considerable for different types of reflecting surfaces too. Predictions obtained for grass surfaces have been used as the most appropriate for calculations of noise levels around the airports (e.g. to meet environmental impact assessment needs). The impedance of grassland can vary from one place to another and with the season. Values for the warm seasons are used for noise calculations. Sometimes, in particular cases of assessment, the mixed types of reflecting surfaces (inside the aerodrome area) must be observed due to the differences in their influence. Numerical predictions for various ground surfaces, elevation angles and distance computed by means of the Soroka and Chien formulas for the interference effects and a semi-empirical model for the impedance characteristics of the reflecting surfaces are shown in Fig. 4.92. For routine aircraft noise assessment a 'ground-effect' generator may be used such as the LATER-generator in ISOBELL'a software, designed in the Ukraine and mentioned elsewhere[44]).

In the same way, the influence of screens on noise levels can be investigated by means of the appropriate software (e.g. SCREEN-generator).

The noise from a landing ground roll is not as important to the total noise exposure as the noise from a takeoff ground roll. If thrust reversal is not used, the engines are normally idling during ground roll, and the noise will therefore be insignificant. There are two possible approaches for this part of

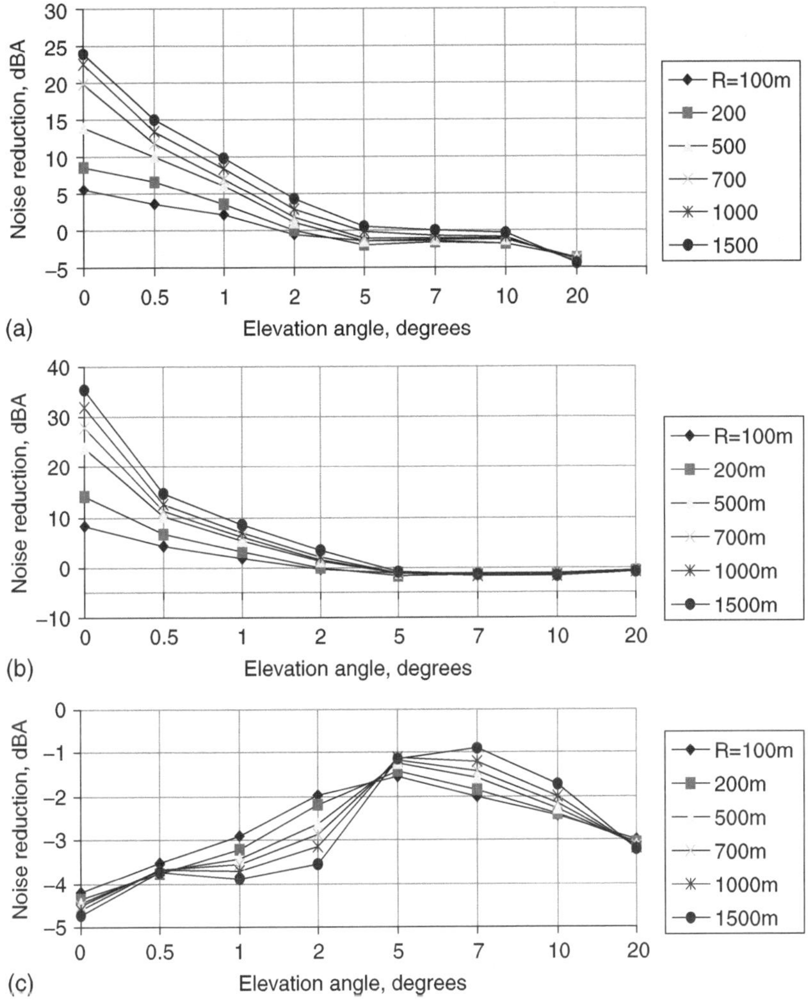


*Figure 4.92* The predicted influence of ground surface type on the lateral attenuation for aircraft fitted with turbojet engines as a function of elevation angle and distance: (a) grass covering (flow resistivity $\sigma = 350\,\text{kPa s/m}^2$); (b) snow covering (flow resistivity $\sigma = 50\,\text{kPa s/m}^2$); and (c) concrete covering (flow resistivity $\sigma = 100,000\,\text{kPa s/m}^2$).

a landing. Ground roll can be ignored or the same equations can be used as for a takeoff. In the latter case, $L_{AE}$ and $L_{Amax}$ are assumed to correspond to a low-power setting, normally idle power, and $\Delta L$ is assumed to be 0 (the contour is ended by a half circle).

When the thrust reverses are activated, the aerodynamics are changed, resulting in increased noise even with the same power setting. No models are so far available for calculating this, but it is recommended to use the noise data format corresponding to the nominal power setting and add a correction for the change in aerodynamics. For aircraft with external thrust reverses, the correction will normally be between 5 and 10 dB. For 'Chapter 2' aircraft a suitable average value will be 8 dB, whereas it is recommended to use 5 dB for 'Chapter 3' aircraft. For aircraft with internal thrust reverses, the increase of the noise due to the aerodynamics is less important and it is recommended not to make any correction in this case. It must be emphasized that these estimations are approximate and based on a very limited data.

For propeller aircraft that are able to use the propellers for reversal, it has not been possible to devise similar general corrections. The effect of propeller reversal has to be estimated in each individual case. If the traffic is a mix of jet and propeller aircraft in which the jet aircraft noise dominates, it is possible to ignore propeller reversal. Furthermore, in connection with thrust or propeller reversal, it is recommended that $\Delta L$ are set to 0, in which case the contours become parallel to the runway and are closed by half circles behind the end-of-roll.

Noise contours are calculated on the assumption that all aircraft departure and approach ground tracks follow exactly the nominal routes, so they may be liable to localized errors of several decibels. It is recommended that, for greatest reliability, the forms and parameters of the distributions of approach and departure ground tracks should be measured on each route at particular airports.

If measurements are not available, nominal departure routes may be assumed or a judgment made about the route. In this case, standard deviations about the route should be used, derived from the following expressions:

(a) routes involving turns of less than 45°

$$\sigma(y) = 0.055x - 0.150 \text{ for } 2.7\,\text{km} \le x \le 30\,\text{km} \tag{4.81}$$

$$\sigma(y) = 1.5\,\text{km for } x > 30\,\text{km}$$

(b) routes involving turns of more than 45°

$$\sigma(y) = 0.128x - 0.42 \text{ for } 3.3\,\text{km} \le x \le 15\,\text{km}$$

$$\sigma(y) = 1.5\,\text{km for } x > 15\,\text{km}$$

In equation (4.81) $\sigma(y)$ is the standard deviation and $x$ is the distance from start of roll. All distances are expressed in kilometers. For practical reasons $\sigma(y)$ is assumed to be zero between the lift-off point and $x = 2.7$ km or $x = 3.3$ km depending on the amount of turn.

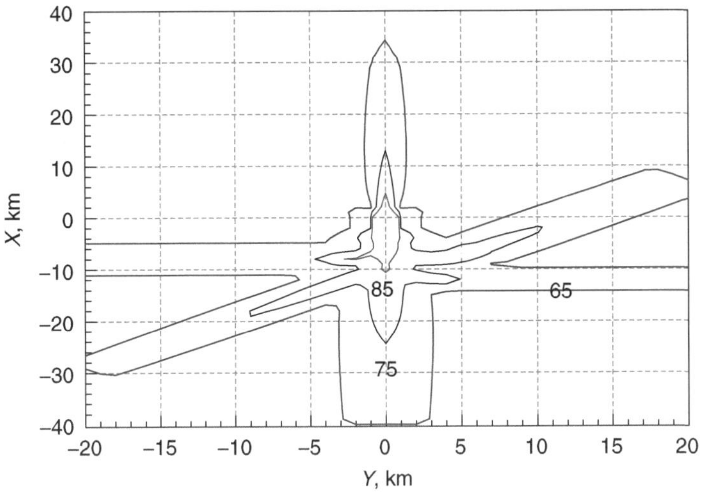


*Figure 4.93* Noise contours predicted by means of the improved calculation method.

For arrivals, the lateral dispersion can be neglected within 6 km of touchdown. Otherwise dispersion depends upon each individual runway and aircraft type.

Calculated values of noise indices are not particularly sensitive to the shape of the lateral distribution. The Gaussian form gives the best fit to many observed distributions. Although continuous distributions can be simulated, an approximate model is preferable on the grounds of computing cost. The accuracy of the five-point discrete approximation given in Table 4.31, where $y_m$ is the mean track or nominal track as appropriate, generally gives values within 1 dB of those obtained from a continuous (Gaussian) distribution, and is recommended.

The proportions given in Table 4.31 are taken into account before adding the contributions together. The effect of lateral attenuation could be taken into account for the discrete positions of the aircraft, or an overall

*Table 4.31* Proportions of aircraft to be assumed following different ground tracks spaced about a nominal track

| *Spacing* | *Proportion* |
|---|---|
| $y_m \pm 2.0\,\sigma(y)$ | 0.065 |
| $y_m \pm 1.0\,\sigma(y)$ | 0.24 |
| $y_m$ | 0.39 |

$y_m$ is the mean track or nominal track as appropriate.

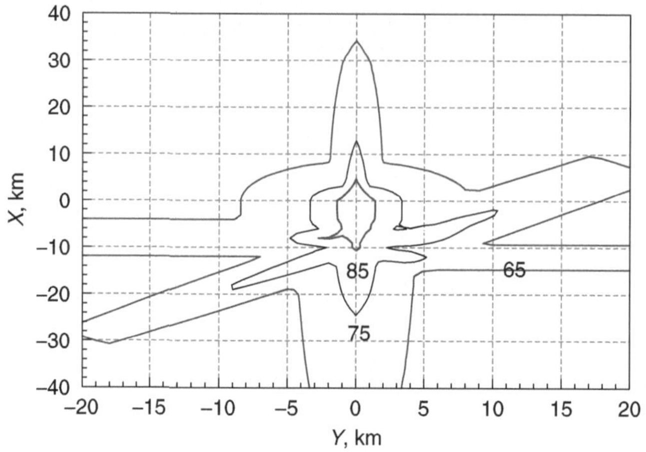


*Figure 4.94* Noise contours predicted by means of an existing calculation method.

effect could be calculated corresponding to zero lateral dispersion from the nominal flight path.

As well as being dispersed laterally, the traffic will also be dispersed vertically. The vertical dispersion is mainly due to the variation in takeoff weight, the headwind (or tailwind) component, the takeoff procedure, and how the pilot is executing the procedure. It is in general sufficient to choose a typical flight profile which is normally the average profile when calculating equivalent sound levels and the flight profile corresponding to the largest takeoff weight in the case of maximum levels. FLIGHTPATH generator is used for this purpose. If the vertical dispersion is very large due to large differences in takeoff weight, it may be necessary when calculating equivalent sound levels to divide the traffic on two or more takeoff profiles corresponding to different stage lengths.

Examples of noise contour predictions are shown in Figs 4.93 and 4.94. They are calculated for identical initial operating conditions but with (Fig. 4.93) and without (Fig. 4.94) an account of the factors and improvements described in this chapter. The comparison shows that the previous method overpredicts noise levels compared with the improved method.

# 5 The influence of operational factors on aircraft noise levels

## 5.1 Aircraft on the ground

Common sources of noise from ground-based operations at airports include engine maintenance run-ups, taxiing aircraft, operation of ground and auxiliary power units, preflight run-ups, takeoff roll and thrust reverses, movement of surface vehicles and some airport power plants. Noise levels have been measured by numerous researchers for a variety of airports and conditions. Of course, the takeoff roll produces the maximum levels of noise in the working area of the airport and in its vicinity because of the maximum engine thrust necessary for this stage. Table 5.1 compares differences between the average peak values of noise for characteristic operations in airport and the noise level during takeoff.

Bearing in mind the short duration of runway operations, it is possible to foresee that, in particular cases, other types of aircraft operations inside the airport may produce a higher noise impact than that due to aircraft takeoff. For example, this might be true of engine maintenance run-ups or taxiing, if they are performed during the night close to residential areas. Engine maintenance run-ups include a half-minute of maximum thrust, so the peak noise level may be the same as during aircraft takeoff roll. If the location of run-ups is near residential areas, several technical means must be used for reduction of noise impact, including acoustic screens or engine noise mufflers.

The basic operational factors that have an influence on noise impact include the state and slope of the runway, meteorological factors (temperature, pressure, relative humidity, wind direction and velocity), flight (takeoff or landing) aircraft weight, thrust-to-weight ratio, the aerodynamic configuration of the aircraft and topographic conditions in the vicinity of the ground operations. Other operational conditions which influence noise impact include frequency of movements, aircraft fleet composition and flight procedures for separate aircraft types. The results of research show that the type of engines in an aircraft power plant has a dominant influence.

Although the acoustical impact in an airport area and its environment is important, air pollution is considered more frequently as a priority for

*Table 5.1* Differences between the average peak values of noise during characteristic operations in an airport and those during takeoff (the reference distance of the data is 150 m)

| *Type of airport operation* | *Noise level difference [dBA]* |
|---|---|
| Engine maintenance run-up | 0 |
| Engine pre-flight run-up | 5–10 |
| Landing run with reverse thrust | 5–8 |
| Taxiing | 16–24 |
| Auxiliary power unit | 20–30 |

*Table 5.2* Effectiveness of operational procedures for air pollution reduction during taxiing

| *Operational procedures for air pollution reduction* | *Air pollution reduction (% by mass)* | | |
|---|---|---|---|
| | *Products of incomplete burning* | *Nitrogen oxides* | *Fuel consumption* |
| Tug use | 35–50 | 5–10 | 26 |
| Taxiing with one working engine | 30 | 0 | 3 |
| Reduction of queuing aircraft | 7 | 1–3 | 1–2 |
| Optimum use of runways (if more than one) | 15 | 5 | 10 |

analysis and optimization of operation modes if an aircraft is moving on ground. The operational procedures that can diminish the total mass of air pollution by aircraft on the ground are listed in Table 5.2.

If an aircraft is taxiing with non-operating engines, the resulting reduction of air pollution depends on two factors. Although there is a decrease in the number of sources of air pollution, the operating engines have to operate at higher thrust than is necessary if all of the engines are used to move the aircraft on the ground.[1,2] The optimum modes and number of operating engines have been analyzed and defined for the taxiing of Tupolev-134A, Tupolev-154 and Il'ushin-62M aircraft (with two, three and four engines in their power plants, respectively). For the latter aircraft, an optimum number of working engines during aircraft taxiing consistent with providing the necessary thrust may be defined (Fig. 5.1).

Chronometrical analysis of the taxiing stages for Tupolev-134A, Tupolev-154 and Il'ushin-62M, which was carried out in real operation conditions in airports, shows that the most important engine operation mode during taxiing is the idle conditions. Figure 5.2 plots the temporal proportion of idling mode use during taxiing of the aircraft $\Delta t_{id} = \Delta t_{id}/t_{tax}$, where $t_{tax}$ is the total duration of the taxiing, in dependence from aircraft thrust-to-weight ratio $\tau_{id} = T_{id}/W$.

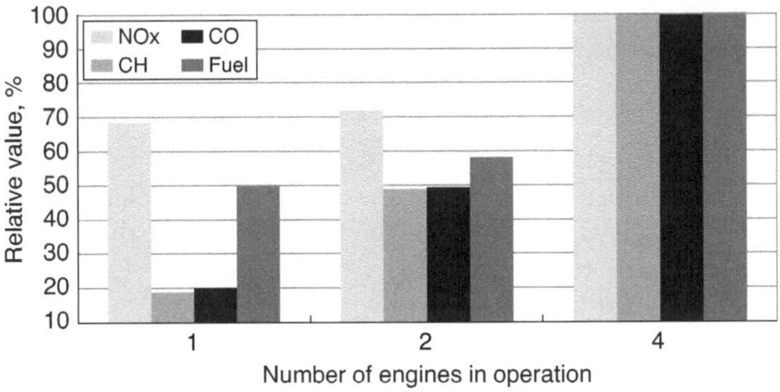


*Figure 5.1* The influence of the number of operating engines on emissions and fuel burn reduction for a four-engined Il'ushin-62M aircraft.

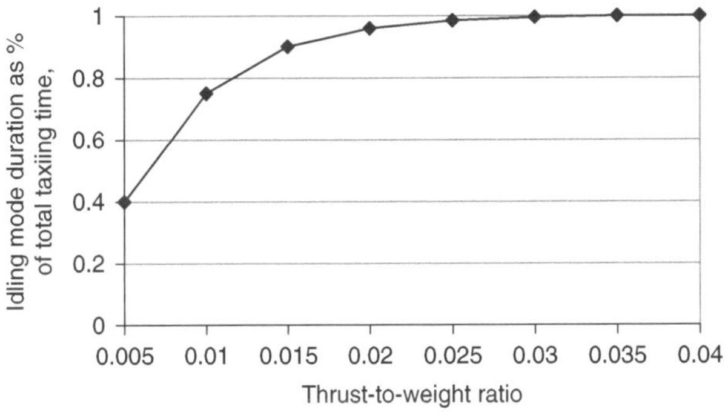


*Figure 5.2* The influence of aircraft thrust-to-weight ratio on the proportion of taxiing time in idling mode.

Decreasing the number of working engines and increasing the aircraft mass (weight) decreases $\Delta t_{id}$. If the thrust-to-weight ratio $\tau_{id}$ is $> 0.025$, during almost the entire taxiing stage the engines are operated at idle mode. This is mostly typical for the taxiing stages after the aircraft landing. For taxiing of aircraft before takeoff with all operating engines and for the thrust-to-weight ratio $\tau_{id} < 0.02$, a small part of taxiing is executed with higher engine modes, which may provide thrust $T = (0.15\ldots0.3)T_{max}$, where $T_{max}$ is the thrust of the engines at maximum operation mode. Taxiing with a smaller number of working engines requires an increase in the duration of higher engine operation modes, which are providing the higher necessary thrust. For one engine in operation during aircraft taxiing, the necessary thrust may reach $T = (0.2\ldots0.4)T_{max}$ for aircraft with two or three engines and $T = (0.5\ldots0.7)T_{max}$ for aircraft with four engines in the power plant.

Taxiing with fewer engines in operation may increase the noise inside the airport area.[3] On the one hand, there is a decrease due to the reduction of noise sources in their number, which can be calculated from

$$\Delta PNL = 10\,\lg n/n_{CY},$$

where $n_{CY}$ is a total number of engines in the aircraft power plant and $n$ is the real number of engines in operation. On the other hand, there is an increase in noise levels due to the increased required thrust of engines in operation. This is calculated from

$$\Delta PNL = 10\,\lg \frac{n}{n_{CY}}\left(\frac{R}{R_{CY}}\right)^{28.6},$$

where $R$ is the thrust cut-off degree for the engines. The total noise level reductions for fewer engines in operation are shown in Fig. 5.3.

Another important aspect for the use of fewer engines in operation during the aircraft taxiing is the choice of the appropriate location for starting the engines that are not used during the taxiing and for making their pre-flight run-ups.

Other important ground activities for the aircraft on the aerodrome are the rolls along runway. The state and slope of the runway can have a significant influence on the noise impact aside of the runway and under the aircraft flight path. In particular, calculations show that a friction coefficient of runway surface changing from 0.02 (dry concrete covering) to 0.03 increases the aircraft running start length by only 3–4 per cent. The aircraft running length also increases if there is an upward runway slope in the direction

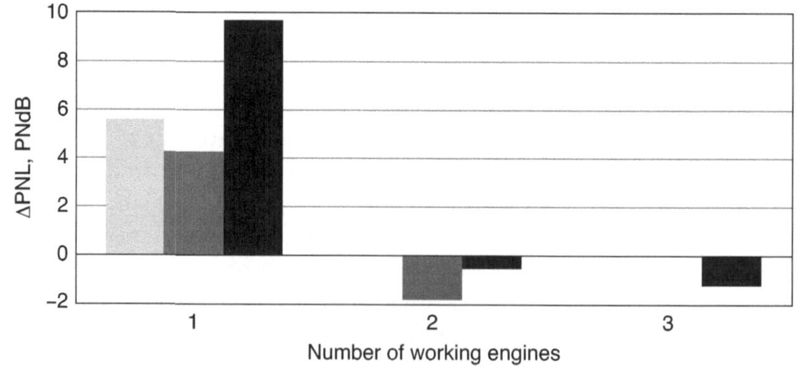


*Figure 5.3* Dependence of noise level reductions on the number of working engines: white – Tupolev 134 with two engines in the power plant; gray – Tupolev 154 with three engines in the power plant; black – Il'ushin-62M with four engines in the power plant.

of takeoff. The relationship between the approximate running length $L_p$ and the running length $L_{p0}$ for the runway without slope ($i = 0$, where $i$ is the runway slope angle in radians) is given by

$$L_p \cong \frac{L_{po}}{1 - \left(2igL_{po}/v_{otp}^2\right)}.$$

where $v_{otp}$ is the speed of the aircraft unstick from the runway.

The reduction of the aircraft takeoff weight in comparison with maximum weight leads to shorter running lengths and a shorter duration for the takeoff run. Although the peak values of noise level alongside the runway must be the same, the equivalent or effective noise levels will be slightly smaller because of the shorter duration of the takeoff run.

The thrust-to-weight ratio and the aerodynamic configuration of the aircraft depend on the aircraft flight procedure (piloting method) during the landing–takeoff cycle. If the takeoff weight and runway length provide the necessary conditions, it may be possible to use the nominal (not the maximal as usually) engine operation mode at the takeoff run. This also has the benefit of increasing the lifetime of the engines. In this case, the length of run should be greater, but the peak and equivalent noise levels are smaller than if the takeoff is performed with maximum engine thrust.

Aerodynamic configuration of the aircraft influences the takeoff run distance mainly. Changes in lift coefficients due to various flap retractions define the speed of aircraft unstick from the runway and therefore the length of run. Accordingly, they influence equivalent noise levels rather than peak levels.

The use of a 'rolling-start', that is, when the aircraft begins to roll as soon as it appears on the runway without stopping at the point of an executive start, has the same consequences as the runway condition. It does not affect the peak noise levels, but decreases the duration and length of the run.

Wind has an influence on takeoff run parameters, but its following influence on noise levels is quite small. The main influence of wind is found on sound propagation effects. This depends on the wind direction and wind profile in the atmosphere between the noise source and the control point. Effects such as the formation of 'shadow zones', which diminish noise levels, are very specific, therefore they are not taken into account in decisions on operational factors. Any other sound propagation effects like atmospheric absorption that depend on meteorological factors are not considered here.

Topographical conditions may have an important influence on sound propagation. Besides the effects of wavefront spreading between noise source and receiver, the interference between direct sound and reflection from the soft covering of the ground surface will influence the rate of the noise level decrease, which, in total, may be as much as 8–12 dBA per doubling of distance depending on the type of covering and the geometry of sound propagation paths. There can be seasonal peculiarities; for example, snow covering has a different effect from grass covering. On other hand, hard 'ground effect' may result in a noise level decrease of less than 6 dBA per distance doubling.

## 5.2 Under the flight path

The advantage of modeling methods is that they enable more thorough analysis of the influence of various factors (separately or in combination) on aircraft noise levels. The reference conditions assumed in the usual methods for calculating aircraft noise levels are the following:

- wind – 4 m/s (8 knots) headwind constant with height above ground;
- runway elevation – mean sea level;
- runway gradient – none;
- air temperature – 15°C;
- aircraft takeoff gross weight – 85 per cent of maximum takeoff weight;
- aircraft landing weight – 90 per cent of maximum landing weight;
- number of engines supplying thrust – all;
- atmosphere – standard atmosphere (SA).

These reference conditions are subject to change for particular noise events, so the changes in flight paths and in noise generation/propagation effects may occur too.

The effective perceived noise levels (EPNL) at noise monitoring points under the flight trajectory [No. 2 for takeoff and No. 3 for landing, both defined in accordance with International Civil Aviation Organization (ICAO) requirements[4]] and the area of the 90 EPNdB noise contour $S_{90}$ have been used as the noise impact criteria in the analysis. Only the influence of factors such as speed and direction of wind on trajectory parameters has been considered. A model for wind speed profile effects for sound propagation, such as 'shadow zones', is considered in Chapter 4. The important factors that influence the noise levels are predicted to be the flight (takeoff or landing) weight, thrust-to-weight ratio, flight speed, aerodynamic configuration of the aircraft and the ambient meteorological parameters including temperature, pressure and humidity.

A special category of operational conditions that may affect noise impact consists of parameters such as flight (movement) intensity, aircraft fleet composition and the flight procedures of separate aircraft types. Research has shown that the type of engines in the aircrafts' power plant and the bypass ratio $m$ for turbofan engines have dominant influences in combination with other operational factors.

Certain of the various operational factors cause no more than 1 dB deviations from noise levels corresponding to normal flight operations. For example, the state and slope of a runway have an insignificant influence on noise under the aircraft flight path, as shown in the previous section.

For the same aircraft types, the use of 'rolling-start' speeds of between 5 and 15 m/s decreases the length of roll along the runway between 3 and 6 per cent. The corresponding noise level at the control point 6500 is predicted to decrease by 0.3 and 0.5 EPNdB and the noise contour area $S_{90}$ is predicted to change only by 2 per cent on average.

An increase in flight acceleration $a$ at initial climbing enables a decrease in the vertical component of flight speed of the aircraft associated with a decrease in the climbing angle by $\Delta\Theta$ given by $\sin \Delta\Theta = a/g$. This means that the flight path is located closer to the point (surface) of noise control.

Noise–power–distance (NPD) curves ('noise radii' in the Ukrainian calculation method[5]), are defined in the same manner for reference conditions, which are those mostly required for noise zoning assessment (as in ICAO Circular 205[6]). In particular cases, which may be different from the reference flight conditions, noise radii must be recalculated. So, for exposure indices, the flight speed influence may be considered by using the known functional dependence between noise radius $R_N$ (i.e. the distance for a particular mode and index value) and flight speed $v$ (see Section 4.2). Considering only the wind effects on flight trajectories, the noise level in the upwind direction at the point of noise control may increase by 0.5 and 1 EPNdB compared with zero wind speed. Downwind, the levels may decrease by a similar amount. The corresponding noise contour changes are less than 2 per cent on average for the aircraft considered.

For stable climbing, the flight path angle $\Theta$ is related to the thrust-to-weight ratio $\tau$ by:

$$\sin\Theta = \tau - K^{-1},$$

where $K$ is a parameter indicating the aerodynamic quality of the aircraft in a climbing configuration.

The decrease in takeoff weight of an aircraft, in comparison to the maximum permitted operational weight, leads to a steeper takeoff/climbing flight path. So the height of the flight path above the control point/area becomes greater and the *EPNL* at the control point decreases. The predicted decrease can be up to 6 EPNdB for aircraft with low bypass ratio ($m < 2$) turbofans, and up to 4 EPNdB for aircraft with higher bypass ratio engines. The corresponding noise contour area $S_{90}$ is predicted to be decreased by 15 to 30 per cent. The maximum values correspond to lower bypass ratio engines (Fig. 5.4). The parameter used is $S_{90} = (S - S_{max})/S_{max} \times$ 100 per cent, where $S$ corresponds to the weight under consideration and $S_{max}$ corresponds to the maximum weight.

Similar considerations apply to noise levels during landing. Decreasing aircraft weight reduces the noise because of the decreased thrust required by the appropriate operation mode and the insignificant decrease of landing speed.

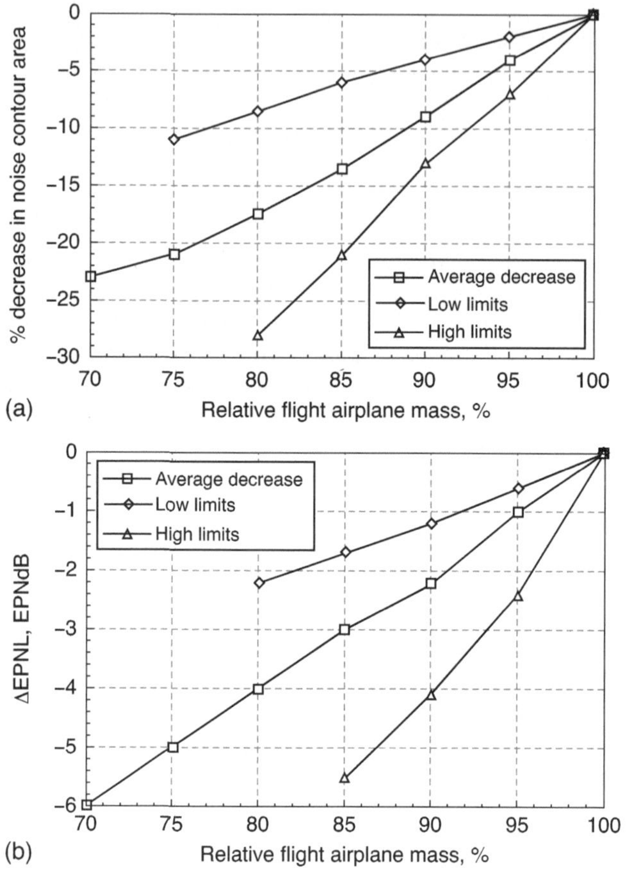


*Figure 5.4* Predicted influence of aircraft flight mass on (a) the area of the 90 EPNdB noise contour and (b) the noise level at the point of noise control 6.5 km from the brake release on the runway.

The landing aircraft weight dominates the necessary operation mode of the engine, but the rate of noise contour area decrease $dS/dW_{landing}$ for landing flight modes is higher than that for climbing modes. Landing of aircraft with maximum landing weight compared with the usual weight, which is approximately 20 and 25 per cent below maximum, is predicted to lead to an increase in the noise contour areas $S_{90}$ by 45–75 per cent, and to increase noise levels at the control point No. 3 by 4–6 EPNdB. The upper limits correspond to aircraft with turbofan engines having bypass ratios $m \leq 2.5$. The predicted noise contour gradient values for landing, $dS/dW_{landing}$, are between 3 and 3.5 and for takeoff/climbing, $dS/dW_{takeoff}$, values are between 0.5 and 1.4.

For some aircraft types during the takeoff run, there are engine life advantages, as long as the takeoff weight and runway length allow the use of engine modes other than the regulatory maximum. Since the thrust-to-weight ratio is less in this case, the length of run is greater and flight path heights are smaller. Levels of noise radiated by the engines are reduced between 3 and 4 EPNdB for turbofan engines with a bypass ratio $m \leq 2.5$ and between 1 and 2 EPNdB for turbofans with $m > 2.5$. Moreover, at the lateral noise control point No. 1 at takeoff (a maximum of 450 m from the side of the runway at the beginning of the running start), the noise levels become lower too. Although for turbofan aircraft with a bypass ratio $m \leq 2.5$ the noise contour area $S_{90}$ is predicted to decrease between 20 and 40 per cent, for turbofan aircraft with $m > 2.5$, it is predicted to increase by between 10 and 20 per cent.

The influence of atmospheric temperature and pressure on aircraft noise levels is the result of the change in flight path parameters, the change in the absorption parameters of the atmosphere and the change in noise generation by the main acoustical sources. An increase in temperature or decrease in pressure means that the corresponding air density decreases and causes a decrease in the air consumption ratios within the engine flow tracts. As a result, the thrust decreases but the engine jet velocity increases. The aircraft takeoff velocities and lengths increase and the takeoff gradients and heights under the control surfaces decrease. The importance of ambient air temperature for noise impact under the takeoff path is shown in Fig. 5.5.

In cold conditions, the noise levels at the control point may decrease by up to 10–12 EPNdB compared with International Standard Atmosphere (ISA) conditions. For example, in Integrated Noise Model (INM, version 6) the noise levels in the NPD database are corrected to reference-day conditions (temperature, 77°F; pressure, 29.92 inches of mercury; and altitude, mean sea level), in a way that noise levels can be adjusted to airport temperature and pressure using:

$$\Delta L_{imp} = 10\lg(\rho c/\rho c_{ref})$$

$$L = L_{ref} + \Delta L_{imp},$$

where $\Delta L_{imp}$ is the acoustic impedance adjustment to be added to noise level data in the INM NPD database (dB), $\rho c$ is the acoustic impedance at observer altitude and pressure (N s/m$^3$).

The noise level contributions from each segment of the flight path are derived from the NPD data stored in the international Aircraft Noise and Performance (ANP) database. However, it must be noted that these data have been normalized using average atmospheric attenuation rates defined in SAE AIR-1845.[7] Those rates are averages of values determined during aircraft noise certification testing.

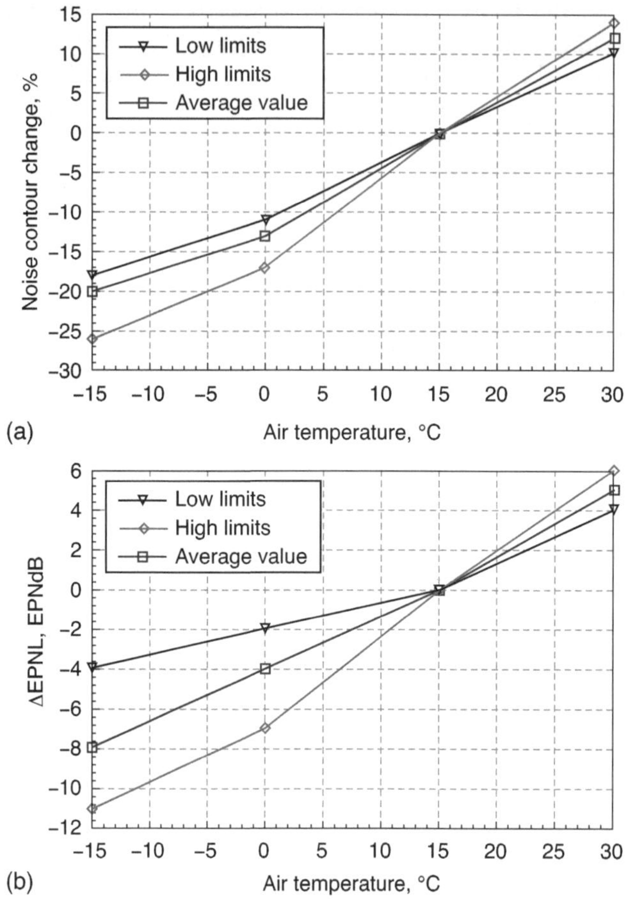


*Figure 5.5* The predicted influence of ambient temperature on (a) the noise contour area and (b) the *EPNL* at control point No. 2 during takeoff.

Adjustment of the NPD curves to user-specified conditions – temperature, humidity and pressure – is performed in three steps:

(1) First, the reference spectrum is corrected to remove the SAE AIR-1845 atmospheric attenuation $\alpha_{n,ref}$:

$$L_n(d_{ref}) = L_{n,ref} d_{ref} + \alpha_{n,ref} d_{ref},$$

where $L_n(d_{ref})$ is the unattenuated spectrum at $d_{ref} = 305$ m and $\alpha_{n,ref}$ is the coefficient of atmospheric absorption for the frequency band $n$.

(2) Next, the corrected spectrum is adjusted to each of the ten standard NPD distances $d_i$ using attenuation rates for both the SAE AIR-1845 atmosphere and the user-specified atmosphere in accordance with it.[8] For the SAE AIR-1845 atmosphere:

$$L_{n,ref} d_i = L_{n,1} - 20\,\lg(d_i/d_{ref}) - \alpha_{n,ref} d_i.$$

For the user atmosphere:

$$L_{n,866A} T,RH,d_i = L_{n,1} - 20\,\lg(d_i/d_{ref}) - \alpha_{n,866A} T,RH\, d_i,$$

where $\alpha_{n,866A}$ is the coefficient of atmospheric absorption for the frequency band $n$ (expressed in dB/m) calculated using SAE ARP866A with temperature $T$ and relative humidity $RH$.[8]

(3) At each NPD distance $d_i$ the two spectra are A-weighted and decibel-summed to determine the resulting A-weighted levels $L_{A,866A}$ and $L_{A,ref}$, which are then subtracted arithmetically:

$$\Delta L(T,RH,d_i) = L_{A,866A} - L_{A,ref}$$

Here, the reference spectrum is one of the possible unique spectral classes. A spectral class is the grouping of aircraft considered similar based on the combination of the aircraft and engine types. Considerations for grouping aircraft include the airframe, type of engine, number of engines, location of engine and bypass ratio. Similarity is based on the shape of the spectrum and the relative location of any tones below 1000 Hz.

The spectra limits for class 104 aircraft are presented in Fig. 5.6a.[9] Figure 5.6b presents the normalized spectra limits along with the spectrum proposed to represent this spectral class. The representative spectrum for this spectral class is the weighted arithmetic average of the individual third-octave band spectral data. The weighting was based on a recent annual survey of the number of departures for each aircraft type.

Given that the ground effect curves for each individual aircraft spectrum fall within the ±1 dB limit curves for all elevation angles, the proposed representative spectrum is considered to represent the individual spectra used to derive the spectral class adequately. Figure 5.6c presents the final spectral class.

Under different atmospheric conditions, such as air temperature, noise radiation and air absorption may contribute significantly to NPD recalculation. Therefore, aircraft/engine reference spectra are used, just as in the new European Civil Aviation Conference (ECAC) approach or the current version of INM. For noise radiation, the recalculation is performed taking account of the dominant noise source for the aircraft and the flight mode

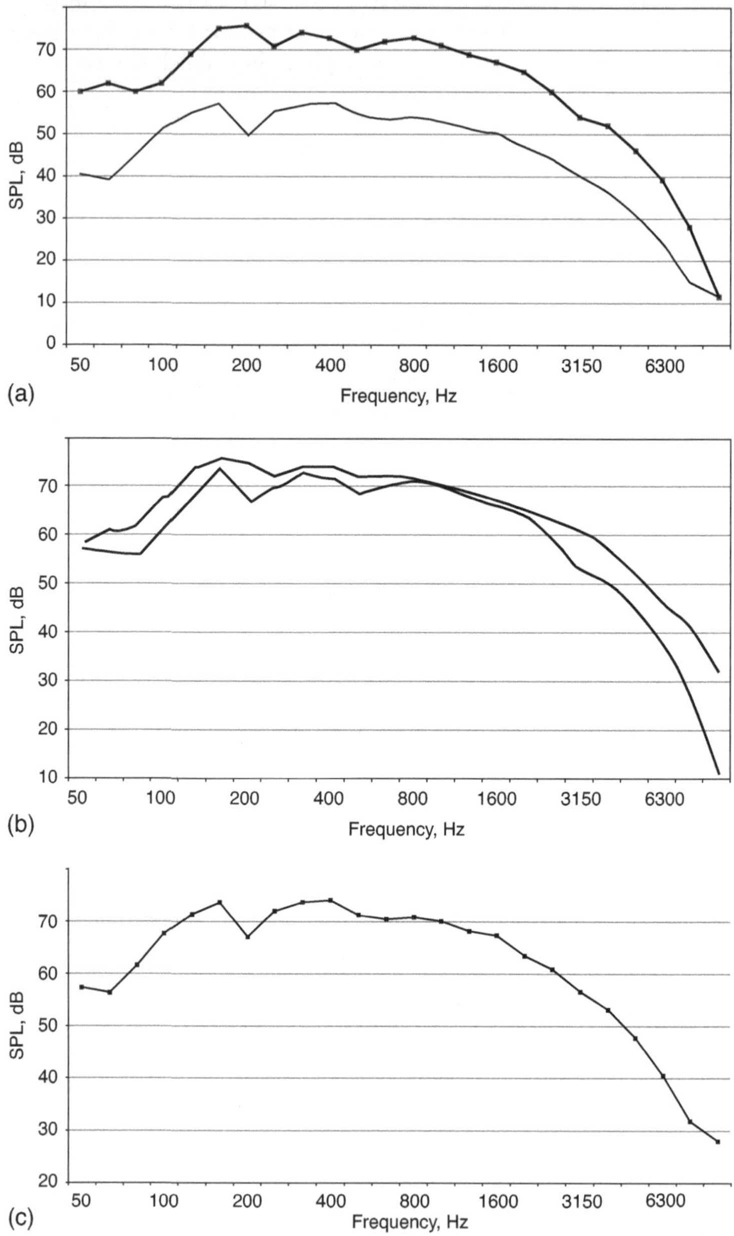


*Figure 5.6* Departure spectral classes for 104 aircraft: (a) actual spectra limits (maximum value – upper curve; minimum value – lower curve) for class 104 aircraft; (b) normalized spectra limits (maximum value – upper curve; minimum value – lower curve); (c) the final spectrum for 104 aircraft class.[9]

(stage) under consideration. There are three possible dominant sources considered in the current national techniques: the jet (usually, bypass jet), the fan (in forward and upward directions of noise propagation) and the airframe. Specific corrections for noise indices are included in the functional dependence on dominant source.

For example, the Ukrainian software *NoBel* enables recalculation of the noise spectrum for specific atmospheric conditions. An example calculation for an Il'ushin-86 aircraft is shown in Fig. 5.7. Even for the reference distance of 1 m, the differences between the overall sound pressure level (OASPL) in the direction of maximum noise generation are quite different from the simple impedance adjustment used in INM (Table 5.3).

It is not obvious that the dependence of the OASPL increment on ambient temperature must be uniform because, for some engines, the laws governing their control may be different for various temperatures. Figure 5.8 shows that, for the D-36 engines, installed on the Yakovlev-42 aircraft, the change in the law on control of engines at temperatures higher than 15°C disrupts the uniform dependence. The temperature adjustment due to the sound propagation effects is higher than simple impedance adjustment too (Fig. 5.9).

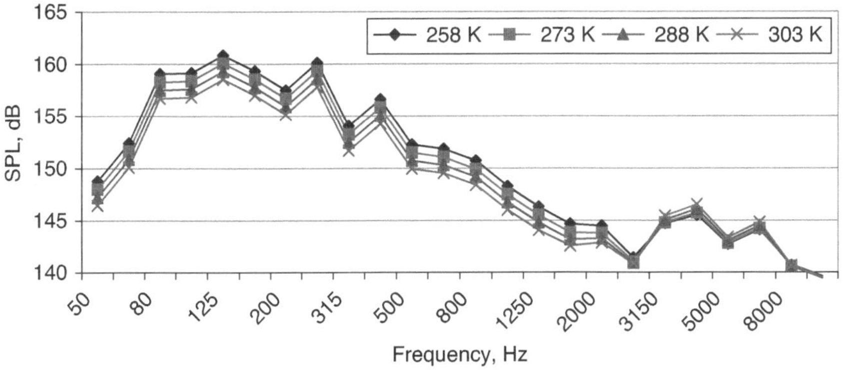


*Figure 5.7* Predicted noise spectra in the direction of maximum noise generation for an Il'ushin-86 aircraft at a reference distance of 1 m for four ambient air temperatures.

*Table 5.3* Calculated overall A-weighted sound pressure level (OASPL) increments (ue to ambient air temperature in the direction of maximum noise generation with impedance adjustment (an example for an Il'ushin-86 aircraft)

| $T_{atm}$ (K) | 258 | 273 | 288 | 303 |
|---|---|---|---|---|
| $\Delta L_{T-288}$ | 1.5 | 0.7 | 0 | −0.7 |
| $\Delta L_{AT-288}$ | 1.2 | 0.6 | 0 | −0.5 |
| $\Delta L_{imp}$ | 0.55 | 0.27 | 0 | −0.25 |

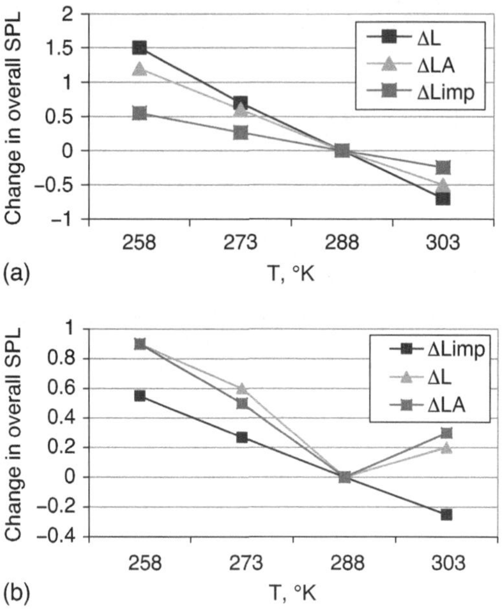


*Figure 5.8* Predicted change in overall SPL due to air atmosphere temperature in the direction of maximum noise generation with impedance adjustment (INM): (a) NK-86 engine (Il'ushin-86); (b) D-36 engine (Yakovlev-42).

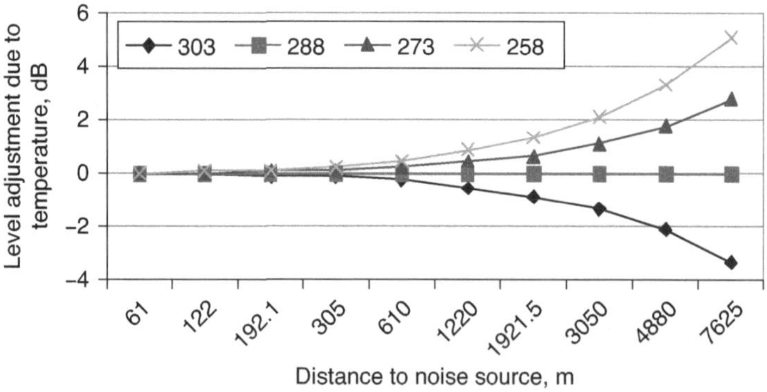


*Figure 5.9* Predicted dependence on distance of the adjustment for sound propagation effects at four temperatures.

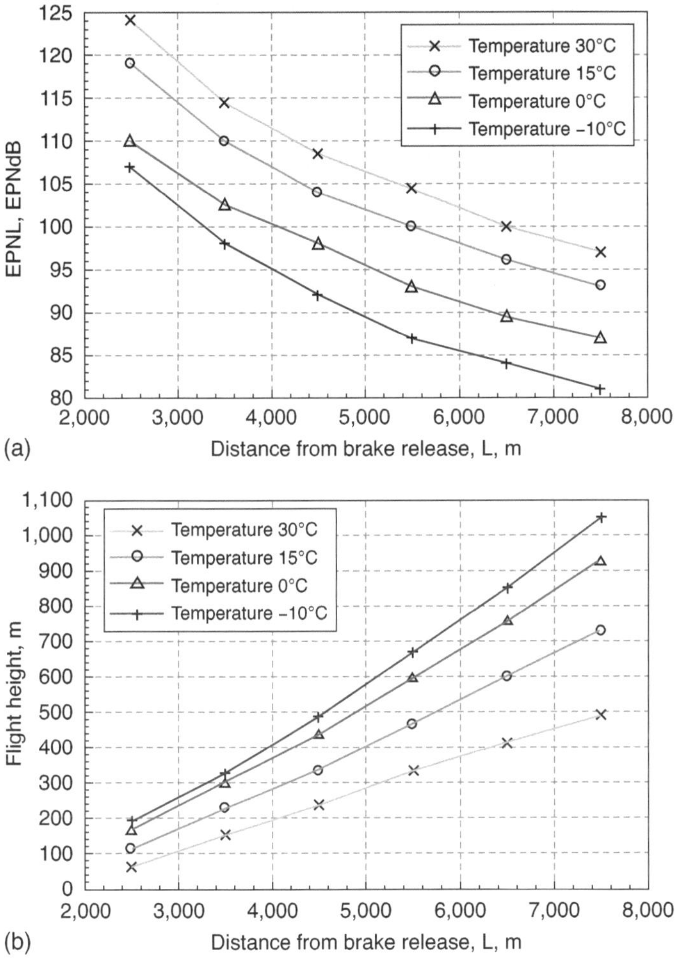


*Figure 5.10* Predicted influence of ambient temperature, for an aircraft with high bypass turbofan engines on (a) noise levels under the trajectories and (b) flight trajectory parameters.

A more thorough analysis of temperature influence for an aircraft with high bypass ratio ($m = 4$) turbofans is shown in Fig. 5.10. As the atmospheric temperature varies between –20 and +30°C, the *EPNL* at the monitoring point under the takeoff path is predicted to change between 8 and 14 EPNdB, and the noise contour area $S_{90}$ is predicted to change between 20 and 30 per cent [relative to the area $S_{90}$ for standard atmosphere (SA) conditions].

The explanations of the predicted changes in *EPNL* at the takeoff control point (6.5 km from the brake release) can be deduced from Figs 5.11–5.13.

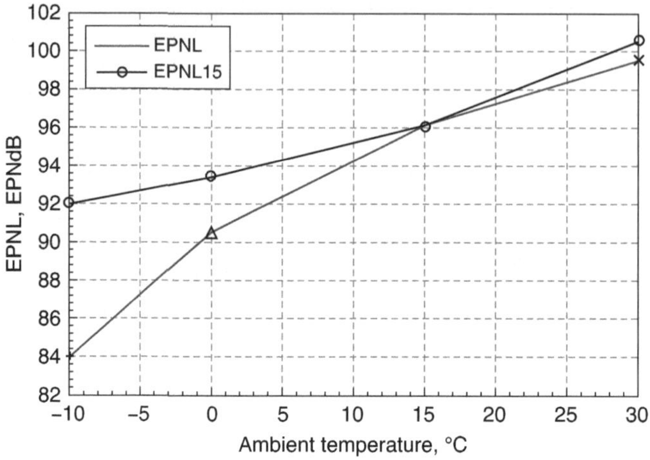


*Figure 5.11* Difference between the effective perceived noise levels (EPNL) predicted at the control point No. 2 (6.5 km) for varied ambient temperature conditions and for International Standard Atmosphere (ISA) conditions ($EPNL_{15}$), but for trajectories computed for varied temperature (Fig. 5.3).

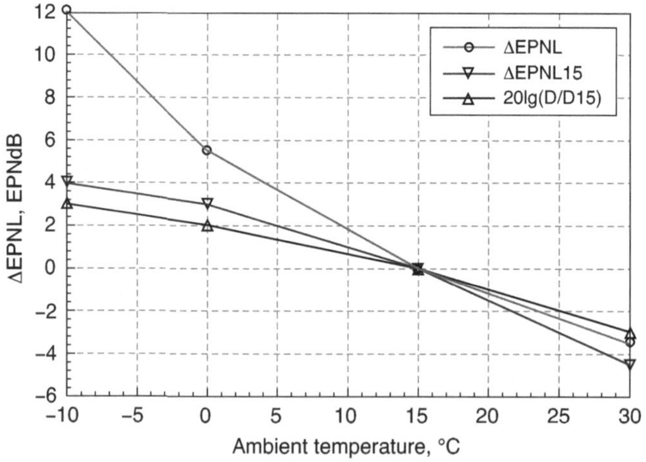


*Figure 5.12* Predicted effective perceived noise levels (EPNL) differences at control point No. 2 in comparison to International Standard Atmosphere (ISA) conditions. The curve $20\lg(D/D_{15})$ is explained in the text.

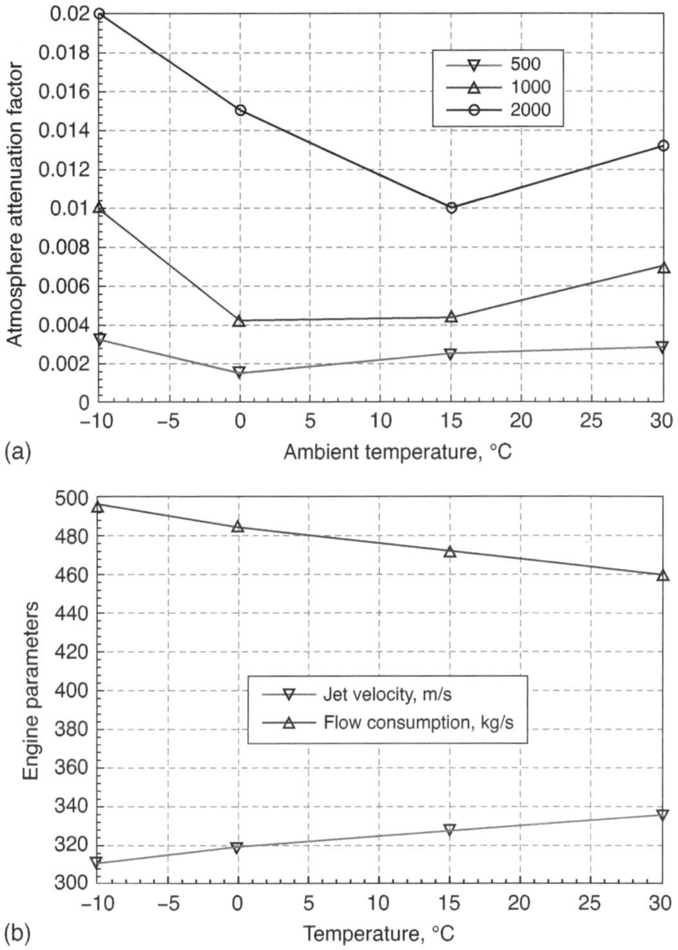


*Figure 5.13* The predicted influence of ambient temperature on (a) atmospheric attenuation and (b) engine mode parameters responsible for noise generation.

The differences between *EPNL* and $EPNL_{15}$ (the value of *EPNL* at temperature 15°C) are partly the result of the temperature influence (shown in Figs 5.11 and 5.12). The corresponding values of engine air flow consumption $G_0$, engine jet velocity $V_j$ and coefficients for air absorption $\alpha$ at the main frequencies 500–2000 Hz (calculated according to ICAO requirements[10]) are shown in Fig. 5.13.

While landing, the noise levels decrease with decrease in air temperatures, but only due to engine thrust changes, since the landing flight path over the control point No. 3 is defined by the glide slope requirements at a specific airport. Usually the angle of the glide slope is equal to 3 degrees.[4]

A decrease in ambient atmospheric pressure (between 101,325 and 90,000 Pa) corresponding to a change of the height of an airport between 0 to 1000 m (assuming that the change of pressure with height is in accordance with the ISA laws) leads to an increase in predicted noise. The increase is between 4 and 8 EPNdB in *EPNL* and between 10 and 20 per cent in $S_{90}$.

The main thrust of this section has been to point out the necessity of accounting for the influence of certain operational factors when calculating noise levels around an airport. Moreover, if these calculations are connected with noise zoning and land-use planning, the worst possible operational conditions must be considered. These will correspond to the highest intensity of aircraft movement, their highest possible in-flight weights and the highest ambient temperatures (i.e. the warmest season of the year). Nevertheless, SA conditions must be used when comparing the calculation results for different airports and for other operational circumstances.

## 5.3 Takeoff and climbing

Various operational factors may influence the effectiveness of low-noise procedures. Typically, the 90 EPNdB noise contour area, $S_{90}$, is used as the criterion of effectiveness for noise control. In addition, estimates of noise level *EPNL* at control points Nos 2 and 3,[4] the fuel consumption $M_{fuel}$ and the equivalent emitted mass of air pollution $M_{eq}$ during the flight stage of interest are considered.

Models for the dependence of noise impact area and noise radius $R_n$ on flight parameters (Chapter 4) enable predictions of the change in noise impact resulting from changes in flight trajectory parameters. Analysis of these predictions is used for determining optimum aircraft climbing trajectories to reduce noise impact. Primarily, the extent of noise reduction is determined by the degree and instant of engine cutback, the height, flight speed and the aerodynamic configuration of the aircraft. The rotation frequency, $n$, of the engine fan rotor is important in considering the influence of engine cutback.

The influence of the aerodynamic configuration (i.e. the wing flaps retraction angle) depends on the following factors:

- The clean (in-flight) aerodynamic configuration (flap angles are equal to zero): the in-flight aerodynamic quality $K$ of aircraft is higher than during takeoff (almost twice for the aircraft types considered). Therefore, the flight trajectory inclination angle $\Theta$ and flight height $H$ increase quicker with the in-flight aerodynamic configuration, even after taking into account the fact that the minimum safe flight speed is greater by between 30 and 35 per cent.
- The degree of engine cutback is higher during flight than during takeoff.

Therefore, to achieve greater noise reduction effectiveness at a point or zone of noise control under the flight path, it is necessary to use a 'cleaner' aerodynamic configuration, while at the same time meeting the flight safety requirements.

On the assumption that the parameters of aircraft motion are approximately constant along the considered section of the trajectory, then expression (4.15) in Chapter 4 for determination of surface area $\Delta S$ along the trajectory section ($X_{begin}$, $X_{end}$) leads to:

$$\Delta S_k = \frac{R_n^2}{\sin\Theta}\left[\frac{\pi}{2} - \arcsin\left(\frac{X_H \sin\Theta}{R_n}\right)\right] - X_H\sqrt{R_n^2 - X_H^2 \sin^2\Theta} \qquad (5.1)$$

The final coordinate $X_{end}$ is determined by the formula

$$x_k = \frac{(R_k v_k)_{nom}}{\sin\Theta_k v_k},$$

where $(R_k V_k)_{nom}$ is the product of the noise radius and flight speed during the current stage of flight (in normal conditions the engine operation mode is nominal).

The change in area $S$ as a result of the change in the flight control parameter $n$ is estimated from the derivative $dS/dn$ according to the following formula:

$$\frac{dS}{dn} = \left(\frac{2R_N}{\sin\Theta}\frac{dR_N}{dn} - \frac{R_N^2}{\sin^2\Theta}\frac{d\sin\Theta}{dn}\right)\left(\frac{\pi}{2} - \arcsin\frac{X_i \sin\Theta}{R_N}\right)$$
$$- \left(1 - \frac{X_i^2 \sin^2\Theta}{R_N^2}\right)^{-1/2}\left(\frac{R_N X_i}{\sin\Theta} - \frac{X_i^3 \sin\Theta}{R_N}\right)\frac{d\sin\Theta}{dn} \qquad (5.2)$$

In expression (5.2), the derivative is given by $d\sin\Theta/dn = dT/dn(1/G)$, where $G$ is the aircraft weight. The value of the derivative $dR/dn$ is determined from the dependence of noise radius on engine rotation speed similar to that shown in Fig. 4.9. Since $dR_N/dn >> d\sin\Theta/dn$, the value of the derivative is positive (i.e. $dS/dn > 0$). Thus, following engine cutback, the area $S$ must decrease.

Values of the derivative $dR_N/dn$ (see Fig. 4.9) greatly exceed the values of derivative $dR_N/dv$ (Fig. 4.10 in Chapter 4) for takeoff and climbing flight modes. For aircraft with in-flight aerodynamic configuration, the absolute values of the noise radius $R_N$ are less than those for aircraft with declined wing flaps (Fig. 4.9). Besides noise radius estimation, it is necessary to consider the dependence on conditions of surrounding atmosphere, for example, on temperature (Section 5.2).

The duration of the engine cutback procedure is determined by $\Delta t_{cutback}$, which is defined by the formula:

$$\Delta t_{cutback} = \left( \frac{R_{Nnom}}{\sin \Theta} - X_i \right) \frac{1}{v}, \tag{5.3}$$

where $R_{Nnom}$ is the noise radius for the nominal setting of the engine, which is used for further aircraft climbing.

The permissible degree of engine cutback $\check{T}_{cutback}$ is defined by the aerodynamic efficiency of the aircraft $K$and by the safe value of angle of climbing, $tg\Theta_{\delta} = 0.04\ldots0.05$:

$$\check{T}_{cutback} = \left( tg\Theta_{\delta} + \frac{1}{K} \right) G. \tag{5.4}$$

The derivative $dS/dv$ (where $v$ is aircraft flight speed) is negative; therefore as aircraft flight speed increases, the value of noise contour area $S$ decreases, and the numeral value of this derivative is higher than the derivative on engine rotation frequency $dS/dn$ (i.e. $|dS/dv| > |dS/dn|$. At the considered stages of height climbing, the range of possible aircraft flight speeds is quite small, $\Delta V$ does not exceed 20–40 m/s along the stage. Therefore, the decrease of the noise contour area $\Delta S_v$, resulting from increase in flight speed, is considerably less than that resulting from engine cutback $\Delta S_n$. Moreover, the increase in flight speed is usually continuous and defined by averaged acceleration $a$ along the flight stage. On this account, for achievement of maximum engine cutback effectiveness, in order to decrease noise under the flight trajectory, climbing is necessary to realize the minimum permissible flight speed (which is limited from flight safety conditions).

Research has been carried out on the influence of the start of the engine cutback on the noise level at the control point under the climbing flight path.[11] This has shown the existence of an optimum moment of engine mode transfer to cutback setting. Analogous research can be done using validated models to determine the influence of the beginning of engine cutback on area $S$. An example of the predicted dependences for the considered criteria on the longitudinal distance to the control point is shown in Fig. 5.14 for a Tupolev-154.

A considerable decrease in noise level (*EPNL*) at the control point is achieved by optimum implementation of the given procedure (more than 10 EPNdB with the maximum takeoff weight of the aircraft). The optimum moment of engines cutback is determined from analysis of the dependence of the perceived noise level (*PNL*) in the point of noise control from the aircraft location on flight path (expression 4.19 in Chapter 4). If $l_{10}$ represents the distance of the flight path, at which the noise level diminishes by 10 dB (10 dB in $L_A$ or 10 PNdB in *PNL*) and assuming that the maximum value of noise level is observed approximately in the middle of the distance $l_{10}$,

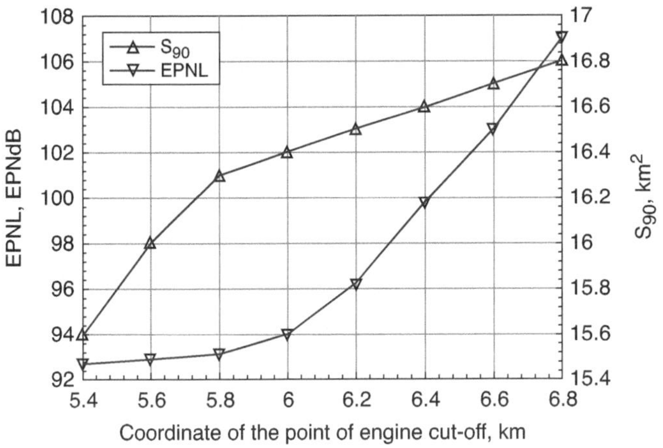


*Figure 5.14* Predicted influences on noise impact criteria of the longitudinal distance from the point of engine cutback to the control point during takeoff and climbing.

then its boundaries are determined from the following condition, which is reproduced from equation (4.19):

$$\left(\frac{l_{10}}{2R_n}\right)^2 = 10^{\frac{1}{1+0.5\delta}}, \tag{5.5}$$

where $\delta$ denotes attenuation due to sound absorption in air.

The temporal boundaries $t_1, t_2$ corresponding to the boundaries of $l_{10}$ are determined by formula (4.19). The optimum target time $t_1^*$ for beginning of engines cutback, based on the influences of the engines transition time in cutback mode and of flight height on the value of $l_{10}$, is within the interval $(t_1, t_2)$. If the purpose is to decrease the area $S$, the engine cutback must be performed just after the retraction ot the flaps. However, compared with the optimum moment of cutback for minimizing *EPNL* at the point of noise control, this will mean that the *EPNL* at the point of noise control No. 2 (6500 m) will be up to 6 EPNdB greater for the aircraft with turbofans $m \leq 2.5$ and between 2 and 3 EPNdB greater for the aircraft with turbofans $m > 2.5$.

Aircraft noise reduction under the flight path depends on the takeoff weight. The influence of aircraft weight on noise levels at takeoff is analyzed in Section 5.1. Decreasing of takeoff weight is equivalent to increasing the thrust-to-weight ratio of the aircraft, which in turn results in a decrease in the running distance along the runway and an increase in climbing gradients. An increase in climbing rate reduces the climbing time and increases the aircraft

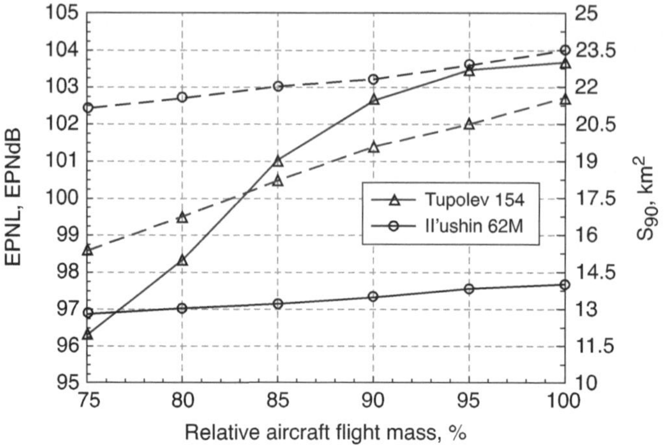


*Figure 5.15* Predicted influences of takeoff weights on noise impact criteria under the flight path during climbing (Tupolev-154 and Il'jushin -62M aircraft) assuming nominal operation modes.

flight height over the control point. The influences of aircraft takeoff weight on noise impact criteria are shown in Fig. 5.15.

This assumes that the aircraft climbs in accordance with the requirements of the operational manual for each given aircraft type. For all the aircraft considered, there is a typical decrease in the considered criteria corresponding to decrease in their takeoff weight. For the maximum weights of the Tupolev-154 and Il'ushin-62M, there are sharp increases in the *EPNL* at the control point, because cutback of the engines to the nominal mode in this case is performed in the immediate proximity to the point of noise control. Generally, reduction of aircraft weight in comparison to the maximum value can result in reductions of up to 6 EPNdB for aircraft with turbofans of $m \leq 2.5$, and up to 4 EPNdB for aircraft with turbofans of $m > 2.5$ (see Fig. 5.15). These results show that limiting aircraft takeoff weight can be effective for noise reduction under the flight path.

Reduction of aircraft weight means an increase in the climbing gradient $tg\Theta$, therefore the degree of permissible engine cutback can be greater [see equation (5.4)], so there can also be a decrease in the noise level under the flight path. Changes of aircraft weight within the operational range can lead to a reduction of thrust-to-weight ratio $\Delta\tau$ during climbing of up to between 25 and 30 per cent. The additional degree of engine cutback, consequent on the growth of thrust-to-weight ratio, results in values of relative rotor frequency $\Delta n$ between 0.07 and 0.09. In accordance with the relationship $R_N = f(n)$, an example of which is shown in Fig. 4.9, the associated decrease of noise radius $\Delta R_N$ is between 400 and 500 m. Based on an analysis of the relationships in Fig. 4.8, there will be a reduction of *EPNL* between 4 and

5 EPNdB for turbofan aircraft with a bypass ratio $m \leq 2.5$, and between 2 and 3 EPNdB for turbofan aircraft with a bypass ratio $m > 2.5$.

The derivative $d\sin\Theta/dn = /dT/dn(1/G)$ increases corresponding to the decrease in aircraft weight. Accordingly, the second (negative) item in expression (5.4) increases, and the first decreases. Similarly, the derivative $dS/dn$ decreases as takeoff weight decreases and the relative reduction of the noise contour area $S$ also decreases. As a consequence, the resulting noise contour for minimum aircraft weight and maximum engine cutback will be minimal.

Operational manuals for Tupolev-154 and Il'ushin-62M aircraft recommend the use of nominal engine operation modes corresponding to lowered aircraft takeoff weight. Figure 5.15 shows the predicted influences on noise impact criteria of the takeoff weights of Tupolev-154 and Il'ushin-62M aircraft while using nominal settings of their engines during aircraft climbing.

The noise contour area $S_{90}$ is predicted to increase between 10 and 15 per cent, fuel consumption is predicted to increase between 15 and 20 per cent, the *EPNL* at the control point is predicted to increase between 2 and 3 EPNdB, the equivalent mass of the engine emission is predicted to reduce by 5 per cent, and the noise contour area $S_{100}$ is predicted to reduce between 40 and 50 per cent. These results show that use of the nominal operation mode of the engines at takeoff, although effective for increasing the lifetime of the engines, slightly reducing emissions and reducing the $S_{100}$ contour area, is not an effective method for decreasing of aircraft noise impact, since the noise contour area $S_{90}$ increases as does the fuel consumption.

Combinations of aircraft acceleration $a$ and cutback of the engines reduce the effectiveness of the noise contour area reduction $\Delta S$ compared with aircraft moving at a constant minimum speed $v = v_2 + 5.5$ m/s. In this case it is necessary to increase engine thrust, which means there is an increase in the noise radius according to $\Delta R_{Nn} = dR_N/dn\Delta n$. The time of the aircraft flight with the engines in cutback mode in this case is determined by equation (5.6), which differs from equation (5.3):

$$t = \frac{1}{a}\left[\sqrt{\left(v_0 + \frac{dR_N}{dv}\frac{a}{\sin\Theta}\right)^2 + 2a\left(\frac{R_N}{\sin\Theta} - x_H\right)} - \left(v_0 + \frac{dR_N}{dv}\frac{a}{\sin\Theta}\right)\right] \tag{5.6}$$

Although increasing flight speed means reduction of the noise radius according to $\Delta R_{NV} = dR_N/dV\Delta V$, the resulting overall noise radius change will be positive $\Delta R_N = \Delta R_n + \Delta R_V > 0$. As a consequence, acceleration reduces the effectiveness of engine cutback. Moreover, flight acceleration provides a reduction of the rate of climb because of the reduction of climbing angle $\Theta$, which is determined by acceleration directly: $\sin\Delta\Theta = a/g$. The area

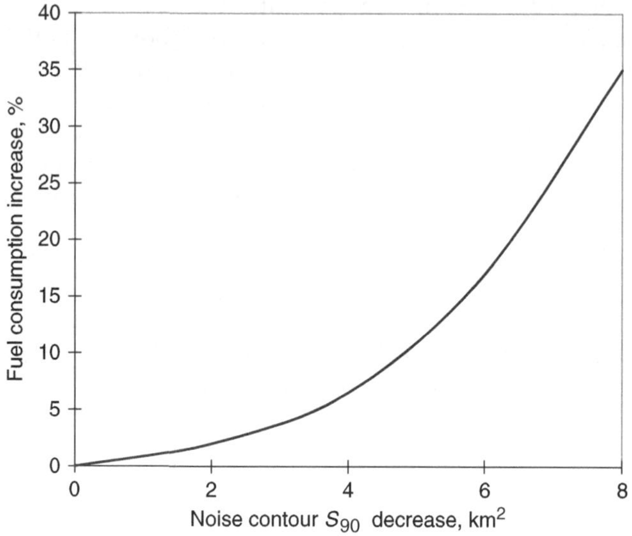


*Figure 5.16* Increase in fuel consumption corresponding to decreasing the noise contour area by climbing with engine cutback.

of noise contour $S_{90}$ is predicted to increase between 10 and 20 per cent depending on takeoff aircraft weight in comparison with minimum value $S_{90}$, which is reached during the climbing trajectory with constant flight speed $v = v_2 + 5.5$ m/s, assuming that the aircraft acceleration to speed $v = v_3 + 20$ m/s is performed with the minimum permissible engine thrust.

Similarly, it has been established that the most effective method of reducing aircraft noise impact (turbofans with small bypass ratio $m \leq 2.5$) is for the aircraft to climb with a constant flight speed $v = v_2 + 5.5$ m/s and with cutback of the engines to minimum permissible thrust. Use of optimum aircraft flight parameters corresponding to these climbing conditions implies an increase in fuel consumption on this flight stage (Fig. 5.16).

These predictions have been obtained assuming meteorological conditions in accordance with ISA requirements. Analysis of the influence of operational factors (Sections 5.1 and 5.2) shows that air temperature has an essential influence on aircraft performance and noise characteristics. Consequently, the dependence of noise reduction during climbing due to engine cutback on air temperature is of practical interest.

Predictions show that the dependence of aircraft noise on temperature is not simply a consequence of the variation of sound attenuation with atmospheric temperature (Fig. 5.13). Therefore the gradient of noise level change for a given flight trajectory in the temperature range between 15°C and 30°C is less than that corresponding to the temperature range between

0°C and 15°C, notwithstanding an approximately identical aircraft flight height change over the control point ($\Delta D_0$ in Fig. 5.12).

Since aircraft aerodynamic efficiency does not depend much on the air temperature, the permissible degree of engine cutback is determined, above all things, by the safe value of climbing angle $\Theta_\delta$ (5.4). For the aircraft types corresponding to the predictions in Figs 5.10–5.13, the degree of engine cutback or change of thrust-to-weight ratio of the aircraft $\tau$ will differ from thrust-to-weight ratio with nominal engine modes. Thrust-to-weight ratios of 0.138, 0.127, 0.105 and 0.041 conform to the following temperatures and engines modes: 40 per cent nominal for temperatures between −10 and 0°C; 60 per cent nominal for 15°C, maximum cruise for 30°C. The noise radii, $R_N$, for the listed modes are equal to 343, 392, 513 and 740 m, respectively. For nominal settings of the engines and the listed temperatures, the appropriate noise radii are equal to 687, 810, 1017 and 913 m. Similarly, with increasing air temperatures, engine cutback effectiveness reduces to 18 per cent in comparison with the maximum effectiveness 55 per cent corresponding to 0°C, if the effectiveness is determined only by noise radius. Taking into account all trajectory parameters and corresponding flight heights over a controlled surface, the maximum effectiveness is observed at the minimum considered temperature (i.e. −10°C –the noise contour area reduction $\Delta S$ is predicted to be 46 per cent. The minimum effectiveness is predicted to be at 30°C (the noise contour area reduction $\Delta S$ is predicted to be 16 per cent).

## 5.4 Descent and landing

During flight along a glide slope the aircraft flight parameters are approximately constant, therefore the noise impact area $S$ can be defined in a simpler way than (4.15), that is

$$S = \frac{\pi}{2} \frac{R_N^2}{\sin|\Theta|}. \tag{5.7}$$

Analysis of the expression (5.7) shows that noise levels under an aircraft descent trajectory can be reduced by: (1) increasing of angle of glide slope $\Theta$; (2) changing the aircraft aerodynamic configuration; and (3) increasing the flight speed on landing. Change in an aircraft's aerodynamic configuration enables decrease of the necessary engine thrust and, hence, decrease in the noise generation by aircraft power plant and in the airframe noise.

Landing with a low drag configuration is recommended as one of the methods for decreasing noise.[12] Its main feature is the use of intermediate wing flap deflection angles. This requires less engine thrust and permits an increase in the flight speed, hence decreasing the flight time along the glide slope.

Another method for decreasing noise impact while the aircraft descends before landing is based on finding the optimum glide slope angle Θ. A maximum glide slope angle of about 6°, limits noise impact during landing for all aircraft types (see Fig. 5.17).

Another possibility for noise control is the use of a two-segment glide slope during which the external (outer) segment of the glide slope has a maximum

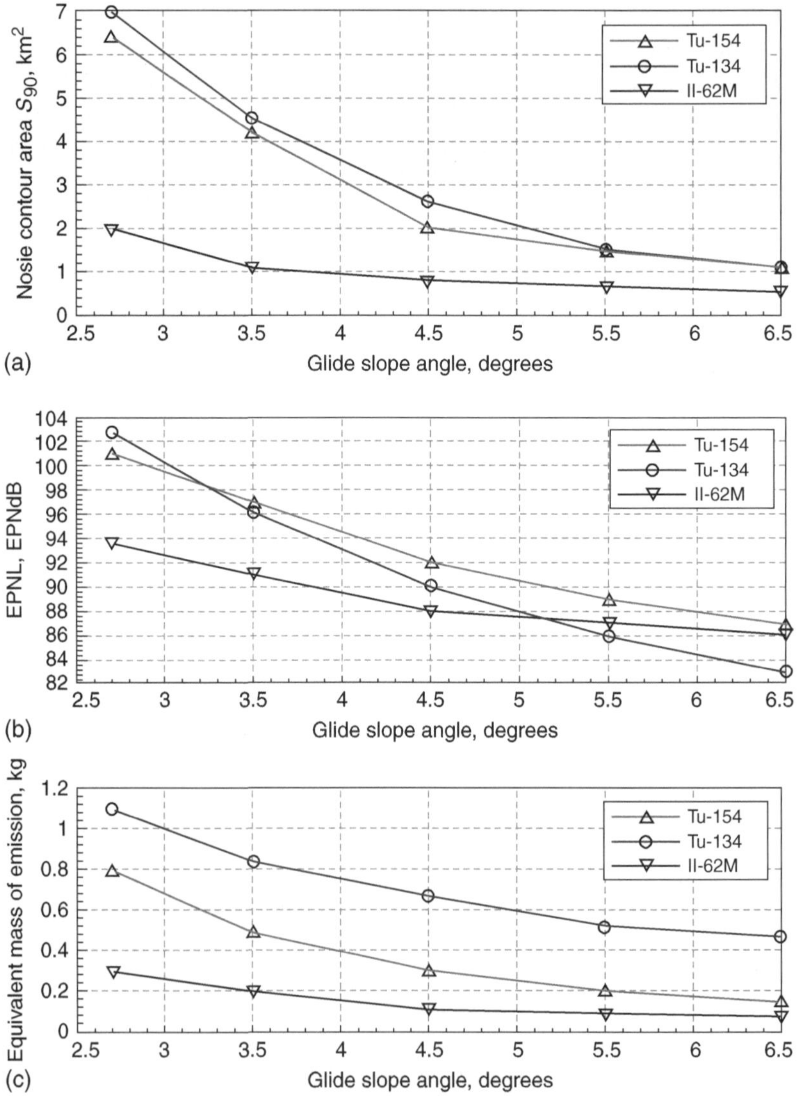


*Figure 5.17* Dependence of noise impact criteria on glide slope angle (a) $S_{90}$ and (b) effective perceived noise levels (EPNL).

angle of 6° and the internal (inner) segment has a glide slope angle of 2.7°. The noise reduction depends on the height of the transition $H_{int}$ between the two segments. A relationship between noise contour area $S$ and transition height $H_{int}$ is given by the following formula:[12]

$$S = H_{int}\left(ctg\Theta_1\sqrt{R_1^2 - H_{int}^2\cos^2\Theta_1} - ctg\Theta_2\sqrt{R_2^2 - H_{int}^2\cos^2\Theta_2}\right) + \frac{R_1^2}{\sin\Theta_1}\arcsin\frac{H_{int}\cos\Theta_1}{R_1} - \frac{R_2^2}{\sin\Theta_2}\arcsin\frac{H_{int}\cos\Theta_2}{R_2} \quad (5.8)$$

where indices 1 and 2 correspond to the internal and external segments of the glide slope.

Recommendations about the transition height during the two-segment approach are determined by safety requirements, which include the requirement for a near-constant flight speed at heights beneath 120 m before landing. Computations using (5.8) show that, for $H_{int} = 200$ m, the noise contour area $S_{90}$ is three times larger than the corresponding area for a one-segment glide slope of 6°.

Another potential contribution to noise control during the approach and landing is optimization of the engine operational mode used during the flight along the glide slope as the flight speed decreases. Compared with 'normal' flight at constant speed, an increased flight speed can be used at the initial point of glide slope and then the necessary deceleration is used to ensure the required speed at heights below 150–120 m before landing. The deceleration is provided by an appropriate value of engine thrust, which must be less than the thrust necessary for a 'stationary' aircraft approach with constant speed. It can be shown that, by control of engine rotation frequency $n$ (the parameter representing the engine thrust) and angle of attack $\alpha$, it is possible to obtain an approximately constant flight deceleration $a$ along the section of interest of the considered flight path.

The initial flight speed $v$ on entrance into the glide slope must be determined so that the resulting aircraft deceleration and flight speed over the front edge of the runway $v_{fe}$ are in accordance with the operational manual requirements (i.e. it should be such that $v = v_{fe} + at$).

The change in the noise radius $R_{Ndec}$ for a flight mode that includes deceleration compared to that for flight at constant speed $R_{Nst}$, is determined from:

$$\Delta R_N(t) = \frac{dR_N}{dn}\Delta n_j + \frac{dR_N}{dv}\Delta v. \quad (5.9)$$

Curves 1 and 2 in Fig. 5.18 show the predicted values of noise radius for a Tupolev-154 aircraft ($G_{noc} = 75$ t) for different engine modes $n_{dec} = n_{st} - \Delta n_j$ (line 3 corresponds to the value for $R_{Nst}$).

A search for the optimum glide slope for aircraft descent for this case has been performed by exploring the space of permissible engine rotor rotation

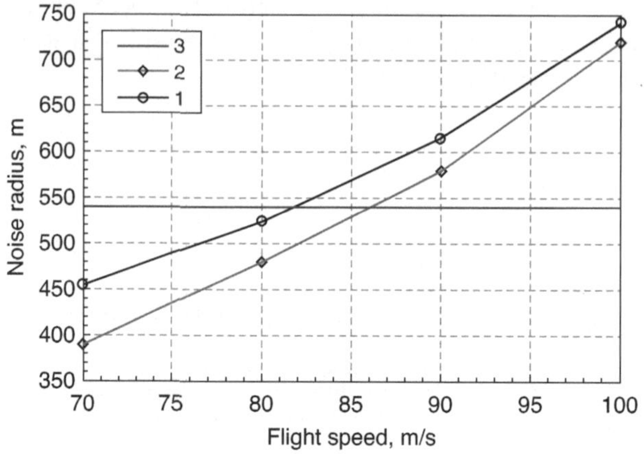


*Figure 5.18* Predicted dependence of noise radius (curves 1 and 2) on the engine modes for a Tupolev-154 aircraft ($G_{noc} = 75$ t) compared with the value for constant flight (line 3).

frequencies $\{n_i\}$ inside the range of permissible operation modes. The angles of attack $\alpha$ have been chosen so that the current engine operational mode value and angle of attack provide the necessary deceleration during the flight stage considered. The result of the search shows that the optimum value $n_{opt}$ is in the range of values $[n_0, n_{st}]$, for which the predicted noise radius is less than $R_{Nst}$ (see Fig. 5.18). The engine rotor rotation frequency $n_0$ corresponds to the engine operational mode for which the average noise radius increase along the aircraft flight trajectory is zero. The predicted noise radius for the optimum engine operational mode does not exceed $R_{Nst}$, and its initial value is $R_{Nst}$ at flight speed $v_n$ (curve 1 in Fig. 5.18). The optimum engine operational mode is determined by the empirical relationship:

$$N_{opt} = n_{st} - \Delta n_{opt};\ \Delta n_{opt} \cong 0.7(n_{st} - t_0).$$

Figure 5.19 shows the dependence of the noise contour area on the engine operational mode during the descent of a Tupolev-154 aircraft ($G_{desc} = 80$ t) along the glide slope. The deceleration value for the optimum flight mode is about $-0.1\ \mathrm{m/s^2}$.

However, the effectiveness of optimum flight speed deceleration procedure during descent along the glide slope is determined by the decrease of noise contour area between 10 and 25 per cent (depending on aircraft type and landing mass). This procedure may be effective only when the rotation frequency for constant speed descent $n_{st}$ is located in the domain of engine modes, for which $dR_N/dn$ has the maximum value (i.e. at $n > n_{tr}$; see Fig. 4.9). At $n < n_{tr}$ the value of $\Delta R_{Ndec}$ will be determined mainly by flight speed. Accordingly, it has positive values, so the considered noise

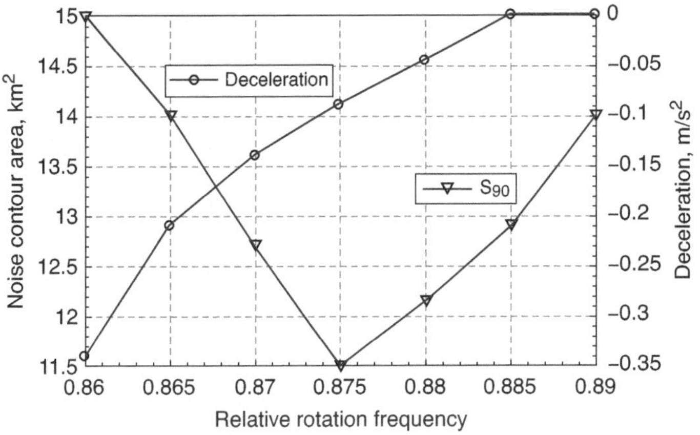


*Figure 5.19* The relationship between flight deceleration and relative engine rotation frequency (i.e. relative to the rotation frequency for the maximum operation mode) and the corresponding influence on the $S_{90}$ noise contour area for a Tupolev-154 aircraft ($G_{desc} = 80$ t).

contour area for a given engine operational mode must always increase in comparison with the value for constant speed descent. Computations show that, for Tupolev-154 with $G_{landing} < 65$ t, Tupolev-134 with $G_{landing} < 35$ t and Il'ushin-62M with any landing mass (as is the case for other turbofan aircraft with a high bypass ratio), descending with deceleration before landing is not effective for noise reduction. In addition, the use of deceleration procedures is not effective for noise reduction during descent with an intermediate aerodynamic configuration of the aircraft; e.g. that used for takeoff).

The optimum aircraft flight procedures, described above, for the descending stage before the landing, are applicable to segments of the trajectories, located between 10 and 12 km away from the runway threshold. These segments determine fully the 90 EPNdB noise contour (i.e. the area $S_{90}$). Nevertheless, it is important to be able to assess aircraft noise and consider optimization of flight procedures for noise reduction at greater distances from the runway (e.g. up to 20 km).[12] The results of relevant investigations may be considered using the example of a Tupolev-154 aircraft.

During normal flight operations, landing gear and wing flaps extraction, and decrease of flight speed to regulated values for landing are performed during the flight 'circle' at heights between 400 and 600 m and at distances between 8 and 20 km away from the runway threshold. The engine operation modes along the horizontal flight stages during the 'circle' are higher than during descent along the glide slope. For example, the Tupolev-154 makes use of a 0.7 nominal mode rather than a 0.6 nominal mode for engines in accordance with the operational manual for the aircraft. Noise levels under these horizontal flight stages are between 80 and 95 EPNdB, depending on aircraft type and landing mass. For the purposes of the following investigations, a 'circle' with height $H = 600$ m was taken as the initial value.

The simplest noise reduction procedure, in comparison with this initial trajectory, is to start the glide slope at a greater flight height.[12] Currently, such a procedure is called a continuous descent approach. For example, consider $H_{circle} = 900$ m. Assume that the landing configuration is fully set at this height and that further aircraft descent is performed at constant speed ($v = 1.3v_S$). The noise levels under the aircraft flight path are reduced between 6 and 8 EPNdB. But the $S_{80}$ noise contour (the contour for $EPNL =$ 80 EPNdB is considered because the contour 90 EPNdB is quite small) becomes insoluble and bounds a greater surface than for the initial variant. On the other hand, use of the aircraft in-flight configuration down to a height of 400 m and consequent deflection of the wing flaps into a landing position along the trajectory segment between heights of 400 and 200 m enables a decrease of the noise levels under the trajectory by up to 20 EPNdB. Moreover the $S_{80}$ noise contour area is reduced by between 60 and 65 per cent, the $S_{90}$ noise contour area is reduced by 25 per cent and the equivalent mass of air pollution is reduced by between 50 and 55 per cent. These results are quite important for determining the operational recommendations for reduction of the equivalent sound level at the point of noise control or in the total zone under consideration.

# 6 Methods of aircraft noise reduction

## 6.1 Reduction of noise at source

### *6.1.1 Power plant*

The noise produced by an aircraft depends on the type of aircraft and the acoustical performances of its engine. The basic parameters affecting aircraft noise characteristics are the mass, the thrust–weight ratio, the aerodynamic load on the wing, the airplane aerodynamics and the type of the power plant.

There is an interrelationship between the different methods of noise reduction, which include the reduction of noise at source, the changing of aircraft performance characteristics and the optimization of operational procedures. The fan and turbine are important elements of the power plant on the turbojet aircraft that generate tonal and broadband noise. The noise of the turbofan engine with a high bypass ratio is typical for other aircraft propulsion systems. When the tip speed of the turbofan is subsonic, the tones in the noise spectrum correspond to the blade-passage frequency and its harmonics. At supersonic relative blade velocities, additional multiple pure tones appear in the noise spectrum due to shock waves on the blades. The broadband component of the noise spectrum is the result of non-stationary flow, including inflow turbulence and strut-induced unsteady flow. The resultant far-field noise spectrum of a turbofan engine is the sum of several effects, including the unsteady aerodynamics and acoustic generation on a blade row, propagation upstream and downstream of noise in the engine duct, reflecting and scattering of the sound waves from other blade rows and inlet and aft radiation.

Modern methods of reducing noise at source on turbofan engines include cycle optimization (optimum bypass ratio, fan pressure ratio, speed reduction), advanced fan blade and fan exit guide design, aeroacoustic design of the air intake and fan, low noise nozzle design, passive and active liners, active noise control and boundary layer control. The more traditional methods of noise reduction at source include blades/vanes spacing optimization, exclusion of inlet guide vanes, selection of the lowest fan tip speeds, design of the scarf inlet, aerodynamic 'cleaniness' of the engine impellers and acoustic lining design. Passive and active liners increase the transmission loss in engine ducts. Jet noise suppression through chevron nozzle design and optimization of the exhaust nozzle ejector are effective methods of jet noise reduction. The application of devices for noise reduction depends on the weight of the devices, and their manufacturing constraints, cost and maintenance requirements.

Propellers are the basic sources of the noise on a turboprop engine. Because propeller blade-tip speed is an important parameter, reducing tip speed decreases the noise. Augmentation of the number of blades decreases propeller noise for small blade-tip speeds of rotation (less than 240 m/s). Reducing the thickness of the blades on a propeller reduces the thickness noise. A large diameter with many blades diminishes loading noise and a large blade sweep also reduces noise. Increasing the propeller diameter for a given thrust combined with a relatively small tip speed reduces loading noise. On an aircraft with few propellers, the noise levels in the aircraft cabin can be decreased by phasing the rotation propellers on opposite sides of the fuselage. This effect is called synchrophasing. It is an example of the positive effect of acoustical interference whereby there is noise cancellation in the domains where there are opposite phases of waves.

As was mentioned in Chapter 1 (item 8 in the list of methods of reducing aircraft noise on page 4 of Chapter 1), noise abatement methods can be realized at any of the stages of design, manufacturing, operation and repair. Figure 6.1 shows the system for reducing the noise of an aircraft during the design stage. The objective of such activity is to achieve minimum noise consistent with cost–benefit assessment and safe operation of the aircraft.

As a result of noise considerations during the design phase of an aircraft, information is available for corresponding jet noise, fan noise, turbine noise, combustor noise and core noise. This information enables calculation of noise levels for certain aircraft configurations and the evaluation of noise levels during flight. The next step is to account for installation effects, the influence of the operational flight mode and to estimate the propagation effects, and hence predict the airframe noise and airplane noise for the three certification points. After that, an assessment of the various measures of noise reduction may be made with regard to the economics (manufacturing

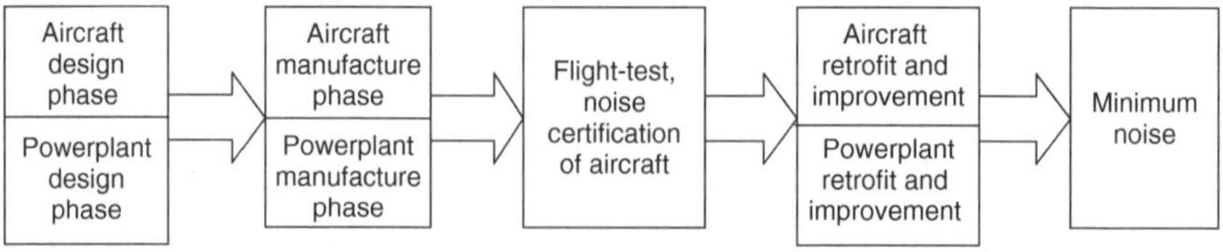


*Figure 6.1* An aircraft noise reduction methodology.

and maintenance cost, weight, drag, fuel consumption), the power plant and aircraft technologies, aircraft maintenance performance and flight safety. Noise reduction must remain compatible with many other aircraft requirements. Flight safety has always been of prime importance for the development of air transport.

Since an aircraft with a turbofan is a complex acoustic source, integrated methodology must be used to reduce noise. The main aim of this methodology is to balance the effects of noise reduction at source against the achievable noise abatement under the flight path. As an example, we calculate the effectiveness of the acoustic treatment of the fan inlet turbofan for an Il'ushin-86 aircraft.

Suppose that an acoustically absorbent material installed in a fan inlet turbofan reduces the tonal and broadband levels frequency by 5 dB. Figure 6.2a shows the directivity patterns of the approach noise of an Il'ushin-86 without acoustic treatment and Fig. 6.2b shows the corresponding patterns with an acoustically absorbing liner on the duct walls of the fan inlet.

Reduction of noise radiation in the forward direction of the engine is predicted. The predicted effect on noise spectra can be seen in Fig. 6.3. The predicted noise reduction is the result of increasing the transmission loss of fan noise in the duct. The high-frequency band noise without treatment (Fig. 6.3a) is larger than that with acoustic treatment (Fig. 6.3b). The predicted decrease in noise level, $\Delta EPNL$, for an Il'ushin-86 aircraft at control point No. 3 (2000 m from the runway) after the acoustic treatment of the fan inlet turbofan is 2.1 EPNdB.

On high bypass ratio engines, acoustic treatment can be used in the fan inlet and bypass ducts and as stream liners. The predicted directivity patterns of turbofan ($m = 4$, bypass jet) of a Yakovlev-42 aircraft with a noise-absorbent lining in the fan inlet, bypass ducts and installation of nacelle treatment are shown in Fig. 6.4. The predicted effectiveness of each of the noise-reducing treatments (the inlet duct, the outlet bypass duct lining and the hot stream liners) is 5 dB. The separate effects of the acoustic treatment of the inlet duct (method 1; Fig. 6.4b), the installation of noise absorbent lining in the inlet and bypass ducts (method 2; Fig. 6.4c) and the installation of a noise absorbent lining in the inlet and the total effects of treating the bypass ducts and incorporating the stream liners (method 3; Fig. 6.4d) have been predicted.

The acoustical spectra predicted for an Yakovlev-42 aircraft with high bypass turbofan engines during the approach flight mode for each of the methods 1–3 of noise abatement are shown in Fig. 6.5. Method 1 is predicted to decrease the fan noise radiation for forward direction (Fig. 6.4b). Method 2 is predicted to decrease the fan noise radiation in both forward and backward directions (Figs 6.4c and 6.5c). Method 3 is predicted to decrease the noise radiation from the fan and turbine (Figs 6.4d and 6.5d).

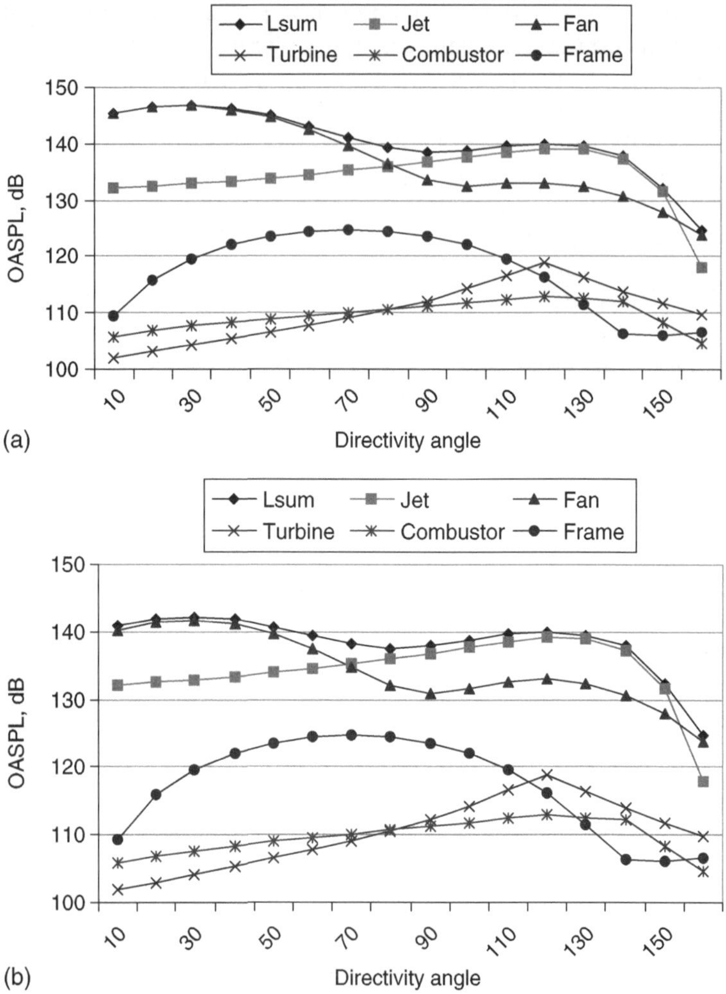


*Figure 6.2* The directivity patterns for approach noise from an Il'ushin-86 aircraft (a) without acoustic treatment of the fan inlet and (b) with acoustic treatment of the fan inlet.

The predicted decrease in noise level from a Yakovlev-42 aircraft at control point No. 3 (2000 m from the runway) after introduction of methods 1, 2 and 3, respectively are: for method 1 – $\Delta EPNL = 1.7$ EPNdB; for method 2 – $\Delta EPNL = 2.7$ EPNdB; and method 3 – $\Delta EPNL = 2.9$ EPNdB. The overall noise levels (effective perceived noise levels: EPNL) are summarized in Fig. 6.6 for a Yakovlev-42 aircraft with high bypass turbofan engines during the approach flight mode.

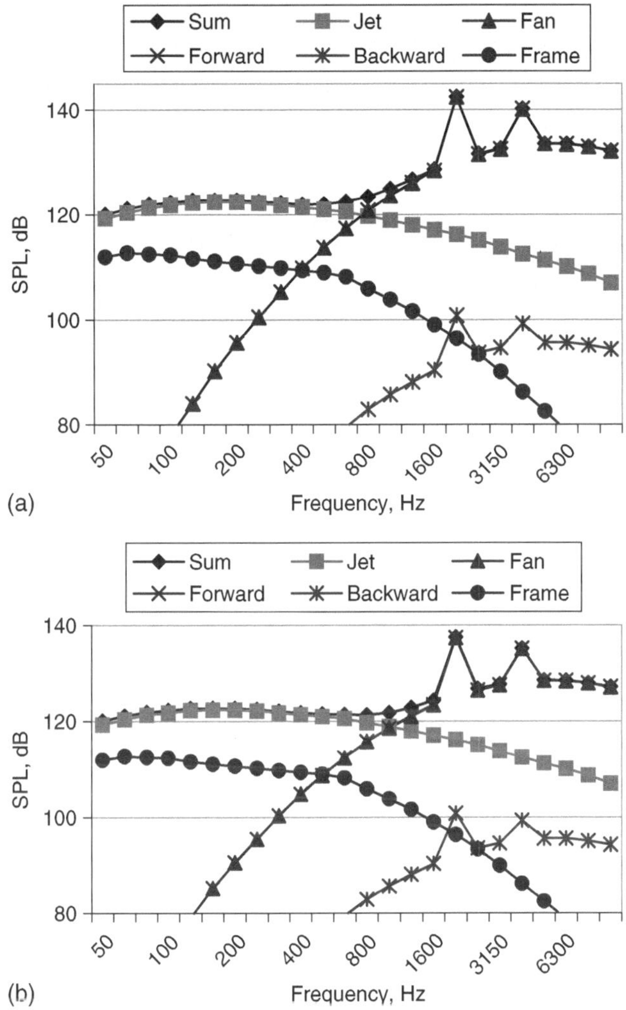


*Figure 6.3* The predicted spectrum of approach noise (for the forward direction of acoustical radiation) from an Il'ushin-86 aircraft in the direction of maximum inlet fan noise radiation (a) without acoustic treatment of the fan inlet and (b) with acoustic treatment of the fan inlet.

### *6.1.2 Simultaneous noise reduction under the flight path and inside the aircraft cabin*

The reduction of the noise at source results in noise reduction under the flight path and inside aircraft cabin. If the distribution function $P_j$ is the relative contribution of the $j$th method of noise abatement (in some zone under the flight path and inside the aircraft cabin), than in the zone under flight path, the equivalent continuous sound level can be represented by

$$L_{Aeq} = 10\lg\left(T_0^{-1}\sum_{j=1}^{N}\tau_{ej}10^{0.1L_{AMAXj}}\right) \tag{6.1}$$

where $T_0$ is the observation time period; the effective time $\tau_{ej}$ is given by $\tau_{ej} = K(D_{0j}/v_{Fj})$, where $K = 3.4$ for a turbojet aircraft and $K = 2.5$ for a turboprop aircraft; $D_{0j}$ is the shortest distance to the airplane for the $j$th

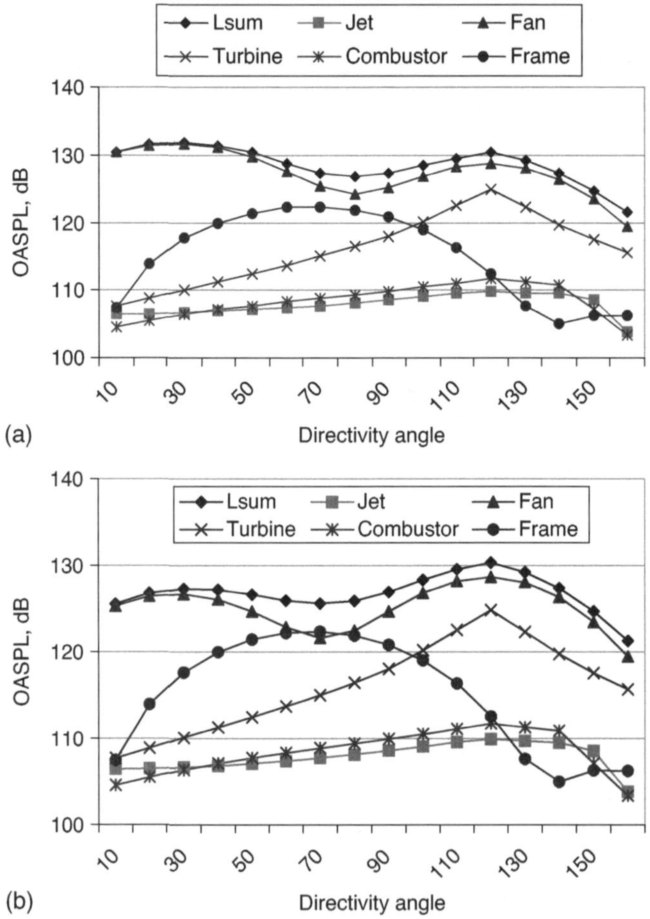


*Figure 6.4* The directivity patterns for a Yakovlev-42 aircraft ($m = 4$, bypass jet) at approach flight mode: (a) without acoustic treatment; (b) after acoustic treatment of the inlet duct; (c) after acoustic treatment of the inlet and outlet ducts for fan noise abatement; (d) after acoustic treatment of the inlet and outlet ducts and the installation of hot stream liners.

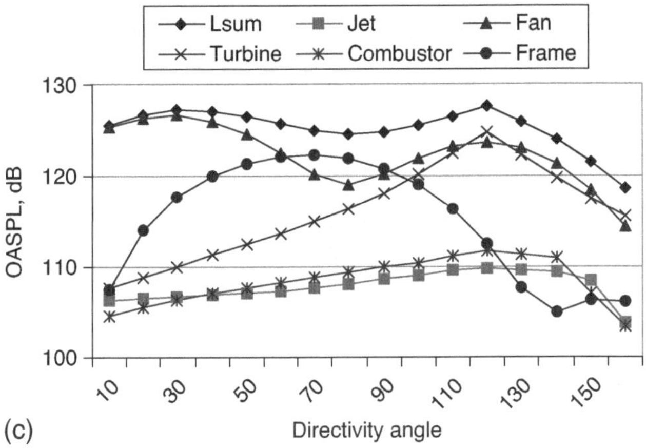


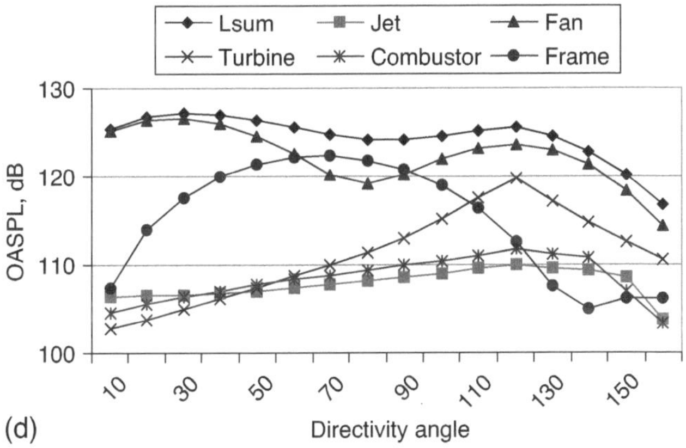


*Figure 6.4* Cont'd

noise reduction method; $v_{Fj}$ is flight velocity; $L_{AMAX_j}$ is the maximum noise level in zone under flight path for the $j$th noise reduction method; and $N$ is the number of methods of noise abatement. In the chosen zone, equation (6.1) can be rewritten as

$$\sum_{j=1}^{N} \tau_{ej} P_j 10^{0.1\Delta L_j} = 1, \tag{6.2}$$

where $\Delta L_j = L_{AMAXj} - L_{Aeq}$ and $L_{Aeq}$, the target equivalent noise level, is given. Suppose that the effectiveness of the $j$th method of noise abatement is represented by a noise reduction inside aircraft cabin of $\delta L_j$. The required

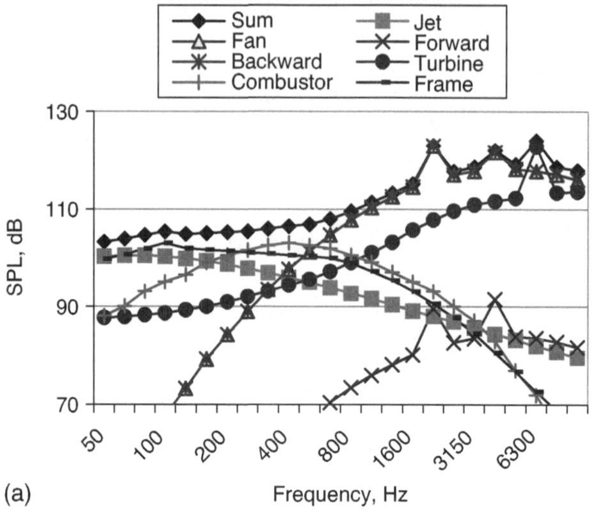


(a)

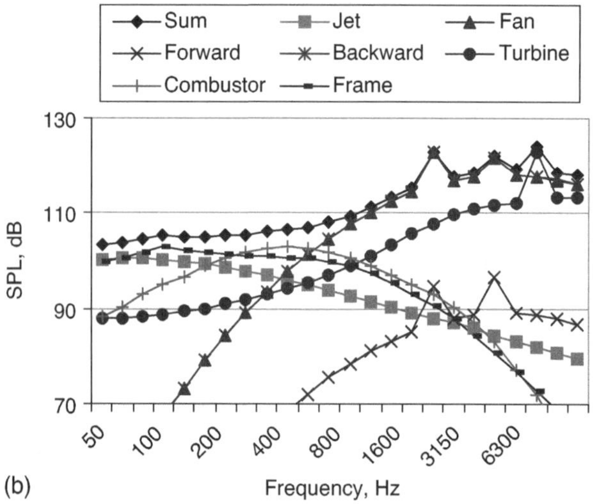


(b)

*Figure 6.5* The predicted spectrum (in the backward direction of acoustical radiation) for a Yakovlev-42 aircraft ($m = 4$, bypass jet, in the direction of maximum fan noise radiation) at the approach flight mode: (a) without acoustic treatment; (b) with acoustic treatment of the inlet duct; (c) with acoustic treatment of the inlet and outlet ducts for fan noise abatement; and (d) with acoustic treatment of the inlet and outlet ducts and the installation of hot stream liners.

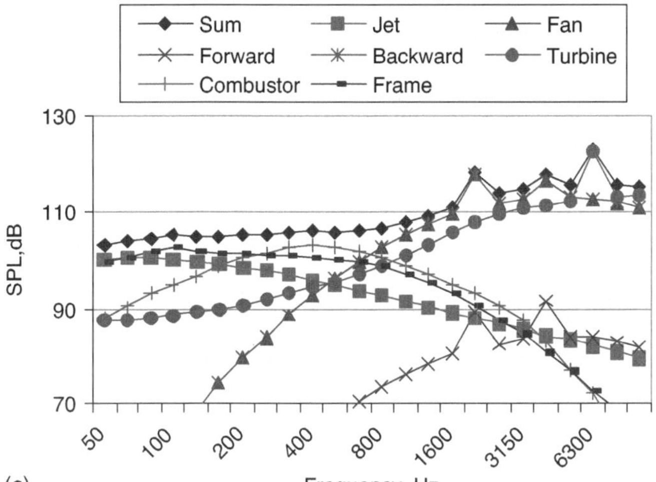


(c)

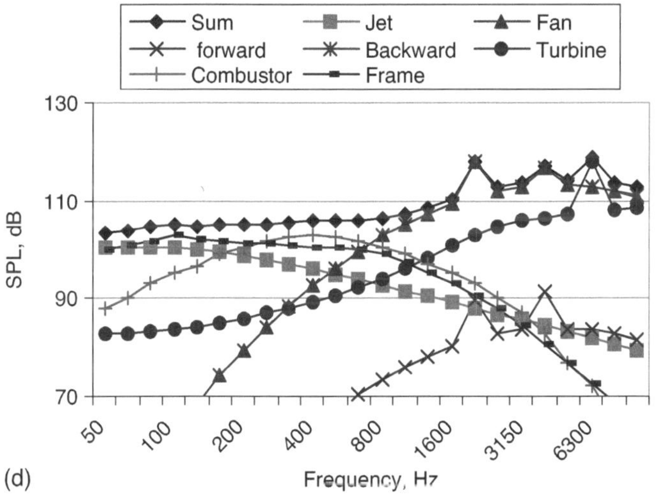


(d)

*Figure 6.5* Cont'd

noise reduction inside the aircraft cabin is given by

$$\delta L = \sum_{j=1}^{N} P_j \delta L_j. \tag{6.3}$$

We introduce the entropy of the system in following form:

$$S_P = \sum_{j=1}^{N} P_j \left(1 + \ln \frac{\nu_j}{P_j}\right), \tag{6.4}$$

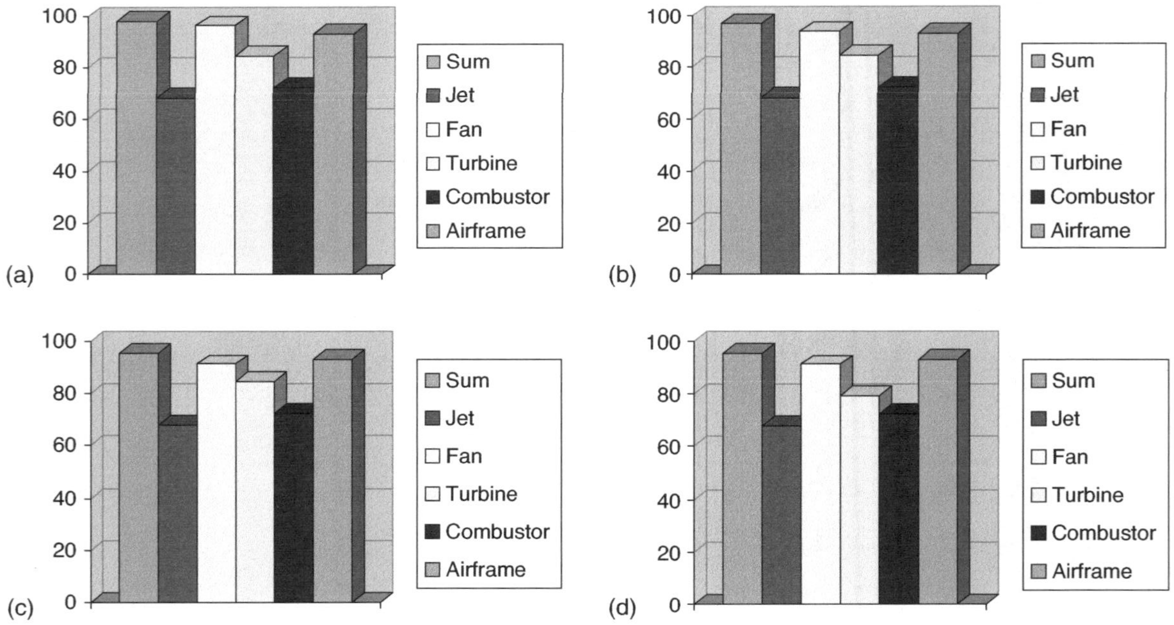


*Figure 6.6* The contribution of every source to the total effective perceived noise levels (*EPNL*) in accordance with proposed acoustic treatment of the turbofan engine: (a) no treatment; (b) with treatment of the inlet fan; (c) with treatment of the inlet and outlet fan; and (d) with treatment of the inlet and outlet fan and installation of hot stream liners.

where $\nu_j$ is the *a priori* probability evaluated for the $j$th method of noise reduction. $\nu_j$ might be evaluated, for example, from the increasing takeoff mass $\Delta m_j$ corresponding to the $j$th method of noise reduction:

$$\nu_j = \frac{\Delta m_j^{-1}}{\sum_{j=1}^{N} \Delta m_j^{-1}}.$$

The distribution function $P_j$ is defined by the relative extreme of entropy according to equation (6.4) with constraints (6.2) and (6.3):

$$P_j = \nu_j \exp(-\lambda \tau_{ej} T_0^{-1} 10^{-0.1\Delta L_j} - \beta \delta L_l),$$

where the multipliers $\lambda$, $\beta$ are defined from equations (6.2) and (6.3). As an example, Table 6.1 shows the results of calculation of the normalized distribution function $\overline{P}_j = P_j / \sum_j P_j$ for a heavy turboprop aircraft.

The distribution function $\overline{P}_j$ ranks the methods of noise reduction for achievable noise efficiency. It is necessary to fulfill the demand of the equation (6.2) so that $\delta L_j^* = P_j \delta L_j$ provides a given target equivalent noise level ($L_{Aeq}$) inside the cabin of the aircraft.

### 6.1.3 *Use of noise mufflers during engine testing*

To protect local dwellings from the adverse impacts of intensive noise, the engine test facilities at airports should provide a reduction in noise levels from engines up to the test levels corresponding to their stable operational parameters. The mufflers used for this purpose have to deal with high jet speeds, high temperatures in exhaust gases, the extent of the sound source

*Table 6.1* Evaluation of methods of reducing noise in heavy turboprop aircraft ($\delta L = -31.9$ dB)

| *Method of noise reduction* | $\delta L_j$ | $\Delta L_j$ | $\overline{P}_j$ |
|---|---|---|---|
| Decrease in propeller blade tip speed | −2 | −1 | 0.0126 |
| Decrease in the airload on the propeller blade | −4 | −3 | 0.0257 |
| Decrease in the airload on the blade tip | −2 | −1 | 0.1257 |
| Decrease in the blade tip thickness | −4 | −3 | 0.0257 |
| The blade twist | −2 | −1 | 0.1257 |
| Increasing the distance between propeller and fuselage | −4 | −1 | 0.5097 |
| Changing the panel stiffness | −4 | 0 | 0.0254 |
| Rationalizing the cabin interior arrangement | −4 | 0 | 0.0508 |
| Changing the mass distribution | −2 | 0 | 0.0063 |
| Increasing the sound absorption inside the cabin | −4 | 0 | 0.0101 |
| Synchrophasing of engine | −5 | −3 | 0.0312 |
| Using active absorbers | −5 | 0 | 0.0511 |

(usually 8–10 times the jet diameter at the engine outlet nozzle), various spectra and noise levels that depend on the location of the sources in the jets. Typically, the noise muffler used on a turbofan jet has a cylindrical form with a diameter that is usually twice that of the jet diameter. Figure 6.7 shows mufflers of the following types: 1 – ejector type; 2 – including vane devices; 3 – including forced mixing of the additional air mass; 4 – chamber mufflers; and 5 – multi-jet mufflers.

The characteristics of the sound absorbing material are chosen so that the maximum of its sound absorption coefficient lies close to the frequency of maximum jet noise. The muffler also achieves reduction of noise from a turbofan jet engine because of the diminished jet speed gradients during interaction with ejected air. In an optimum design of the engine test muffler, the jet noise energy at the exhaust must not exceed the noise energy of the primary jet and the jet noise should be attenuated in the muffler. Because the jet speed at the outlet of the muffler is the determining parameter for noise generation, its decrease is achieved by means of a diffuser. In some mufflers, reticulated screens are used, mounted at some distance from the nozzle. The effect of a reticulated screen is to break large eddies into smaller ones. This means that the jet spectrum is transformed from low frequencies to high frequencies, which are attenuated well by the muffler lining. Tubular diffuser mufflers combining the principles of active influence on noise generation, decreasing flow speed in the diffuser, and sound-absorbing elements can achieve noise reductions on the order of 30 dB.[1–3]

## 6.2 Noise reduction under the flight path

A balanced approach to environmental protection includes protection from excessive aircraft noise. One of the most important components of this is implementation of low noise flight procedures. Their efficiency is very dependent on the ambient and other operational conditions. For this reason, routine noise control procedures have been suggested.[4,5] Their realization depends upon the presence of accurate and reliable information about the prevailing aircraft noise situation.[6] A decision-support system is required that will enable choice between the possible solutions for aircraft noise control around the airport. There are several procedures that have been implemented in aircraft operation: low noise approach and takeoff procedures under the flight path. Methods for optimization and methods for selecting optimal strategies can be used in such cases.

### *6.2.1 The mathematical formulation*

The mathematical formulation for minimizing the noise impact of a single aircraft includes the trajectory model for the aircraft, an acoustical model accounting for various influences on noise levels under the flight paths and the criteria for optimization. The mathematical trajectory model of the

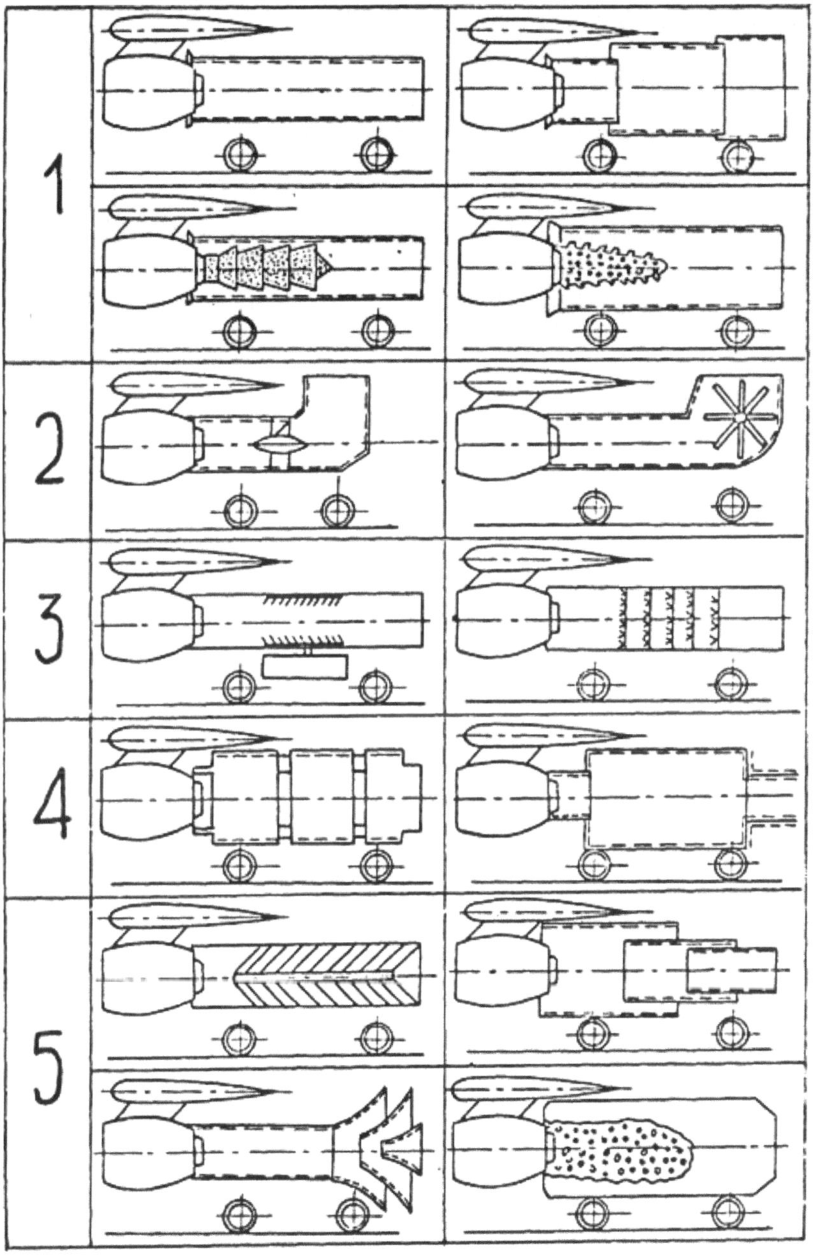


*Figure 6.7* Typical aerodrome noise mufflers.

airplanes is represented by a set of ordinary differential equations derived for the center of mass of the aircraft in a flight velocity coordinate system. In matrix form, it is expressed as:[7]

$$\dot{X} = F(X, U), \quad X_0(t_0) = x^0, \quad t \in [t_0, t_1], \tag{6.5}$$

where $X = (x_0, \ldots x_n)$ is a vector of trajectory coordinates in an $(n+1)$-sized Euclid volume; $U = (u_1, \ldots u_m)$ is the flight control parameters vector within the defined region of constraints $\Omega \subset E^m$; $t$ is time in the interval $[t_0, t_1]$, where $t_0$, $t_1$ represent the beginning and the end of trajectory segment; $F = (f_1, \ldots f_n)$ is an $n$-sized vector-function at region $E^n \times \Omega$.

We consider differential equations (6.5) derived for the center of mass of the aircraft in the velocity coordinate system without sideslip motion (sideforce is absent):

$$\begin{aligned}
& x_1 = v, \; x_2 = \theta, \; x_3 = \varphi_p, \; x_4 = x, \; x_5 = y, \; x_6 = z \\
& f_1 = g\left(\frac{T\cos\alpha_e - X_a}{mg} - \sin\theta\right) \\
& f_2 = \frac{g}{v}\left(\frac{T\sin\alpha_e + Y_a}{mg}\cos\gamma - \cos\theta\right) \\
& f_3 = -\frac{g}{v\cos\theta}\left(\frac{T\sin\alpha_e + Y_a}{mg}\sin\gamma\right) \\
& f_4 = v\cos\theta\cos\varphi_p \\
& f_5 = v\sin\theta \\
& f_6 = -v\cos\theta\sin\varphi_p \\
& U_1 = T, \; U_2 = \alpha, \; U_3 = \delta_f \ldots,
\end{aligned} \tag{6.6}$$

where $v, \theta, \varphi_p$ are, respectively, the velocity, flight path angle and yawing angle; $T$is the total engine thrust of the aircraft; $x$, $y$, $z$ are the Cartesian coordinates with origin at the center of mass of the aircraft; $X_a$, $Y_a$ are, respectively, the aerodynamic drag and the lift carried by the wing; $\alpha_e = \alpha + \varphi_e$, $\alpha$ is wing angle of attack, $\varphi_e$ is the engine tilt angle; $\gamma$ is the rolling angle; and $\delta_f$ is the flaps angle, $g$ is acceleration due to gravity.

Since the rotational motion around the center of mass of the aircraft is not included, the trajectory model is valid only in the vertical plane and can be represented by a set of four equations with the following phase parameters: flight velocity $v$, flight path angle $\theta$, longitudinal coordinate $x$, and vertical coordinate $y$. The flight control parameters in a trajectory model of the airplane are the total engine thrust of the aircraft $T$, the flap angle $\delta_f$ and the pitch angle $\vartheta = \alpha_e + \theta$.

The region of permissible changes of coordinates and control parameters is limited by inequalities of the form:

$$g(X, U) \geq 0,$$

which are defined along the trajectory.

In addition, some constraints may be imposed on the phase parameters at the end of the trajectory segment:

$$\Psi[X(t_1)] = 0$$

and some constraints of permissible changes of control parameters must be included in accordance with the aircraft or engine technical characteristics:

$$U_{\min} \leq U \leq U_{\max}.$$

Therefore, the domain of permissible changes of the parameters along the aircraft trajectory segment considered can be represented by

$$\Omega = \{U \,|\, g(X, U) \geq 0, \Psi[X(t_1)] = 0\} \tag{6.7}$$

The optimization criterion may be any of the known aircraft noise control criteria discussed in Chapter 1 (see Table 1.6). These include:

- noise levels, accounting for the psycho-physiological human response to noise impact, and modified by means of any type of spectrum correction, for example, weighted sound levels $L_A$, $L_D$ or perceived noise levels *PNL*, *PNLT*, etc.;
- effective noise levels, accounting for the time duration of a particular aircraft noise event, for example, *SEL* or *EPNL*;
- equivalent noise levels or noise indices accounting for the number of noise events during the particular time interval, diurnal mostly, for example, units like $L_{Aeq}$, *ECPNL* or *NEF*, *NNI*, etc.;
- the area inside noise contour.

After analyzing all possible types of noise impact criteria, it has been established that the most useful form of criterion is given by

$$I = \frac{1}{T_0} \int_{t_0}^{t_1} f_o[X(t), U(t)]dt. \tag{6.8}$$

All of the models intended for aircraft noise assessment are used to define noise impact criteria for the optimization tasks and connect them with the phase and control parameters of the flight (trajectory) model of the aircraft. The numerical gradient method has been used to solve this task.

Besides this, another method has been developed for obtaining the optimum trajectory.[8–10] The method consists of searching for the optimum control parameters $\Omega$. The starting points for the search are defined by means of the $\Lambda\Pi_\tau$-series, which have the most uniform distribution in any considered space. This approach is useful and necessary to estimate not only the solution according to given noise impact criteria, but also to combine them with other types of operational efficiencies like fuel consumption, air pollution, etc. They can be represented as additional criteria of the task or as additional constraints. So for every examined trajectory or particular segment of the flight trajectory, the values of control parameters $U_{kl}$ are defined by the relationship:

$$U_{kl} = U_{k\min} + q_{kl}(U_{k\max} - U_{k\min}),$$

where $q_{kl}$ is an element of the $\Lambda\Pi_\tau$ series and $l$ is the ordinal number of investigations.

Tables of results of the search must be made up to define the optimal solution. If there are multiple criteria, the Pareto-set is constructed and hence the optimal solution is found.

### 6.2.2 *The approach and landing stage*

A general structure of the optimum solution of aircraft trajectory for achieving minimum noise under the flight path has been published elsewhere.[4] It is supposed that, for flight trajectories, the true condition is represented by $\theta =$ *constant* (constant flight path angle).

The differential equations governing the aircraft approach are

$$\begin{aligned} m\frac{dv}{dt} &= T - X_a mg\sin\theta \\ 0 &= Y_a - mg\cos\theta \\ \frac{dx}{dt} &= v\cos\theta \\ \frac{dy}{dt} &= v\sin\theta. \end{aligned} \tag{6.9}$$

The criterion for optimizing the noise level at the point of noise control under the flight path is the integral form of *EPNL*:

$$I = \int_{t_1}^{t_2} 10^{0.1PNL(t)}dt, \tag{6.10}$$

where the integrand is determined by the engine behavior, the landing configuration and the distance from the aircraft to the observation point.

Equation (6.10) can be rewritten for a specific aircraft type in the form

$$I = K_0 \int_{t_1}^{t_2} \frac{T^s}{r^p} dt, \tag{6.11}$$

where $K_0, s, p$ are some empirical constants, $r = \sqrt{(x - x_0)^2 + (y - y_0)^2}$ is the distance from the aircraft to the observation point $(x_0, y_0)$. It is necessary to determine the solution of equations (6.9), which minimizes the criterion (6.11) and the resulting reduction in aircraft noise impact at the point $(x_0, y_0)$. For example, Fig. 6.8 shows the domain OAKBO of permissible solutions of equations (6.7) for the approach flight trajectory.

The point O is determined by the height of the aerodrome holding circle and the point K $(x_K, y_K)$ is the end point $(x_0 > x_K)$. From the permissible changes in the approach flight trajectory, one can determine the flight path which minimizes the noise impact of a single aircraft. From equation (6.9), the transform of noise impact criteria (6.11) gives

$$I = K_0 \int_{x_1}^{x_2} T^s \varphi dx, \tag{6.12}$$

where $\varphi = (r^p v \cos\theta)^{-1}$. From the point O to point K one can use different trajectories, for example, OPK or OQK (Fig. 6.8). The difference of the criteria (6.12) along the trajectories is given by

$$\Delta I = KT^s\left(\int_{OQK} \varphi dx - \int_{OPK} \varphi dx\right) = KT^s \oint_{OQKPO} \varphi dx.$$

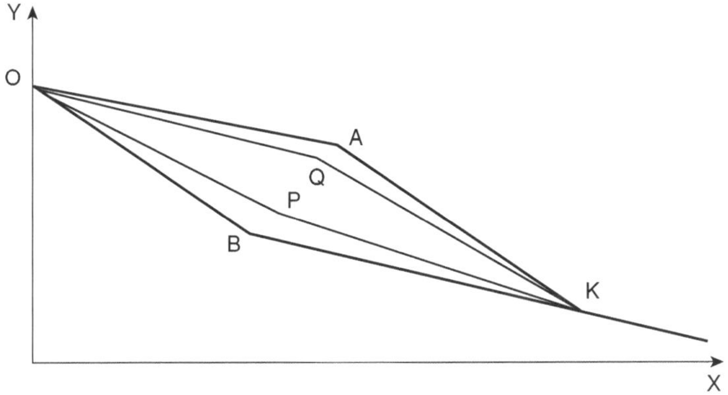


*Figure 6.8* The domain of permissible approach trajectories.

The continuity property of the function $\varphi$ representing permissible changes of the approach flight trajectory allows the use of Green's theorem

$$\Delta I = KT^s \int_S W dx dy, \quad W = -\frac{\partial \varphi}{\partial y},$$

where $S$ is the area OBKAO (the positive direction is the counter-clockwise direction). If $v = constant$

$$W = \frac{p(y - y_0)}{v r^{p+1} \cos\theta} > 0. \tag{6.13}$$

From inequality (6.13) it follows that $I_{OQK} > I_{OPK}$. The functional (6.1) is minimum and the minimum noise impact [determined by criteria (6.12) in point $(x_0, y_0)$] is achieved along a two-segment trajectory OAK. Examples of a preferred minimum noise trajectory and that based on the evaluation of the noise efficiency of the two-segment trajectory for a Tupolev-154M aircraft are shown in Fig. 6.9a and b, respectively.

Analysis of the influence of the landing flap deflection in noise reduction can be achieved by using the same method. On the assumption that the wing angle of attack and the engine tilt angle are small, aircraft may be considered to land along a polar parabola. One can use a dimensionless variable in equation (6.9)

$$u = \frac{v}{v_p}, \quad \overline{T} = \frac{TK_M}{mg}, \quad \overline{x} = \frac{gx}{v_p^2 K_M}, \quad \overline{y} = \frac{gy}{v_p^2 K_M}, \quad v_p = \sqrt{\frac{2mg}{\rho S \sqrt{\pi \lambda c_{x_0}}}},$$

$$K_M = \frac{1}{2}\sqrt{\frac{\pi \lambda}{c_{x_0}}},$$

where $S$ is the wing surface, $\lambda$ is the effective wing aspect ratio and $c_{x_0}$ is the air wing drag coefficient for zero lift. The solution of the first equation (6.9) can be represented in the form ($\overline{T}, \theta = constant$):

$$\overline{x} + C = -0.5\cos\theta \ln\left| -\frac{u^4}{\cos\theta} + 2\left(\frac{\overline{T}}{\cos\theta} + K_M tg\theta\right) u^2 + \cos\theta \right|$$

$$+ \frac{\cos\theta}{2\sqrt{\sigma}} \ln \left| \frac{-\frac{2u^2}{\cos\theta} + 2\left(\frac{\overline{T}}{\cos\theta} + K_M tg\theta\right) - \sqrt{\sigma}}{-\frac{2u^2}{\cos\theta} + 2\left(\frac{\overline{T}}{\cos\theta} + K_M tg\theta\right) + \sqrt{\sigma}} \right|, \tag{6.14}$$

where $\sigma = 4\left[\left(\frac{\overline{T}}{\cos\theta} + K_M tg\theta\right)^2 - 1\right] > 0$ and C is a constant of integration. In the plane $(v, x)$ the solution of (6.14) gives the domain of permissible

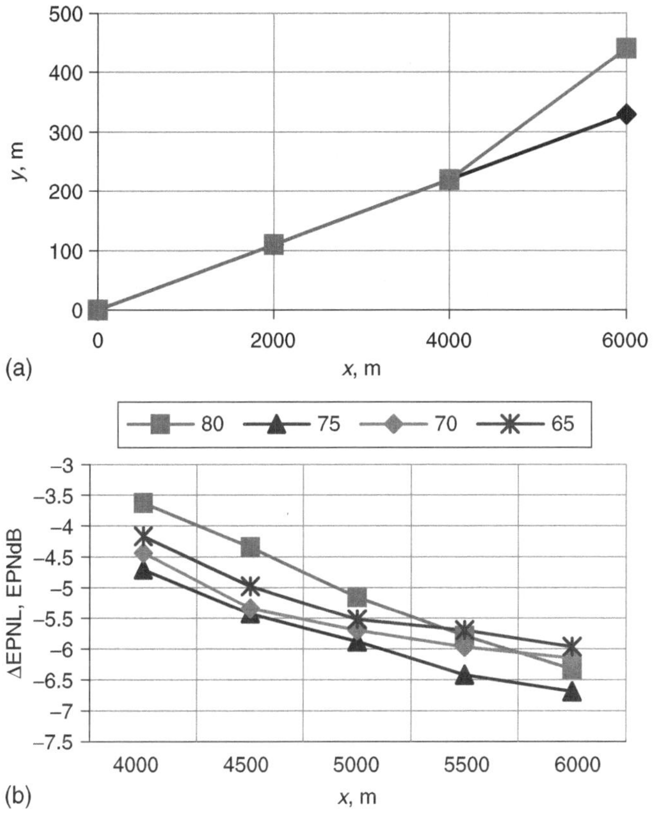


*Figure 6.9* (a) The form of a two-segment glide slope and (b) the decrease in *EPNL* corresponding to various two-segment glide slopes for aircraft landing mass in the range between 65 and 80 tons.

changes for the parameters for the approach of aircraft as shown in Fig. 6.10. Point M is the reference point and K is the end point. The arcs MA, BK correspond to the aircraft approach with intermediate flap deflection and the arcs MB, AK represent the aircraft approach with maximum landing flap deflection.

If the noise impact criterion is written in the form of (6.12), then it is possible to define the function

$$W = -\frac{K_0}{r^p \cos\theta}\left(-\frac{T^s}{v^2} + \frac{sT^{s-1}}{v}\frac{\partial T}{\partial v}\right). \tag{6.15}$$

To fulfill the condition $\partial T/\partial v \leq 0$, in the domain MAKBM of permissible changes of the parameters, the function $W > 0$. As a result, the minimum aircraft noise impact is achieved along the path MAK: the line segment

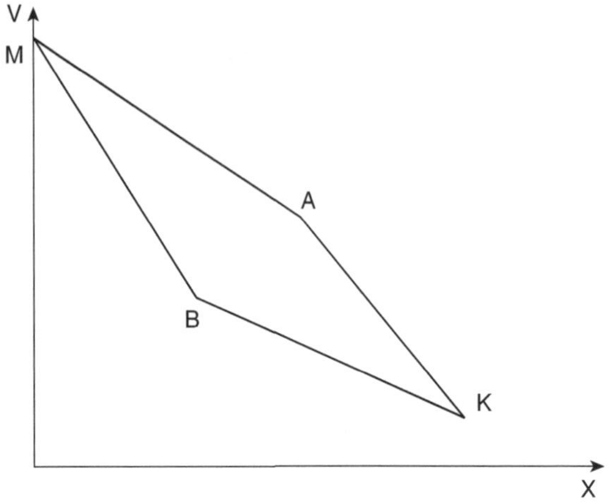


*Figure 6.10* The domain of permissible flight parameters along the airplane trajectory in the $(v, x)$ plane.

MA corresponding to the aircraft approach with intermediate flap deflection followed by segment AK corresponding to maximum landing flap deflection. The noise level reduction achieved at the control point is between 2 and 4 EPNdB.

Some low noise approach procedures involve programming the landing flap deflection and the wing angle attack. For such an approach, the domain of permissible changes of parameters is similar to the area MAKBM in Fig. 6.10. In this region the function (6.15) $W > 0$, therefore a minimum of aircraft noise impact is achieved along the path MAK: the segment MA corresponds to the aircraft approach in a low drag configuration and segment AK corresponds to the regime of deceleration of the aircraft between the heights of 350 m and 200 m along the standard glide path. The *EPNL* at points under the flight path is reduced between 5 and 11 EPNdB.

Since the realization of two- or multi-segment glide poses considerable technical problems, other low noise approach procedures have been investigated. In Fig. 6.11 the predicted results of programmed deceleration of a turbofan aircraft are shown. This is analogous to the low drag/low power method. For a particular value of the flight weight (Tupolev-154, $mg = 80$ t) the optimum value of deceleration predicted to be $-0.1$ m/s$^2$ and the relative reduction of noise contour area is predicted to be between 10 and 15 per cent.

By combining the continuous approach methods of low drag/low power and programmed deceleration, a considerable decrease in noise levels can be achieved. For example, if the gears and flaps are deflected between the

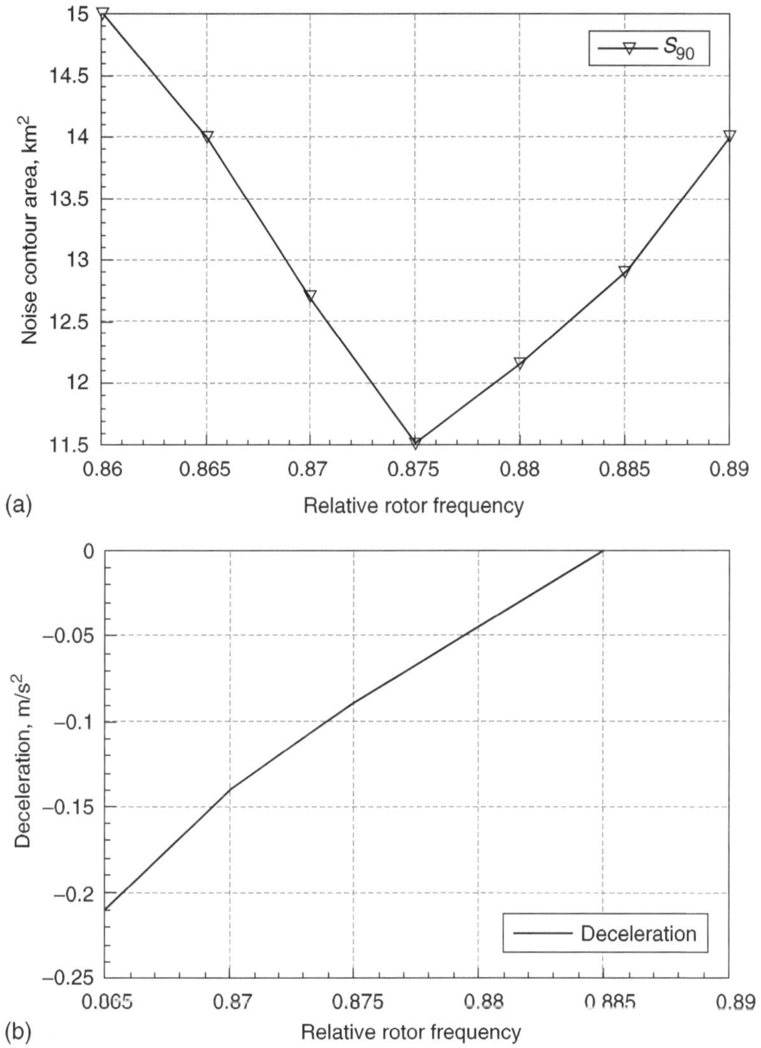


*Figure 6.11* (a) Predicted influence of the deceleration mode on the area of the $S_{90}$ contour area and (b) deceleration corresponding to the rotor frequency for a turbofan aircraft.

flight path heights of between 400 and 200 m, the area of the $S_{90}$ noise contour (defined by the $EPNL = 90$ EPNdB) is predicted to be decreased by 25 per cent. The *EPNL* at points far from the runway are reduced by almost 20 EPNdB and the *EPNL* at points near the runway are reduced between 7 and 10 EPNdB.

### *6.2.3 The takeoff stage*

The initial control parameters for takeoff procedures are the engine power setting (for simplicity defined as the relative value of rotor frequency $\overline{n}$), the pitch angle $\vartheta$ (the solutions for pitch angle are more stable than for angle of attack) and the angle of flap deflection $\delta$. The constraints for control variables $\Omega$ are defined by the flight safety requirements and by the aircraft and engine characteristics:

$$\delta = \delta_{to} \quad \text{for } H \leq 120\,\text{m},$$

$$\vartheta_{\min} \leq \vartheta \leq \vartheta_{\max}$$

$$T_{\min} \leq T \leq T_{\max}$$

$$\theta \geq \theta_{\min}$$

$$v \geq v_2 + 20$$

where $\delta_{to}$is the flap angle takeoff value, $v_2$ is the safe velocity established for a given takeoff aerodynamic configuration (km/h). The criterion for optimizing the noise level decrease at the point of noise control is represented by equation (6.10).

The most typical noise solutions involve the following:

- Takeoff and initial climbing require maximum values of thrust $T$ (maximum rotor frequency $\overline{n}$) and a flight path angle $\theta$ achieved with the minimum permissible value of the flight velocity $v$; therefore, the flight path should be as high as possible above the control point of interest.
- At a distance of between 200 and 1000 m before the control point the engines are throttled back to the values of the thrust $T$, which are defined for the permissible safe value of flight path angle $\theta_{min}$.
- On passing the noise control point, the engine thrust and other control parameters are reset to those for normal climbing.
- If the point of noise control is placed near the runway, the flaps are preserved in a takeoff position. In other cases, the flap deflection may be changed to a clean aerodynamic position or to any intermediary position between takeoff and clean deflections for decreasing the drag forces, and to increase the flight path angle and height above the control point.

These features are demonstrated in Fig. 6.12 for an aircraft with two low-bypass-ratio engines. The corresponding values for normal takeoff and climbing are shown as dotted lines for comparison.

If the optimization criterion is represented by the area of the noise contour around the flight paths, then the optimal solution has some differences from the previous one. First of all, the cutback of the engine operation mode should begin earlier at the end of the takeoff stage (i.e. at a height

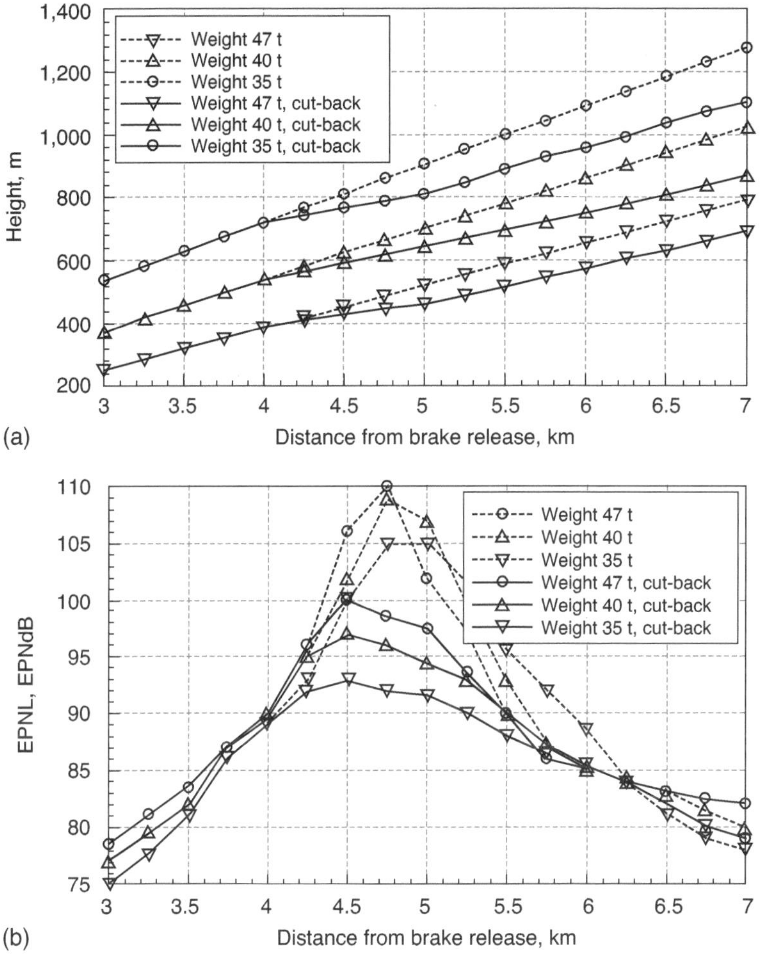


*Figure 6.12* (a) Flight trajectories and (b) predicted *EPNL* at the noise control point 4.5 km from takeoff.

of $H = 120$ m for most aircraft; see Fig. 6.13). The engine is reset to the normal climbing mode somewhat later.

The reset point on a trajectory may be defined by employing the concept of noise radius according to the following formula:

$$H = R_{Nnom} \cos \Theta,$$

where $R_{Nnom}$is the noise radius defined for normal climbing and nominal engine mode.

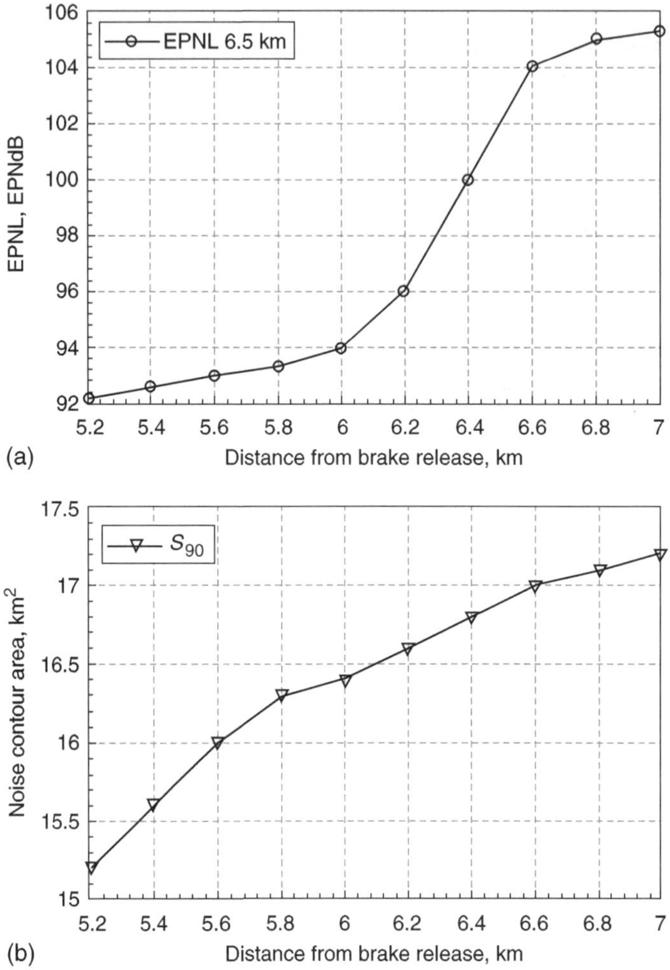


*Figure 6.13* (a) Predicted *EPNL* at the 6.5-km control point and (b) the corresponding $S_{90}$ contour area.

The concept of noise radius is very useful for analyzing the resulting solutions. For these purposes, the results of noise radius investigations are discussed in relation to noise abatement. The influence of the flaps deflection angle can be explained by the following factors:

- In a clean aerodynamic configuration of the aircraft, its aerodynamic quality $K$ is higher by nearly a factor of 2 compared with the normal takeoff aerodynamic configuration, so the flight path angle $\theta$ and the flight height $H$ must increase more quickly in a clean-up configuration, even though the minimum safe clean-up velocity is between 30 and 35 per cent higher.

- The degree of throttling-back of the engine thrust in the clean-up configuration is higher (throttling is deeper) than in the takeoff configuration of the aircraft.

If it is possible from the point of view of flight safety, the clean-up configuration should be used for more efficient noise reduction at a control point under the flight path.

The derivatives $dR_N/d\bar{n}$ within the set of takeoff–climbing modes are much greater than the derivatives $dR_N/dv$ and the absolute values of $R_N$ for the clean-up configuration are smaller than with deflected flaps. Therefore, the results of the optimization accord with the noise radius model of the aircraft.

## 6.3 Noise reduction in the vicinity of an airport

The concept of noise radius $R_N$ (see Section 4.2) is useful when investigating the noise footprints formed under the flight of an aircraft in the vicinity of an airport. The aircraft noise impact is minimized by minimizing the surface area contained within the contour defined by $EPNL = constant$, $\theta = constant$. The equation of the contour is

$$(x \sin\theta - y_0 \cos\theta)^2 + z^2 = R_N^2,$$

where $x, z$ is the coordinate of the contour on the ground surface and $y_0$ is the ordinate of the chosen coordinate system relative to the ground coordinate system. For part of the flight path ($\theta = constant$) the surface area is defined by

$$S = -\int_l \sqrt{R_N^2 - (y - y_0)^2 \cos^2\theta}\, dx, \qquad (6.16)$$

where the minus sign accounts for the convention used in path-tracing $l$. To minimize the noise contour area during the approach–landing flight stage, we investigate the function $W$ [corresponding to (6.16)] in the domain of permissible changes of parameters inside the area OAKB shown in Fig. 6.8:

$$W = -\frac{(y - y_0)\cos^2\theta}{\sqrt{R_N^2 - (y - y_0)^2 \cos^2\theta}} < 0. \qquad (6.17)$$

In the present case, the minimum noise impact can be realized along the flight path segment OBK shown in Fig. 6.8. The solution is valid as long as

$$R_N > (y - y_0)\cos\theta.$$

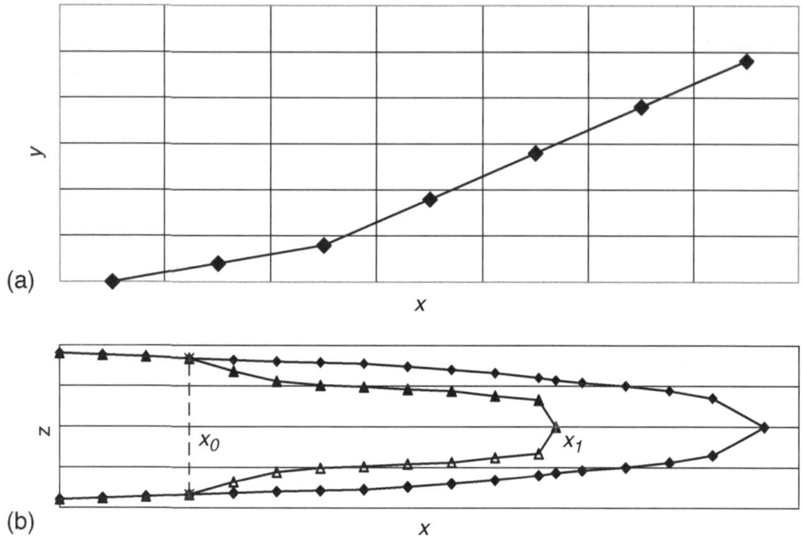


*Figure 6.14* Noise contour area effects of a two-segment glide slope: (a) the trajectory of the two-segment glide slope; and (b) the contour noise area.

Analysis of noise contour alterations due to a multi-segment glide slope can be done through a noise radius analysis. The flight path parameters along the segment that is nearest to the runway are equal to normal conditions, so the part of the noise contour $\Delta S_1$ in this segment is not changed and corresponds to the initial procedure (see Fig. 6.14a). For a higher slope segment the derivative $\partial \Delta S_2/\partial v$ is given by ($R_N v = C_v$; $\theta_2 = constant$):

$$\frac{\partial \Delta S_2}{\partial v} = -\frac{2C_v R_{N2}}{v^2 \sin|\theta_2|}\left[\frac{\pi}{2} - \arcsin\frac{q_2 \sin|\theta_2|}{R_{N2}}\right], \tag{6.18}$$

where $q_2 = x_0 tg|\theta_1|\, ctg|\theta_2|$, $x_0$ is the coordinate transition of an aircraft from the standard glide-path (Fig. 6.14a and b).

This derivative is negative, so with increase in velocity, the noise contour area $S_2$ is decreased. The influence of coordinate $x_0$ corresponding to the transition from a higher slope to the normal one can be analyzed using the derivative in a form:

$$\frac{\partial \Delta S_1}{\partial x_0} = -2\left(\sqrt{R_{N2}^2 - 0.25x_0^2 \sin^2 2|\theta_2|\, tg^2|\theta_1|} \times tg|\theta_1|\, ctg|\theta_2| \right. \tag{6.19}$$

$$\left. - \sqrt{R_{N1}^2 - x_0^2 \sin^2\theta_1}\right), \quad x_0 < \frac{R_{N1}}{\sin|\theta_1|}, \quad |\theta_2| > |\theta_1|.$$

This derivative is positive, so if $x_0$ is decreasing, the noise contour is decreasing too, and $x_0$ must be as small as possible to be consistent with

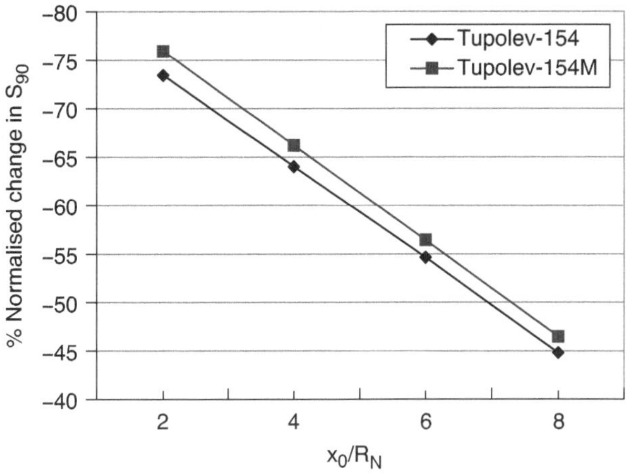


*Figure 6.15* The percentage normalized noise level reduction under a two-segment glide slope: $\Delta\overline{S}_{90} = (\Delta S_{90}/S_{90})$ 100 per cent, where $\Delta S_{90}$ is the reduction in the $S_{90}$ noise contour areas through the use of a two-segment glide slope.

safety requirements. Figure 6.15 shows the predicted influence of $x_0$ on reductions of the $S_{90}$ noise contour areas for Tupolev-154 and Tupolev-154M aircraft.

Let us now investigate low noise takeoff procedures using (6.19). The problem of the minimization of aircraft takeoff noise can be formulated after making the aircraft climb to some height (e.g. 400 m is assumed in Fig. 6.16). The trajectory MOK is typical for aircraft with turbojet engines. The line segment OA corresponds to the takeoff thrust trajectory of the aircraft. The trajectories OB and AK correspond to a reduced thrust. The line segment BK corresponds to the nominal engine mode flight thrust.

The domain of permissible flight paths ($\theta$ = *constant*) can be described by the differential equations derived for the center of mass of the airplane in the velocity coordinate system

$$\begin{aligned} m\frac{dv}{dt} &= T\cos\alpha - X_a mg\sin\theta \\ 0 &= T\sin\alpha + Y_a - mg\cos\theta \\ \frac{dx}{dt} &= v\cos\theta \\ \frac{dy}{dt} &= v\sin\theta. \end{aligned} \tag{6.20}$$

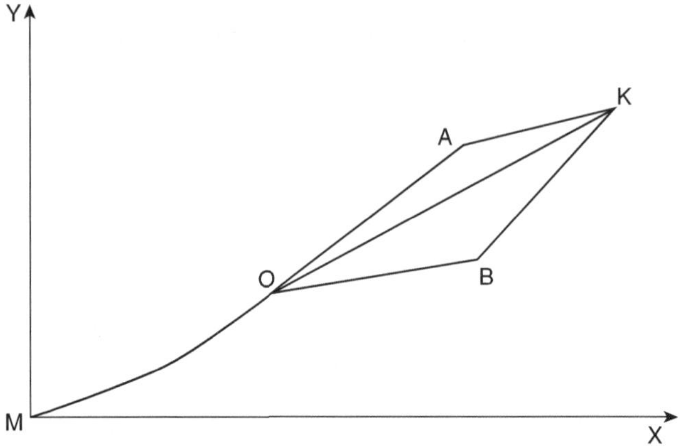


*Figure 6.16* The domain of permissible changes of parameters for the takeoff trajectory.

In the domain of permissible flight paths, defined by solutions of the equations (6.20), the function $W < 0$ [see inequality (6.17)]. The minimization of aircraft noise impact can be achieved along trajectories of the form of OBK: a segment of reduced thrust in the line segment OB in Fig. 6.16 is followed by use of the nominal mode of the engines for aircraft ascent (line segment BK).

To minimize the noise contour area, we introduce the initial coordinate $x_i$ (Fig. 6.17), so that part of the noise contour area is given by:

$$\Delta S_i = \frac{R_{Ni}^2}{\sin\theta_i}\left(\frac{\pi}{2} - \arcsin\frac{q_i \sin\theta_i}{R_{Ni}}\right) - q_i\sqrt{R_{Ni}^2 - q_i^2 \sin^2\theta_i}, \tag{6.21}$$

where $q_1 = x_0, q_2 = x_0 tg\theta_1 ctg\theta_2, q_3 = R_{N2} ctg\theta_3 (\cos\theta_2)^{-1}$ and $x_0$ is the point at which the engine is throttled down.

The influence of aircraft mass on the noise contour area is defined by the derivative $\partial\Delta S_i/\partial m$

$$\frac{\partial\Delta S_i}{\partial m} = \frac{\overline{T}_i ctg\theta_i}{mg}\left\{\frac{R_{Ni}^2}{\sin\theta_i}\left[\frac{\pi}{2} - \arcsin\left(\frac{q_i \sin\theta_i}{R_{Ni}}\right)\right] + q_i\sqrt{R_{N1}^2 - q_1^2 \sin^2\theta_i}\right\}, \tag{6.22}$$

$$q_i < \frac{R_{Ni}}{\sin\theta_i}, \ \theta_i > 0,$$

where $\overline{T}_i = T_i/mg$ and $T_i$ is the thrust on the $i$th flight path (such as in Fig. 6.17a). Since the derivative $\partial\Delta S_i/\partial m > 0$, decrease in aircraft mass will

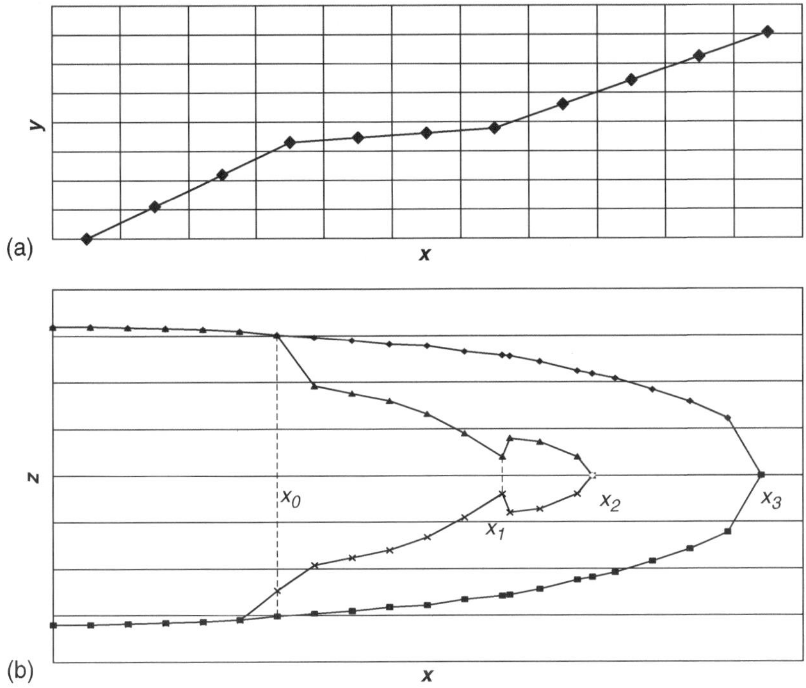


*Figure 6.17* (a) A form of reduced noise trajectory and (b) noise contours corresponding to different throttle-back points.

ensure decrease in the noise contour area. The influence of the engine mode is defined by the derivative $\partial \Delta S_i / \partial n$ (where $n$ is the rotor frequency)

$$\frac{\partial \Delta S_i}{\partial n} = A_i \frac{\partial R_{Ni}}{\partial n} + B_i \frac{\partial \overline{T}_i}{\partial n}, A_i = \frac{2R_{Ni}}{\sin\theta_i}\left[\frac{\pi}{2} - \arcsin\left(\frac{q_i \sin\theta_i}{R_{Ni}}\right)\right],$$

$$B_i = -\frac{ctg\theta_i}{mg}\left\{\frac{R_{Ni}^2}{\sin\theta_i}\left[\frac{\pi}{2} - \arcsin\left(\frac{q_i\sin\theta_i}{R_{Ni}}\right)\right] + q_i\sqrt{R_{Ni}^2 - q_i^2\sin^2\theta_i}\right\}, \quad (6.23)$$

$$\theta_i > 0, q_i < R_{Ni}(\sin\theta_i)^{-1}.$$

where $mg$ is the flight weight of the aircraft. Analysis shows that

$$A_i > 0,\ B_i < 0,\ \left|A_i \frac{\partial R_{Ni}}{\partial n}\right| >> \left|B_i \frac{\partial \overline{T}_i}{\partial n}\right|,$$

therefore $\partial \Delta S_i / \partial n > 0$ and throttling-back of the engines (decrease in rotor frequency) provides a decrease in the total noise contour area. The coordinates of the noise contour area such as in Fig. 6.17b are determined from:

$$x_1 = x_0(1 - tg\theta_1 ctg\theta_2) + R_{N2}(\sin\theta_2)^{-1},$$

$$x_2 = x_0(1 - tg\theta_1 ctg\theta_2) + R_{N2}(\sin\theta_2)^{-1} \cdot 1 - tg\theta_2 ctg\theta_3) + R_{N3}(\sin\theta_3)^{-1},$$

$$x_3 = R_{N1}(\sin\theta_1)^{-1}.$$

The duration $t_2$ of the flight stage with engines throttled-back during takeoff is defined by

$$t_2 = \left(\frac{R_{N2}}{\sin\theta_2} - x_0 tg\theta_1 ctg\theta_2\right)\frac{1}{v_{m2}}, \tag{6.24}$$

where $v_{m2}$ is middle flight velocity for the second flight stage of climbing.

The derivative $\partial \Delta S_1/\partial\theta_1 < 0$ (for the first flight stage during climbing; see Fig. 6.17a), so if the flight path angle increases, the noise contour area decreases.

The derivative $\partial \Delta S_i/\partial v < 0$, so if the flight velocity increases, the noise contour area decreases. The possible range of velocity changes is between 20 and 40 m/s during the flight stage considered, so the resulting decrease $S$ due to the increased velocities is much smaller than the possible decrease $S_i$ due to throttling-back of the engines.

In accordance with equation (6.21), the derivative $\partial \Delta S_3/\partial v < 0$ (for the third flight stage during climbing; see Fig. 6.17 a), so acceleration of aircraft is possible provided the noise contour area decreases. A program for the flight velocity can be determined by using the investigation function $W$ [for criteria (6.16)] in the domain of permissible changes of parameters inside the area BCKD shown in Fig. 6.18. In the plane $(x, v)$, the domain of permissible changes of the parameters includes the arcs BC and DK, which correspond to climbing with acceleration, and the arcs CK and BD, which correspond to climbing with constant velocity.

For the trajectories with $\theta_3 = constant$, the function

$$W = -\frac{C_v R_{N3}}{v_m^2\sqrt{R_{N3}^2 - (y - y_0)^2\cos^2\theta_3}} < 0,$$

where $v_m$ is the flight velocity during the flight stage of the climbing (Fig. 6.18) and $C_v = R_{N3}v_m$ is constant. In the domain BCKD shown in Fig 6.18 the function $W < 0$, therefore a decrease in the total noise contour area is achieved along path BDK consisting of the arc BD corresponding to climbing with constant velocity and the arc DK corresponding to climbing with acceleration.

Several manuals suggest operational procedures for reducing the impact of aircraft noise. For example, implementation of the continuous descent

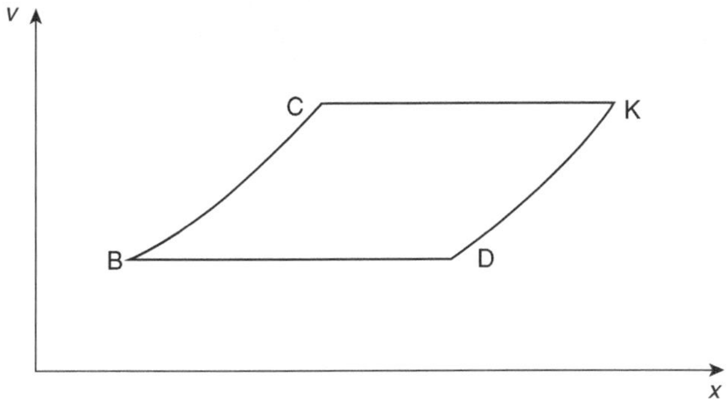


*Figure 6.18* The domain of permissible changes in the flight parameters along the airplane trajectory in the $(v, x)$ plane.

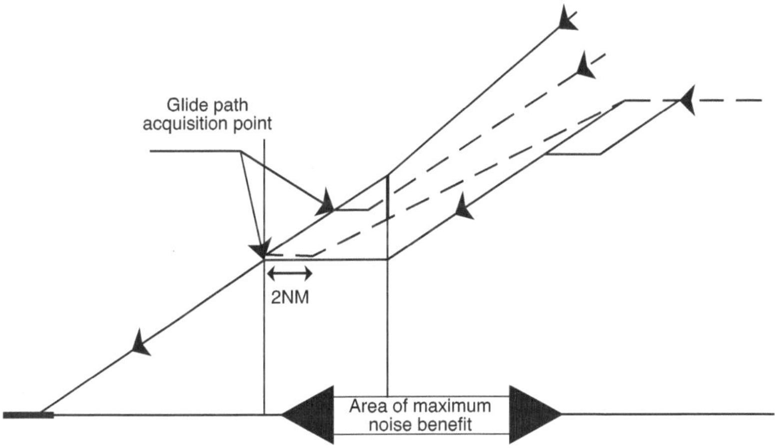


*Figure 6.19* Conceptual diagram of a continuous descent approach.

approach (CDA) has the potential to reduce noise during the approach phase of flight as well as reduce the fuel used between 50 and 150 kg per flight.[13] CDA is an aircraft operation technique in which an arriving aircraft descends to an optimal position with minimum thrust as shown in Fig. 6.19. The broken lines are typical of an aircraft using CDA in order to reduce the noise impact on the ground, and reduce fuel use and atmospheric emissions.

The basic aspects of CDA include:[13]

- providing accurate and timely distance to go (DTG) information to pilots in order achieve CDA;

- avoiding giving descent clearance prior to the point at which a CDA would naturally occur and giving estimated distance from touchdown to the pilot to allow the aircraft to intercept the approach glide-path with the minimum of level flight;
- provision of approach speed requirements to facilitate a continuous descent profile without the need for segments of level flight;
- avoiding unnecessary early deployment of flap and undercarriage where this does not conflict with the safety requirements and existing operational procedures.

## 6.4 The efficiency of acoustic screens for reducing noise from airport ground operations

One of the important methods of mitigation of noise from ground operations at airports is the use of an acoustical screen (noise barrier) to decrease the noise along with sound propagation path. Prediction of the acoustical efficiency of such a screen must include consideration of the types of noise sources and their acoustical characteristics, the location of the noise sources relative to the point of noise impact assessment, the type and the characteristics of screen and the influence of ground reflecting surfaces.

The acoustical efficiency or performance of a screen includes the shielding effects of line-of-sight sound waves. However, in addition, sound waves propagate over the screen because of wave diffraction at its top and sides. Investigations of the influence of the height, width and thickness of screens for airport noise problems show that their performance depends on the acoustical characteristics of aviation noise sources, the interference of direct and reflected waves at the receiver point and the sound transmission characteristics through the screens. Since all of the effects depend on frequency, a decision-making process on the use of screens at airports requires information about the spectral characteristics of noise sources and the noise reductions along the propagation path with and without acoustic screens. Specialized software has been developed for this purpose and its main features and some results are presented here.

The barrier height is of fundamental importance to the attenuation produced by the barrier. The higher the barrier, the more the line-of-sight is blocked, the larger is the path difference (difference in length between the line-of-sight and the sound path over the barrier top or side) and, hence, the greater is the attenuation. So, in respect of the aircraft-related noise, acoustic screens are suitable only for abating noise from aircraft operations on the ground. These operations include engine ground run-ups, runway takeoff and landing operations and taxiing. Other airport noise sources for which screens may be appropriate include vehicles used for maintenance of airport ground surfaces and the aviation engines, and other ground transport used in and around the airport.

Ground run-ups are aircraft engine tests that are performed to ensure that they are ready for operation after the servicing. The duration of the ground run-up can be from a few seconds to many minutes and usually it includes the testing of all engine settings – from idle up to maximum operating mode. Usually run-ups do not follow a predictable time pattern. Indeed run-ups may be conducted at night. Mechanics and other aviation personnel, working inside the airport, frequently recognize the run-ups as the most annoying noise source.

Since the noise emissions from aircraft engines are directional, the orientation of the aircraft during a ground run-up is very important and there might be a particular problem if the distance to residential land is less than a kilometer. Thus noise barriers or various enclosures have been constructed at many airports to reduce the noise of run-up operations. If the walls of the screen are installed around the aircraft, then screens with sloping sides are used to avoid multiple reflections. In some cases sound-absorptive material may also be proposed for use.

Long sound propagation distances give rise to issues that need to be considered in the acoustical design of screens. The primary issues are atmospheric effects (i.e. wind- and temperature-gradient effects) on screen performances. Downwind receiver conditions or temperature inversions reduce screen efficiency, since the curved propagation paths may decrease the path difference and accordingly reduce the insertion loss of the screen. Acoustical screen insertion loss may also be reduced by the loss of ground effect in the vicinity of the screen.

In many airports, the runways are located a few kilometers from the nearest homes and often the intervening terrain is flat and unobstructed. A-weighted sound levels of around 90 dB have been measured at homes during the start of the takeoff roll and landing with reverse thrust of the engines. Since areas around runways must be clear of obstruction from safety point of view, screens cannot be located within about 200 m of a runway. This means that barriers intended to mitigate noise at the start of the takeoff roll must be constructed near the receivers to be effective.

Unless all ground surfaces between the source and the receiver are acoustically hard, the ground-effect attenuation must usually be considered to assess properly the expected insertion loss of barriers.

Earth berms are sometimes used to control runway noise at airports. They are less expensive than walls and they may be less unsightly. Moreover, the necessary land is often available on the airport, especially if the berms are constructed at the same time as the runway.

The nature of the noise wall is also a factor. The forms and types of acoustical screens used at airports include barriers, walls, buildings and artificial and natural structures associated with the local topography. There are modifications of performance due to shape of the noise wall, the nature of the diffracting edge, the finite length of the noise wall, such as at access gaps (for example, between buildings) and the addition of absorptive material.

The acoustical efficiency of a screen $\Delta L_{ef}$, otherwise known as insertion loss *IL*, is defined as the difference between the noise reductions with $\Delta L_{scr}$ and without screen $\Delta L$ along the noise propagation path from noise source (characterized by sound level $L_s$) to receiver (characterized by sound level $L_r$). Specifically

$$\Delta L_{ef} = \Delta L_{scr} - \Delta L; \Delta L = L_s - L_r; \Delta L_{scr} = L_s - L_{r'},$$

where the sound levels at the receiver points $L_r$ and $L_{r'}$ differ according to the presence and influence of screen.

The noise generated by aircraft on the ground is determined most of all by the type of engine, its nominal power and the settings on the aircraft. Aircraft engines are complex noise sources including one or more jets, fans, turbines, compressors, combustion chambers or propellers. To investigate the influence of noise source characteristics on barrier efficiency, four examples of noise sources have been used: (1) turbojet noise; (2) turbofan noise; and (3) turboprop noise spectrum. The corresponding third-octave bands are plotted in Fig. 6.20. The aircraft source spectra have been averaged for specific groups of engines and normalized to a reference distance $R_0 = 1$ m.

The noise levels at the receiver point located on various heights above the ground are reduced relative to noise levels at a source by a number of effects. The level of noise reduction at every spectrum band with central frequency $f$ is defined by:

$$\Delta L(f) = \Delta L_R + \Delta L_{atm} + \Delta L_{int} + \Delta L_B + \Delta L_{scr},$$

where $\Delta L_R = 20\,\lg(R/R_0)$ is the result of spherical spreading, $R$ is the distance between the source and the receiver; $\Delta L_{atm} = \alpha(R - R_0)$, $\alpha$ is the sound absorption coefficient in air as a function of frequency and

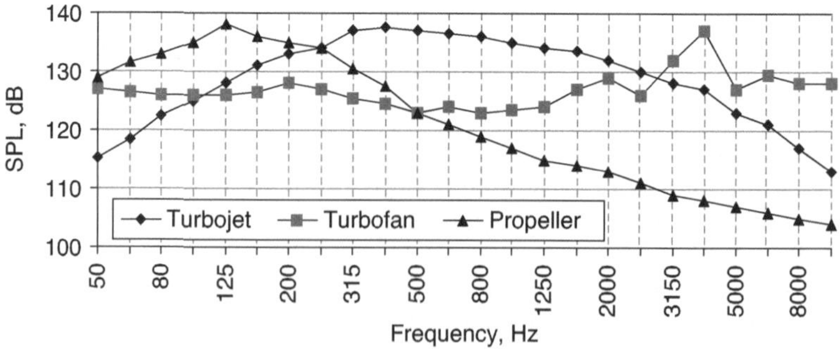


*Figure 6.20* Characteristic engine noise spectra referenced to 1 m.

atmospheric parameters (temperature, pressure, relative humidity); $\Delta L_{int}$ is the result of interference between direct and ground-reflected waves at various frequencies; $\Delta L_B$ is the transmission loss through green belts along the sound propagation path; and $\Delta L_{scr}$ is the noise reduction by screens, other than green belts.

Sound absorption coefficients for any atmospheric conditions are defined in accordance to recommendations of ICAO Annex 16:[9]

$$\alpha = 10^{\left[2.05\lg\left(\frac{f}{1000}\right)+1.14\bullet 10^{-3}T-1.917\right]} + \eta 10^{\left[\lg(f)+8.43\bullet 10^{-3}T-2.756\right]},$$

where $T$ is air temperature and $\eta$ is a parameter, which depends upon the sound frequency, and the relative humidity and temperature.

Transmission losses $\Delta L_B$ in green belts, which may consist of various kinds of trees and bushes, are defined by empirical formulas:[10]

$$\Delta L_B = 20\lg\left(\frac{d+\sum_i B_i+\sum_i A_i}{d}\right)+1.5Z+\beta\sum_i B_i,$$

where $d$ is the green belt width, $B_i$ is the width of the $i$th tree crown, $B_i$ is the distance between particular crowns of the $i$th and the $(i+1)$th tree rows (usually about 5 m), $Z$ is the number of rows and $\beta$ is the relative sound absorption by green belts. The influence of interference (frequently called excess attenuation or lateral noise attenuation) and diffraction effects are considered in detail elsewhere.[10]

Special account must be taken regarding propagation paths around the screen edges – at the top and the sides of the screen. Geometrical propagation paths over the top edges of thin and thick screens are shown in Figs 6.21a and b.

The algorithm used in a software tool developed for screen characteristics assessment is based on the model by Maekawa[11] discussed previously in Chapters 3 and 4.

Typically, in controlling the noise from aircraft ground operations the sound waves propagate along the surface, and the length of the propagation path is much larger than the wavelength while the dimensions of the noise source are shorter than the wavelength, corresponding to compact sources (Fig. 6.22).

The mathematical expression for the total sound pressure assessment at a receiver point consists of two parts: the direct sound wave and the reflected sound wave. Since usually the reflecting surfaces are not rigid, the amplitude of the reflected wave is decreased and the phase of the reflected wave is altered as a result of the sound–wave–surface interactions. Both effects are described by the complex acoustic impedance $Z$ of the reflecting surface. The relation between the acoustic impedance and the plane wave reflection

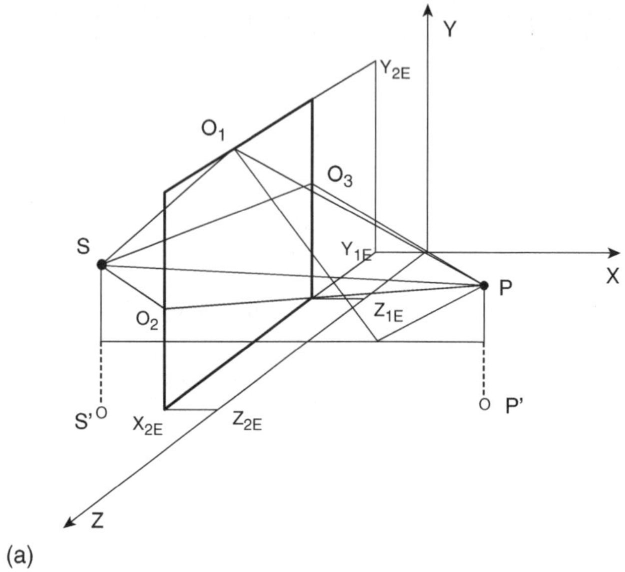


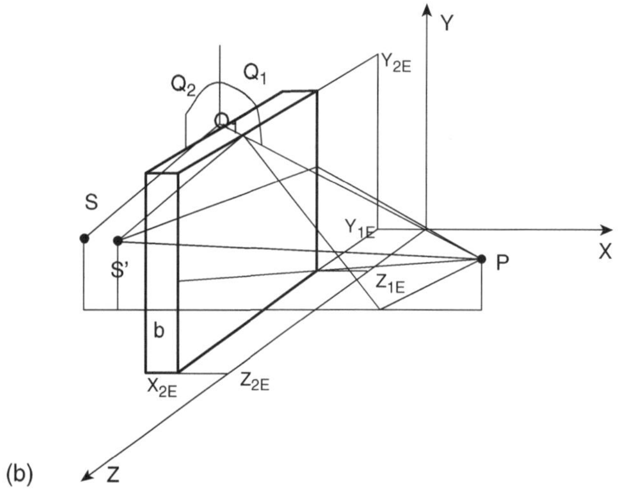


*Figure 6.21* Propagation paths around (a) thin and (b) thick screens.

coefficient $R_p$ is shown in the following:

$$R_p = \frac{Z\sin\phi - [1-(k/k_2)^2\cos^2\phi]^{1/2}}{Z\sin\phi + [1-(k/k_2)^2\cos^2\phi]^{1/2}},$$

where $Z$ is normalized by the characteristic acoustic impedance of air; $\phi$ is the angle of incidence; and $k$, $k_2$ are the wave numbers for air and for

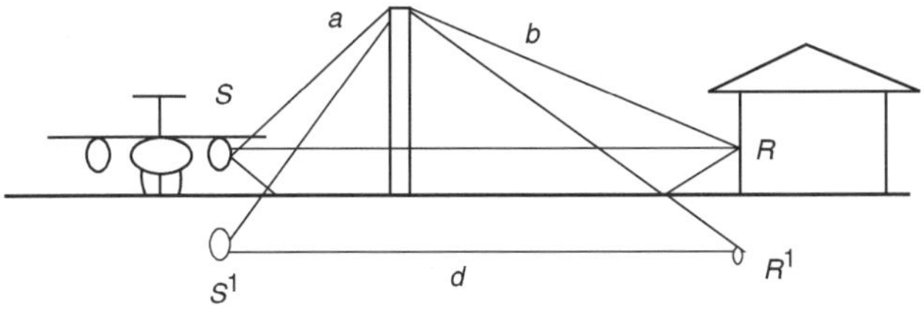


*Figure 6.22* The source–receiver geometry for ground operation noise reduction by a screen: $S$ is the real noise source, $S^1$ is the image noise source, $R$ is the actual noise receiver and $R^1$ is the image noise receiver.

the surface, respectively. Usually $\phi > 0$, $k_2 > k$, so the last formula can be transformed into a simpler form:

$$R_p = \frac{Z\sin\phi - 1}{Z\sin\phi + 1}.$$

In the equations above, the reflection coefficient corresponds to that for a plane wave, but in practice reflection of a spherical wave should be taken into account (see Chapters 3 and 4). The spherical wave reflection coefficient $R$ is given by:[10]

$$R = R_p + (1 - R_p)F(w),$$

where function $F(w)$ is the 'boundary loss factor', which describes the interaction of the curved wave front with the flat surface; $w$ is the 'numerical distance', which is directly proportional both to the propagation distance $R_2$ and to the frequency $f$ through the wave number $k$:[10,11]

$$F(p_e) = 1 + ip_e^{1/2}\exp(-p_e)erfc(-ip_e);\ p_e = (ikR_2/2)^{1/2}(\beta + \cos\theta).$$

Phase and amplitude differences between direct and reflected waves result in complicated interference patterns either decreasing or increasing sound pressure levels in particular frequency bands. The interference effects in any spectral bands are calculated from:[10]

$$\Delta L_{int} = 10\lg\left\{1 + S^2|R|^2 + 2S|R|[(\sin\alpha\Delta R/\lambda)/(\alpha\Delta R/\lambda)]\cos([\beta\Delta R/\lambda + \delta)]\right\},$$

where $S = R_1/R_2$, $\Delta R = (R_2 - R_1)$, $\alpha = \pi(\Delta f/f)$, $\Delta f$ is the spectral bandwidth, $\beta = 2\pi[1 + (\Delta f/f)(\Delta f/f)/4]^{1/2}$; for third-octave bands $\alpha = 0.725$ and $\beta = 6.325$.

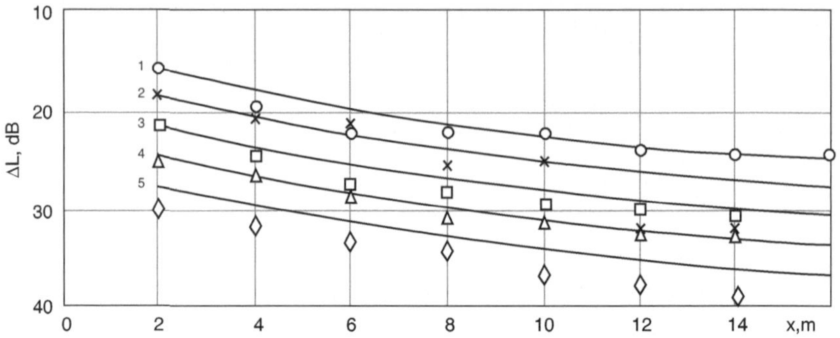


*Figure 6.23* Comparison of calculated and measured screen efficiencies for various distances to noise receiver for five frequencies: 1 – 250 Hz, 2 – 500 Hz, 3 – 1000 Hz, 4 – 2000 Hz, 5 – 4000 Hz.

*Table 6.2* Comparison of measurements and predictions of the efficiency of the screen with finite length (dB)

| *Calculated point (X,Y,Z)* | *Insertion loss of the screen* | *Frequency (Hz)* | | |
|---|---|---|---|---|
| | | *2500* | *5000* | *10,000* |
| 1, 0.5, 1.28 | Predicted | 19.7 | 22.6 | 25.7 |
| | Measured | 19.9 | 21 | 26 |
| 1, 1, 1.28 | Predicted | 17 | 20 | 21.5 |
| | Measured | 15.3 | 17.2 | 19.6 |
| 1, 0.5, 0 | Predicted | 23 | 25 | 27 |
| | Measured | 20.9 | 23.9 | 26.9 |
| 1, 1, 0 | Predicted | 19 | 20.5 | 21 |
| | Measured | 15.7 | 17.5 | 19.8 |

These formulas have been implemented in software and have been compared with measurements. Results of such comparisons ($\sigma^2 = 0.63$) are shown in Fig. 6.23 and in Table 6.2.

The adequacy of these models and algorithms is shown by comparison of the calculated and measured noise levels from an Il'jushin-86 aircraft in the presence and absence of the screen installed at Sheremetjevo airport to control ground operation noise from this type of the aircraft. Calculated values are presented in Fig. 6.24 and in Table 6.3.

The ground (interference) effect contribution for this example has been calculated and the results are presented in Fig. 6.25.

The influence of interference effects on a screen's acoustic efficiency has been calculated. The geometrical scheme for location of the noise source, screen and receiver has been chosen to represent a case of aircraft noise control near to the hotel in the vicinity of the airport. The length of the

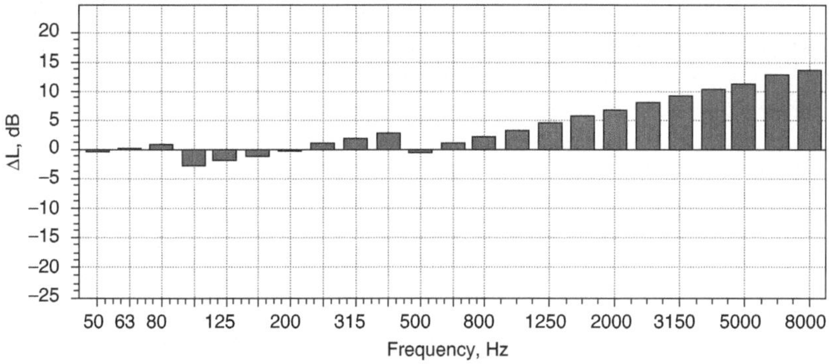


*Figure 6.24* Insertion loss of the screen installed at Sheremetjevo airport for the abatement of noise from Il'jushin-86 aircraft on the ground.

*Table 6.3* Measured insertion loss of the screen installed at Sheremetjevo airport for abatement of noise from an Il'jushin-86 aircraft on the ground

| Frequency (Hz) | 32.5 | 63 | 125 | 250 | 500 | 1000 | 2000 | 4000 | 8000< |
|---|---|---|---|---|---|---|---|---|---|
| $\Delta L_{ef}$ (dB) | 2 | 5 | 0.5 | 5 | 5 | 8.5 | 14 | 16 | 14 |

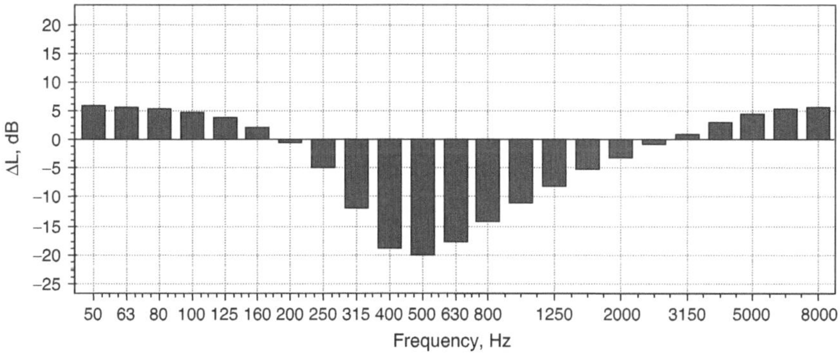


*Figure 6.25* The interference effect for distance 270 m, source height 2.5 m, receiver height 1.2 m; the surface covering is assumed to be grass.

screen is about 500 m and its height is 12 m. The center of the coordinate system is on the left of the screen. The axis OZ is directed along the screen and the axis OX is perpendicular to the screen. The coordinates of noise source are (−175, 325, 5); the coordinates of the receiver are (113, 175, 5), using ($x$, $z$, $y$)-coordinates, respectively.

Three types of reflecting surfaces have been considered: (1) one where reflection is absent; (2) one where reflection is perfect (absolutely rigid surface); and (3) one where reflection is from an impedance surface (grass). The types of noise sources were described earlier. The results of spectral efficiency of the screen are shown in Fig. 6.26.

Compared to the absence of reflection, the presence of a perfectly reflecting ground causes a decrease in the acoustical efficiency to an equal extent throughout the frequency spectrum. Reflection from an impedance surface gives results that lie between those of the two previous variants, but in some frequency bands there is a much decreased efficiency (even negative insertion loss for the screen), for example, between 160 and 250 Hz (Fig. 6.26). The frequencies of poor insertion loss depend on the geometry of the scheme and the impedance characteristics of the surface.

Table 6.4 shows predictions of spectral efficiency of this barrier performed using noise indices $\Delta L_{Aef}$ and $\Delta L_{ef}$ for three types of noise sources and the three types of ground. These results are interesting because A-weighted noise levels are used frequently in national standards and regulations for noise control, and they show that the influence of noise source spectral characteristics is important.

The influence of source spectrum on barrier efficiency during ground operations might be particularly significant for propeller aircraft since their main acoustic energy is located in the low-frequency bands of the spectra, where the interference effect tends to be strong.

Predicted insertion loss spectra for various combinations of surfaces either side of the screen are shown in Fig. 6.26. The impedance characteristics for these surfaces (1 – concrete/concrete; 2 – concrete/grass; 3 – freshly raked soil/grass; 4 – sharp sand with moisture < 10 per cent/grass) are based on values published elsewhere.[10]

The differences in spectral efficiencies cause the differences in $\Delta L_{Aef}$ between 1 and 2 dBA.

Seasonal changes in the acoustical properties of reflecting surfaces may mean that the screen efficiency changes with the season in the same manner. Measured efficiencies may differ from calculated ones (made at the stage of screen construction) due to the influence of this factor.

*Table 6.4* Comparison of screen efficiency for various types of noise sources

| *Noise source type* | *Type of sound wave reflection* | | |
|---|---|---|---|
| | *None* ($\Delta L_a/\Delta L_{in}$) | *Complete* ($\Delta L_a/\Delta L_{in}$) | *Impedance* ($\Delta L_a/\Delta L_{in}$) |
| Turbojet | 12.9/10.9 | 8.3/6.1 | 12.0/7.9 |
| Turbofan | 15.5/8.8 | 10.8/3.8 | 13.7/5.7 |
| Pink noise | 13.8/8.5 | 9.1/3.5 | 12.6/5.9 |

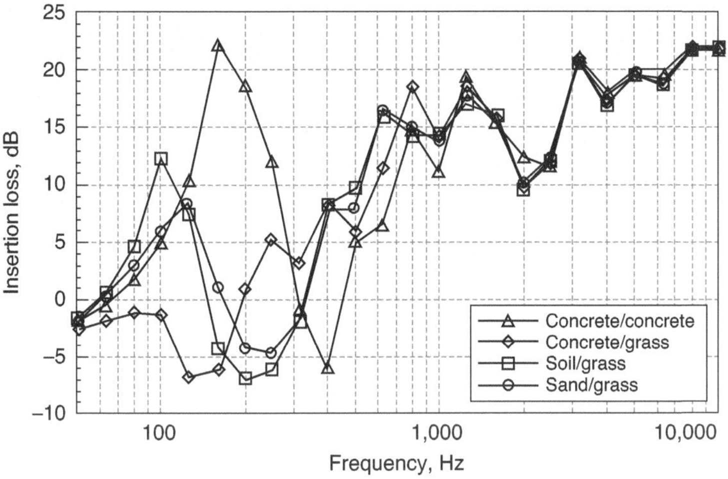


*Figure 6.26* Predicted insertion loss spectra for various combinations of surfaces either side of the screen.

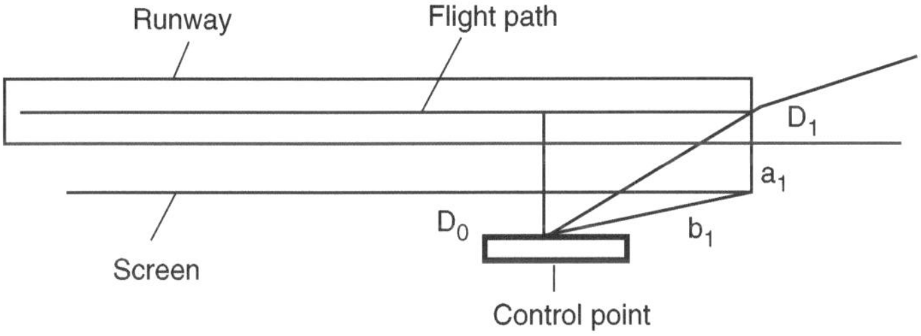


*Figure 6.27* An example of a noise attenuation screen design.

However, meteorological influences at the time of measurement may be equally if not more important.

A design calculation has been carried out for the installation of a screen alongside the runway or near some other stationary or moving point sources. The situation is illustrated in Fig. 6.27.

Analysis of noise generation by aircraft engine (Chapter 2) and of sound wave propagation conditions (Chapter 3) allows us to observe that the maximum level of noise expected from this source is not associated necessarily with the shortest distance from control point to the source $D_0$. In Table 6.5 and in Fig. 6.28 the predicted effects of geometrical spreading, directivity and interference of sound waves are shown, and also the summation of the results for concrete and grass covering of the reflecting

*Table 6.5* Predicted dependence of the various noise propagation criteria on the directivity angle from the source to the point of noise control (data were obtained for Il'jushin-86 aircraft for the shortest distance, $D_0 = 500\,\text{m}$)

| *Directivity angle (degrees)* | $\Delta L_{int}$ *(dBA), grass* | $\Delta L_{int}$ *(dBA), concrete* | $\Delta L_{int}$ *concrete* $-\Delta L_{int}$ *grass (dBA)* | $\Delta L_{\Theta}$ *(dBA)* | $\Delta L_R$ *(dBA)* | $\Delta L_{\Sigma}$ *(dBA) grass* | $\Delta L_{\Sigma}$ *(dBA) Concrete* |
|---|---|---|---|---|---|---|---|
| 10.0 | 1.8 | 1.3 | −0.5 | −7.0 | −15.2 | −20.4 | −20.9 |
| 20.0 | 1.8 | 1.1 | −0.7 | −9.5 | −9.3 | −17.0 | −17.7 |
| 30.0 | 1.6 | 0.8 | −0.8 | −7.0 | −6.0 | −11.4 | −12.2 |
| 40.0 | 0.25 | −1.0 | −1.25 | −4.5 | −3.85 | −8.1 | −9.3 |
| 50.0 | −0.7 | −2.0 | −1.3 | −3.0 | −2.3 | −6.0 | −7.3 |
| 60.0 | −1.1 | −3.2 | −2.1 | −2.0 | −1.25 | −4.25 | −6.35 |
| 70.0 | −1.0 | −1.9 | −0.9 | −1.0 | −0.54 | −2.5 | −3.4 |
| 80.0 | 0 | 0.1 | 0.1 | 0 | −0.13 | −0.1 | 0 |
| 90.0 | 0 | 0 | 0 | 0 | 0 | 0 | 0 |
| 100.0 | 0.4 | 6.4 | 6.0 | 1.5 | −0.13 | 1.8 | 7.8 |
| 110.0 | 1.0 | 4.3 | 3.3 | 3.0 | −0.54 | 3.5 | 6.8 |
| 120.0 | 1.2 | 4.6 | 3.4 | 5.0 | −1.25 | 5.0 | 8.4 |
| 130.0 | 1.5 | 3.4 | 1.9 | 7.5 | −2.31 | 6.7 | 8.5 |
| 140.0 | 3.2 | 1.4 | −1.8 | 10.0 | −3.84 | 9.4 | 6.6 |
| 150.0 | 5.4 | −0.1 | −5.5 | 12.0 | −6.0 | 11.4 | 5.9 |
| 160.0 | 6.9 | −1.0 | −7.9 | 9.5 | −9.3 | 7.1 | −0.8 |
| 170.0 | 7.0 | −1.0 | −8.0 | 1.0 | −15.2 | −8.2 | −15.2 |

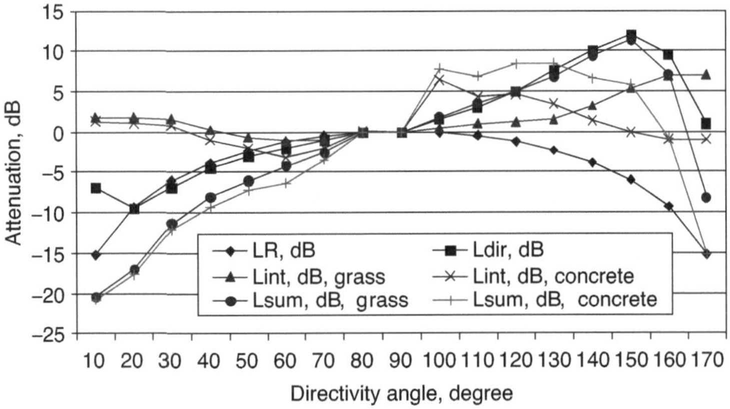


*Figure 6.28* Results of attenuation calculations for the screen shown in Fig. 6.27.

surfaces around the flight path of the aircraft. All of the listed effects have an influence on the noise level at the point of noise control, but the most important is the effect of the interference–ground effect.

For complete concrete covering, the maximum is observed at angles between 120 and 130 degrees. For complete grass covering, the maximum is observed at 150 degrees from the direction of aircraft motion. Therefore, it is necessary to define the screen's geometrical parameters and its acoustic efficiency for these directions.

## 6.5 Reduction of noise impact by optimum scheduling of aircraft operations

In Section 5.1, a special category of operational conditions was mentioned, which defines noise impact around an airport directly. It includes characteristics of air traffic such as aircraft fleet composition, flight intensity and the flight procedures used by individual aircraft types along separate routes.

Typically, a fleet is composed of aircraft with various acoustic performances, including International Civil Aviation Organization (ICAO) 'Chapter 2 aircraft', 'Chapter 3 aircraft' or/and 'Chapter 4 aircraft' requirements. This composition changes over time. The various consequences for the noise footprints are investigated here using the example of a regional airport in the Ukraine.

In the 1980s, two main types of aircraft, the Yakovlev-40 and Antonov-24, operated from the airport (Antonov-2, Antonov-14, Antonov-28 and L-410 also aircraft were used, but there were much fewer of them and they were much quieter). For noise-zoning purposes, noise contours were calculated using a graph-analytical method developed in the 1970s. The contours were recalculated in 1991 when the number of flights reached a maximum, which was quite close to airport operational capacity. Later, as a result of a transitional period for the national economy, the demand for aviation transportation decreased and it still had not reached the 1991 level. Predictions obtained using INM version 5.2 are shown in Figs. 6.29 and 6.30. The actual noise and aerodynamic performances of the Yakovlev-40 and Antonov-24 aircraft and flights in 26R direction from the airport have been assumed to obtain for these predictions. In reality, the noise contours are likely to be larger, not only because of louder noise performances, but because real flight paths differ from those assumed in the INM database.

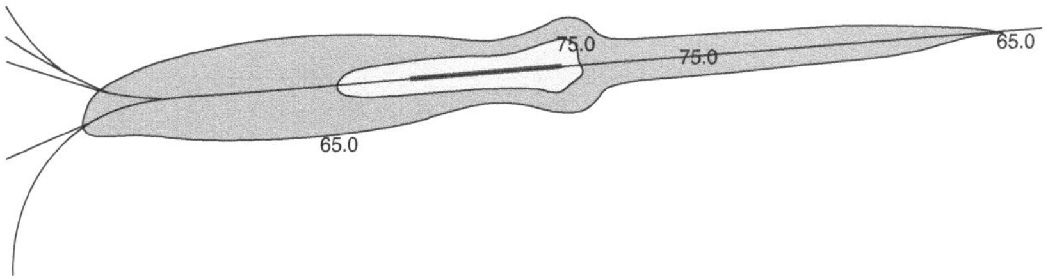


*Figure 6.29* Predicted daytime $L_{Aeq}$ contours for flights in 1980 in the 26R direction from a regional airport in the Ukraine. The assumed parameter values correspond to the noise and aerodynamic performances of the Yakovlev-40 and Antonov-24 aircraft.

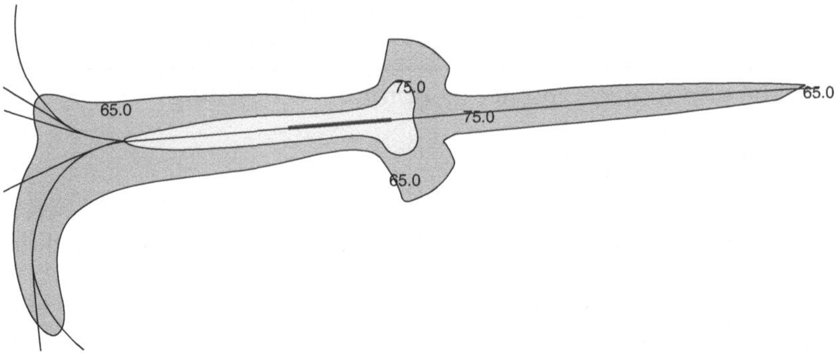


*Figure 6.30* Predicted $L_{Aeq\ day}$ contours for flights in 1990 in the 26R direction. The assumed parameter values correspond to the noise and aerodynamic performances of the Yakovlev-40 and Antonov-24 aircraft.

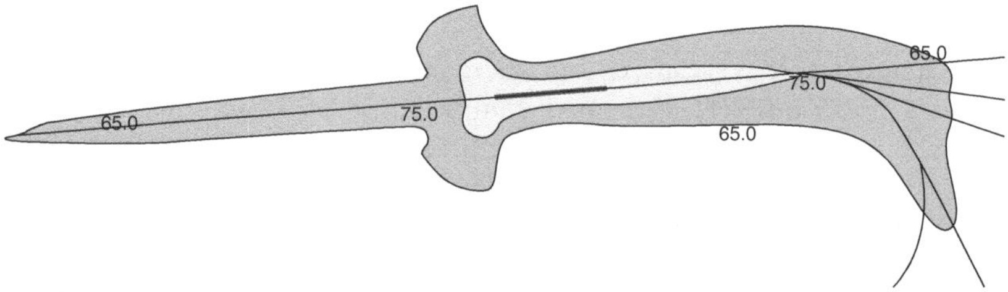


*Figure 6.31* Predicted $L_{Aeq\ day}$ contours for flights in 1991 in the 08L direction. The assumed parameter values correspond to the noise and aerodynamic performances of the Yakovlev-40 and Antonov-24 aircraft.

Predicted $L_{Aeq\ day}$ contours for flights in the 08L direction are shown in Fig. 6.31. The scale used in these and subsequent figures in this section is the same.

According to the 1991 results, the 65 dB $L_{Aeq\ day}$ contour (used as the limit for building restrictions) includes huge residential areas on both sides of the runway. For this reason, night flights (noise limit $L_{Aeq\ night} = 55$ dB) were prohibited in the middle of the 1980s – one of the first operating restrictions implemented at the airport under consideration. Of course, such a restriction constrains the operational capacity of the airport.

Currently, quieter aircraft are in use. For example, the Antonov-140 aircraft has been substituted for the Antonov-24 and the Antonov-72 aircraft has been substituted for the Yakovlev-40. Both of the substitutions represent 'Chapter 3' types of aircraft. Nevertheless, Antonov-24 and Yakovlev-40 aircraft are still flying. So the aircraft fleet is mixed, containing both 'Chapter 2' and 'Chapter 3' aircraft types. The noise contours predicted for operation at the full capacity of the airport are shown in Fig. 6.32.

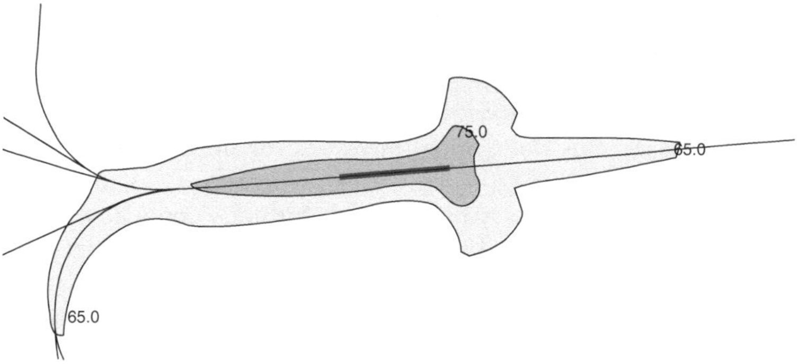


*Figure 6.32* Predicted daytime $L_{Aeq}$ contours for flights in the 26R direction from a regional airport in the Ukraine at full airport capacity. The assumed parameter values correspond to the noise and aerodynamic performances of the current mix of ICAO 'Chapter 2' and 'Chapter 3' aircraft.

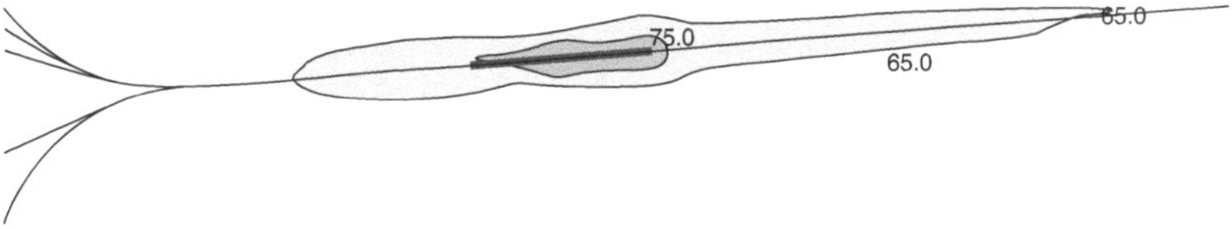


*Figure 6.33* Predicted daytime $L_{Aeq}$ contours for flights in the 26R direction from a regional airport in the Ukraine at full airport capacity. The assumed parameter values correspond to the noise and aerodynamic performances of only 'Chapter 3' aircraft.

If all of the aircraft fleet were to be changed to 'Chapter 3' aircraft, then considering the same flight intensity, the results of noise contour calculation would be as shown in Fig. 6.33. The $L_{Aeq\ day} = 65$ dB contour areas are much less for this scenario, but once again they cover the residential areas in the vicinity of the airport. So today, at some airports, implementation of 'Chapter 3' aircraft throughout the fleet is not a sufficient solution – they need even quieter aircraft types.

For the control of aircraft noise impact around an airport, it is necessary to implement all of the possible measures including low noise flight procedures, sound insulation of buildings, sound screening, etc. Low noise flight procedures include:

(1) flight route optimization in the airport vicinity;
(2) low noise takeoff and approach flight procedures;
(3) optimal distribution of the aircraft among the routes, etc.

To consider the noise impact of the intensity of movements by various types ($i = 1, \ldots, L$) of aircraft in operation (total number of $N$), it is necessary to distribute the aircraft between the routes ($j = 1, \ldots, M$) of departure–arrival according to the noise level requirements in critical zones. The vicinity of the airport is divided into several zones with various grades of noise impact. Besides the various operational measures for noise control, there are a number of other factors $Q(m = 1, \ldots, Q)$, where $m = 1$ is the initial operational procedure without any modification for noise control, $m = 2$ is the preferential runway usage and so on. So the quantity $N_{ijm}$ defines the number of aircraft of type $i$ on route $j$ subject to implementation of noise control measure of kind $m$. This quantity $N_{ijm}$ is subject to the restrictions[12]

$$\sum_{j,m} N_{ijm} = C_i, \quad \sum_{i,m} N_{ijm} = T_j, \quad \sum_j T_j = \sum_i C_i = N,$$

where $C_i$ is the number of aircraft of type $i$ in operation and $T_j$ is the number of aircraft of all types employing the $j$th route. The noise levels at the control points are calculated as equivalent continuous levels by means of the formula:

$$L_{Aeq} = 10 \lg \left[ \frac{T_0}{T} \sum \left( N_{ijm} \tau_{ijm} 10^{0.1 L_{A\max ijm}} \right) \right],$$

where $L_{Amax\ ijm}$ is the maximum noise level due to aircraft of type $i$, flying along route $j$ using flight procedure $m$; $T_0$ is a time constant; $T$ is the observation time period; and $\tau$ is the duration of the noise impact of a particular aircraft noise event. It should be noted that

$$\sum_{i,j,m} (N_{ijm} P_{ijm}) = 1,$$

$$P_{ijm} = \frac{\tau_{ijm} T_0}{T} 10^{(0.1 L_{A\max ijm} - 0.1 L_{Aeq})}.$$

The system is determined by any quantity of distributions $N_{ijm}$ and the number of possible states of the system with distribution $N_{ijm}$ equal to $W\{N_{ijm}\}$, where

$$W\{N_{ijm}\} = \frac{N!}{\prod_{i,j,m} N_{ijm}}$$

ln $[W\{N_{ijm}\}]$ is the equivalent of the entropy of the system, which may be defined as the logarithm of the probability of any distribution being fulfilled.

Hence, the entropy of the system can be introduced in the following form:[12]

$$S = \ln\left[\frac{N!}{\prod\limits_{i,j,m} N_{ijm}} \times \prod_{i,j,m}(V_{ijm}N_{ijm})\right] = \ln(N!) + \sum N_{ijm}\left(\ln\frac{V_{ij}}{N_{ijm}} + 1\right),$$

$$\sum_{i,j} V_{ij} = 1$$

where $V_{ij}$ is the normalized frequency of aircraft of type $i$ using route $j$ determined by the restrictions on maximum noise levels in critical zones.

The probable distribution of aircraft $N_{ijm}$ for any one point of noise control is defined by the relative maximum of entropy with the additional constraints:

$$N_{ijm} = V_{ij}A_iC_iB_jT_j \exp(-bP_{ijm})$$

$$A_i = \left[\sum_{j,m} V_{jm}B_jT_j \exp(-bP_{ijm})\right]^{-1}$$

$$B_j = \left[\sum_{i,m} V_{im}A_i\ C \exp(-bP_{ijm})\right]^{-1}$$

where the multiplier $b$ is a parameter defined from the initial restrictions and $A_i$, $Bj$ are 'balancing' multipliers. Multiplier $A_i$ decreases the number of flights in accordance with the attraction of route $j$.

For any number of points of noise control, the probable distribution of aircraft $N_{ijm}$ is defined by

$$N_{ijm} = V_{ij}A_iC_iB_jT_j \exp\left(-\sum_l b_lP_{ijm}\right)$$

$$A_i = \left[\sum_{j,m} V_{jm}B_jT_j \exp\left(-\sum_l b_lP_{ijm}\right)\right]^{-1}$$

$$B_j = \left[\sum_{i,m} V_{im}A_iC \exp\left(-\sum_l b_lP_{ijm}\right)\right]^{-1}$$

where the value of $l = 1 \ldots i$ identifies points of noise control. These formulae are used in an iteration procedure which solves for the value of $N_{ijm}$ at which the criterion on noise control is fulfilled.

*Table 6.6* Example distributions of three types of aircraft between two routes for optimum noise control. Each aircraft of type $i$ flies along route $j$ following procedure $m$, where $m = 1$ represents the original procedure without any noise control modification and $m = 2$ represents a procedure including engine cut-off after takeoff for noise control

| $i$ | $j$ | $m$ | $L_{Amax\,ijm}$ (dB) | $N_{ijm}$ |
|---|---|---|---|---|
| 1 | 1 | 1 | 70 | 32 |
| 1 | 2 | 1 | 80 | 49 |
| 2 | 1 | 1 | 67 | 14 |
| 2 | 2 | 1 | 79 | 21 |
| 3 | 1 | 1 | 71 | 14 |
| 3 | 2 | 1 | 75 | 21 |
| 1 | 1 | 2 | 67 | 32 |
| 1 | 2 | 2 | 79 | 49 |
| 2 | 1 | 2 | 70 | 14 |
| 2 | 2 | 2 | 80 | 21 |
| 3 | 1 | 2 | 68 | 14 |
| 3 | 2 | 2 | 72 | 21 |

A particular example is the definition of $N_{ijm}$ for limits expressed in terms of $L_{Aeq}$ (or in terms of another measure of noise). Results for $L_{Aeq} = 63$dB in the control zone are presented in Table 6.6 as the optimal distribution of aircraft of three types with constraints $C_1 = 162$, $C_2 = 70$, $C_3 = 70$. In this case, $m = 2$ defines a procedure with engine cut-off along the takeoff trajectories.

The proposed method for calculating the optimal distribution of aircraft between the routes for arrival and departure may be used to compose a flight schedule for up to a year at the airport that takes account of noise level restrictions in critical zones around the airport. The normalized frequency for usage of an aircraft fleet can be assumed to satisfy the form:

$$V_{ij} = Ca(i, k)b(k),$$

where $C$ is a normalizing constant; $a(i, k)$ is the prior assessment of frequency for exploitation of aircraft of type $i$ during the month $k(k = 1, \ldots, 12)$ and $b(k)$ is the frequency of using a particular arrival and departure route defined by analysis of the wind direction in the airport region for this period. So, the optimal distribution is defined by:

$$N_{ijm} = \frac{V_{ik} C_i \exp\left\{-\sum_{s}\left[b(s) P_{ijks}\right]\right\}}{\sum_{j,k}\left\{V_{ik} \exp\left\{-\sum_{s}\left[b(s) P_{ijks}\right]\right\}\right\}},$$

where the parameters $b(s)$ are defined for noise level restrictions at the considered control points, $s = 1, \ldots, S$ and $S$ is the number of critical zones.

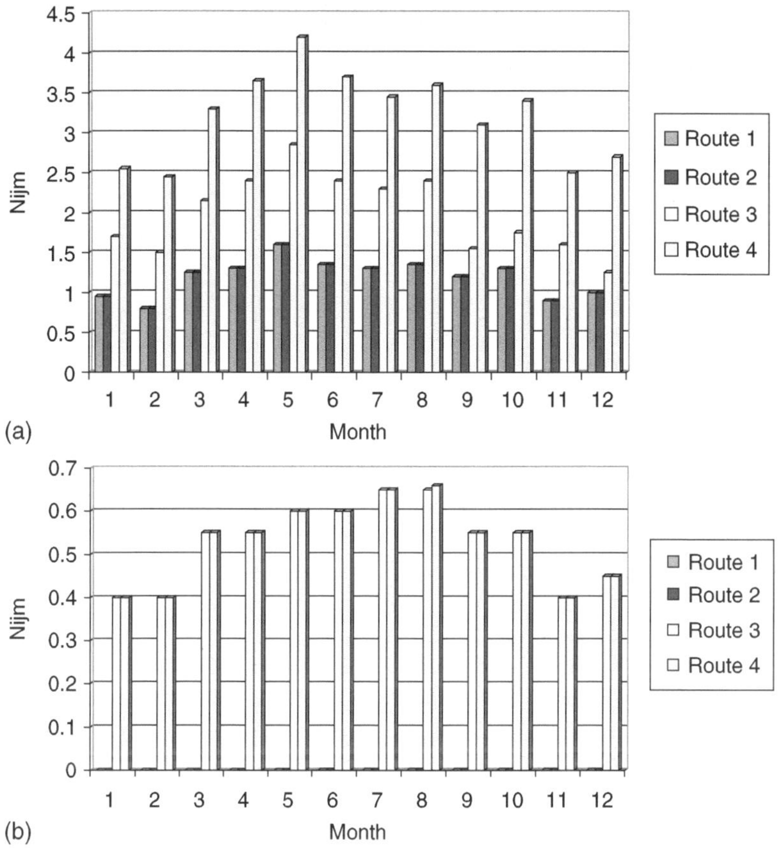


*Figure 6.34* Optimal schedules for restricting noise impact to $L_{Aeq} = 65$ dB in critical zones for two types of aircraft using four routes: (a) Antonov-24; and (b) Yakovlev-40.

Examples of the optimal non-dimensional distribution of aircraft $\overline{N}_{ijm}$ between four routes for arrival and departure for two types of aircraft to meet a noise level requirement in the critical zone of $L_{Aeq} = 65$ dB are shown in Fig. 6.34a (Antonov-24) and Fig. 6.34b (Yakovlev-40).

# 7 Monitoring of aircraft noise†

## 7.1 Reasons for noise monitoring

There are several reasons why it is important that information on aircraft noise exposure is gathered and/or generated and then released in a form that describes the noise in a way that is both 'accurate' and comprehensible to the user. 'Accurate' in this context means much more than pure technical accuracy. The information must give a realistic 'picture' of the noise exposure patterns around an airport.

Monitoring involves a system of continued observation, measurement, forecasting and evaluation for defined purposes, and is the basic tool in underpinning responsible environmental management. It is now generally accepted that the pollution levels generated by any significant activity in society should be monitored and published so that the public can be aware of the potential adverse effects which may be generated by that activity. Also, it is important that the results of environmental monitoring are archived so that long-term trends in pollution levels can be checked and published. On a more detailed level, the information can demonstrate the changing nature of aircraft noise exposure around an airport (as well as trends in air pollution, crash risk and exposure to electromagnetic fields). Specifically, it reveals the gains in terms of significantly quieter aircraft and the losses in terms of increasing numbers of aircraft movements.

Monitoring and evaluation depend on one another. Some consider the distinction between the two to be that monitoring is the collection of data, whereas evaluation contains a value judgment necessary to interpret the results of monitoring. Others appear to take the next step of analysis, which makes it parallel to evaluation. For example, the purposes of monitoring are described elsewhere as: (1) to assess the current status of the resource to be managed or to help determine the priorities for management; (2) to determine if the desired management strategies were followed and produced the desired consequences; and (3) to provide a greater understanding of the system being managed. These are also the purposes of evaluation.

† The authors wish to express their thanks to SINTEF Telecom and Informatics (Norway) for permission to include Figures and Tables from their report **STF40 A02032** *Aircraft Noise Measurements at Gardermoen Airport, 2001. Part 1: Summary of results* in Chapter 7.

So the steps for evaluation and for a monitoring program are the same: (1) problem definition; (2) inventory; (3) experimental design and indicator selection; (4) sampling; (5) validation of the model; and (6) data analysis and management adjustment.

There are two main types of monitoring systems: trend-based and predictive monitoring. Predictive monitoring is also referred to as a model based approach to monitoring. Like the model-based approach, goal oriented monitoring is also used for establishing progress towards a predetermined objective.

All monitoring systems are subdivided by their operational responsibility in following ways:[1]

- by geographical area or jurisdiction – local, city, county, country, continent, global; by medium – air, water, soil, vegetation, population, etc;
- by method of monitoring – direct instrumental sensing, remote sensing, modeling, indirect indicators, questionnaires, diaries;
- by the strength of the effect or process being monitored – impact monitoring (in cities), regional monitoring (in rural areas), baseline monitoring (in remote areas);
- by type of impact – physical, biological, chemical, human health, socioeconomic, institutional (governmental responses to environmental problems); and by objective.

Monitoring systems may be divided into two categories – instrumental and non-instrumental. For aircraft noise, instrumental systems include a set of sensing terminals, usually installed round the airport, their number being determined mainly by the tasks to be solved for routine noise management purposes. Non-instrumental systems operate from airport information data and statistics, real (historical) and forecast, which are usually reported for many different purposes. Both types of system must be operated jointly with the aim to reach the solutions appropriate for routine and strategic management of the 'airport–environment' system.[2]

Environmental noise in a wider context, caused by transport, industrial and recreational activity, is one of the main local environmental problems, because noise exposure is produced close to the source of its radiation. But regional (nationwide or even worldwide) distribution of noise exposure distribution is a subject of strategic solutions, including political and economic regulation of the problem.

In particular, the objectives for monitoring include the following: to determine present conditions; to determine trends; to understand phenomena; to validate and/or calibrate environmental models; to make short-term predictions; to make long-term assessments; to optimize the utility and/or cost-effectiveness of any of the above; and to control. If the control objective is dominant, the monitoring system becomes a management sub-system based on observation.

*Table 7.1* Strategies for the optimizing monitoring system design according to its goals

| *Goal* | *Strategy* |
|---|---|
| Environmental factor control | Place monitors near hotspots |
| Environmental factor description | Place monitors over a wide area |
| Health-related studies | Place monitors at sites representative of noise sensitive exposure |
| Trend analysis | Place monitors in places of largest trends and smallest variability |
| Environmental factor modeling | Place monitors according to model considerations |

Related to objectives is the question of tolerances or confidence limits within which the system is to operate. These considerations are crucial in the design of a useful monitoring system and they are analyzed in detail later (see Section 7.3).

The optimum design of a monitoring system depends greatly on the objectives established for the system, for example, following the guidelines shown in Table 7.1

When the geographical area is to be monitored, the preferred sampling strategy depends on:[1]

- the funds available and the costs of equipping and operating stations;
- the objectives (e.g. is the goal to estimate the area mean, the long-term trend or the highest value occurring anywhere in the area?);
- the required tolerance levels (e.g. is it necessary to estimate the area mean within ±5 per cent with 95 per cent confidence, or will some coarser value be satisfactory?); and
- the complexity of the field being monitored (gradients, variances, etc.).

There are three general approaches to the problem of optimizing network density. The statistical approach uses information on gradients, variances and space correlation fields, and the site that is expected to provide the greatest amount of new information is selected. Conversely, the station that is of least use could be determined and removed from the network.

The modeling approach uses knowledge of the environmental behavior of the element/process being studied to estimate the areas where data are most likely to be useful. For example, a climatological dispersion model could be used to estimate sites, where high values of air pollution are most likely to occur. For aircraft noise, sites must be chosen to avoid the influence of other possible environmental noise sources.

A number of systems and philosophies of monitoring are available at the moment. In this chapter, the current issues for aircraft noise monitoring are explained. Usually a monitoring system is designed to estimate change in both space and time. In most cases, however, networks of monitoring stations are established, each measuring the same elements at the same times. Frequently, an aircraft noise monitoring system is designed to meet multiple objectives.

As for global (nationwide, regional) issues, local aircraft noise assessment and control requires consideration of other dominant factors of the impact of aviation on the environment. As shown in Section 1.1, the environmental capacity of the airport is determined by several factors: aircraft noise, local air quality, third party risk, water and soil pollution, surrounding land use and habitat value, and wastes. Recently estimation of the impact on environment is distributed between local and regional levels. Currently, noise and air pollution mapping of large industrial centers and urban areas are of prime concern.

For example, Directive 2002/49/EC (Environment Noise Direction; END) deals with noise from road, rail and air traffic, and from industry. The END requires: (1) the determination of exposure to environmental noise for the values of the noise indicators $L_{den}$ and $L_{night}$, through noise strategic mapping, using noise measurements or computer-based noise modeling; (2) provision of information on environmental noise and its effects on the public; and (3) adoption of action plans, based upon noise-mapping results, which should be designed to manage noise issues and effects, including noise reduction if necessary.

For example, in England, in accordance with the Environmental Noise Regulations 2006, strategic noise maps have been developed estimating noise levels from: (1) roads with more than 6 million vehicle passages annually; (2) railways with more than 60,000 train passages annually; (3) airports with more than 50,000 aircraft movements annually (excluding training and light aircraft); and (4) urban areas with populations greater than 250,000 and a certain population density.

Capturing sufficient information at the necessary resolution through measurement alone would involve an extremely large survey that would be prohibitively resource intensive. Many measurements would have to be carried out over long periods and the subsequent analysis and collation work would be extensive and expensive. However, unattended measurements are generally indiscriminate, recording the total noise present at a location including extraneous noise, so it can be very difficult to differentiate between sources. Thus, although noise measurements are likely to have a role to play in the noise mapping process, for example, for verification purposes, it will not be possible to provide the level of detail required by the END through measurement alone.

Modeling noise requires information on the noise sources, the transmission path and the receptor as shown in Table 7.2.[3] Computer software models are used to determine the noise level at the various receiver locations from the particular noise sources (e.g. airports), taking account of the various features described in Table 7.2. Modeling tools and software, described in

*Table 7.2* Noise modeling input information for noise mapping

| *Source* | *Pathway* | *Receiver* |
|---|---|---|
| The different elements within the source<br>The noise generated by the different individual elements<br>The number of each of the individual elements that occurs<br>The location of the individual elements<br>The time of day that these sources occur | The distance between the source and receiver<br>The effect of atmospheric and climatic conditions<br>For land-based sources, the type of ground cover between the source and receiver<br>The existence or otherwise of obstacles that will affect the transmission of sound from the source to the receiver | The location of the receiver<br>The effect of localized acoustic reflections<br>The type of receiver (dwelling, school, hospital, open space, etc.) |

detail in Chapter 4, satisfy most of the END requirements and must be used for the aircraft noise-mapping purposes. The task is how to combine the measurement results, made locally and at specific times (including day and seasonal variance, if important), with global/regional assessments of the impact of noise (and other factors).

In its White Paper on the future development of the Common Transport Policy, the European Commission stressed the need to ensure that areas surrounding airports are adequately protected against an increase in noise volume due to the growth in air transport and that no new noise sensitive activities are allowed near airports. To that effect, measures were announced including *implementing noise monitoring* around airports.

Polluters have a responsibility to monitor and report on the pollution they are generating and the public has a right to know environmental pollution levels. It is self-evident that, if pollution levels are reported in a manner that is unintelligible to the non-expert, there has not been effective disclosure of what is happening.

If a meaningful picture is to be painted of aircraft noise exposure patterns around an airport, a person needs, at the least, to have access to the following information: where the flight paths are; at what times aircraft use a flight path (in particular, at sensitive times – night/early morning, evenings and weekends); how often the aircraft use the flight path; variations in activity levels from hour to hour, day to day, week to week, etc; and noise levels from individual flights.

Any aircraft noise abatement policy should include instrumental aircraft noise monitoring to provide information to the public on the actual noise around an airport and to assess complaints about aircraft noise. When combined with flight data from the airport surveillance radar, the noise monitoring system allows compliance with prescribed standard flight procedures and tracks to be checked. Such an integrated flight track and aircraft noise monitoring system makes it possible to detect immediately any violations of standard procedures and to trace offenders against established noise limits.

The noise monitoring systems, which are installed at many airports around the world, range from simple systems measuring the noise levels of individual flights to complex systems, which accumulate and analyze noise data and monitor the flight tracks of aircraft, weather information, complaints of the neighboring population and so on. It must supply factual analysis of aircraft operations and their consequences for the environment and information useful to a surrounding community of the activities taken for environmental noise control. The data and analysis available from the system may improve planning efforts for noise control, such as in aircraft operation, or the best location for residential zones around the airport.

There are many national rules and guidelines that govern the installation of monitoring systems around international airports. During the last two decades, around ten airports per year have provided new or improved installations. Initial systems were very simple in design and operation compared with more recent ones.

According to a special Work Programme of the Committee on Aviation Environmental Protection (CAEP) of the International Civil Aviation Organization (ICAO) an airport noise-monitoring effort should:

(1) compile data on methods used to describe aircraft noise exposure and applications of the data;
(2) determine the contribution (general and/or specific by type, route, arlines, etc.) of aircraft to the overall noise exposure;
(3) collect data on the characteristics of airports with noise and/or flight path monitoring systems;
(4) collect details of airport noise monitoring systems such as capabilities, data stored and technical support;
(5) compare calculated and monitored noise levels for a suitable sample of airports;
(6) compare measured noise levels with certificated noise levels for a range of aircraft types and operating conditions;
(7) examine changes in measured noise exposure over a representative time period; and
(8) update advisory documents on methodologies and applications of noise contouring and monitoring, supplemented, for environmental noise management, by the elements of expert and decision-making systems.

This collection of information:

(1) enables determination of the contribution of aircraft to overall noise exposure;

(2) enables detection of occurrences of excessive noise levels from aircraft operations;
(3) enables assessment of the effects of operational and administrative procedures for noise control and compliance with these procedures and/or assess alternative flight procedures for noise control (the tool of objective assessment of efficiency of the proposed operational and administrative procedures for noise control in the vicinity of the airport);
(4) assists in the planning of airspace usage issues;
(5) increases public confidence that airport-related noise is being monitored to protect the public interest;
(6) enables validation of noise forecasts and forecasting techniques and their methodologies over an extended period of time (collection of data for noise contouring, system noise exposure forecasting and contouring with compiled data);
(7) assists relevant authorities in land-use planning for developments and noise impact on areas in the vicinity of an airport;
(8) enables assessment of a Quota Count system (special mitigation procedure which defines an appropriate number of flights of the aircraft of specific types during a specific period of the day without violation of noise limits), among other possible noise-mitigation measures; and
(9) indicates official concern for airport noise by its jurisdiction and its governing bodies, and enables provision of reports to, and responses to questions from Government and other Members of Parliament, industry organizations, airport owners, community groups and individuals.

Also, an airport-monitoring system can assist:

(1) in answering noise complaints about aircraft operations from the general public and their enquiries;
(2) detecting unusual flight events (measurement and verification of noise levels by aircrafts, air companies);
(3) educating pilots, airlines, airport proprietors, the public (detection of operations which have not complied with flight corridor requirements;
(4) obtaining statistical data using an objective resources (aircraft types, operating times, usage of flight tracks and routes, of runways, complaints, etc.);
(5) applying research tools to assist the airport in performing certain tasks as required and mandated (planning of airspace usage around the airport, detection of operations which have not complied with flight corridor requirements, determination of the contribution of aircraft to overall noise exposure); and
(6) assessing compliance with mandatory noise levels, established by a governmental entity, etc. (measurement and verification of noise

levels and flight procedures by aircraft types, air companies, detection of occurrences of excessive noise levels from aircraft operations, etc.).

The most recent phase of aircraft noise monitoring improvements began at the end of the 1980s toward the development in a single complex system of the data from noise monitors, flight tracks, weather information and population complaints. The requirements for noise source identification, contribution assessment of the separate noise events, supervising flight control of the aircraft, and combining measurement and modeling tools have become urgent.

The benefits of operating a noise and flight path monitoring system are substantial. However, these benefits may not be fully realized and the operating agency's credibility may be reduced if insufficient resources are provided to oversee the system's operation and ensure its accuracy. Points to note include the following:

(1) The noise and flight path monitoring system generates vast quantities of data, and a methodical process of summarizing and reporting the data is vital. This may take the form of standardized report formats, produced at predetermined intervals.
(2) It is essential to check the accuracy of the data carefully before it is issued publicly. This relates particularly to the noise data, where the system may be performing extensive mathematical calculations on data which has been gathered automatically from unattended instrumentation. With the logarithmic processing, which is basic to sound level calculations, it does not require many incorrect inputs to distort a summarized average severely.
(3) It is essential to keep records of system outages, particularly in regard to flight track information, to avoid the circumstance where a complainant may be told there was no aircraft operation at the time and place corresponding to that complained about, when in fact there was an operation, but it was not recorded due to a system outage.
(4) There is a need to run a preventative maintenance and calibration program and this will be an ongoing cost. If the system is used to detect violations of noise limits and/or of flight corridor boundaries, for the purpose of prosecution of offenders, then the records of maintenance and calibration data may become evidentiary material in legal proceedings.
(5) The process of installation of a noise and flight path monitoring system may be seen as a service to the community in helping to deal with the adverse affects of aircraft operations. However, while the system will provide for a more informed discussion, it is not in itself a solution to those adverse effects.

## 7.2 Instrumentation for aircraft noise monitoring

Most of the airports in the world are planning or already have installed aircraft noise monitoring systems (ANMS). Each ANMS has special characteristics. These range from systems used occasionally by individuals to systems with 15 or more permanent staff. Associated noise control or monitoring offices range from those that work on only one or two complaints about noise per month to a complete sub-division or *unit* dedicated to noise in a large international airport dealing with thousands of complaints per week. The number of noise-measuring terminals (NMTs) may range from only two at a small airport to a network of more than 50 (as at Narita airport in Japan). The number of NMTs around an airport will depend on the runway layout at the airport, the situation of the airport in relation to the noise sensitive locations and the number of such locations to be monitored. The important objectives for the number and location of the NMTs around an airport were discussed in the previous section.

Usually a three-phase study must be undertaken to satisfy the main objective of noise monitoring. This involves: (1) developing a preliminary set of contours using the analytical model and the best available database; (2) using contours to select noise monitor locations and acquire monitor data; and (3) normalizing monitor data to typical flying conditions and updating noise contours.

An ideal system configuration includes: (1) hardware for the measurement of aircraft and background noise at selected locations around the airport; (2) the recording of the aircraft flight traffic schedule from air traffic control; (3) the correlation of measured noise events with aircraft flights; (4) the identification of the aircraft correlating with noise events; (5) the storage of data; and (6) operation and management of a database of operational and permanent information for the production of displays and reports of measured results.

Original systems often included only noise monitoring. A notable disadvantage of such an arrangement is the reduced ability to attribute noise to particular aircraft. Today, many of the systems enable noise data to be identified with the operation of individual aircraft. The identification is achieved either through the correlation of noise events with radar flight track data or through the correlation of noise events with flight schedule data. The ability to correlate the two types of data – noise and flight path – makes the system much more useful. This subject is described in Section 7.4. If flight track monitoring is required, then it is necessary to include the recording of aircraft flight tracks via an interface to the airport radar and the correlation of measured noise events with aircraft movements (position) on the flight tracks.

The system must have a connection with a source of weather information. It is necessary for both noise data interpretation and flight track analysis. The information commonly recorded includes wind speed, wind direction,

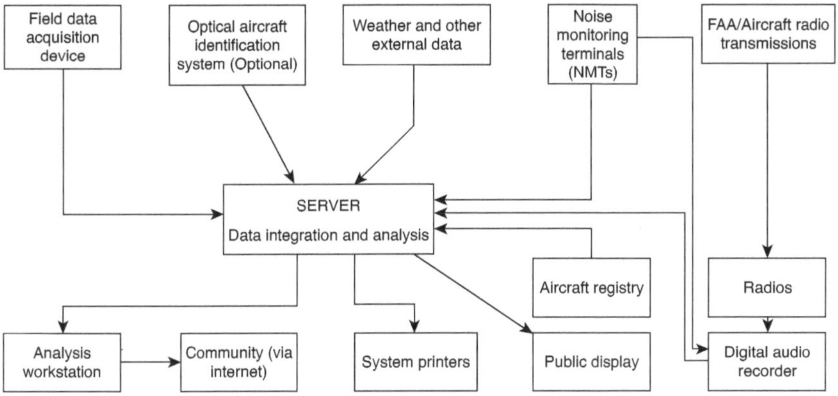


*Figure 7.1* The structure of an aircraft noise monitoring system.

temperature and relative humidity. The meteorological data are recorded at either one of the NMT monitoring sites or from the airport meteorological station.

To assess the number of people annoyed by noise it can be necessary to provide a sub-system for handling complaints and messages.

Two airport ANMS, including flight track control units, are shown in Fig. 7.1.

The basic components are the noise monitoring terminals and the central noise monitoring station (NMS) with the necessary software for storage, processing, analysis and presentation (enabling generation of reports in different forms) of information about noise impact estimation in airport region. The noise and track-keeping (NTK) systems, at each airport, match air traffic control radar data (i.e. aircraft flight paths) to related noise measurements from noise monitors at prescribed ground positions. Typical uses of the NTK system at airports include the detection of departure/arrival noise limit infringements and the monitoring of aircraft track-keeping performance. In this case, a communication network with airport radar is realized autonomously.

It is important that the noise data and the information about routes and flight tracks, provided by air traffic control system of the airport, are accurate. Accuracy can be provided by supervision of the noise-measuring process, and by comparing tracks obtained from board flight information registration equipment and from control equipment with the tracks given by the air traffic control system.

ANMS architecture: (1) should provide for reliable and fast presentation and analysis of long-term data; (2) allow for cheap and effective fit to changes in amount and type of users, number of terminals (monitors) and the number and type of peripheral equipment; and (3) be open and allow

*Table 7.3* System descriptions

| *Noise control monitoring tasks* | *Requirements for noise control terminals (portable and permanent)* |
| --- | --- |
| Central computerized noise monitoring station (NMS) | Requirements for NMS, usually specific to apparatus including computer, data storage, printers, number of users, etc. |
| Functional peculiarities of the software | Requirements for data-processing software |
| Reports | Requirements for reporting formats |
| Unessential possibilities | Supervising and storage of weather data, records of crew-dispatcher radio connections, remote displays, etc. |
| System maintenance | Details on every required maintenance, calibration, check, etc. |

for simple integration into airport infrastructure and nearby operational environment.

The system descriptions are provided by realization of specific system applications. These descriptions can be divided into logical blocks (Table 7.3).

Airport noise control systems are founded on noise control terminals (NMTs), which allow for flexible noise data agglomeration and transmission. The system usually works in real time transferring data to the central station of the system (NMS) for display of a noise event. In real time, typical data can also be recorded for special analysis and display (reconstruction). Typical reconstructions of noise data can be shown on a computer and used for detailed visual and mathematical analysis of noise events. The NMS provides for real-time transmission and storage of noise events and periodical data. NMTs usually work from an Alternating Current (AC) network. During power interruptions, a terminal makes use of a reserve battery, which can allow for several days of continuous data collection. An NMT includes: (1) a microphone (see Fig. 7.2b) fixed on a high mast (Fig. 7.2a); (2) a weather-protected box with equipment for analysis of sound signals and storage of the associated analysis results (Fig. 7.2a); (3) a modem for data transmission to a central station; and (4) a power system (net and/or batteries). Usually the modem link is realized through a dedicated data line, or through a 'dial-up' telephone mode.

Important aspects of environmental noise include: (1) data reduction methods and clarity of displays; (2) dynamic range; (3) impulse response; (4) environmental performance (temperature extremes and wind); (5) shock resistance; (6) ease of operation; and (7) likelihood of errors in operation.

It is necessary to install the microphone of a noise monitoring system as far away as practical from any sound-reflecting objects and to protect it from weather conditions and from birds (Fig. 7.2b). All reflecting surfaces other

(a) (b)

*Figure 7.2* Elements of a noise monitoring system: (a) mast including meteorological sensors; and (b) microphone mounting and protection arrangement.

than the ground should be at least 10 m away from the microphone. The entire microphone assembly, as used in normal operation (e.g. microphone, preamplifier, rain protection, windscreen, microphone device support, anti-bird devices, lightning conductor and any calibration device) must meet the directional requirement of class 1 of The International Electrotechnical Commission (IEC) 61672-1 (according to IEC651 and American National Standards Institute (ANSI) S1.4 standards) for the reference direction. These directional requirements must be met with every accessory used in service mounted in their normal place.

All of the noise-measuring equipment in NMT should meet the requirements for type 1 of the international standards IEC 651 and IEC 804; this should be independently verified by an accredited organization.

Sites for unattended measuring microphones must be chosen to minimize the effect of spurious sound from non-aircraft sound sources. To provide reliable event detection using a system based on sound level discrimination it is recommended that sites should be selected such that the A-weighted maximum sound pressure level of the least noisy aircraft to be detected is at least 15 dB above the residual long time average sound pressure level.

The standard microphone height is at least 6 m above the ground. However, to minimize the interference effects due to ground reflections, microphone heights higher than 6 m are recommended, up to a height of 10 m. For monitoring jet aircraft with high bypass-ratio engines then, if it can be shown that the elevation angle of each aircraft to be monitored (relative to the ground plane) is always greater than 30 degrees during the time that the sound pressure level is within 10 dB of the maximum, microphone heights as low as 1.2 m might be used.

The range of noise levels during environmental noise monitoring is pretty large, especially around airports. To estimate the noise climate at the control point, the equipment should be able to register the highest levels associated with separate noise events produced by aircraft and the much lower background noise. If the background levels are very low, for example, levels near to 30 dB may occur particularly at night, the airport must be assured that the apparatus is suitable to measure such low levels. Exact measurements according to international standards require that the measured levels are at least 6 dB higher than the noise floor of the instrumentation. Indeed, the noise floor should be 10 dB less than the lowest level to be measured.

Noise monitoring instrumentation ought to have a dynamic range at least 63 dB (usually more than 80 dB), to meet the demands of standards. The dynamic range is the range of noise levels that the equipment can register without overload. Although large system ranges can be handled by switching between particular sub-ranges, this can cause mistakes in measurements, particularly in conditions when the levels change quickly (e.g. during an aircraft noise event). The majority of equipment that is currently available has a dynamic range of more than 110 dB.

Usually noise measurements are frequency corrected (weighted) to conform with human reaction to noise. Measuring equipment should at least make use of the 'A-weighted' frequency correction. The complexity of the noise analysis carried out in the NMT will vary depending on the requirements of the operator, and may include dBA levels at regular intervals (say one per second), average noise levels ($L_{Aeq}$) over each hour, day and/or other period and statistical noise levels ($L_n$ values) over those same periods.

Usually, measurements required for control of external noise are broadband. For more detailed analysis, it is necessary to provide data in discrete frequency bands: usually third-octave bands. Analyzers distribute noise between many individual bands (24 bands in the frequency range from 50 to 10,000 Hz) and indicate in which spectral bands noise energy is found and at what value. Frequency analysis is relatively new in airport noise control systems, but is particularly important for countries, where legislation dictates the measuring of indices like those recommended by

ICAO standards [perceived noise level (*PNL*) and/or effective perceived noise level (*EPNL*)] and where measurements are used for validating noise models. The required accuracy of frequency analysis is specified by international standard IEC 225 and conforms with type 1.

Every NMT should provide for local storage of different reports as well as registration of noise levels every second. Storage is most important for systems of 'dial-up' type, but also such storage is necessary for systems that communicate to a central computer in real time, if the central computer is inaccessible for any reason.

In addition to permanently placed NMT, a system can include several portable NMTs located in additional residences according to the requirements of a given situation. Alternatively, for a program of short-term control (from 2 to 3 months) in particular locations, they can allow regular noise levels estimation in many other locations, which can be technically or economically more expedient, rather than using permanently placed NMTs. Portable NMTs can have the same capabilities as permanent ones and be fully compatible with stationary system terminals. Flexibility in placing portable NMTs is enhanced by the use of cellular telephone technology for communication between the portable NMTs and the central computer.

If portable terminal installations do not have automatic data collection, then they should be capable of keeping data for at least five days.

Based on worldwide experience, there are several criteria to be satisfied in the selection of an area for noise monitoring and in selecting NMT locations:

(1) the area should be predominantly residential, and preferably not more than about 10 km from the nearest end of the closest runway;
(2) the area should be near regularly used flight tracks, to minimize the chances of mis-identification of ambient noise sources as aircraft;
(3) the area must be relatively quiet so that aircraft noise is generally detectable above the background noise. This means it should not be in close proximity to noise sources such as main roads, railway lines, factories and air-conditioning systems. Depending on the local climate, the close proximity of metal roofs may also need to be avoided because of the noise generated during heavy rain;
(4) there must be a direct line-of-sight from the NMT microphone to the aircraft, and it is best to avoid proximity to large buildings or steep terrain which could cause echoes or shielding of noise;
(5) there must be access to electrical power and telephone lines;
(6) there must be ready access for maintenance;
(7) the site should be secured from malicious damage.

Also, it is necessary to run a program of preventive maintenance, including regular audits of the measuring performance of every part of the NMT. As indicated by standards, the microphone should be replaced twice a day and at fixed time intervals by a calibrated laboratory device such as a pistonphone. It is important that the calibration results are kept in the central station as a guarantee of measurement accuracy.

The heart of the central NMS is a computer with the necessary software installed. The general structure of the software depends on the objectives of the monitoring and consists of three modules:[2] inventory, functional and analytical.

The inventory module is concerned with collection of data about pollution levels and sources, including the volumes of the source activities and the various protection measures. The inventory module determines the requirements for the sensing terminals of the system (their number and their location).

The functional module is the part of the software that includes all necessary descriptions of the sources and environment. Most important are the performances of the sources and the environment which contribute directly to the levels of impact, for example, the flight route scheme, the flight traffic along a particular route and the type of the aircraft.

The software in the analytical module serves both the instrumental and non-instrumental system modules. It aids the assessment of environmental impact in terms of ecological and economical circumstances.

Software should provide for full noise data collection from terminals placed around an airport, together with the corresponding information from radar about flights and flight plans from air flight control. Additional inclusion in the software of weather data and flight information makes the system more complete for management of the acoustic environment.

The processes of noise and flight tracks control call for processing of a great amount of data, and realization of the full potential of airport noise monitoring relies on recent advances in the speed and memory of computation systems. The ability to correlate between two data types makes the modern systems far more useful than the earlier systems that were dedicated only to noise control.

A weakness with the use of aircraft noise contours is that they can give the impression that there is no noise outside the contours. Therefore, the inclusion of 'distant' flight paths is particularly important to demonstrate that aircraft noise does not stop at the outermost noise contour. For some airports, community interest in the location of flight paths can extend to distances between 30 and 40 km from the airport. In addition, 'distant' flight path information shows that the shape of noise contours does not necessarily reflect where a significant number of the aircraft fly – the shape of some noise contours is particularly influenced by the loudest aircraft types and by landing aircraft.

An NTK system provides information on: (1) which aircraft are flying; (2) where they fly to and from; (3) where they are in the air; (4) how high and how fast they are; (5) which runways and routes they are using; (6) how much noise they make on the ground; and (7) the corresponding weather conditions, and so on.

Flights plan or radar data collection can be real time or involve 'dial-up' or by parties (e.g. using aircraft onboard magnetic information carriers for transmission, which include all the necessary information for flight path reproduction). Also, flights planning information can be added separately by hand.

The radar interfaces can be different: from simple textual ASCII format, sent by asynchronous interfaces, to complexes of synchronous interfaces with strict temporal demands.

Given the magnitude of the safety and aircraft separation functions that are demanded of the airport surveillance radar (e.g. ASR-9), it is not surprising that the radar provides extremely accurate and reliable positional information. ASR-9 provides a range accuracy of ~0.05 km and the azimuth accuracy is within 0.09 degrees.

An evaluation of flight track location information was conducted relative to the location of observed aircraft over flights. It was determined that the most effective way to verify the radar flight track data was to determine the location of an over-flight in the area of concern and compare it to the available flight track location data.

Several methods of identifying observed over-flight locations were considered. The goal of the location assessment methodology was to determine a point on the ground where the aircraft was observed passing directly overhead. Triangulation of the aircraft altitude and distance from a measured point was contemplated, but no accurate measurement of distance from the ground to the passing aircraft was identified. Although this method seems simplistic, the theory was that with fewer measurements being made, there would be less total error compounded during the procedure. A member of staff visually observed the arrival flight path of aircraft over-flights and used a global positioning system (GPS) receiver to identify the location perceived to be directly underneath the aircraft as it passed overhead. When the location of the aircraft was identified, the aircraft type and time were noted to link that event to its associated radar flight track location information. The geographic location of the over-flight was also referenced and measured from street intersections. The location of the observed arrival over-flight was then compared to the flight track data to ascertain the accuracy of the flight track data.

The processes outlined above resulted in the determination of the distance between physically monitored tracks and tracks depicted in ANMS. By doing this, the member of staff was able to determine the accuracy of the flight tracks depicted in ANMS, with respect to the physically monitored flights. The results of the analysis showed the average deviation of flights depicted in ANMS from the physically monitored tracks to be approximately 8 m. Considering the multi-faceted elements of error introduced in such an aircraft positional evaluation, the above findings are extremely favourable with regard to the unprecedented aircraft positional accuracy considerations that have been part of the INM contour generation process.

The question of accuracy relative to noise and operational data collected at major airports around the world is not new. This question is analyzed in Section 7.3.

Collection of weather data is essential for exact estimation of noise impact. Weather data are provided either as reports from specialized control systems about weather through a computer or are introduced by hand.

System administration functions must include: (1) reserving (daily and monthly, automatic, regular); (2) archiving (i.e. the keeping of historical data for use, when necessary, without restarting); (3) ensuring system safety (defining and controlling access levels for individual use of the system, to prevent unrepresentative or illegal access and use of the system); and (4) reporting (i.e. the registering and reporting of relatively important cases about unloading of data, changes in important parameters and interruptions in delivery of data from external sources).

System administration functions are vital because they generate an enormous amount of data, the accuracy of which must be verified carefully before feeding reports on results of noise monitoring to the public.

To reduce the need for continuing investments, it is important that ANMS can be modified and extended without having to replace existing components. Requirements for expansions of an ANMS might arise from: (1) the need to introduce extra functional possibilities to meet the demands of new laws or new operational airport demands; (2) developments in the functional possibilities of NMT; (3) increase in the number of users; (4) increase in the area to be monitored; and (5) the development of additional airports.

In normal operation, the system will collect noise and flight path data continuously; 24 hours per day and 7 days per week. Timing is of critical importance to the system's operation, and it is essential that all parts of the system are synchronized with each other and with the external data sources to which the system is connected. Regular synchronization of the clocks in the various parts of the system is as necessary as the sensitivity checks carried out on the microphones at the NMTs.

As the NMTs measure all the noise to which they are exposed, not just the aircraft noise, there is a need to distinguish aircraft noise from the noise of other sources. This process may use the concept of the 'noise event'. A noise event occurs each time that the noise level exceeds a preset threshold for more than a preset duration. The noise level and time duration thresholds defining a noise event will be determined by experience with the conditions at a particular site. The selection of thresholds will generally be a compromise between ensuring that as many aircraft as possible are measured and identified, and avoiding overloading the system's storage capacities with large numbers of extraneous, non-aircraft, noise events.

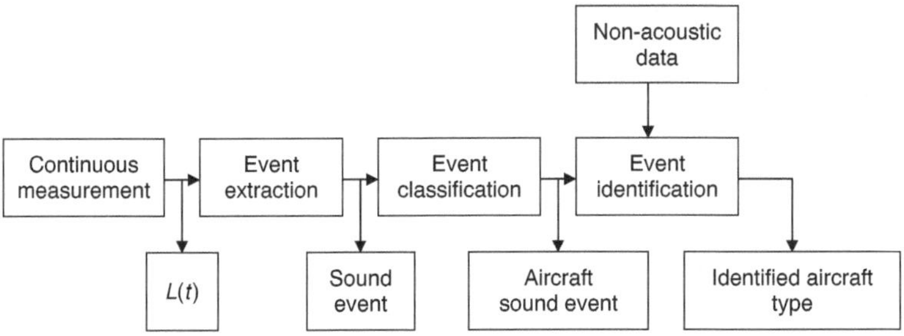


*Figure 7.3* Data processing for sound event identification.

The noise events can be extracted from the continuous measurement based on purely acoustic properties of the measured signal. Then, with the aid of additional acoustic properties that are specific to aircraft sound events, the system is able to classify the event either as an aircraft sound event or a non-aircraft sound event (Fig. 7.3). If there is also non-aircraft information available, the aircraft sound events may be further processed to link them with specific movements and aircraft types.

Automatic long-term aircraft monitoring is possible only when the aircraft are reliably and precisely detected and identified. Different identification techniques can be used to detect the aircraft sound event depending on the situation. It may be necessary to use different techniques for different periods of the day.

An aircraft technique must fulfill the following two criteria:

(1) The level difference between the measured cumulated exposure of all aircraft sound events and the true value of the exposure shall not exceed 3 dB. The true value shall be estimated by also accounting manually for sound contributions of unmeasured aircraft movements and of mistakenly included non-aircraft sound events.
(2) The count of aircraft sound events shall not be different from the true value of the total count of aircraft operations by more than ±50 per cent of the true count.

An estimation of the quality of detection should include aircraft-position data, such as air traffic radar data and will be done by the comparison of the results of the automatic identification with the acoustic identification made by the operator. The latter identification can be carried out either by *in-situ* identification of aircraft events or by listening to recorded signals. The test period will include at least 20 aircraft sound events of the same type of aircraft operation.

The noise data measured and stored for the noise event may include the maximum noise level during the event ($L_{Amax}$), the time history of the noise of the event, the sound exposure level (*SEL*), the equivalent continuous sound level ($L_{Aeq}$) and the *EPNL*. In addition, the event data will also include the time at which the event starts and finishes, and the time at which the maximum sound level during the event occurs.

The information received from the radar includes not only the aircraft positional information but also the time that each radar return was received. The computer correlates the noise event and radar track information. When the time of a noise event coincides with the time that an aircraft's track was in the general vicinity of the NMT, then the noise event is attributed to the aircraft. Naturally, since it is possible that another noise source, such as a truck, may have produced a noise coincident with the presence of an aircraft, it is not possible to be completely certain that the noise event was generated by aircraft. With frequency analysis capabilities included in the NMT, the certainty that the source of the noise was actually an aircraft can be enhanced.

Geographic information systems (GIS) is a computer system that includes tools for processing spatial data into information that can be used for making decisions about areas on the Earth. The GIS system is a powerful analytical tool that supports decisions affecting thousands of residents. GIS is continually being utilized in new and innovative ways to map and analyze aircraft noise impacts around the monitoring system of airports.

There are several types of maps which are used as informational tools for aircraft noise impact evaluation. Their strengths and weaknesses are shown in Table 7.4.

*Flight path maps* are the basic aircraft noise information tool. They provide people with an indication of where aircraft fly and effectively underpin all other aircraft noise information. They clearly show where the aircraft is going and also may show the height of the aircraft (a colour coding system is used to give an indication of the heights of aircraft) in the vicinity of the enquirer's home.

An enhanced flight path map – a track density plot for periods longer than a few days (for a busy airport) – can tend to have too much information and present a confusing picture.

*Flight path movements charts* show a picture of aircraft noise distribution around an airport and that aircraft do not all follow the same track but tend to spread to generate distinct flight path zones. A flight path movement chart therefore contains, in addition to average day information, data on the busiest and quietest days during the period covered by the chart to give an indication of how noise varied over that period.

*Single event noise data* are generally either provided by showing computer generated 'noise footprints' or by providing data on the noise levels registered by individual flights at the noise monitors around airports. The contours allow the person to readily compare the noise levels generated by different aircraft types and particularly for one type of aircraft landing versus takeoff.

*Table 7.4* Strengths and weaknesses of the particular types of noise/track maps

| *Type of mapping* | *Advantages* | *Disadvantages* |
|---|---|---|
| Flight path maps | Show the aircraft flight paths over inhabited areas | Do not normally contain information on the number of movements and the times at which they occurred |
| | Based on actual monitoring | Do not give direct information on 'noise' |
| Flight path movement charts | Monitored flight-path location and aircraft movement numbers are combined | Provide information on aircraft movements, not aircraft noise |
| Single event noise data | Good site indication of the sound pressure level from an aircraft of a specified type on a specified flight path | Contour is generated solely by computer modeling and relates to one flight path only |
| | | Can give the impression that there is no noise beyond the outer contour |
| N70 contour maps | Showing noise exposure in a meaningful way for short time periods | Can give the impression that there is no noise beyond the outer contour |
| Measured N70 charts | Information is derived from noise monitoring | Provide information for those locations at or very near to the noise monitoring terminals |
| | Rapid examination of the noise generated by particular operations or aircraft types | Not readily apparent what a sound pressure level of 70 dBA will be like in practice |
| Noise contour maps | Showing noise contours for day–night periods averaged during the year or few months | Contours are generated by computer modeling using a set of realized flight paths – difficult to relate to measured noise levels on NMTs |
| Noise zoning maps | Showing noise zones for permitted human activities support | Contours are generated by computer modeling using a set of realized flight paths – difficult to relate to measured noise levels on NMTs |
| Audio/visual tools | Clearly demonstrate the variations between different types of noise events and what the noise is like | Not able to revisit the information in order to think about it |
| Website viewers | Can receive almost real-time information | Do not explain why it is happening |

NMTs, noise monitoring terminals.

*N70 contour maps* summarize single event data for a specified time period over the area surrounding an airport and indicate the number of aircraft noise events louder than 70 dBA, which occurred on the average day during the period covered by the chart. The term 'N70' has also commonly become used as a generic expression for the family of 'Number Above' descriptors. Contours showing the number of events above 60 dBA and/or 80 dBA are also frequently generated for specific applications.

*Measured N70 charts* report actual noise measurements, in contrast to N70 contour maps, made around an airport. The measured N70 not only provides information from a source that is more trusted by some people, it also provides a good tool for checking the accuracy of N70 contours.

*Noise contour maps*, usually calculated for day–night noise indices using real flight tracks and meteorological conditions, may provide information for noise management programs.

*Noise zoning maps*, usually calculated for day–night noise indices averaged during the year, are necessary tools for the support of permitted human activities.

A number of *audio/visual products* have been developed in recent years to demonstrate aircraft noise to non-expert audiences via the medium of public 'noise simulation' presentations by generating very high-quality sound and visual images. Typically, they would be used as a part of the consultation process on a major airport project such as the construction of a new runway. The demonstrations are likely to include, for example, showing the differences in noise generated by different aircraft types, the changes in noise generated by moving a flight path, the reduction in noise achieved by insulating a house, etc. The tools can also be used to help an audience understand the technical noise descriptors used in noise assessment reports.

A number of airports provide *website flight path viewers*, which are updated as the aircraft move, usually delayed between 10 min and 1 day, from the real time of the move. An individual aircraft can be selected and tracked as it traverses through the airspace. The sites also allow the user to replay the flight tracks for any selected time period contained in the flight track database. This GIS technology allows public access to the GIS and ANMS databases through interactive web mapping applications. Actual noise contours are created on a map (an example is shown in Fig. 7.4), according to the tracks actually realized by the aircraft and corrected by noise measurements (ANMS monitors), enables definition of the real noise impact of the aircraft operation and to prove (or not) the validity of the boundaries of the noise zoning.

The connection of instrumentally measured noise data with predictions from noise modeling is one of the basic directions for improvement in aircraft noise monitoring systems. This approach enables estimation of noise levels in the whole area around the airport and not only at the location of the measuring stations. An additional advantage is that the locations of the

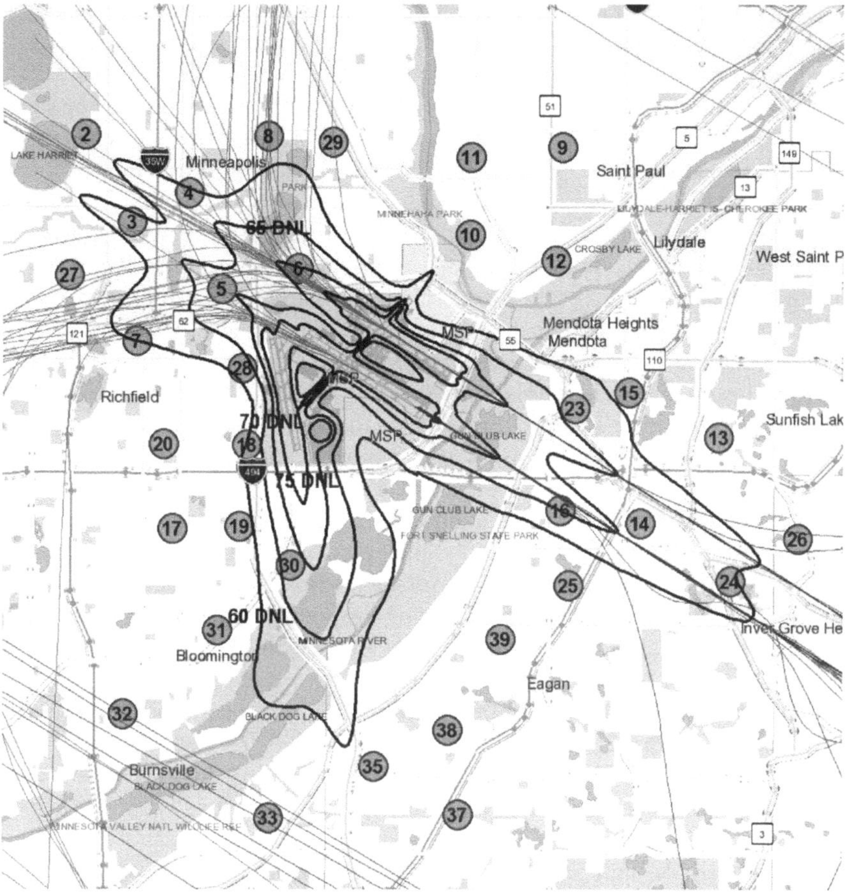


*Figure 7.4* Typical noise map created on a web viewer: 38 monitors are installed for instrumental noise measuring; departure flight paths are to the West; arrival flight paths are to the East (created by http://maps.macnoise.com/interactive/ on March 11, 2008 3:53. This information is to be used for reference purposes only).

measuring stations can be chosen more freely. Traditionally, the measuring stations must be located near residential areas, where the influence of other noise sources cannot be eliminated. Better locations can be found that yield more reliable data and therefore more reliable noise levels.

There may be large differences between noise level predictions and measurements at specific sites because most of the calculation/modeling tools are based on averaged descriptions or on summer season conditions (the most unfavourable season for aircraft noise exposure). The influence of diurnal or seasonal changes of the metheorological parameters, and their variation with height over the ground surface, and the influence of topographical characteristics (ground covering, plant belts, screens, etc.), which may have an important impact on NMT data, may not be taken into account by current calculation tools. To update the sound propagation model frequently, the measuring stations must not only measure noise levels, but also other parameters, like temperature and wind profiles. The sound emission of the aircraft, according to the type of the aircraft and the environmental factors which influence noise levels, is derived from an aircraft noise emission model. Chapter 4 describes methods of connecting measurements with calculated ones, enabling improvement in the calculation procedures for noise levels on the ground. Appropriate algorithms for assessment, identification and optimization were proposed and realized. An aircraft noise emission model uses a point source with a directivity characteristic in contrast to many other models, which use line segments and correction terms. With the powerful model of a moving point source, the level-time history of a single flight can be reconstructed. The model is validated from time-to-time for single flight events by comparing the level-time history of measurements with calculations. The power of the tool is in combining data and models in this model-based monitoring system. In addition, the use of models opens possibilities for: (1) short-term noise forecasts; (2) scenario studies with various aircraft distributions; and (3) enforcement of noise limits.

To optimize the net of NMTs (their number and locations in vicinity of the airport), they are defined by the aims and objectives of the monitoring (see Table 7.1 and Section 7.1). All the functions of the system must be provided instrumentally, particularly for: control of flight routes; control of noise levels at specific sites from specific types of the aircraft or for all the types; control of flight mode procedures by analysis of the measured noise under specific routes; and control of noise levels at population sites and their comparison with specific limits depending on the activities performed by the local population.

The most general task – defining the noise contours – allows for other approaches to designing monitoring systems, particularly where it is necessary have a wide area coverage with NMTs (for this reason, there are 56 NMTs installed around Narita airport, Japan!). Bearing in mind the costs for each NMT and for the system as a whole, the task does not have to be solved by increasing the number of NMTs. This is true not only for aircraft noise, but for other environmental factors also. Two concepts have been proposed to deal with this task.[4]

The first of these is that of the total representative assessment of the noise level (alongside other factors of environmental impact) at the site of continuous monitoring, over a period that may cover quite a few different scenarios of noise exposure (and exposure by other factors). This is realized by means of the models outlined in Chapter 4.

The representative assessment $C_i$ is calculated as weighted averaged value over all scenarios:

$$C_i = \sum_j c_{ij} p_j, \tag{7.1}$$

where $p_{ij}$ is a probability of a given scenario during the interval. An alternative is to use:

$$C_i = \sum_j q_{ij} p_j, \tag{7.2}$$

where $q_{ij} = 1$, if $c_{ij}$ is higher than normative limit for noise level (or for another environmental factor), $q_{ij} = 0$, if $c_{ij}$ is less than the limit.

The second concept is that of a representative zone (i.e. a zone around the terminal) inside the boundaries of which data on environmental factor assessment may be extrapolated with a predefined sufficient accuracy and reliability.

This concept is realized in two stages. The first stage involves the definition of representative points of the possible location of the monitors due to the representative assessments of the environmental factors $C_i$ – the points of maximum values of these assessments. In the second stage, the representative zones $A_i$ are defined for all monitors with bounds defined by predefined confidence levels of the assessments. The optimal solution consists in minimization of the total area, which is covered by a few (or all in consideration) particular representative zones.

The task of reconciling the modeling (calculation) results with those measured by monitors is important for the purposes of validating and referencing predictions with measurements.[4] The solution consists in searching for the models (aircraft noise models) for every type of the aircraft $L_P$, which give predictions closest to the measured results $L_O$. A possible criterion for the choice of the model (identification criterion) is:[4]

$$D = \left[ \sum_i (L_{Pi} - L_{Oi})^2 W_i \right]^{1/2}. \tag{7.3}$$

The solution of the task involves minimizing the criterion. The solution depends on the extent of the set of initial data – flight routes and profiles, meteorological and topographical parameters. The identification task is solved for every type of the aircraft (every event) separately.

It is relatively easy to produce information on the location and usage levels of current flight paths around an airport; it is more complex to generate such information for the future. The level of confidence in such future information will reduce as time horizons or the distances from the airport increase.

When developing long-term land-use planning contours, the time horizon can be up to 50 years and hence any flight path assumptions have to be conservative and only extend as far as is necessary to generate the contours. Irrespective of the restricted nature of the flight paths used in developing long-term planning contours, it is important, for reasons of transparency, that details of the flight paths used to generate the contours be made available to the public if requested.

## 7.3 Uncertainties in measurements and predictions

The noise abatement requirements at some airports require measurement accuracy to within 0.1 dB. However, this is more accurate than the current microphone tolerances. Since they could affect the spending of millions of US$, such requirements signal the need for a new approach. The measurement error can be considered to result from a combination of electronic, mechanical and acoustic errors. The accuracy and linearity of electronic circuits are better than those in the past and, also, there have been improvements in compensation for temperature, humidity and static pressure effects. In addition, field calibration devices and acoustic limitations must be considered. Considering all of these factors, it might be possible to reduce the system tolerances.

When providing data to the public, every effort should be made to ensure its accuracy. In practice, data are unlikely ever to be perfect and will nearly always have some form of minor deficiency or be based on certain assumptions. It is important that these 'imperfections' should not be used as an excuse for withholding information. The preferred approach is to recognize the shortcomings in the data and to indicate to the recipient the level of confidence that may be placed on the information.

There are a number of potential sources of error in the process of noise mapping, which can combine to produce large errors in predicted noise levels. These include errors in calculation methodologies, errors in computer implementations of methodologies, errors in input data, errors introduced in processing data for noise mapping and errors introduced in the software calculation of noise levels perhaps as result of efficiency techniques.

The accuracy of measurements and data for assessing aircraft noise around the airports is quite an important issue, bearing in mind that noise contours, which are the basis for noise zoning, may have an important influence on the number of contracts awarded for the sound insulation of homes inside particular zones. Over thousands of homes, this can amount to hundreds of millions of US$. So the accuracy of the estimated contours and boundaries of the noise protection zones are subjects for cost evaluation, particularly in respect of sound-proofing measures, which can result in huge additional costs for the airports and their operators.

Both for measurement and estimation, the uncertainties in single flights of specific aircraft types have to be estimated. Noise estimation is based on algorithms that are subject to experimental error. Both measurements and calculations are estimates of the unknown true sound level. There is no way to predict *a priori* if measurement or calculation is closer to the true sound level.[5,6] There are two uses that are greatly affected by uncertainty in measurement: (1) the determination of sound power (by measurement or by algorithm) to estimate sound immission based on algorithms; and (2) the measurement of noise compared with community noise regulations. Uncertainties in the first case compound the uncertainties in the second and, without attention to these, there may be unhappy local residents.

Errors in noise estimation may either result from limitations in the modeling algorithms or errors in the input data. Possible sources of errors in modeling include the following.[7,8]

- Effects of the atmosphere on sound propagation – many models assume a standard homogeneous atmosphere. Errors occur when ambient conditions depart from the standard conditions (15°C, 1013.25 hPa, 70 per cent humidity). The main reasons are changes in atmospheric absorption (temperature and air humidity), refraction effects (wind speed and temperature gradients) and atmospheric turbulence.
- Terrain effects – due to variations in local terrain and topography (complex ground reflections and scattering).
- Aircraft noise sources – engines, aerofoil (flaps) surfaces.

The total modeling uncertainty $u_m$ may be expressed as, following (4.72),

$$u_m^2 = u_{source}^2 + u_{div}^2 + u_{atm}^2 + u_{add}^2 \tag{7.4}$$

where $u_{source}$ is the uncertainty in modeling the sound source (directivity dependent emission levels at various power settings, aerodynamic configurations and flight velocities); $u_{div}$ is the uncertainty of geometrical divergence due to uncertainty $u_r$ in distance; $u_{atm}$ is the uncertainty of atmospheric attenuation due to uncertainty $u_r$ in distance and uncertainty $u_\alpha$ of air absorption coefficients; and $u_{add}$ is the uncertainty that accounts for variations in propagation (meteorological effects, ground properties, atmospheric and topographic conditions).

Considering the modeling of aircraft noise levels in third-octave spectra, the errors (inaccuracy) of the modeling can be shown by means of the vector of total spectral differences $E$ [see (4.27)] and/or the relative error index of agreement $d$ [see (4.28) The uncertainty of source modeling $u_{source}$ for a specific type of the aircraft, using the sum of contributions from the main aircraft noise sources and solution of identification tasks (Section 4.3), is quite small, less than 1 dBA for the overall *SPL*. Sound propagation (atmosphere and terrain) effects may contribute to inaccuracy much more, especially when aircraft are at greater heights (far enough from the runway). This uncertainty is important for noise contours with values less than 65 dBA ($L_{eq}$, $L_{DN}$ or $L_{DEN}$) around airports with a traffic intensity which is close to airport capacity limit.

Input data errors are caused by inaccuracy in: (1) aircraft position information – the use of theoretical flight paths or errors in the measured position from radar data; (2) flight procedures for departure and arrival; (3) standard profiles, including standard instrument departures (SID) or standard arrival routes (STAR); (4) takeoff weights and thrust cutback procedures; (5) estimation of engine thrust; (6) knowledge of the atmospheric and topographic conditions; and (7) operational data, including aircraft type and engine type.

The radar assistance, performed by the different air traffic control (ATC) in current ANMS, is aimed to verify each aircraft position in different flight states: takeoff, cruise and landing. Consequently, the radar tracks give integrative information for many parameters needed for aircraft-generated noise evaluation.

Usually radar tracks are determined by a series of spatial–temporal data $(x, y, h, t)$, correlated with more generic flight information. For the profile analysis, it is necessary to compute the distance covered from the beginning of the takeoff phase. This permits a useful comparison between the real profile and the modeled ones. In the majority of cases, the takeoff initial point coincides with the runway start point, but in some cases, there may be a displaced threshold. The weather conditions have to be acquired contextually to radar tracks to verify their influence on profiles. An important variable is the wind speed and its direction. In particular, the latter component is dominant for runway orientation. Finally, it is necessary to obtain detailed information regarding runway use as a function of wind direction.

Accurate data for the sound emission of individual aircraft is crucial. If the sound level emitted by the aircraft is estimated incorrectly, all related calculations are incorrect by the same amount. Depending on the application and the requirements, some factors are considered to be boundary conditions for a given measurement set-up and others are considered to be uncertainties. For monitoring measurements, which are representative for a certain area, local effects of reflection are considered as uncertainties.

Measurements have to be corrected by location dependent systematic deviations of the order of 0.4–0.9 dB due to the influence of the detection threshold and residual noise.[7,8] The remaining standard uncertainty of yearly averaged measurements is nearly 0.9 dB.

Two types of evaluation of standard uncertainties have been defined, as follows:[9]

*Type A:* the standard uncertainty $u_A$ can be evaluated using statistical methods and decreases with increasing numbers of measurements. Type A uncertainties include

- $u_{dist}$ – the uncertainty in level caused by variable shortest distances from the flight paths to the receiver point;
- $u_{add}$ – the uncertainty due to variations in propagation (atmospheric and topographic conditions);
- $u_{ac}$ – the uncertainty resulting from variations in sound emission of the aircraft (the engine power setting, the position of the flaps, the slats and gear, individual differences in the aircraft and engine, etc.)

*Type B:* the standard uncertainty $u_B$ depends on factors influencing the result and it does not decrease with increasing numbers of measurements. Type B uncertainties may be estimated using analytical investigations, and include

- $u_{inst}$ – the uncertainty of the measurement equipment (calibration and sound level meter, e.g. the uncertainty of a sound level meter class 1 according to IEC 61672-1);
- $u_{set}$ – the uncertainty resulting from the influence of varying parameters (e.g. threshold level) on the aircraft event detection and on calculated exposure level $L_{AE}$ of a single event;
- $u_{res}$ – the uncertainty resulting from the influence of residual (i.e. 'non-aircraft' sound levels during an aircraft event measurement);
- $u_{env}$ – the uncertainty resulting from the influences at the measurement site of reflections of the sound wave (on the ground and the walls) and of the microphone height (these uncertainties are also contributed by components of type A (e.g. the varying angles of sound incidence due to the moving source on varying flight paths within a flight corridor).

Each of these uncertainties, either of type A or B, are between 0.2 and 1 dB. The standard uncertainties of type A and type B can be combined in the combined standard uncertainty $u_c$ using:[10]

$$u_c = \sqrt{u_A^2 + u_B^2}. \tag{7.5}$$

Errors in measured position, as determined by the radar[11] are functions of distance from the radar, the radar resolution and trajectory orientation relative to the radar (radial speed). Radar systems measure position in terms of range and azimuth from the radar location. Before use, the data are projected on to a suitable coordinate reference (e.g. on to Cartesian coordinates with the airport datum as origin). All commercial aviation noise (AN) calculation models assume a flat Earth. The horizontal position error induced due to projection has been estimated to be negligible.[12] According to radar performance specification, and depending on lateral distance (from the aircraft to the radar station) and altitude of the aircraft, the standard uncertainty $u_r$ is estimated to be between 90 and 130 m.

Errors may also be incurred as a function of target elevation and sensor configuration. In detection, there are two categories of possible missing data: blanking and scan-to-scan detection losses. Blanking is used to avoid saturation of the radar system with replies from targets close to the radar and in areas where the radar data are not used. Noise monitoring systems are normally connected to the approach radar, thus blanking is mainly a problem for targets on the runway. However, if the radar is optimized for ATC use (to survey targets in the approach and departure phases of flight) targets on the runway may be blanked out. Scan-to-scan losses cause most errors during target maneuvers. Rectilinear flight segments, such as takeoff roll, present less of a problem, as the model can use linear interpolation from the start of the runway.

Current ATC secondary surveillance radar provides aircraft height information as flight level with a precision of 30 m. Aircraft on-board altitude may be derived from the static pressure. To avoid errors due to correction of local pressure it is usual to derive the height of the target above the airfield by using the on-board flight level information.

The results from comparisons[12] indicate that the accuracy of track keeping (the accuracy of radar tracking), on average, is no worse than ±7–8 m in aircraft height and no worse than 40 m in position. These are less than the estimated errors of the track keeping (i.e. ±25 m in height and ±60 m in position). From the sample of individual radar readings analyzed, the typical range of measured height errors at each airport was found to be within ±30 m, and the average '*x*' and '*y*' errors of individual data points were found to be within ±100 m.

Most current noise models use time-independent position information to produce noise contours. This makes it difficult to correlate the noise model data with conventional surveillance and on-board information, both of which rely on time-stamped information. Careful manual correlation of noise events will be required to ensure the events correlate with the position of the target. The flight data recorder and noise monitor data will be time-stamped in Universal Time Coordinated (UTC) to a precision of 1 s or better. Radar data are generally time-stamped with a precision of 0.1 s or better.

A sensitivity analysis[13] of the INM fundamental computational algorithms has checked two main variables groups and evaluated their influence on simulated acoustical levels. The variable groups include:

- local conditions: meteo-climatic variables, especially pressure that influences operation; and
- aeronautical parameters: the aircraft performance characteristics, especially the takeoff weight value and the aerodynamic coefficients related to flap configuration.

Some approximations, even if mandatory for a statistical model, lead to discrepancies with measured data, particularly considering the *SEL*, for every value set in a generic day. For every flight segment, the major sensitivity is to aircraft weight for takeoff, to database aerodynamic and thrust coefficients for initial climb and to flight speed for acceleration.

With a variation of 1 per cent in the input parameters, the corresponding variation in output parameters is determined up to a maximum of 6 per cent. More interesting is that, for the *SEL* value, the errors are less than 0.5 dB. However, noise–power–distance (NPD) interpolation errors lead to a minimum error of 0.8 dB and 'noise fraction adjustment' errors are less than 1.4 dB.

The results of preliminary analysis of aircraft takeoff weight data have been provided in a study at Heathrow airport.[14] Each flight was matched against the corresponding event in NTK. Figure 7.5 plots the highest laterally adjusted references for each flight against takeoff weights (in good correspondence with Figs 5.4 and 4.1a). A best-fit straight line has been drawn through the data, although the true relationship (especially at lighter weights) is unlikely to be linear. The line has a slope of 3.8 dB per 100,000 kg.

One basic assumption for state-of-the-art models that never has been validated is that sound radiation from the aircraft is cylindrically symmetric about the fuselage center line. However, some investigations have shown that this is not the case.[12,14] None of the current noise models take account of this result.

Both measurements and subsequent simulations show that lateral attenuation is less than predicted by Society of Automotive Engineers (SAE),

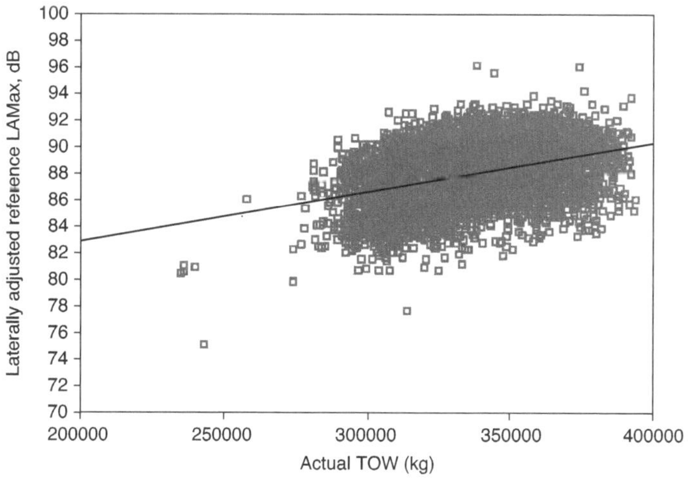


*Figure 7.5* The effect of takeoff weight (TOW) on laterally adjusted reference level.

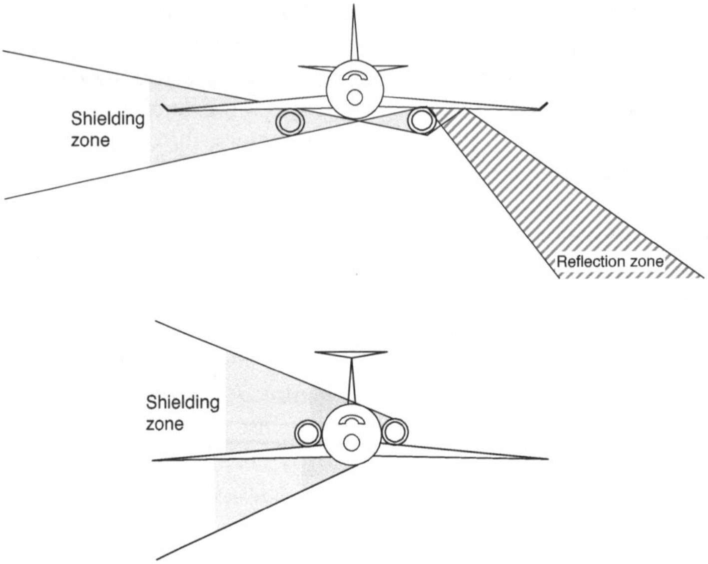


*Figure 7.6* Comparison of the database (noise–power–distance; NPD) and measurements of (a) the maximum noise level ($L_{Amax}$) and (b) the sound exposure level (*SEL*) for B737-600 aircraft.

Aerospace Information Reports (AIR) 1751.[12] This trend is true for all aircraft types at both departure and landing operations. The results also indicate that the aircraft family B737 exhibits less lateral attenuation than the aircraft family MD80. The reason for this is related to the effects of engine position. Engines that are wing mounted (Fig. 7.6a), as for the B737 aircraft family, and tail-fuselage mounted (Fig. 7.6b), as for MD80 aircraft family, produce different installation effects, which may lead to over-estimation of the total lateral attenuation and, in consequence, to underestimation of predicted noise levels of the order of between 2 and 4 dBA (depending on where the calculation point is located relative to the flight path; Fig. 7.6)[14].

The analyses show that there are statistically significant differences between measured noise and the database NPD curves. The A-weighted *SEL* and $L_{Amax}$ levels were compared to the corresponding noise curves from the NORTIM database NPD tables. The measured levels at specific thrust values during the descent of aircraft and flight along the glide slope were normalized to the curve value by linear interpolation between the database noise levels (Fig. 7.7).

The deviations, in dBA, between the NPD curves from the database and the normalized data have been analysed[14] and some statistics on the differences are given in Tables 7.5 and 7.6 for *SEL* and $L_{Amax}$ values,

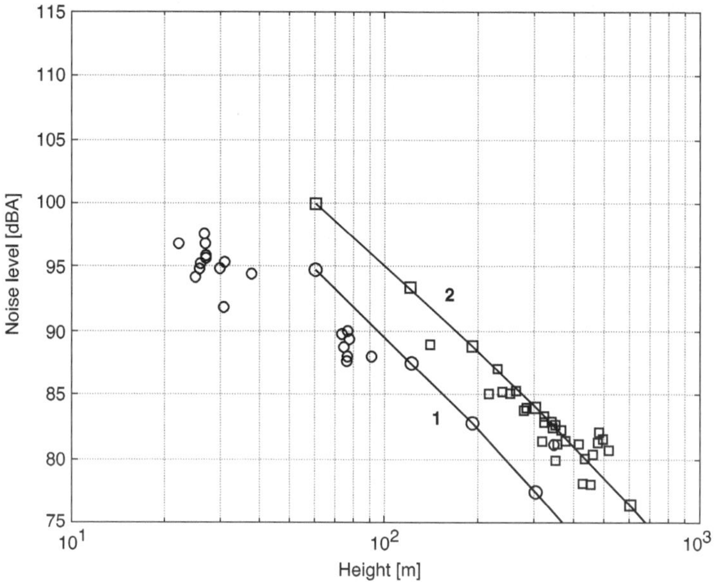


*Figure 7.7* Altitudes from the flight recorder (thick lines) and Noise and Track Monitoring System (NTMS) (thin lines) as a function of time for three randomly selected arrivals.[14]

respectively. The number of observations, mean value, standard deviation and limits for the 95 per cent confidence interval are shown for each aircraft and operation. A positive mean value suggests that the database NPD curves give too high a noise value. It is clear from these results that there are statistically significant differences between measured noise data and the database values for similar thrust settings and distance for some aircraft.

Quite meagre samples (i.e. with a low number of observations) for B738, MD90 at landing and for MD83, MD90 at takeoff indicate that the mean differences for these aircraft are not significant. Perhaps the differences are so huge for the B737 types of aircraft because their typical total lateral attenuation effects (Fig. 7.6a) are higher than those calculated by the model for this effect in NORTIM. A second possible reason is that the thrust adjustments typically used from 5 Nautical Miles (NM) to touchdown are not taken into account in the standard approach profiles. The effect on maximum levels can be up to 10 dB and this correction may be implemented on a short to intermediate term. For $L_{Amax}$ the differences are higher than those for *SEL* because *SEL* is an integrative value for the event, and usually integration over the flight event period decreases the uncertainty defined by lateral attenuation effects. So *SEL* data are generally more stable with higher confidence than $L_{Amax}$ data.

*Table 7.5* Statistical analysis of the sound exposure level (*SEL*) value differences between the noise–power–distance (NPD) curves from the database and the normalized measured data

| *Type* | *Operation* | *No. of observations* | *Mean* | *SD* | $CI_{low}$ | $CI_{high}$ |
|---|---|---|---|---|---|---|
| B736 | Landing | 20 | 4.1 | 1.6 | 3.4 | 4.9 |
| B737 | Landing | 22 | 3.8 | 1.4 | 3.1 | 4.4 |
| B738 | Landing | 3 | 1.4 | 2.1 | −3.8 | 6.6 |
| MD81 | Landing | 13 | −2.6 | 0.7 | −3.0 | −2.2 |
| MD82 | Landing | 25 | −1.6 | 1.6 | −2.3 | −1.0 |
| MD87 | Landing | 8 | −2.4 | 1.4 | −3.5 | −1.2 |
| MD90 | Landing | 2 | 8.9 | 6.5 | −49.5 | 67.2 |
| B736 | Takeoff | 31 | −0.9 | 1.2 | −1.4 | −0.5 |
| B737 | Takeoff | 36 | −0.8 | 1.8 | −1.4 | −0.2 |
| MD81 | Takeoff | 29 | −0.8 | 1.1 | −1.3 | −0.4 |
| MD82 | Takeoff | 58 | −1.4 | 1.5 | −1.8 | −1.0 |
| MD83 | Takeoff | 2 | 0.1 | 1.6 | −14.5 | 14.8 |
| MD87 | Takeoff | 11 | −1.1 | 1.2 | −1.9 | −0.3 |
| MD90 | Takeoff | 4 | −2.0 | 1.0 | −3.5 | −0.5 |

SD – standard deviation; CI – confidence interval.

*Table 7.6* Statistical analysis of the maximum noise level ($L_{Amax}$) value differences between the noise–power–distance (NPD) curves from the database and the normalized measured data

| *Type* | *Operation* | *No. of observations* | *Mean* | *SD* | $CI_{low}$ | $CI_{high}$ |
|---|---|---|---|---|---|---|
| B736 | Landing | 20 | 6.0 | 2.5 | 4.9 | 7.2 |
| B737 | Landing | 22 | 6.3 | 2.6 | 5.1 | 7.4 |
| B738 | Landing | 3 | 4.2 | 2.7 | −2.4 | 10.8 |
| MD81 | Landing | 13 | −1.8 | 1.1 | −2.4 | −1.1 |
| MD82 | Landing | 25 | −0.4 | 2.5 | −1.5 | 0.6 |
| MD87 | Landing | 8 | −1.6 | 1.8 | −3.1 | −0.1 |
| MD90 | Landing | 2 | 12.3 | 10.9 | −85.9 | 110.5 |
| B736 | Takeoff | 31 | 0.2 | 1.6 | −0.4 | 0.8 |
| B737 | Takeoff | 36 | 0.4 | 2.0 | −0.3 | 1.1 |
| MD81 | Takeoff | 29 | −0.5 | 1.9 | −1.2 | 0.2 |
| MD82 | Takeoff | 58 | −1.2 | 2.4 | −1.8 | −0.6 |
| MD83 | Takeoff | 2 | 0.7 | 2.0 | −17.1 | 18.6 |
| MD87 | Takeoff | 11 | −1.0 | 1.6 | −2.1 | 0.1 |
| MD90 | Takeoff | 4 | −3.8 | 1.0 | −5.4 | −2.2 |

SD – standard deviation; CI – confidence interval.

The noise- and track-monitoring system at airports shows results in good agreement with the aircraft recorder data. Actual flight profiles could therefore be taken directly from the Noise and Track Monitoring System (NMTS) into program GMTIM, which calculates the aircraft noise exposure (Norwegian software using the INM database with respect to

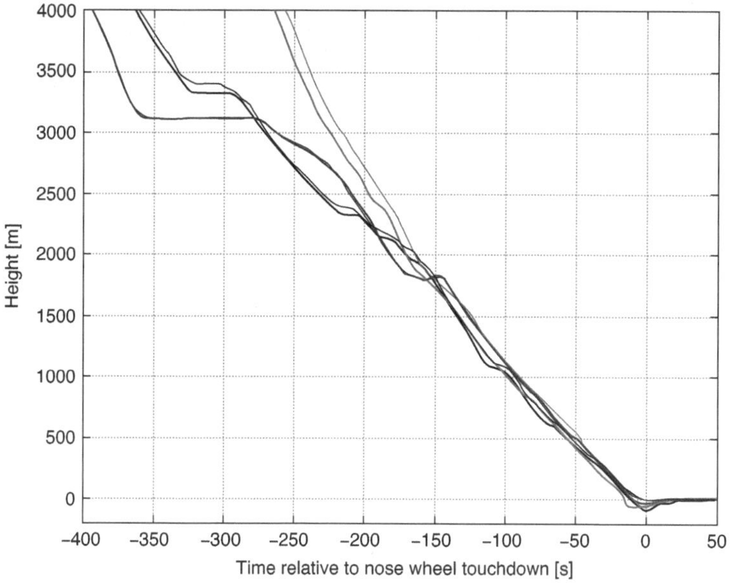


*Figure 7.8* Altitudes from flight recorders (thick lines) and NTMS (thin lines) as a function of time for three randomly selected departures.[14]

noise and flight profiles).[11] The implementation calls for the development of algorithms to estimate the applied thrust for each flight.

Of the 155 flights investigated, 70 flight records included GPS position data.[14] The pressure altitude is also reported in the flight recorder. A comparison has been made between the flight recorder data and the data reported from the radar tracking part of the noise- and flight-monitoring system. Figures 7.8 and 7.9 (three randomly selected flights) show this comparison between the altitudes given in the flight records and reported by the radar tracking system for arrivals and departures, respectively. The altitudes of both sources have been adjusted by setting the altitude at the runway to zero.

The average deviation may be due to a time delay in the altitude reported by the radar tracking system. A time adjustment of between 1.5 and 3 s applied to the radar data results in better agreement. However, the deviation is much smaller than sensitivity analysis[13] of the INM fundamental computational algorithms defined for input parameter uncertainties, which may lead to corresponding discrepancies between the calculated results and the measured data.

A flight profile is defined as altitude, velocity and engine thrust settings as a function of distance from a reference point along the flight track. For a takeoff profile, the reference point is the brake release point. For approach

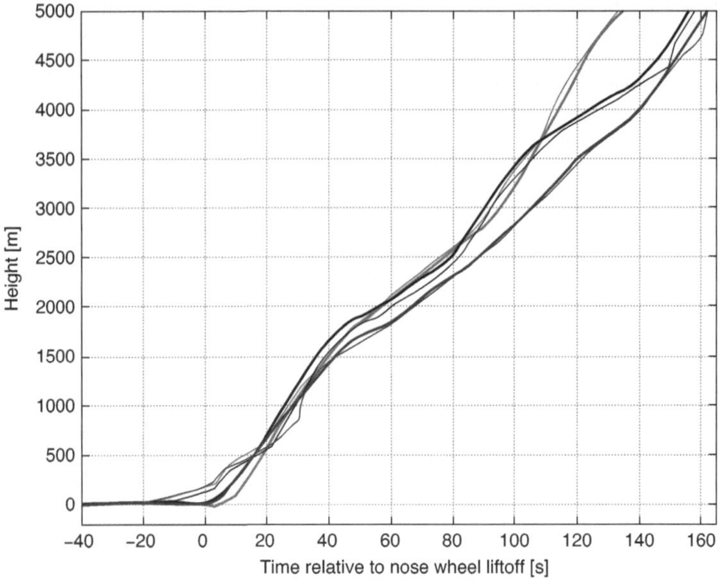


*Figure 7.9* Comparison of MD82 database approach profiles and 15 measured profiles for: (a) height; (b) velocity; and (c) thrust.[14]

profiles, the reference point is the runway threshold at the touchdown end of the runway.

Flight recorder data allows construction of actual flight profiles. All 155 profiles have been extracted for valid flights. Figures 7.10 and 7.11 show height, speed and net thrust as a function of distance from the reference point for the arrivals and departures, respectively.

For approach and landing there is little deviation from the 3-degree glide slope (which begins at an altitude of 600 m) with a few exceptions, where the aircraft descent is steeper (e.g. approaches performed in the mode of continuous descent approach). Differences in speed and net thrust are evident (Fig. 7.10). Outside 5000 m, the measured speed is almost higher than the database profile. Error increases with distance along the flight path. From sensitivity analysis,[13] a maximum error in speed of the order of 1.5 knots (i.e. about 1 per cent of the input) relates to an error in speed adjustment of about 0.4 dB. This means that, for approach and landing (Fig. 7.10b), the discrepancies in noise levels between calculation results and measured data are up to ±3–4 dBA. An error of 1 per cent in the power level (around 50–100 kg) relates to a maximum error in interpolated sound exposure level of the order of 0.2 dB, and corresponding discrepancies in noise levels (Fig. 7.10c) of up to ±2–3 dBA.

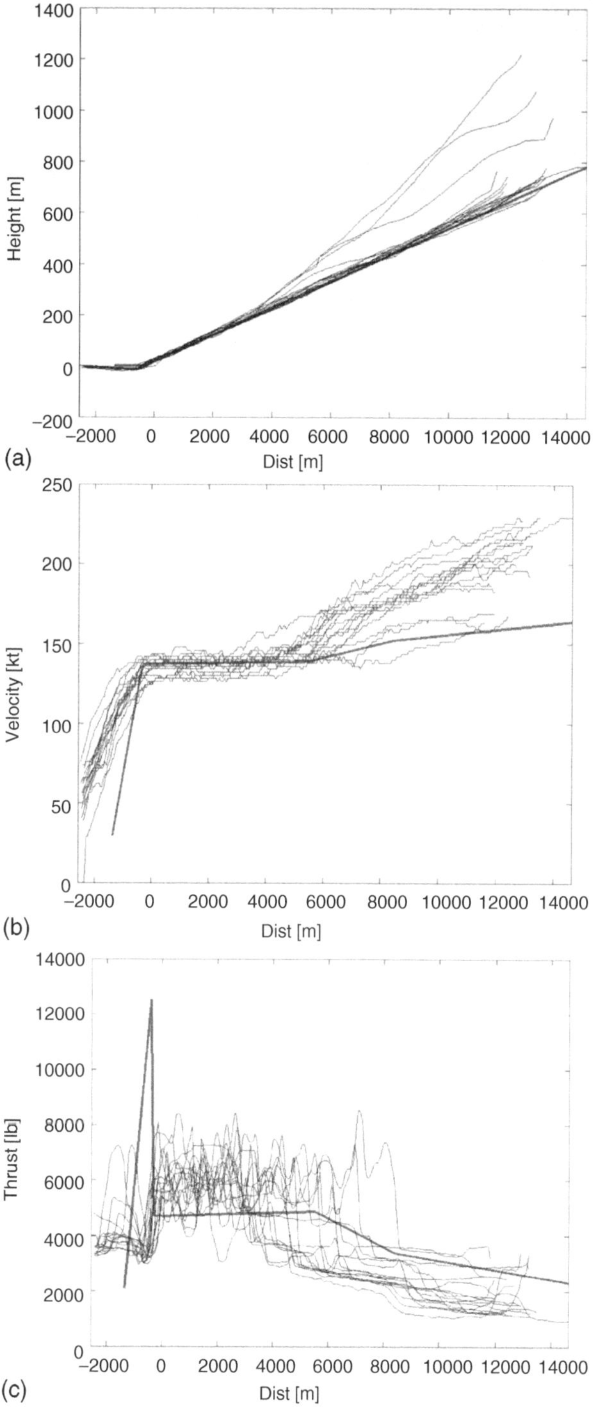


*Figure 7.10* Comparison of MD82 database takeoff profiles and 29 measured profiles.[14]

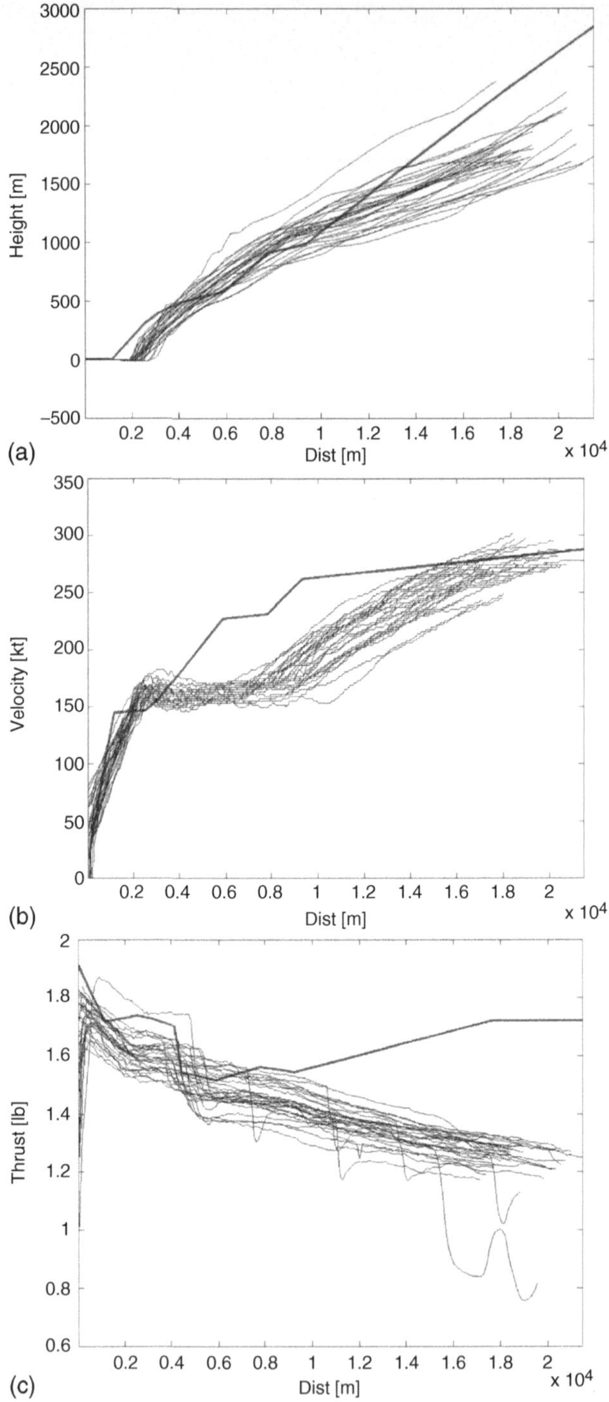


*Figure 7.11* Comparison of position data from NTMS (dots) and flight recorder GPS-based flight tracks (lines).[14]

Outside 9000 m the thrust is also generally lower, which results in at least 2–3 dBA less noise produced than the database profile would suggest. Between 9000 and 5000 m from threshold there is an altitude shift in thrust enforcement, so a noise level should be between 2 and 3 dBA higher than the database input. The noise level will fluctuate and the prediction of maximum levels will be uncertain. For the shown variations along the glide slope flight between net thrust 1000 and 2000 kg, the database NPD gives a difference in maximum noise level of up to 10 dBA for this aircraft type.

The measured takeoff profiles (Fig. 7.11) show that the INM database profile is based on different climb-out procedures. Lower thrust is applied both during the takeoff run (thus the distance of running is longer than from database input) and throughout the climb (thus the altitude at distances over 10 km is smaller than that from database input). Power cutback can only be seen in around half of the events, most likely due to the frequent use of derated thrust procedures. The effects on noise level are mixed and will vary both along the flight path and to the sides. Lower altitudes (Fig. 7.11a) and velocities (Fig. 7.11b) result in greater noise levels. Smaller thrusts contribute to lower noise levels – for distances over 10 km (Fig. 7.11c) of up to 0.5–1.0 dBA less than database inputs.

Figure 7.12 compares the position data given in the flight recorder and reported by the radar tracking system.[14] The continuous lines represent the GPS position (smoothed). The dots represent the positions given by the radar. The crosses mark the positions of the five measurement sites. The coordinates are given relative to measurement position 1. Figure 7.12 shows that near the ground (i.e. close to the runway) the radar data are subject to significant random errors. At greater altitudes, there are no large random errors, although the deviation is on the order of ±50–100 m in the north–south direction, and ±20–60 m in the east–west direction.

New technologies and analysis methodologies for developing flight-track operation statistics, which is a main subject of the track-keeping sub-system of the ANMS, are used in the development of the actual noise contours. In most cases of noise zoning around the airports, the noise contours are calculated, both for current and forecasted scenarios of flight operation. Using the track keeping data (from the radar included in ANMS), noise contours may be assessed more accurately. For example, in Fig. 7.13, actual noise contours are found to be smaller than the forecast mitigated contours (developed under the requirements of the Consent Decree.4): by a 9.0 per cent reduction in the contour day–night sound level ($DNL$) = 60 dBA and by a 11.6 per cent reduction in the contour $DNL$ = 65 dBA.

These differences are caused by differences between the forecast or assumed and actual: (1) number of flights during the day (being less than forecast) and night (being more than forecast); (2) fleet mixture (e.g. hushkitted Stage 3 aircraft average daily operations in actual statistics were down 42.1 per cent from the forecast mitigated number); (3) runway use percentages; and (4) flight procedures.

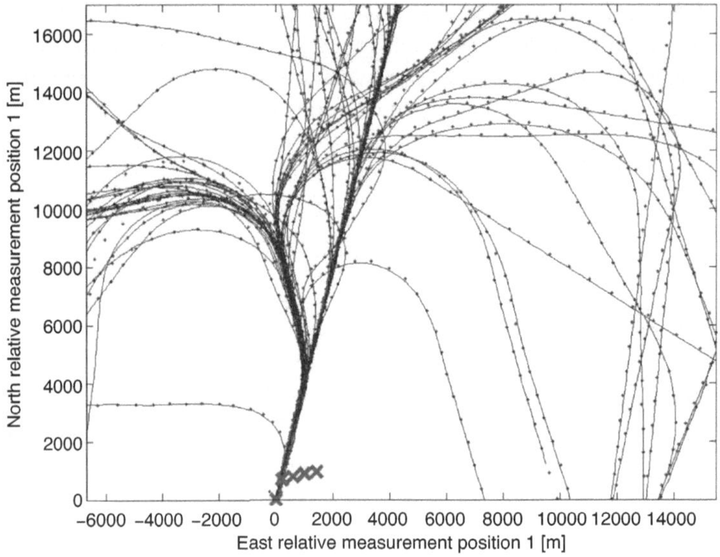


*Figure 7.12* Comparison of position data from NTMS (dots) and flight recorder GPS-based flight tracks (lines) near the runway. The crosses mark the position of the five measurement sites. The western runway is shown as a thick line.[14]

## 7.4 Identifying sources of noise events

The pattern recognition built into each terminal of the noise monitoring system enables identification of noise events due to aircraft flying over in a routine mode at the point of noise measurement and is based on a set of algorithms developed for use in noise monitoring systems. Correct classification performance and misclassification errors vary as a result of the different decision rules applied to the outputs of the algorithms.

Noise source identification (NSI) is a vital step prior to a successful noise control program.[15] Among the NSI methods, sound field visualization techniques are particularly useful in estimating the source position and the source strength. In addition to NSI, sound field visualization techniques also find application in non-destructive evaluation,[16,17] underwater imaging[18,19] and machine diagnosis.[20,21]

Generally, noise assessment is about evaluating the impact of one specific noise source, for example, the noise from a specific type of the aircraft. This is not always an easy task. In practically every environment a large number of different sources contribute to the ambient noise at a particular location.

The terminology, commonly used and derived from ISO 1996, defines *ambient noise*, which is the noise from all sources combined, and *specific*

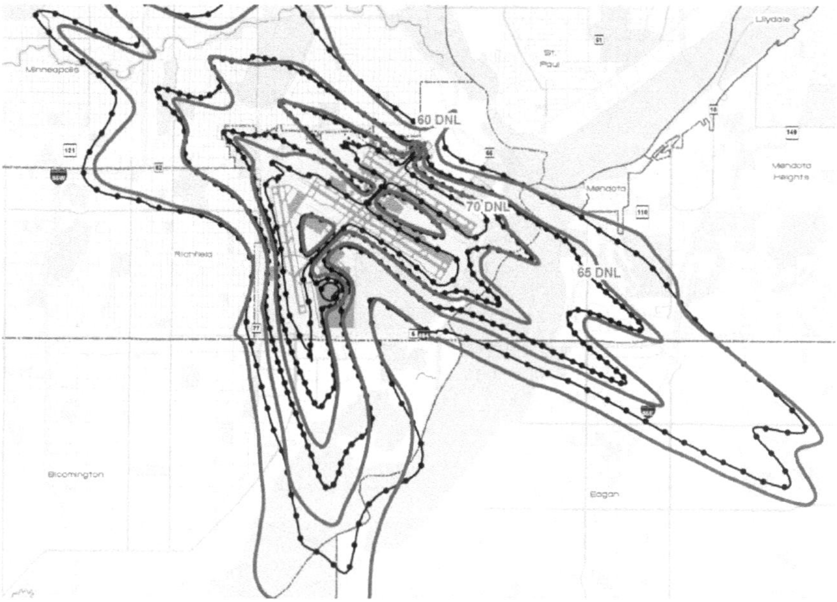


*Figure 7.13* A typical comparison of actual (joined dots) and forecast (continuous) noise contours.

*noise*, which is the noise from the source under investigation. The specific noise is the component of the ambient noise that can be identified and associated with the specific source (i.e. an aircraft of a specific type).

The *residual* noise is the noise remaining at a point, under certain conditions, when the noise from the specific source is suppressed. The term *background noise* is not used in ISO 1996 but is also common, and should not be confused with residual noise. It is sometimes used to mean the level measured when the specific source is not audible and sometimes it is the value of a noise index, such as the $L_{A90}$ (the level exceeded for 90 per cent of the measurement time).

In the context of building planning, the term *initial noise* is used to denote the noise at a certain point before changes, for example, the extension of a production facility or the building of barriers, are implemented.

Noise generated by an aircraft and identified as such is called a *noise event*. The variation of noise level with time is characteristic for each type of noise and allows clear identification of the noise events linked to air traffic, as a result of their highly specific signature. An example is shown in Fig. 7.14a. Generally, a distinction can be made between the temporal profiles for the three different types of transportation noise: aircraft noise, road traffic noise, and train noise (see Fig. 7.14b).

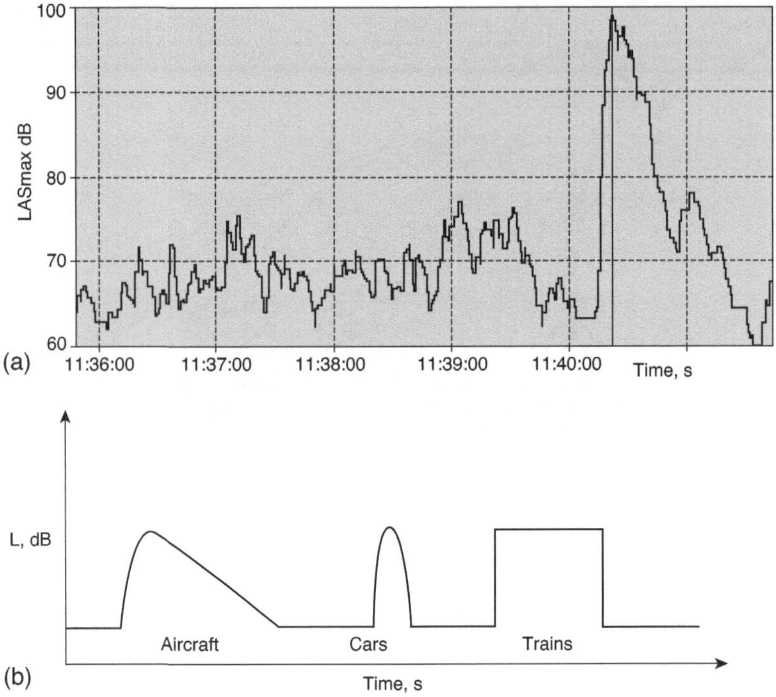


*Figure 7.14* Aircraft noise events: (a) an actual variation of noise level with time measured during an aircraft at takeoff, (the maximum appears at 11:40:22); (b) a comparison of the idealized temporal noise profiles for aircraft, road traffic and trains.

For example, an airplane over-flight can be differentiated from a car pass by simply by comparing the rise and decay times from and to the background level. While both events are characterized by a more or less Gaussian rise to a maximum and fall to a residual level, typically an aircraft over-flight can last up to the order of a minute, while noise from a car passing by at 35 mph may take only 15 s to rise from and fall to the background level. In addition, an aircraft is heard from a much longer distance than a car, so its C- to A-weighted noise level ratio changes as it approaches as a consequence of the fact that atmospheric absorption increases with frequency. As a result, the C to A ratio of an airplane over-flight declines as an aircraft approaches and increases as it moves away. Indeed, during an aircraft over-flight, there is a large difference between C and A levels that widens as the plane gets further away. In analyzing the events, it is important to subtract out the background C- and A-weighted levels from the event. In this way, it is much easier to see the temporal changes towards the beginning and end of the event that help to characterize a source.

An important step in object identification is to obtain information suitable for modeling the object within the automated recognition system. Objects can often be identified with samples as short as 20 ms, which for convenience is called a 'frame' of the signal. It is desirable to reduce these data for object recognition. This process of reducing the amount of data while retaining the ability to recognize the object is called feature extraction. The features will be represented as vectors and, if each object is to be identified, it is necessary to have distinguishable feature vectors.

There are several factors that complicate the object recognition process. One problem is that the communication channel may vary even if the object is the same. That is, sensors placed in different environments will behave somewhat differently and will produce different signals for the same aircraft. Another complication is that the objects in the same class (e.g. aircraft of the same type) will produce somewhat different signals. Signal classification is the process of identifying the object associated with a given input signal. Once the features have been extracted, an $n$-dimensional feature vector will be input to a classifier.

Finally, noise and multiple signals from different sources will distort the signals. A goal is to develop models of the audio signal in a feature space that can be used for object identification. The features will be in the form of an n-dimensional vector, and object identification is accomplished by analyzing the feature vector. There are a number of useful methods for extracting time domain and spectral domain features.

As shown in Fig. 7.15, the shape of the time-domain signal is characteristic of the different types of aircraft events.[15] This is true irrespective of what kind of descriptor, including any of the usual sound level indices ($L_{eq}, L_{max}, SEL$, etc.) is used. So the temporal sound level signal can be processed to reduce the number of measurements and extract features useful for classification. One measurement of relevance is the maximum value. Some sound events such as those due to jet aircraft are louder than others. Contrastingly, single-engine propeller aircraft landings are very quiet. Other identifying features are related to the shape of the curves. A fast aircraft such as a jet will have a sound level versus time curve that is steeper as the plane approaches than that due to a propeller aircraft.

The basis for a generalized analysis of the temporal signal during an aircraft over-flight is shown in Fig. 7.16.[15]

Besides the maximum (peak) value, three measures are useful for describing the signal. The measure related to the portion of the signal including the rise time is called alpha ($\alpha$). If $a^2$ is the coefficient of the second-order term in a polynomial fit to a temporal signal from the beginning of the sound event to $t_{max}$, which is the time of the maximum sound level, then $\alpha = 1/a^2$. This measurement should reflect the way in which the sound builds up at the microphone. The two other measurements, $b_1$ and $c_1$, are the slopes of the curve from the beginning of the event to $t_{max}$ and from $t_{max}$ to the end of the event, respectively. The first of these measurements again reflects the rate at

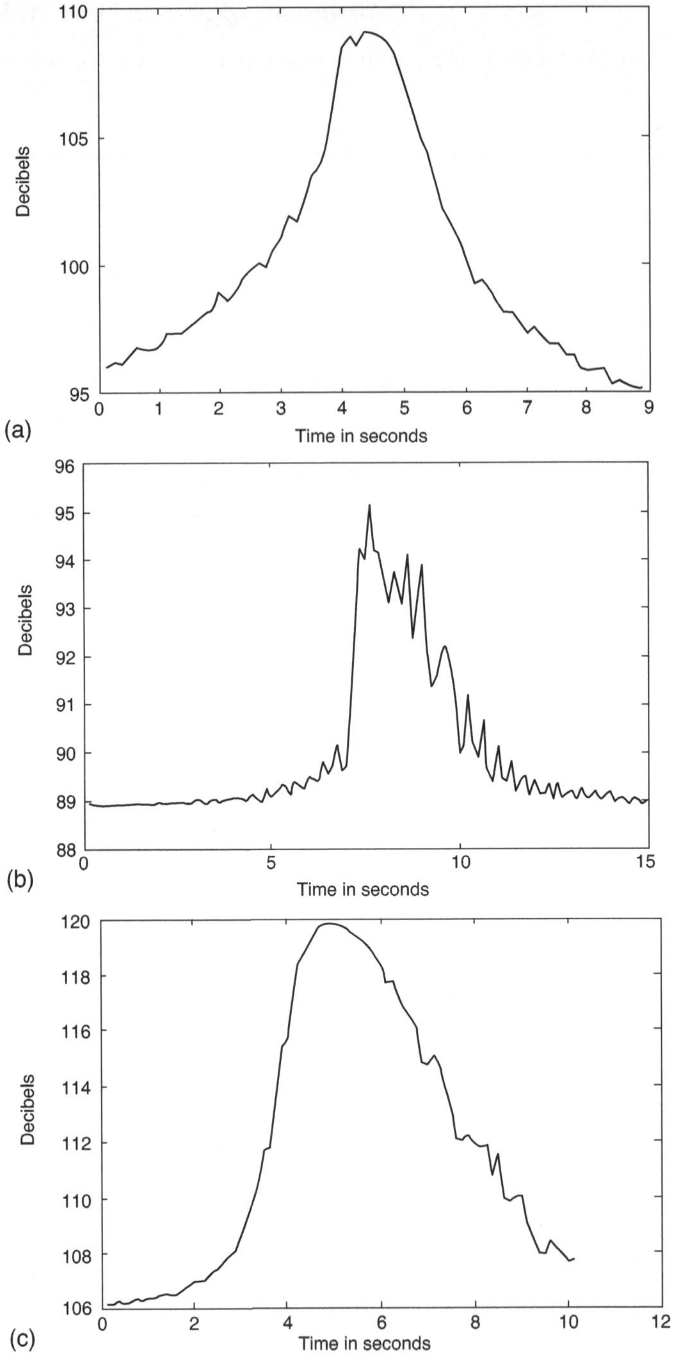


*Figure 7.15* A sound pressure level (*SPL*) signal under the flight path from: (a) a single engine aircraft on takeoff; (b) a multi-engine aircraft on takeoff; and (c) a jet aircraft on takeoff.

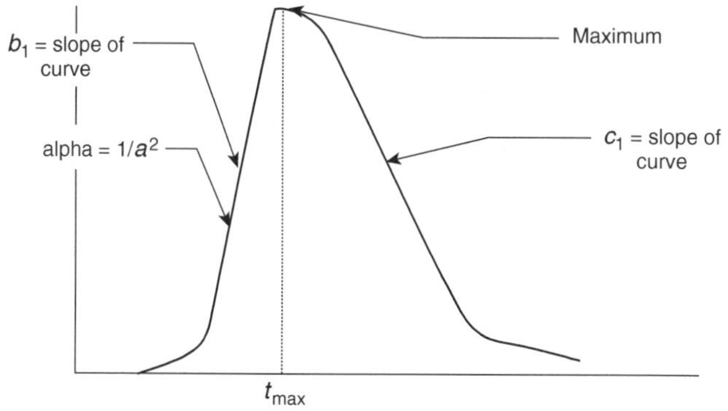


*Figure 7.16* A generalized analysis of the noise–time signature during an aircraft over-flight showing the definitions of $\alpha$, $b_1$ and $c_1$.

which sound builds up at the microphone. Another measurement extracted is a sound event skewness $S_{SE}$, which is the obliqueness of the curve about $t_{max}$, defined by

$$S_{SE} = \mu^3/\sigma^3$$

where $\mu^3$ is the third central moment. This term reflects the difference in the fitting function to the left and right of $t_{max}$. Another measure involves the symmetry of the function about $t_{max}$. The measure is a ratio of the area to the right of $t_{max}$ to the area to the left of $t_{max}$.

Measures of frequency spectra reflect differences in the aircraft. For instance, analysis of the helicopter time signatures reveals that the frequency content is concentrated in frequency bands (e.g. third-octave bands) numbered from 3 to 10. The single-engine propeller aircraft frequency content is concentrated in frequency bands from 6 to 13, whereas the jet aircraft frequency content is concentrated in frequency bands from 7 to 18.

A standard supervised pattern recognition paradigm is described in Fig. 7.17. A pre-processor uses signal processing techniques to generate a set of features characteristic of the signal to be classified, for example, a sequence of short-time spectra or the average spectrum, in noise recognition. These features form a pattern (or feature vector).

Subsequently, the classifier utilizes decision logic and a binary tree classification system (Fig. 7.18) to assign the pattern to a particular class. A useful concept is that of fuzzy modeling,[22–24] which is based on the collections of IF-THEN rules with both fuzzy antecedent and consequent predicates. The advantage of such a model is that the rule base is generally provided by an expert and the main paradigm of a fuzzy model is that the fuzzy algorithm is a knowledge-based algorithm, the essential concepts of which are derived from fuzzy logic. The fuzzy system is an expert

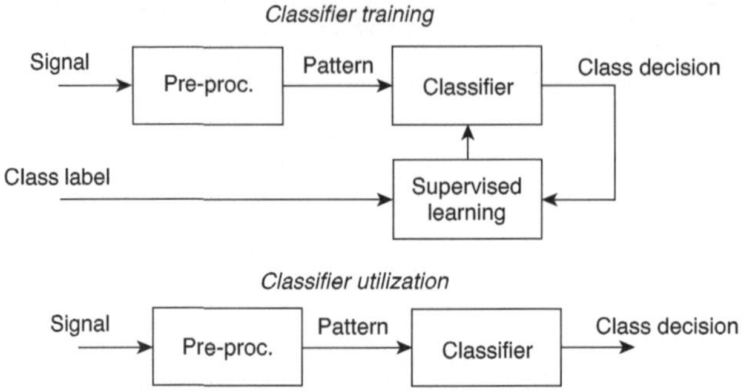


*Figure 7.17* Standard supervised pattern recognition paradigm.

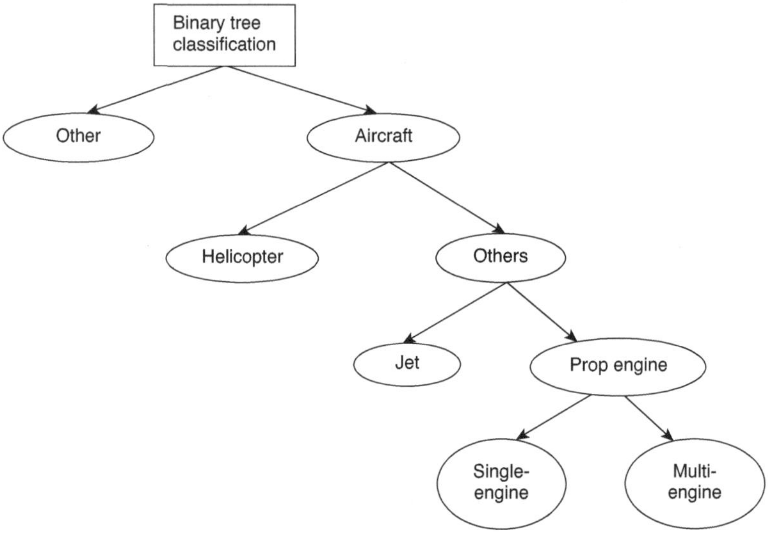


*Figure 7.18* A binary tree classifier for aircraft noise.

knowledge-based system that contains the fuzzy algorithm in a simple rule-base.

During the supervised training or learning phase, class labels identifying the training patterns or training samples are provided to the system so that it can adjust the parameters of the classifier to obtain optimum performance according to some criterion, usually the minimization of the error rate. Once the system has been trained for a particular pattern-recognition application, no more modifications are performed, and the classifier is put into service.

It is theoretically possible to train a classifier for a large number of noise sources and observation conditions, but such training is not practical.

It requires an enormous amount of training data if the training data are to be representative of the variability of the patterns. Insensitivity to the variations in observation conditions will often be obtained at the cost of a loss in classification power because features that are less sensitive to variations are often less discriminating too. Possible solutions are either *adaptable classifiers*, which are trained in a particular situation but can be adapted to a different one by tuning some parameters, or *adaptive classifiers*, which can perform the adaptation automatically. In some noise monitoring applications, it may be more important to detect one particular type of event (e.g. aircraft) precisely against all the other types of events than to provide a classification for all types of noise sources.

The desirable properties of a pattern-recognition system for noise-monitoring application include: (1) adaptability to different situations; (2) flexibility of the performance criterion; and (3) integration of acoustical domain expert knowledge.

In the present state of pattern-recognition technology, it is clear that a monolithic 'black box' approach to the design of noise recognition system will not meet all these requirements. To attain this objective, a better approach is the constitution of a noise classification 'toolbox' with a library of classifier elements that can be easily selected and tuned by a noise control expert to provide an *ad hoc* system for noise control.

The statistical paradigm for pattern recognition provides a framework for the realization of the noise classification 'toolbox' and for the implementation of the desirable adaptation mechanisms. The statistical approach is quite powerful and rigorous yet flexible. Let us assume that the only feature available for the classification of a noise event is its *SEL*.[25,26]

A statistical model can be assumed for the data, for example, a Gaussian model. Let $\omega_1, \omega_2, \omega_3$ denote the set of three possible noise source classes, let $P(\omega_i), i = 1, 2, 3$, denote their *a priori* probabilities and let $p(x|\omega_i), i = 1, 2, 3$ denote the class conditional probability distribution functions (pdfs), where $x$ stands for the *SEL*. Under the Gaussian hypothesis, we have:

$$p(x|\omega_i) = \frac{1}{\sqrt{2\pi}\sigma_i} \exp\left[-\frac{1}{2}\left(\frac{x-\mu_i}{\sigma_i}\right)^2\right].$$

The parameters of the distributions (means and variances) can be estimated from the training samples by the usual methods. Once the pdfs are available, classifiers can be easily constructed. For example, it can be shown that the classifier minimizing the error rate (the Bayes classifier) is obtained by assigning to a new pattern $y$ the class maximizing the *a posteriori* probability:

$$P(\omega|y) = p(y|\omega_i)\, P(\omega_i) \Big/ \sum_{i=1}^{3} p(y|\omega_i)\, P(\omega_i).$$

It can be shown that the optimal detector for 'aircraft' events against 'car' or 'truck' events is given by the comparison of the likelihood ratio $p(y|\omega_3)|p(y|\omega_1 V \omega_2)$ to a threshold $T$ selected to obtain adequate 'miss' and 'false alarm' probabilities according to the Neyman–Pearson criterion. Note that both decision tests can be obtained straightforwardly from the set of pdfs.[26]

The change in distance can be taken care of by modifying the parameters of the pdfs accordingly. The effect of the distance variation on the *SEL* can be easily modeled and the means and variances of the pdfs can be correspondingly adapted. Furthermore, on-site fine-tuning of the parameters of the pdfs can be obtained without the intervention of an external supervisor by mixture density estimation methods such as the Expectation Maximization (EM) algorithm.[26]

When submitting aircraft noise complaints to the airport or low-flying aircraft complaints to the Civil Aviation Authority, it is important to identify the offending aircraft as accurately as possible. The observer should try to identify several items: aircraft type (with jet or propeller engines); number of engines (single engine or multiple engines); engine locations [on/under the wings or body of the plane (on fuselage), at the tail or at the front]; type of wing (straight wing or swept back); wing mounting [high wing (on top of the fuselage) or low wing (on the bottom of the fuselage)]; landing gear [retractable gear (typically only visible during takeoff or landing) or fixed gear (visible at all times)]; and the registration number which can sometimes be visible on the tail.

The data captured by the NMT are not sufficient to perform the correlation of the noise event with an aircraft according to the previous list of identification items. The interpretation of the recordings is left to a human expert acting 'offline.' The goal is to develop the tools necessary to build an NMS able to identify automatically the nature of the sources of the noise events in addition to recording their acoustic characteristics. For a further correlation procedure besides noise measurements (provided by NMT), the system needs information about the radar system, the flight plan processing system and the clock system.

The NMT continuously analyzes the incoming noise signal to identify the source of noise. Usually the process of detecting a noise event is based on threshold and time change criteria. The standard noise event detection works on a template defined to pick out the required type of noisiness values perceived in the noise envelope according to the ICAO Annex 16 standards.

A number of measurements of the environmental noise to be analyzed are extracted by a peak-detection process. The peaks cannot be detected unless the trigger level $L_{trig}$ or threshold level $T$ is properly set in advance. The appropriate $L_{trig}$ value also varies according to the kind of target noise, the distances between the target and the receiver and the atmospheric conditions. It must therefore be determined by means of a preliminary measurement. It is easy to determine the value of $L_{trig}$, when the distance between the target

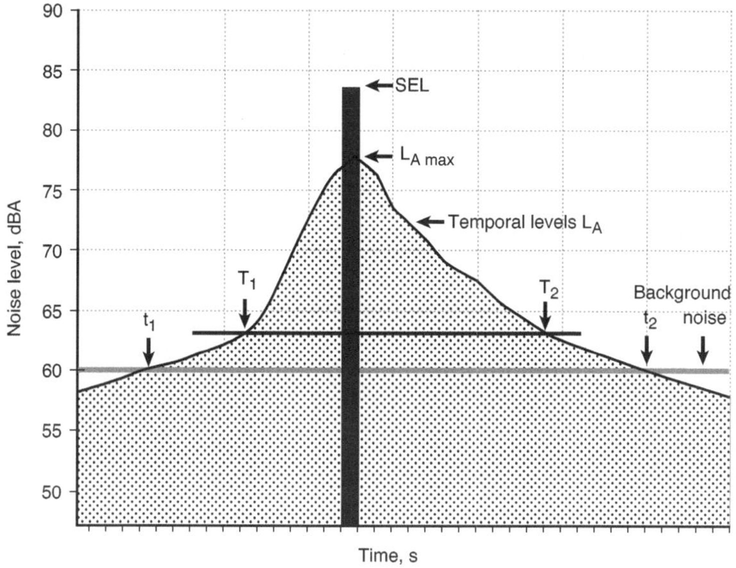


*Figure 7.19* Background noise level threshold criterion for a possible aircraft noise event.

and the receiver is short, and there is no interfering noise source near the receiver.

1. *Threshold criterion:* all information below a primary threshold is ignored. By exceeding the first threshold $T_1$ the subsequent noise data are validated as a possible noise event (Fig. 7.19). The event continues while the noise level passes $L_{max}$ and remains above the secondary threshold $T_2$ (often equal to $T_1$) for a given duration $D$. When the noise signal moves below the threshold $T_2$, it must remain below until a given termination time at which the noise event is closed.

The background noise is the ever-present noise and is recorded throughout the day and night. The threshold values, examples of which are listed in the Table 7.7, are input individually for each site to avoid the requirement for an NMT to check whether every noise is a possible noise event or not. In fact, an NMT can only handle one event at a time.

2. *Duration criterion:* for the final validation of a sound event, the event duration $D$, between the thresholds $T_1$ and $T_2$, must equal or exceed the minimum duration time for the event (data length for a single session) otherwise the event is discarded.

After an event has been identified, the NMT sends the data via a modem to the central station. Identification of a noise event does not necessarily mean that it is due to an aircraft. Thus, it is necessary to refer to the data from peripheral systems (Fig. 7.20). The incoming data: noise events, aircraft

*Table 7.7* Example threshold and background noise values for a noise-measuring terminal (NMT)

| *NMT unit* | *Threshold 1 = threshold 2 [dBA]* | *Background noise [dBA]* | *Minimum duration (s)* |
|---|---|---|---|
| 1 | 65 | 40 | 5 |
| 2 | 65 | 48 | 5 |
| 3 | 75 | 49 | 5 |
| 4 | 66 | 58 | 5 |
| 5 | 64 | 62 | 5 |

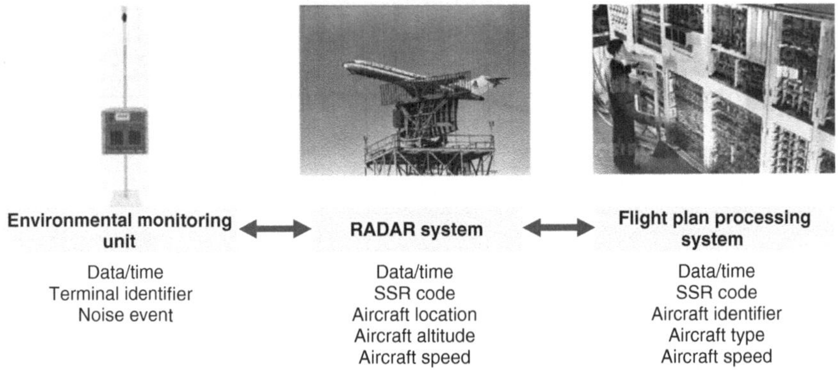


*Figure 7.20* Correlation of data from peripheral systems.

*Table 7.8* Three different basic databases for correlation of a noise event with a specific aircraft flight.

| *Noise monitoring terminal* | *Radar* | *Aircraft flight plans* |
|---|---|---|
| Data/time | Data/time | Data/time |
| Terminal identifier | SSR (secondary surveillance radar) code | SSR code |
| Noise event | Aircraft location | Aircraft identifier |
| | Aircraft altitude | Aircraft type |
| | Aircraft speed | Aircraft speed |

flight plans, radar information, are stored in three different database tables (Table 7.8).

A comprehensive flight data processing system does a lot of work. Aircraft movements have to be monitored continually, resources have to be evaluated and assigned, various types of information from diverse sources have to be interpreted and forwarded to other systems and everything has to be done at the same time. An extensive data collection and processing system is indispensable for modern airport operation.

The correlation algorithm permanently scans these tables for uncorrelated data and allows the following correlation modes:

1. *Track to flight plan correlation:* this associates a radar flight track to a specific flight plan by mean SSR (secondary surveillance radar) code.
2. *Track to noise correlation or distance-based correlation:* the distance based correlation is performed in real time. Each sound event is correlated to the nearest flight track at the time of the $L_{max}$ of the noise event.
3. *Time-based correlation:* for flights without flight tracks, noise events may be correlated to the flight plan based on travel time to the NMTs.
4. *Manual correlation:* the noise system operator is able to correlate manually a noise event after in-depth data analysis.
5. *Measurement availability:* the system is based entirely on automatic event correlation. If one of the components fails – the noise monitoring station, the radar system (no SSR code attribution possible), or the flight plan processing system (no call sign attribution possible) – no automatic correlation is possible.

During the last few years, several acoustic monitoring systems for long term surveys of environmental noise have been developed, including the recording of weather conditions and communication functionalities to allow distance control and diagnostics. Modern monitoring systems are based on dual-channel (Symphonie) or four-channel (Harmonie) PC Card acquisition boards, allowing simultaneous measurements at multiple points for complete and pertinent site characterization.[27]

A classifier for different kinds of noise sources (Fig. 7.21) uses MADRAS (Methods for Automatic Detection and Recognition of Acoustic Sources) methodology,[27] including signal processing (time-frequency and time-scale analysis), mathematical morphology, factorial data analysis and neural networks. The aim of the MADRAS is to develop new noise monitoring instruments with the ability to identity and quantify automatically, in real time, the various acoustic sources which make up a given acoustic environment. The MADRAS database includes high-quality recordings of various types of common environmental noise sources, such as trains, cars, trucks, delivery vans, motorcycles, mopeds, aircraft, chain saws, lawnmowers, industrial plants, etc.

After realizing detection and segmentation in the time domain, a neural network, trained offline to recognize different shapes based on $L_{Aeq}$ evolution, decides the identification between impulse, stationary, pass by, heavy carrier and energy blasts. In each sub-family, by means of appropriate methodology (e.g. wavelet transform for impulsive signals or third-octave statistical spectrum for pass by,) a specialist will be able to associate a shape

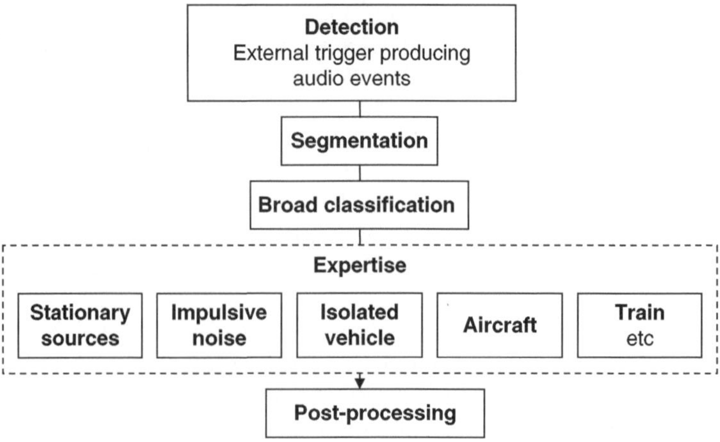


*Figure 7.21* The MADRAS (Methods for Automatic Detection and Recognition of Acoustic Sources) architecture.

to a source. By comparing known shapes from a dictionary with current events, MADRAS will be able to encode in real time the acoustic source, allowing the final measurement file to contain only codes and symbolic information. The encoded file will allow regulation application and source statistics.

This system reaches a success ratio of around 80 per cent in realistic but not excessively perturbed situations (not too many sources merging). For more noisy situations (urban areas), some important adaptations have to be made.[27]

The detection structure adopted in MADRAS methodology is sufficiently robust for monitoring in urban environments. So, rather than increasing the quantity of detected events, the real gain is obtained through precise temporal localization of events and pertinent audio recordings, including complete source signatures. The post treatment of recorded signals by specialists will be more precise, increasing the success ratio in the recognition process. A recursive implementation of this detector is currently under study, which will allow review of the management of different noise states.

A natural idea for determining a detector structure consists in searching for a contrast function in the spectral domain. The method chosen is based on a comparison between two estimators operating on two time sliding windows (Fig. 7.22). The largest will construct an adaptive pattern for background noise in computing a linear average third-octave spectrum on a length depending of the noise fluctuation (typically few seconds in urban context). The pattern consists in calculating average and standard deviation in each spectral channel to realize the classical detector structure.

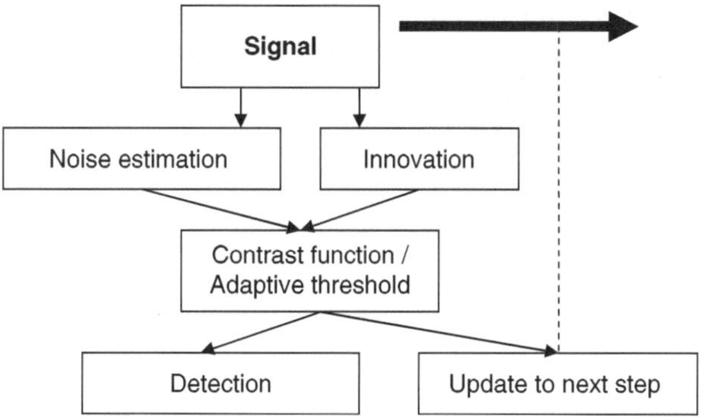


*Figure 7.22* An adaptive detection structure.

Nevertheless, despite new exciting possibilities, two drawbacks have appeared which penalize all monitoring systems. First, in many situations, the acoustical characteristics of all of the noise sources present are not known sufficiently to define release (decision-making) mechanisms. The consequences are non-adapted thresholds for automatic encoding and audio recordings, and low efficiency in using these tools. Secondly, the current situation in long-term measurement is that the work involved in post-processing is more important than the measurement time because of the amount of the data.

So the global challenge for the future is to propose intelligent monitoring tools that are (a) able to extract the necessary information in real time to enable application of regulations (b) thereby identify the acoustic sources violating these regulations and hence (c) fill a dynamic database for final user consultation and referencing.

## 7.5 Interdependencies and tradeoffs between noise and other environmental factors associated with civil aviation

The three basic environmental priorities at an airport are to limit or reduce noise exposure, to reduce local air-quality emissions and to reduce greenhouse gas emissions. There are complex interdependencies between aircraft noise emissions and other types of emissions. So to achieve effective mitigation of aviation impact on the environment, these interdependencies must be taken into account. To solve such a complex task, it is important to have analytical tools and supporting databases that can account for the interdependencies among these factors and that can potentially optimize the environmental benefits of these mitigation measures.

At particular airports worldwide, the priority of local environmental factors may change to take account of the possible dominance of concern over water and soil pollution, land use, wastes or third party risk. Thus, decisions taken about noise control and air pollution, described elsewhere.[28,29] must be analyzed to ascertain whether they assist or hinder the control of other important factors for a specific airport.

The complexity of decisions has increased over time, from a primary concentration on standard setting applied to aircraft, to providing policy advice on operational aspects and consideration of potential market-based options to reduce the impact of aviation on the environment. In seeking to meet the internationally or nationally adopted goals to limit or reduce aviation environmental impacts, an appropriate approach must consider strategic elements, including more stringent noise standards; more stringent landing/takeoff (LTO) operations and new Nitrogen Oxides ($NO_x$) cruise standards; new particulate matter standards; new operational procedures in combination with technological advancements in Communications, Navigation and Surveillance Systems for Air Traffic Management (CNS/ATM); and usage of market-based options and land-use measures to complement more stringent environmental standards.

The main tasks of the combined assessment of the airport impact on environment are: (1) reliable monitoring of the considered environmental factors; and (2) elaboration of adequate models for computation of these factors depending on operating parameters. Here the experience of the Ukraine is considered as an example of complex assessment of influence of the airports infrastructure on the environment.

Analytical tools for existing aircraft noise, third party risk and aviation emissions (local and regional) may be enclosed effectively in a single framework to assess the interdependencies of these factors and analyze the cost/benefit of the proposed actions.

This framework of tools will allow aviation stakeholders such as government agencies, industry and the public to understand: (1) how proposed regulatory actions and policy decisions impact environmental factors on local, national, regional and global levels; (2) how operational decisions affect them and their potential impact on aviation projects; and (3) the cumulative effects of regulatory and non-regulatory actions that affect environmental factors separately and overall.

Anticipated benefits of this framework include the ability to: (1) optimize environmental benefits of proposed actions and investments; and (2) quantify uncertainty associated with complex policy decisions. Other benefits include: (1) improved data and analysis on airport/airspace capacity projects; (2) an increased capability to address noise and emissions interdependencies in the resolution of community concerns; (3) more effective portfolio management; and (4) the ability to analyze and accommodate environmental constraints to capacity growth.

The tools will interact within the strategic policy decision-making environment shown in Fig. 7.23.

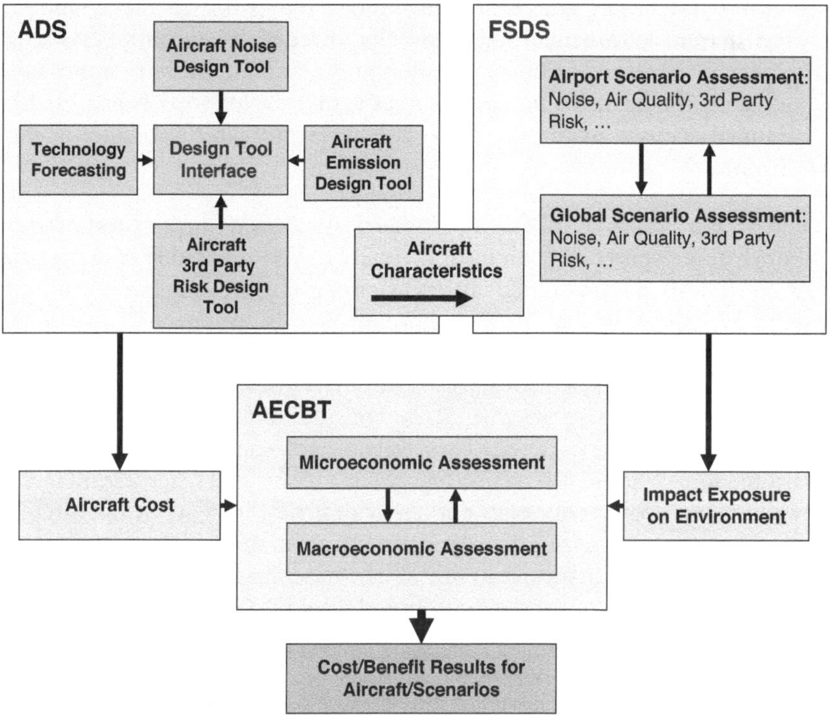


*Figure 7.23* High-level schematic of the components of the new aviation environmental impact tool suite.

The elements of this framework include the following:

- *Aircraft Design Space* (ADS), which will provide integrated analysis of the environmental factors at the aircraft design, maintenance and operation level. Tools like *NoBel* and *NoiTra*,[30] described in Chapter 4, are basic for ADS in respect of noise assessment. In addition, the emission-defined modules of the *PolEmiCa* tool,[31] which were designed for local air pollution tasks in the airports, are the basis for aircraft emission assessment during the LTO cycle of the flight.
- *Flight Scenario Design Space* (FSDS), which comprises ADS and the integration of analytical modules and databases, including airport descriptions, flight routes, fleet structure and operational circumstances, to provide an integrated capability of assessing interdependences and tradeoffs between the environmental factors. Tools, like *INM* and *IsoBell'a* for noise assessment, *PolEmiCa* for airport air pollution, *3Prisk* for third party risk in the vicinity of an airport due to aircraft crash probabilities,[32] *EMISource* for electromagnetic fields assessment from navigation facilities in airports,[33] are the basic elements for FSDS.

- *Aviation Environmental Cost–Benefit Tool* (AECBT), which interacts with FSDS, ADS and economic modules to provide the common, transparent cost/benefit methodology, needed to optimize aviation policy in harmony with environmental policy. The primary approach, which is the basis for AECBT, is a monetary evaluation of the possible damages caused by airport environmental factors in the vicinity of this airport.

The primary focus of ADS is a parametric analysis of aircraft technology and operation aspects on ecological factors. So ADS is capable of analyzing existing aircraft designs under different circumstances and future aircraft designs by capturing high-level technology trends. Additional functional requirements beyond ADS are (1) to provide quantitative estimates of the environmental factors, performance and cost of modifications to existing aircraft as well as future aircraft; (2) to consider different assumptions for technological capabilities, design choices, market scenarios and environmental protection policies; (3) to provide the tradeoffs and interdependencies between technology, economics and environmental impacts at the aircraft level; (4) to explore potential variations within an aircraft class; and (5) to quantify uncertainty associated with all the modeling factors and with the inputs required. The appropriate inputs and the type of information typically required to design an aircraft and its engine as well as to determine its technological impacts are shown in Table 7.9.

*Table 7.9* Inputs and type of information typically required to design an aircraft and its engine as well as to determine technological impacts

| *Inputs* | *Type of information* |
|---|---|
| Vehicle specifications | The parameters to be considered under this category are those typically used to size an aircraft for a particular mission, including class definition, mission definition, material structural selections, aerodynamic inputs and constraints, such as maximum range and maximum approach speed |
| Engine cycle variables | Engines are a sub-system from the vehicle perspective and an environmental impact assessment is not truly possible without a detailed definition of the engine used |
| Economic influences | The economic parameters generally center on a market scenario including things such as production schedule, labor rates and fuel costs |
| Technology impacts | These may be generic in nature, such as factors used to improve aerodynamic efficiency, or they may be introduced to model specific technologies, for example, new materials or cooling techniques that allow for a higher turbine inlet temperature |

*Table 7.10* Influence of engine bypass ratios on their thrust/dimensions and environment factor parameters (ratio of mass of engine $NO_x$ emission during Landing and Take-Off cycle; to take-off thrust of the engine ($F_{oo}$))

| *Bypass ratio* | *Thrust/ dimensions* | *Fan pressure ratio* | *Combustion outlet temperature (K)* | *Certification noise level cumulative (noise margin from Chapter 2, value)* | $NO_x$/ $F_{oo}$ *(g/kN)* |
|---|---|---|---|---|---|
| 5 | 100/100 | 1.8 | 1585 | 20 | 33 |
| 10 | 123/147 | 1.3 | 1450 | 28 | 26 |
| 15 | 146/190 | 1.23 | 1370 | 32 | 22 |

To meet the information requirements, ADS must include at least five seamlessly integrated modules:

(1) the *propulsion system simulator* – calculates the engine thermodynamic cycle analysis;
(2) *the engine weight (dimensions) component estimator* – based on the thermodynamic cycle parameters of the turbine engines;
(3) the *aircraft flight performances optimizer* – calculates aircraft weights and performances based on the mechanical model from the engine estimator and propulsion system simulator;
(4) the *environmental factor predictors (NoiTra* or ANOPP for noise); and
(5) the *economic calculator*.

An analysis within the framework of ADS has been carried out for aircraft noise and its results for aircraft with turbine engines and varied bypass ratio were shown in Chapter 2. Table 7.10 shows the influence of engine bypass ratios on their thrust/dimensions and environment factor parameters.

As the bypass ratio increases, the engine maximum diameter (i.e. the fan outer diameter) increases substantially. The engine diameter is the greatest (i.e. the engine is larger) when the design priority is the cruise thrust and the same moderate thermodynamic heat-cycle loading as in the basic engine is assumed. The impact of these design variants on other engine parameters and the different thermodynamic loading can be characterized by the combustor outlet temperature and the fan pressure ratio. The consequences for a simulated LTO cycle engine operation in terms of thrust-related $NO_x$ emission deposits ($D_p/F_{oo}$) show a decreasing emission characteristic. For this particular technological interdependence, win–win (benefits for both aircraft noise and engine emissions) results are observed. An operational example of the win–win results is the continuous descent approach (CDA) described in Chapter 6.

FSDS requires data about the aircraft source in order to calculate the noise and emissions generated by the aircraft operation at a particular airport. The initial version of FSDS draws upon the existing aircraft and engine databases used by tools such as *IsoBell'a* and *PolEmiCa*.

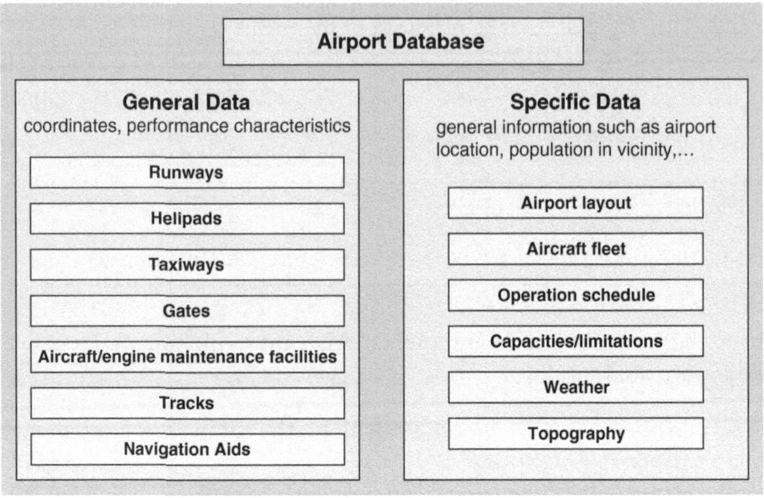


*Figure 7.24* Airports database data sources.

The FSDS system relies on four core input databases for source information with a structure that supports all tools (i.e. *IsoBell'a*, *PolEmiCa*, *3Prisk*, *EMISource*).

The benefit of the approach is that the databases of initial data were generalized for analysis including airport database, fleet database, movements database and aircraft performance database. Details of an airport database are shown in Fig. 7.24. The specifications are the results of a general analysis of the specific databases of the tools *IsoBell'a*, *PolEmiCa*, *3Prisk* and *EMISource*.

AECBT will ultimately use a variety of information provided by ADS to determine the effectiveness of proposed environmental measures. Much of this information will be passed to and acted upon by FSDS. However, there is a set of ADS information that is needed for fleet and operation planning and cost assessment directly within AECBT, including airframe/engine combination costs and aircraft performance.

The detailed structure of AECBT is depicted in Fig. 7.25. The Aviation Economic Module simulates economic flows in the aviation market. The Aircraft Design Space (module ADS) provides environmental factors, flight performance and economic characteristics to AECBT to simulate technology tradeoffs for potential future vehicles when this option is desired (these tradeoffs can be based on either existing technological capability or future technological capability). The FSDS converts aviation activity into quantities of environmental factors distributed in space. The Environmental Impacts Estimation Block converts the quantities of environmental factors into health and welfare impacts, including broad socioeconomic and ecological

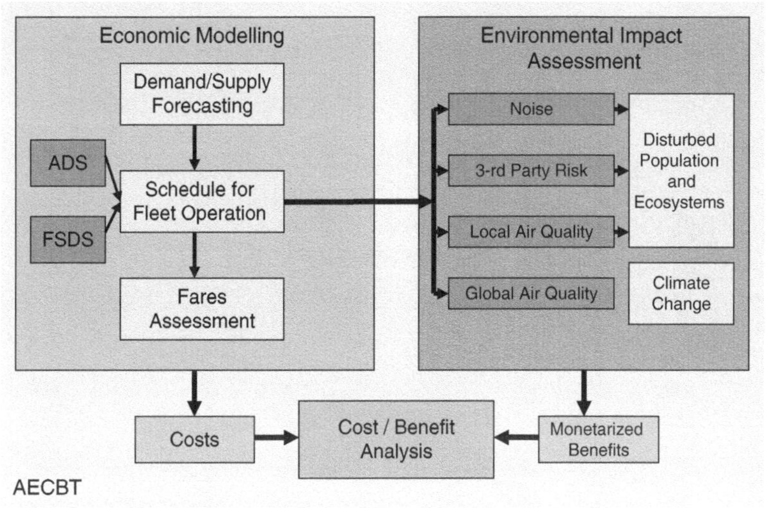


*Figure 7.25* Schematic of the components of the Aviation Environmental Cost–Benefit Tool.

effects. Finally, the costs and benefits block integrated collected costs, environmental inventories and monetary benefits, allows graphical analysis and qualitatively estimated uncertainties.

An analysis in the framework of FSDS has been carried out for an airport which meets the requirements of the END (annual number of flights is more than 50,000). Three types of scenarios have been considered: (1) the current one; (2) a scenario with double the number of flights (since the current rate of increase of flights corresponds to doubling in around 10 years, this represents a 10-year forecasting scenario) with 'Chapter 2' aircraft excluded from the fleet; and (3) a scenario in which the number of flights is at the maximum operational capacity of the airport. The corresponding flight/fleet parameters are shown in Table 7.11.

The noise contours calculated according to these scenarios are shown in Fig. 7.26. A scenario with a predominant number of 'Chapter 2' aircraft has been analyzed as well (Fig. 7.26b), but with an annual flight intensity of less than half of that in the current situation and only one runway in use. For this scenario, the lateral distances of the contours with the same nominal value are greater than for the current scenario between 50 and 70 per cent. Furthermore, the lengths of the contours along the route are greater. In particular, the daytime contour $L_{Aeq} = 65$ dBA is 2 km longer.

There is a different result for the scenario with traffic at the maximum operational capacity of the airport, which may be reached in two decades for the airport under consideration if the current annular flight growth rate remains constant during the forecasting period. In this scenario, the

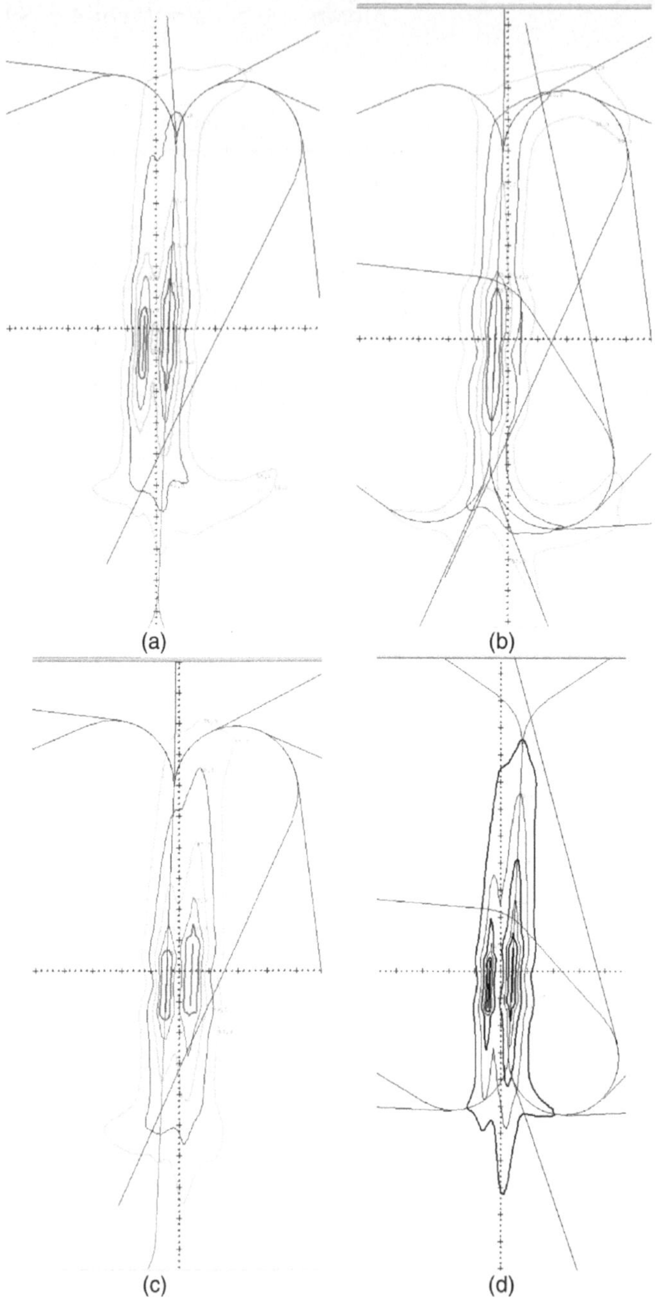


*Figure 7.26* $L_{Aeq}$ noise contours calculated for: (a) the current scenario; (b) an historical scenario with less than half the annual number of flights but with a predominant number of 'Chapter 2' aircraft in fleet and only one runway; (c) a scenario with the number of aircraft doubled (i.e. after 10 years at the current rate of increase); and (d) the scenario with traffic at the maximum operational capacity of the airport (scale = 2 km per main division).

*Table 7.11* Annular flight/fleet parameters for the airport under consideration

| *Aircraft category* | *Flight intensity* | |
|---|---|---|
| | *Current* | *Equal to operational capacity* |
| L1, PAX | 4 | – |
| L2, PAX | 1,101 | – |
| L3, PAX | 3,276 | 85,410 |
| L4, PAX | 18,002 | 98,550 |
| SU, PAX | 6,280 | – |
| T1, PAX | 923 | 20,650 |
| T1, NP | 12 | 1,250 |
| T2, PAX | 7,368 | – |
| EJ, PAX | 972 | 8,760 |
| LT, PAX | 1,659 | 4,380 |

L1–L4 – Western Airliner Jets grouped using the Boeing classes differ by the terms of their production and implementation in operation; SU – aircraft produced in the USSR; T1 – turboprops produced after 1970; T2 – turboprops produced before 1970; LT – light aircraft; EJ – Executive Jets (PAX – passenger, NP – non-passenger).

*Table 7.12* Area of noise contours (in km$^2$), for four flight/fleet scenarios

| *Contour* | *Flight/fleet scenario* | | | |
|---|---|---|---|---|
| $L_{Aeq}$*[dBA]* | *Current* | *Doubling of current, plus 'Chapter 2' phased out* | *Maximum operational capacity of the/ airport – 'Chapter 4'/ 'Chapter 3' mix (1/2)* | *Half of current – 'Chapter 2' dominant* |
| 55.0 | 156.2 | 150.1 | 113.6 | 164.8 |
| 60.0 | 78.3 | 83.2 | 56.8 | 70.6 |
| 65.0 | 34.5 | 41.3 | 22.9 | 28.0 |
| 70.0 | 15.1 | 17.0 | 9.2 | 12.0 |
| 75.0 | 6.0 | 7.6 | 4.0 | 6.1 |

aircraft fleet will be changed considerably, with an increasing dominance of ICAO 'Chapter 3' aircraft and with one-third of the fleet consisting of ICAO 'Chapter 4' aircraft. For this scenario, the areas of the contours are much less than for the other three. The predicted areas are compared in Table 7.12.

For defining noise zones in the vicinity of this airport, the results of the second scenario were chosen, with double the number of flights and improved aircraft fleet ('Chapter 2' aircraft phased out). This scenario is more economically efficient and appropriate, in the sense of the number and total costs of the noise-protective measures, which must be realized inside the zones.

The modeling tools for air pollution, third party risk and electromagnetic field assessment, at airports (*PolEmiCa*[31] or EDMS[34]) may use the same grid/point principles of calculation and the same airport layouts and traffic maps as for aircraft noise. For air pollution, more detailed meteorological and topographical input information are used than for all other factors.

Contamination of the air in the region of an airport is the result of the operation of both moving and stationary sources. Inventories of sources of air pollution in the airports of the Ukraine and other countries have shown that, apart from aircraft, the basic emission sources are (1) airport vehicles; (2) passenger transport; (3) fuel storage; (4) boilers; and (5) aircraft maintenance/repair facilities.

Nevertheless, aircraft are the dominant air pollution sources. The principal pollutants are carbon oxide (CO), hydrocarbons (HC), nitrogen oxides (NOx) and smoke (PM). These are the typical products of fuel combustion of any conventional means of transport. Analysis shows that the emissions of the products of incomplete combustion (CO and HC) are maximum during the stages of engine start, aircraft taxiing before the takeoff and after the landing. These are the stages where the engine operation modes are close to idle. Emissions of nitrogen oxides are maximum during the stages of takeoff and climbing.

A second difficulty for air pollution assessment is the necessity to calculate a huge number of possible daily scenarios with different fleet/flight content and different meteorological/seasonal conditions when defining averaged annual concentrations. For example, the short-term (30 min of averaging) pollution concentrations during the LTO cycles of an aircraft in a particular airport are shown in Fig. 7.27 and their maximum values in Table 7.13.

The plumes and appropriate maximum concentrations are defined mainly by the direction and velocity of the wind. They are produced not only by aircraft, but also by airport vehicles, which are used for aircraft maintenance before and after the flight.

Annual averaged concentrations for the airport as a whole are defined by the annual wind rose and seasonal atmospheric stratification. Examples of these concentration contours are shown in Fig. 7.28. As with aircraft noise, the forecasted scenarios for air pollution show a non-uniform rise with the growth of air traffic, because of the expected constant improvement in aircraft emission performances.

In the same way as aircraft noise zones are defined by aircraft noise contours (Fig. 7.26) and the sanitary zone for air pollution is defined by air contamination contours (Fig. 7.28), third party risk public safety zones (PSZs) are defined by individual risk contours. PSZs are defined around the runways of the busiest airports or those that are located close to residential areas. Their purpose is to eliminate the risk to populations around airports from damage associated with aircraft accidents.

*Figure 7.27* An example of the NOx plume created during the landing/takeoff cycle of an aircraft.

*Table 7.13* Maximum concentrations of air pollution in an airport and their ratio to standard environmental limits

| *Maximum concentrations (mg/m$^3$)* | | | | *Maximum concentrations/limit* | | | |
|---|---|---|---|---|---|---|---|
| *CO* | *$NO_x$* | *$SO_x$* | *PM* | *CO* | *$NO_x$* | *$SO_x$* | *PM* |
| 984.48 | 178.19 | 4.243 | 4.161 | 197 | 2096 | 8.49 | 0.5 |

CO – carbon oxide; $NO_x$ – nitrogen oxides; $SO_x$ – sulfur oxides; PM –smoke.

For such a purpose, the contours of annual individual risk for a population are calculated[32] and, for their standard limits ($10^{-4}$ or $10^{-5}$, depending on the country), the boundaries of the PSZs are defined in a manner similar to those for aircraft noise zones or for reduced emission zones associated with air pollution. For some very busy airports, even the contour $10^{-4}$ may exceed the airport boundaries and reach the closest residential areas. Inside this risky area, all non-airport activities must be fully removed.

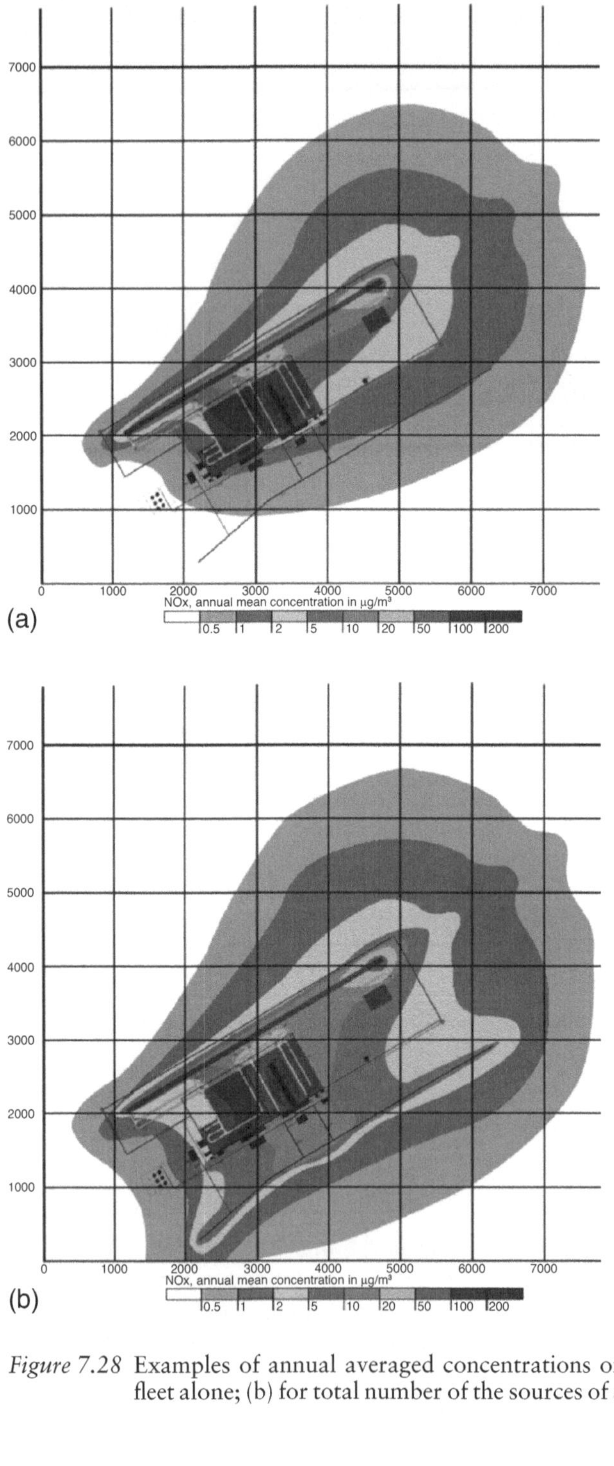


*Figure 7.28* Examples of annual averaged concentrations of $NO_X$: (a) for aircraft fleet alone; (b) for total number of the sources of air pollution in airport.

Calculation of risk contours includes three main stages:[29]

(1) a determination of the total annual number of flights for every aircraft category for which the accident rates are predefined, and calculation of the averaged annual crash rate for the airport (for a particular runway);
(2) use of an accident location probability model, based on probability distribution functions for historical aircraft accidents and calculation of the area of probable accident locations;
(3) determination of contours of individual risk magnitudes $10^{-4}$, $10^{-5}$ and $10^{-6}$, and their representation on a map showing airport facilities and residential areas in the airport vicinity, as is the case with noise or air pollution contours.

Individual risk contours for a particular airport based on the initial data in Table 7.11 are shown in Fig. 7.29.

The lengths of the corresponding contour areas are summarized in Table 7.14 for both runways of the airport.

For all of the considered scenarios, the contour for a risk magnitude of $10^{-4}$, which is the legal prohibition criterion for all types of human activities, lies more or less inside airport area and does not exceed the appropriate prohibition zones for either aircraft noise ($L_{Aeq}$ = 75 dB in the daytime

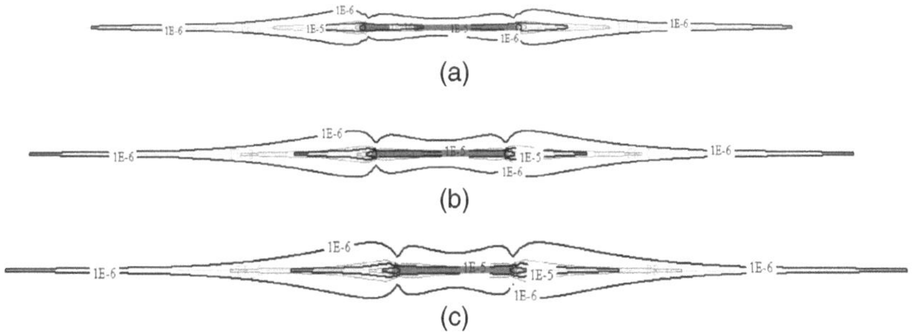


*Figure 7.29* Results of the computation of risk contours for: (a) current aircraft flight intensity; (b) doubled aircraft flight intensity; and (c) forecast for the maximum carrying capacity of a particular airport.

*Table 7.14* Lengths of the individual risk contours for the three scenarios under consideration (Runway 1/Runway 2)

| *Risk* | *Length (m)* | | |
|---|---|---|---|
| | *Current intensity* | *Double intensity* | *Airport capacity* |
| $10^{-4}$ | 80/22 | 217/45 | 394/81 |
| $10^{-5}$ | 1,311/360 | 2,416/795 | 3,507/1,302 |
| $10^{-6}$ | 7,233/3,315 | 10,499/5,335 | 13,261/7,207 |

and $L_{Aeq} = 65$ dB at night) or air pollution (although this depends on the pollutant).

Efficient assessment of environment impact inside and around the airports is based on the following procedure.

(1) Calculation of the number of people impacted by every considered environmental factor inside its corresponding nominal contour.
(2) Assuming that the population density is constant, calculation of a relative area of 'general concern' for every environmental factor under consideration.
(3) Definition of a maximum area of 'general concern' for each environmental factor separately (e.g. for noise alone).
(4) Identification of the methods for reducing the area of concern for the dominant factor.
(5) New calculations of the contours, taking into account the proposed methods of reduction for the considered environmental factors and hence revised areas of 'general concern'.
(6) Definition of a new dominant environmental factor and identification of proposed methods for reduction of the area of concern for this factor.
(7) Continuation of the iteration process until the criteria of minimum impact (impacted population, limited finances, limited resources) are reached.

# Notes

## 1 A review of the aircraft noise problem


1 Callum, T. (2000) 'Environmental capacity of airports – what does it mean?' *Workshop Proceedings 2 Environmental Capacity. The challenge for aviation industry*. London, Heathrow Airport, 8–11.

2 Janic, M. (2003) 'Modeling operational, economic and environmental performance of an air transport network', *Transportation Research*, **8**, 415–32.

3 Nelson, P.M. (ed.) (1987) *Transportation Noise Reference Book*, London: Butterworths & Co. Ltd.

4 Kvitka, V.E., Melnikov, B.N. and Tokarev, V.I. (1980) *Standardization and Noise Abatement for Airplanes and Helicopters* [in Russian]. Kyiv, Vyscha Shkola.

5 Tokarev, V.I., Zaporozhets, O.I. and Straholes, V.A. (1990) *Noise Abatement for Passenger Airplanes in Operation* [in Russian]. Kyiv, Tehnika.

6 Lighthill, M.J. (1952) 'On sound generated aerodynamically: I. General theory', *Proc. Royal Soc. London*, Ser. A, **211** (1107), 564–87.

7 Lighthill, M.J. (1954) 'On sound generated aerodynamically: II. Turbulence as a source of sound', *Proc. Royal Soc. London*, Ser. A, **222** (1148), 1–32.

8 Howe, M.S. (1975) 'Contribution to the theory of aerodynamic sound with application to excess noise and theory of the flute', *J. Fluid Mech.*, **71** (4), 625–73.

9 Crow, S.C. (1970) 'Aerodynamic sound emission as a singular perturbation problem', *Stud. Appl. Math.*, **49** (1), 21–44.

10 Ribner, H.S. (1964) 'The generation of sound by turbulent jets', Volume 8 of *Advances in Applied Mechanics*, Dryden, H.L. and von Karman, Th. (eds). New York, Academic Press, Inc., 103–82.

11 Lilley, G.M. (1954) 'Aerodynamic noise', *J. Royal Aeronaut. Soc.*, 58, 235–39.

12 Hubbord, H.H. (ed.) (1995) *Aeroacoustics of Flight Vehicles – Theory and practice*. Vol. 1: *Noise sources*. Woodbury, NY, Acoustical Society of America.

13 Hubbord, H.H. (ed.) (1995) *Aeroacoustics of Flight Vehicles – Theory and practice*. Vol. 2: *Noise control*. Woodbury, NY, Acoustical Society of America.

14 Crighton, D.G., Dowling, A.P., Ffowcs Williams, J.E., Heckl, M. and Leppington, F.G. (1992) *Modern Methods in Analytical Acoustics*, London, Springer-Verlag.

15 Munin, A.G., Kuznezchov, V.M. and Leontev, E.A. (1981) *Aerodynamic Noise Sources* [in Russian]. Moscow, Mashynostroenie.

16 Mhytaryan, A.M., Enyenkov, V.G., Melnikov, B.N., *et al.* (1975) *Noise Abatement of Aircraft with Turbojet Engines* [in Russian]. Moscow, Mashynostroenie.

17 Ffowcs Williams, J.E. and Hawkings, D.L. (1969) 'Sound generation by turbulence and surfaces in arbitrary motion', *Phil. Trans. Royal Soc. London*, Ser. A, **264** (1151), 321–42.

18 Farassat, F. (1981) 'Linear acoustic formulas for calculation of rotating blade noise', *AIAA J.*, **19**(9), 1122–30.
19 Proposed amendment to procedures for air navigation services. Aircraft operations. Volume 1: Flight procedures, Part V, Noise abatement procedures (2001), PANS-OPS, Doc8168, ICAO, CAEP/5 Recommendations.
20 Von Gierke, H.E. and McEldred, K. (1975) 'Effect of noise on people', *Noise/News International*, **1**(2), 67–89.
21 ENV Noise–Comments–Parliament, Noise policy A40183/97 Resolution on the Commission Green Paper on future noise policy (COM(96), (1997), 0540 C40587/96).
22 Working Group on EU-noise indicators, Commission of the European Communities–Directorate–General Environment, (1999).
23 Zaporozhets, O.I. and Tokarev, V.I. (1998) 'Aircraft noise modelling for environmental assessment around airports', *Applied Acoustics*, **55**(2), 99–127.
24 Position paper on dose response relationships between transportation noise and annoyance, EU's future noise policy, WG2 – Dose/Effect, 20 February 2002. Luxembourg, Office for Official Publications of the European Communities, 24 pp.
25 Lambert, J. and Vallet, M. (1994) *Study Related to the Preparation of a Communication on a Future EC Noise Policy*. Paris, LEN, Final Report, No 9420.
26 ICAO Standard and Recommended Practice, *Environmental Protection. Annex 16 to the Convention on International Civil Aviation. Aircraft Noise*. Montreal, Volume 1 (1993).
27 Technical meeting on exposure–response relationships of noise on health. World Health Organization Regional Office for Europe, European Centre for Environment and Health, Bonn Office, 19–21 September 2002 Bonn, Germany, Meeting report (2002).
28 Langdon, F.J. (1985) 'Noise annoyance', in Tempest, W. (ed.) *The Noise Handbook*, London, Academic Press.
29 Ollerhead, J. and Sharp, B. (2001) 'Computer model highlights the benefits of various noise reduction measures', *ICAO J.*, No. 4, 18–19.
30 Report of the FESG industry response task group. Information Paper CAEP/7-IP/2, CAEP 7th Meeting, Montréal, 5–16 February 2007, 37 pp.
31 *Airport Planning Manual*, Part 2: *Land Use and Environmental Control*, ICAO Doc 9184, 4th edition (2004).
32 OECD Report: Noise abatement policies for the 1990s (1991), Geneva, OECD.
33 Berglund, B., Lindvall, T. and Schwela, D.H. (eds) (2000) *Guidelines for Community Noise*. Geneva, World Health Organization.
34 Procedures for Air Navigation Services – Aircraft Operations, Volume I – Flight Procedures (2005), PANS-OPS, ICAO Doc 8168.
35 Working paper on curfews (2007), Working Papers CAEP/7-WP/15, CAEP 7th Meeting, Montréal, 5–16 February 2007, 37 pp.
36 Airport Noise Compatibility Planning. 14 CFR Part 150. DOT FAA, 1989.

## 2 The main sources of aircraft noise

1 Recommended Method for Computing Noise Contours around Airports (1987), Circular 205 ICAO, International Civil Aviation Organization.
2 Standard Method of Computing Noise Contours around Civil Airports (1997), ECAC Document 29.
3 Procedure for the Calculation of Airplane Noise in the Vicinity of Airports (1986), Society of Automotive Engineers, SAE AIR 1845.

4 Integrated Noise Model (INM) version 6.0 (2000), Report No. FAA-AEE-99-03, Washington, Federal Aviation Administration.
5 The UK Civil Aircraft Noise Contour Model ANCON: Improvements in Version 2 (1999), National Air Traffic Services.
6 Nord 2000. Comprehensive Outdoor Sound Propagation Model (2001).
7 Prediction method for lateral attenuation of airplane noise during takeoff and landing. AIR 1751 (1981), Society of Automotive Engineers. Inc.
8 Gas turbine jet exhaust noise prediction. ARP 876C (1982) Society of Automotive Engineers. Inc.
9 House, M.E. and Smith, M.J.T. (1966) 'Internally generated noise from gas turbine engines – measurement and prediction', ASME Paper 66-GT/N-43.
10 Gas turbine coaxial exhaust flow noise prediction. AIR 1905, (1985) Society of Automotive Engineers. Inc.
11 Heidmann, M.F. (1975) 'Interim prediction method for fan and compressor source noise', NASA TM X-71763.
12 Matta, R.K., Sandusky, G.T. and Doyle, V.L. (1977) 'GE core engine noise investigation, low emission engines', FAA-RD-77-4.
13 Kershaw, R.J. and House, M.E. (1982) 'Sound-absorbent duct design', in White, R.G. and Walker, J.G. (eds) *Noise and Vibration*, Chichester, John Wiley & Sons.
14 Fink, M.R. (1977) 'Airframe noise prediction method', FAA-RD-77-29.
15 Prediction procedure for near-field and far field propeller noise (1977), AIR 1407, Society of Automotive Engineers, Inc.
16 Maglizzi, B. (1977) 'The influence of forward flight on propeller noise', NASA CR-145105.
17 Huff, R.G. and Clark, B.J. (1974) 'Interim prediction methods for low frequency core engine noise', NASA TM X-71627.
18 Munin, A. (ed.) (1986) *Aviation Acoustics*. Part 1: *Aircraft Noise* [in Russian], Moscow, Mashinostroenie.
19 Tokarev, V.I., Zaporozhets, O.I. and Straholes, V.A. (1990) *Noise Abatement for Passenger Airplanes in Operation* [in Russian]. Kyiv, Tehnika.


## 3 Aircraft noise propagation


1 Li, K.M. and Tang, S.H. (2003) 'The predicted barrier effects in the proximity of tall buildings', *J. Acoust. Soc. Am.*, **114**, 821–32.
2 Zaporozhets, O., Tokarev, V. and Attenborough, K. (2003) 'Predicting noise from aircraft operated on the ground', *Appl. Acoustics*, **64**, 941–53.
3 Bass, H.E., Sutherland, L.C. and Zuckewar, A.J. (1995) 'Atmospheric absorption of sound: further developments', *J. Acoust. Soc. Am.*, **97**, 680–83.
4 Larsson, C. (1997) 'Atmospheric absorption conditions for horizontal sound propagation', *Appl. Acoustics*, **50**, 231–245
5 Larsson, C. (2000) 'Weather effects on outdoor sound propagation', *Int. J. Acoustics Vibration*, **5**, 33–6.
6 Zaporozheth, O.I. and Tokarev, V.I. (2002) 'Investigation influence of impedance plate on sound wave propagation', *Proc. Natl Aviation Univ.*, **1**, 240–7.
7 Tokarev, V. and Zaporozhets, O. (2004) 'Calculation of sound wave propagation, radiation and transmission through plate with different boundary conditions'. Inter-noise-2004, *Proceedings of the 33rd International Congress and Exposition on Noise Control Engineering*, Prague, 1–4.
8 Agrest, M.M. and Maksimov, M.S. (1971) *Theory of Incomplete Cylindrical Functions and Their Applications*, Berlin, Springer-Verlag.
9 Attenborough, K., Li, K.M. and Horoshenkov, K. (2007) *Predicting Outdoor Sound*, London, Taylor and Francis, Chapter 2.

10 Abramowitz, M. and Stegun, I.A. (1972) *Handbook of Mathematical Functions with Formulas, Graphs, and Mathematical Tables*, New York, Dover Publications, Inc.
11 Matta, F. and Reichel, A. (1971) Uniform computation of the error function and other related functions, *Math. Comput.*, **25**, 339–44.
12 Attenborough, K., Hayek, S.I. and Lawther, J.M. (1980) 'Propagation of sound above a porous half-space', *J. Acoust. Soc. Am.*, **68**, 1493–501.
13 Banos, A. (1966) *Dipole Radiation in the Presence of Conducting Half-Space*, New York, Pergamon, Chapters 2–4.
14 Donato, R.J. (1978) 'Model experiments on surface waves', *J. Acoust. Soc. Am.*, **63**, 700–3.
15 Daigle, G.A., Stinson, M.R. and Havelock, D.I. (1996) 'Experiments on surface waves over a model impedance using acoustical pulses', *J. Acoust. Soc. Am.*, **99**, 1993–2005.
16 Wang, Q. and Li, K.M. (1999) 'Surface waves over a convex impedance surface', *J. Acoust. Soc. Am.*, **106**, 2345–57.
17 Albert, D.G. (2003) 'Observation of acoustic surface waves in outdoor sound propagation', *J. Acoust. Soc. Am.*, **113**, 2495–500.
18 Li, K.M., Waters-Fuller, T. and Attenborough, K. (1998) 'Sound propagation from a point source over extended-reaction ground', *J. Acoust. Soc. Am.*, **104**, 679–85.
19 Nicolas, J., Berry, J.L. and Daigle, G.A. (1985) 'Propagation of sound above a finite layer of snow', *J. Acoust. Soc. Am.*, **77**, 67–73.
20 Allard, J.F., Jansens, G. and Lauriks, W. (2002) 'Reflection of spherical waves by a non-locally reacting porous medium', *Wave Motion*, **36**, 143–55.
21 Taherzadeh, S. and Li, K.M. (1997) 'On the turbulent jet noise near an impedance surface', *J. Sound Vib.*, **208**, 491–6.
22 Delany, M.E. and Bazley, E.N. (1970) 'Acoustical properties of fibrous absorbent materials', *Appl. Acoust.*, **3**, 105–16.
23 Rasmussen, K.B. (1981) 'Sound propagation over grass covered ground', *J. Sound Vib.*, **78**, 247–55.
24 Attenborough, K. (1992) 'Ground parameter information for propagation modeling', *J. Acoust. Soc. Am.*, **92**, 418–27. [see also Raspet, R. and Attenborough, K. (1992) Erratum: 'Ground parameter information for propagation modeling', *J. Acoust. Soc. Am.*, **92**, 3007.]
25 Raspet, R. and Sabatier, J.M. (1996) 'The surface impedance of grounds with exponential porosity profiles', *J. Acoust. Soc. Am.*, **99**, 147–52.
26 Attenborough, K. (1993) 'Models for the acoustical properties of air-saturated granular materials', *Acta Acust.*, **1**, 213–26.
27 Allard, J.F. (1993) *Propagation of Sound in Porous Media: Modelling Sound Absorbing Material*, New York, Elsevier Applied Science.
28 Johnson, D.L., Plona, T.J. and Dashen, R. (1987) 'Theory of dynamic permeability and tortuosity in fluid-saturated porous media' *J. Fluid Mech.*, **176**, 379–401.
29 Umnova, O., Attenborough, K. and Li, K.M. (2000) 'Cell model calculations of dynamic drag parameters in packings of spheres', *J. Acoust. Soc. Am.*, **107**, 3113–19.
30 Horoshenkov, K.V., Attenborough, K. and Chandler-Wilde, S.N. (1998) 'Padé approximants for the acoustical properties of rigid frame porous media with pore size distribution', *J. Acoust. Soc. Am.*, **104**, 1198–209.
31 Yamamoto, T. and Turgut, A. (1988) 'Acoustic wave propagation through porous media with arbitrary pore size distributions', *J. Acoust. Soc. Am.*, **83**, 1744–51.

32 ANSI S1.18-1999 (1999) *Template Method for Ground Impedance*, New York, Standards Secretariat, Acoustical Society of America.
33 Taherzadeh, S. and Attenborough, K. (1999) 'Deduction of ground impedance from measurements of excess attenuation spectra', *J. Acoust. Soc. Am.*, **105**, 2039–42.
34 Attenborough, K. and Waters-Fuller, T. (2000) 'Effective impedance of rough porous ground surfaces', *J. Acoust. Soc. Am.*, **108**, 949–56.
35 Aylor, D.E. (1972) 'Noise reduction by vegetation and ground', *J. Acoust. Soc. Am.*, **51**, 197–205.
36 Attenborough, K., Waters-Fuller, T., Li, K.M. and Lines, J.A. (2000) 'Acoustical properties of farmland', *J. Agric. Eng. Res.*, **76**, 183–95.
37 De Jong, B.A., Moerkerken, A. and van Der Toorn, J.D. (1983) 'Propagation of sound over grassland and over an Earth barrier', *J. Sound Vib.*, **86**, 23–46.
38 Pierce, A.D. (1974) 'Diffraction of sound around corners and over wide barriers', *J. Acoust. Soc. Am.*, **55**(5), 941–55.
39 Chessel, C.I. (1977) 'Propagation of noise along a finite impedance boundary', *J. Acoust. Soc. Am.*, **62**, 825–34.
40 Parkin, P.H. and Scholes, W.E. (1965) The horizontal propagation of sound from a jet engine close to the ground at Radlett, *J. Sound Vib.* 1, 1–13.
41 Parkin, P.H. and Scholes, W.E. (1965) 'The horizontal propagation of sound from a jet engine close to the ground at Hatfield', *J. Sound Vib.*, **2**, 353–74.
42 Kantola, R.A. (1975) 'Outdoor jet noise facility: a unique approach', *2nd Aero-Acoustics Conference*, Hampton, VA, March, pp. 223–245 (AIAA paper 75-530).
43 Li, K.M., Attenborough, K. and Heap, N.W. (1991) 'Source height determination by ground effect inversion in the presence of a sound velocity gradient', *J. Sound Vib.*, **145**, 111–28.
44 Rudnick, I. (1957) 'Propagation of sound in open air', in Harris, C.M. (ed.) *Handbook of Noise Control*, New York, McGraw Hill, pp. 3:1–3:17.
45 Li, K.M. (1993) 'On the validity of the heuristic ray-trace based modification to the Weyl Van der Pol formula', *J. Acoust. Soc. Am.*, **93**, 1727–35.
46 Zouboff, V., Brunet, Y., Berengier, M. and Sechet, E. (1994) *Proceedings of the 6th International Symposium on Long Range Sound Propagation*, Havelock, D.I. and Stinson, M. (eds), Ottawa, NRCC, 251–69.
47 'The propagation of noise from petroleum and petrochemical complexes to neighbouring communities', (1981) CONCAWE Report no.4/81, Den Haag.
48 Marsh, K.J. (1982) 'The CONCAWE model for calculating the propagation of noise from open-air industrial plants', *Appl. Acoustics*, **15**, 411–28.
49 Monin, A.S. and Yaglom, A.M. (1979) *Statistical Fluid Mechanics: Mechanics of Turbulence*, Vol. 1, Cambridge, MA, MIT Press.
50 Stull, R.B. (1991) *An Introduction to Boundary Layer Meteorology*, Dordrecht, Kluwer, pp. 34–86.
51 Salomons, E.M. (1994) 'Downwind propagation of sound in an atmosphere with a realistic sound speed profile: a semi-analytical ray model', *J. Acoust. Soc. Am.*, **95**, 2425–36.
52 Holtslag, A.A.M. (1984) 'Estimates of diabatic wind speed profiles from near surface weather observations', *Boundary-Layer Meteorology*, **29**, 225–50.
53 Davenport, A.G. (1960) 'Rationale for determining design wind velocities', *J. Am. Soc. Civ. Eng.*, ST-86, 39–68.
54 Huisman, W.H.T. (1990) 'Sound propagation over vegetation-covered ground', Ph.D. thesis, University of Nijmegen, The Netherlands.
55 Salomons, E.M., van den Berg, F.H. and Brackenhoff, H.E.A. (1994) 'Long-term average sound transfer through the atmosphere based on meteorological

statistics and numerical computations of sound propagation', *Proceedings of the 6th International Symposium on Long Range Sound Propagation*, Havelock, D.I. and Stinson, M. (eds), Ottawa, NRCC, 209–28.

56 Heimann, D. and Salomons, E. (2004) 'Testing meteorological classifications for the prediction of long-term average sound levels', *Appl. Acoustics*, **65**, 925–50.

57 Sutherland, L.C. and Daigle, G.A. (1998) 'Atmospheric sound propagation', in Crocker, M.J. (eds) *Encyclopedia of Acoustics*, New York, Wiley, pp. 305–29.

58 Embleton, T.F.W. (1996) 'Tutorial on sound propagation outdoors', *J. Acoust. Soc. Am.*, **100**, 31–48.

59 Stinson, M.R., Havelock, D.J. and Daigle, G.A. (1996) 'Simulation of scattering by turbulence into a shadow zone region using the GF-PE method', *Proceedings of the 6th International Symposium on Long Range Sound Propagation*, Ottawa, NRCC, pp. 283–307.

60 Wilson, D.K. (1996) 'A brief tutorial on atmospheric boundary-layer turbulence for acousticians', *Proceedings of the 7th International Symposium on Long Range Sound Propagation*, Ecole Centrale, Lyon, pp. 111–22.

61 Wilson, D.K., Brasseur, J.G. and Gilbert, K.E. (1999) 'Acoustic scattering and the spectrum of atmospheric turbulence', *J. Acoust. Soc. Am.*, **105**, 30–4.

62 L' Esperance, A., Daigle, G.A. and Gabillet, Y. (1996) 'Estimation of linear sound speed gradients associated to general meteorological conditions', *Proceedings of the 6th International Symposium on Long Range Sound Propagation*, Ottawa, NRCC.

63 Wilson, D.K. (1988) 'On the application of turbulence spectral/correlation models to sound propagation in the atmosphere', *Proceedings of the 8th International Symposium on Long Range Sound Propagation*, Penn State.

64 Ostashev, V.E. and Wilson, D.K. (2000) 'Relative contributions from temperature and wind velocity fluctuations to the statistical moments of a sound field in a turbulent atmosphere', *Acustica/Acta Acustica*, **86**, 260–8.

65 Clifford, S.F. and Lataitis, R.T. (1983) 'Turbulence effects on acoustic wave propagation over a smooth surface', *J. Acoust. Soc. Am.*, **73**, 1545–50.

66 Daigle, G.A. (1979) 'Effects of atmospheric turbulence on the interference of sound waves above a finite impedance boundary', *J. Acoust. Soc. Am.*, **65**, 45–9.

67 Weiner, F.M. and Keast, D.N. (1959) 'Experimental study of the propagation of sound over ground', *J. Acoust. Soc. Am.*, **31**(6), 724–33.

68 Gilbert, K.E., Raspet, R. and Di, X. (1990) 'Calculation of turbulence effects in an upward refracting atmosphere', *J. Acoust. Soc. Am.*, **87**(6), 2428–37.

69 ISO 9613-2 (1996) *Acoustics – Attenuation of Sound during Propagation Outdoors* – Part 2: *General Method of Calculation*, Geneva, International Organization for Standardization,

70 ISO 10847 (1997) *Acoustics – In-situ Determination of Insertion Loss of Outdoor Noise Barriers of All Types*, Geneva, International Organization for Standardization,

71 ANSI S12.8 (1998) *Methods for Determination of Insertion Loss of Outdoor Noise Barriers*, Washington, DC, American National Standard Institute.

72 Daigle, G.A. (1999*) Report by the International Institute of Noise Control Engineering Working Party on the Effectiveness of Noise Walls*, Noise/News International I-INCE Publication.

73 Kotzen, B. and English, C. (1999) *Environmental Noise Barriers – A Guide to Their Acoustic and Visual Design*, London, E&FN Spon.

74 Sommerfeld, A. (1896) 'Mathematische Theorie der Diffraction', *Math. Ann.*, **47**, 31374.

75 MacDonald, H.M. (1915) 'A class of diffraction problems', *Proc. Lond. Math. Soc.*, **14**, 410–27.

76 Redfearn, S.W. (1940) 'Some acoustical source–observer problems', *Phil. Mag.*, 30, 223–36.
77 Keller, J.B. (1962) 'The geometrical theory of diffraction', *J. Opt. Soc.*, **52**, 116–30.
78 Hadden, W.J. and Pierce, A.D. (1981) 'Sound diffraction around screens and wedges for arbitrary point source locations', *J. Acous. Soc. Am.*, **69**, 1266–76.
79 Embleton, T.F.W. (1980) 'Line integral theory of barrier attenuation in the presence of ground', *J. Acoust. Soc. Am.*, **67**, 42–5.
80 Menounou, P., Busch-Vishniac, I.J. and Blackstock, D.T. (2000) 'Directive line source model: a new model for sound diffracted by half planes and wedges', *J. Acoust. Soc. Am.*, **107**, 2973–86.
81 Medwin, H. (1981) 'Shadowing by finite noise barriers', *J. Acoust. Soc. Am.*, **69**, 1060–4.
82 Maekawa, Z. (1968) 'Noise reduction by screens', *Appl. Acoustics*, **1**, 157–73.
83 Tatge, R.B. (1973) 'Barrier-wall attenuation with a finite sized source', *J. Acoust. Soc. Am.*, **53**, 1317–19.
84 Kurze, U.J. and Anderson, G.S. (1971) 'Sound attenuation by barriers', *Appl. Acoustics,* **4**, 35–53.
85 Menounou, P. (2001) 'A correction to Maekawa's curve for the insertion loss behind barriers', *J. Acoust. Soc. Am.*, **110**, 1828–38.
86 Lam, Y.W. and Roberts, S.C. (1993) 'A simple method for accurate prediction of finite barrier insertion loss', *J. Acoust. Soc. Am.*, **93**, 1445–52.
87 Price, M.A., Attenborough, K. and Heap, N.W. (1988) 'Sound attenuation through trees: measurements and models', *J. Acoust. Soc. Am.,* **84**, 1836–44.
88 Kragh, J. (1982) 'Road traffic noise attenuation by belts of trees and bushes', *Danish Acoustical Laboratory Report* no.31.
89 Huddart, L.R. (1990) 'The use of vegetation for traffic noise screening', *TRRL Research Report* 238.
90 Heisler, G.M., McDaniel, O.H., Hodgdon, K.K., Portelli, J.J. and Glesson, S.B. (1987) 'Highway noise abatement in two forests', *Proceedings of the NOISE-CON 87, PSU*, USA.
91 Defrance, J., Barriere, N. and Premat, E. (2002) 'Forest as a meteorological screen for traffic noise'. *Proceedings of the 9th International Conference on Sound and Vibration*, Orlando.
92 Barrière, N. and Gabillet, Y. (1999) 'Sound propagation over a barrier with realistic wind gradients. Comparison of wind tunnel experiments with GFPE computations', *Acustica/Acta Acustica*, **85**, 325–34.
93 Barrière, N. (1999) 'Etude théorique et expérimentale de la propagation du bruit de trafic en forêt', Ph.D. thesis, Ecole Centrale de Lyon.
94 Swearingen, M.E. and White, M. (2007) 'Influence of scattering, atmospheric refraction, and ground effect on sound propagation through a pine forest', *J. Acoust. Soc. Am.*, **122**, 113–19.

## 4 Methods for aircraft noise prediction

1 Tokarev, V.I., Zaporozhets, O.I. and Straholes, V.A. (1990) *Noise Abatement for Passenger Airplanes in Operation* [in Russian], Kyiv, Tehnika.
2 Munin, A. (ed.) (1986) *Aviation Acoustics*, Part 2: *Noise Inside Passenger Airplanes* [in Russian], Moscow, Mashinostroenie.
3 *Manuals for Civil Aviation Impact Assessment on the Environment* [in Russian] (1984) State Institute of Civil Aviation, Moscow, Kyiv Institute of Civil Aviation Engineers, Kyiv.

4 *ICAO Technical Manual for Environment Specified the Usage of Methods for Aircraft Noise Certification* (1995) Montreal, Doc. 9501 AN.
5 *Recommended Method for Computing Noise Contours around the Airports* (1988) Montreal, ICAO Cir. 205-AN.
6 *Relationship between Aircraft Noise Contour Area and Noise Levels at Certification Points* (2003) NASA/TM-2003-212649, Hampton, Virginia, Langley Research Center, 2199–2368.
7 Vlasov, E., Munin, A. and Samokhin, V. (1976) 'Calculation method for noise, produced by the aircraft on the ground surface, based on results of acoustic trials of the engines [in Russian]', *Aviation Acoustics*, Trudy TsAGI, No. 1806, Moscow, 81–88.
8 Stewart, E.C. and Carson, T.M. (1980) 'Simple method for prediction of aircraft noise contours', *J. Aircraft*, **17**(11), 828–30.
9 Mhytaryan, A.M., Enyenkov, V.G., Melnikov, B.N. *et al.*(1975) *Noise Abatement of Aircraft with Turbojet Engines* [in Russian], Moscow, Mashynostroenie.
10 Sean Lynn Summary report for an undergraduate research project to develop programs for aircraft takeoff analysis in the preliminary design phase, Virginia Polytechnic Institute and State University Blacksburg, Virginia, 11 May 1994.
11 Miele, A. (1962) *Flight Mechanics* (Part 1), Reading, MA, Addison-Wesley.
12 Zaporozhets, A.I. (1987) 'Influence of flight safety requirements on optimization results of takeoff parameters with purpose to reduce impact of noise and engine emission', in *Modelling in flight safety provision* [in Russian], Kiev, 1987, pp. 102–8.
13 Kantola, R.A. (1975) 'Outdoor jet noise facility: a unique approach', *2nd Aero-Acoustics Conf.*, Hampton, VA, March, pp. 223–45 (AIAA paper 75–530).
14 Li, K.M. (1993) 'On the validity of the heuristic ray-trace-based modification of the Weyl–Van der Pol formula', *JASA*, **93**(4), 1727–35.
15 Hidaka, T., Kageyama, K. and Masuda S. (1985) 'Sound propagation in the rest atmosphere with linear sound velocity profile', *J. Acoust. Soc. Jpn. (E)*, **6**(2), pp. 117–125.
16 Raspet R., L'Esperance, A. and Daigle, G.A. (1995) 'The effect of realistic ground impedance on the accuracy of ray tracing', *JASA*, **97**(1), 154–8.
17 L'Esperance, A. (1992) 'Modalisation de la propagation des ondes sonores dans un environnement naturel complexe', Ph.D. thesis, Sherbrooke University, Canada.
18 Embleton, T.F.W., Thiessen, G.J. and Piercy, J.E. (1976) 'Propagation in inversion and reflections at the ground', *JASA*, **59**(2), pp. 128–142.
19 Plovsing, B. and Kragh, J. (1998) *Prediction of Sound Propagation in an Atmosphere Without Significant Refraction*, DELTA Acoustics & Vibration Report AV No 1898, 98, Lyngby.
20 Ollerhead, J. (1999) 'CAEP progress on aircraft noise contour modelling', in *Improved Tools for Aircraft Noise and Airport Impact Assessment*, X-NOISE Workshop "Improved Tools for Aircraft Noise and Airport Impact Assessment", Trinity College, Dublin, 1999.
21 Hellstrom, G. (1974) *Noise Shielding Aircraft Configurations, A Comparison Between Predicted and Experimental Results* // ICAS Paper No. 74-58.
22 Maekawa, Z. (1966) *Noise Reduction by Screens*, Memoirs of the Faculty of Engineering, Kobe University, Japan, Vol. 12, pp. 472–9.
23 Lieber, L. (2000) *Small Engine Technology (SET) – Task 13*. ANOPP Noise Prediction for Small Engines., Jet Noise Prediction Module, Wing Shielding Module and System Studies Results, NASA/CR-2000-209706, Phoenix, AZ, AlliedSignal Engines and Systems.

24 Brekhovskikh, L.M. and Godin, O.A. (1989) *Acoustics in Layered Media* [in Russian], Nauka, Moskcow.

25 Ostashev, V.E. (1997) *Acoustics in Moving Inhomogeneous Media*, London, E & FN Spon.

26 Chessell, C.I. (1977) Propagation of noise along a finite impedance boundary, *J. Acoust. Soc. Am.*, **62**, 825–34.

27 Plovsing, B. and Kragh, J. (1999) *Validation of Nordic Models for Sound Propagation – Comparison with Measurement Results*, DELTA Acoustics & Vibration Report No 2006, 99, Lyngby.

28 Ögren, M. and Jonasson, H. (1998) *Measurement of the Acoustic Impedance of Ground*, KFB project 1997-0222, Nordtest project 1365-1997, SP Report 1998:28 Acoustics Borås.

29 Kawai, T., Hidaka, T. and Nakajama, T. (1982) 'Sound propagation above an impedance boundary'. *J. Sound Vibration* **83**(1),125–38.

30 Thomasson, S.-I. (1977) 'Sound propagation above a layer with a large refraction index', *J. Acoust. Soc. Am.*, **61**, pp. 659–674 (1977).

31 L'Espérance, A., Herzog, P., Daigle, G.A. and Nicolas, J. (1992) 'Heuristic model for outdoor sound propagation based on an extension of the geometrical ray theory in the case of a linear sound speed profile', *Appl. Acoust.*, 37, 111–39.

32 Rudnick, I. (1947) 'The propagation of an acoustic wave along a boundary', *JASA*, **19**, 348–56.

33 Daigle, G.A. (1979) 'Effects of atmospheric turbulence on the interference of sound waves above a finite impedance boundary', *JASA*, **65**(1), 45–9.

34 Raspet, R. and Wu, W. (1995) 'Calculation of average turbulence effects on sound propagation based on the fast field program formulation', *JASA*, **97**(1), 147–53.

35 Daigle, G.A., Piercy, J.E. and Embleton, T.F.W. (1978) 'Effects of atmospheric turbulence on the interference of sound waves near a hard boundary', *J. Acoust. Soc. Am.*, **64**(2), pp. 622–630.

36 Plovsing, B. (2000) *NORD 2000: Comprehensive Model for Predicting the Effect of Terrain and Screens in the New Nordic Prediction Methods for Environmental Noise*, Inter-Noise, 2000, Nice, August.

37 Parkin, P.H. and Scholes, W.F. (1965) 'The horizontal propagation of sound from a jet engine close to the ground at Hatfield', *J. Sound Vib.*, **2**(4), 353–74.

38 ISO 9613-2 (1996) *Acoustics – Attenuation of Sound During Propagation Outdoors* – Part 2: *General Method of Calculation*, Geneva, International Organization for Standardization,

39 Delaney, M.E. and Bazley, E.N. (1970) 'Acoustic properties of fibrous absorbent materials', *Appl. Acoustics*, 3(2), pp. 105–116.

40 Zaporozhets, O., Tokarev, V.I. and Shylo, V.F. (1996) 'Influence of impedance characteristics of the reflecting surfaces on reduction of aviation noise by screens', *Proc. 4th Int. Congress on Sound and Vibration*, St Petersburg, Vol. 2, 1135–40.

41 Zwieback, E.L. (1975) *Aircraft Flyover Noise Measurements*, AIAA Paper 75–537, Reston, VA, AIAA.

42 Devis, L.I.C. (1981) *A Guide to the Calculation of NNI*, DORA Communication 7908, 2nd edition, London, CAA.

43 *Recommended Method for Computing Noise Contours around Airports* (1988) Circular 205 AN/1/25, Montreal, ICAO.

44 Zaporozhets, O. and Tokarev, V. (1998) 'Aircraft noise modelling for environmental assessment around airports', *Appl. Acoustics*, 55(2), 99–127.

## 5 The influence of operational factors on aircraft noise levels


1 Tokarev, V., Zaporozhets, O. and Straholes, V. (1990) *Noise Decreasing During the Aircraft Operating* [in Russian], Kyiv, Technika.
2 Vorotyntsev, V., Zaporozhets, O. and Karpin, B. (1981) 'Definition of the task of aircraft description as a source of air pollution' [in Russian], Vol 197: *Problems of Environment Protection from Civil Aviation Impact*, GosNIIGA, Moscow, pp. 14–21.
3 Zaporozhets, O. (1985) 'Determination of the most profitable flight procedures in airport area with purpose of their minimal impact on environment', *Methods and Means of Aviation Impact Reduction on the Environment*, Kiev, KIICA, pp.17–27.
4 ICAO Standard and Recommended Practice (1993) *Environmental Protection. Annex 16 to the Convention on International Civil Aviation. Aircraft Noise*, Vol. 1, Montreal, ICAO.
5 Zaporozhets, O. and Tokarev, V. (1998) 'Aircraft noise modelling for environmental assessment around airports', *Appl. Acoustics*, 55(2), 99–127.
6 ICAO (1988) *Recommended Method for Computing Noise Contours Around Airports*, Cir. 205-AN, Montreal, ICAO,
7 Society of Automotive Engineers (SAE), *Procedure for the Calculation of Airplane Noise in the Vicinity of Airports*, Airspace Information Report (AIR), SAE-1845, Warrendale, PA, SAE.
8 Society of Automotive Engineers (SAE) (1975) *Standard Values of Atmospheric Absorption as a Function of Temperature and Humidity*, Committee A-21, Aircraft Noise, Aerospace Recommended Practice No. 866A, March, Warrendale, PA, SAE.
9 US Department of Transportation, Federal Aviation Administration (FAA) (1999) *Spectral Classes for FAA'S Integrated Noise Model Version 6.0*, Letter Report DTS-34-FA065-LR1, December 7, Washington, DC, FAA.
10 *ICAO Technical Manual for Environment Specified the Usage of Methods for Aircraft Noise Certification* (1995) Doc. 9501 AN, Montreal, ICAO.
11 Mkhitaryan, A.M. (1975) *Noise Reduction in Aircraft with Jet Engines* [in Russian], Moscow, Mashinostrojenije, 264 pp.
12 Melnikov, B. and Zaporozhets, O. (1985) 'Investigation of optimal noise and emission flight modes at descending and landing of the aircraft', *Investigation, Testing and Reliability of Aircraft Engines*, Moscow, GosNIIGA, Vol. 236, pp. 66–74.


## 6 Methods of aircraft noise reduction


1 Munin, A. (ed.) (1986) *Aviation Acoustics*. Part 1: *Aircraft Noise* [in Russian], Moscow, Mashinostroenie.
2 Munin, A. (ed.) (1986) *Aviation Acoustics*. Part 2: *Noise Inside Passenger Airplane* [in Russian], Moscow, Mashinostroenie.
3 Munin, A.G., Kuznezchov, V.M. and Leontev, E.A. (1981) *Aerodynamic Noise Sources* [in Russian], Moscow, Mashinostroenie.
4 Kvitka, V.E., Melnikov, B.N. and Tokarev, V.I. (1984) *Civil Aviation and Environment Protection* [in Russian] Kyiv, Vyscha Shkola.
5 Zaporozhets, O.I. and Tokarev, V.I. (1998) Predicted flight procedures for minimum noise impact. *Appl. Acoustics*, 55(2), 129–43.
6 Mhytaryan, A.M., Enyenkov, V.G., Melnikov, B.N. *et al.* (1975) *Noise Abatement of the Aircrafts with Turbojet Engines* [in Russian], Moscow, Mashynostroenie.

7 Tokarev, V.I., Zaporozhets, O.I. and Straholes, V.A. (1990) *Noise Abatement for Passenger Airplanes in Operation* [in Russian], Kyiv, Tehnika.
8 Sobol, I.M. and Statnikov, P.B. (1981) The *Choice of Optimum Parameters in the Tasks with Many Variables* [in Russian], Moskow, Nauka.
9 International Civil Aviation Organizaton (ICAO) Standard and Recommended Practice (1993) *Environmental Protection, Annex 16 to the Convention on International Civil Aviation. Aircraft Noise*, Vol. 1, Montreal, ICAO.
10 Didkovsky, V.S., Akimenko, V.Y., Zaporozhets, O.I. *et al.* (2001) *Bases of Acoustic Ecology* [in Ukrainian], Kirovograd, Imex Ltd.
11 Maekawa, Z. (1968) Noise reduction by screens. *Appl. Acoustics*, **1**, 157–73.
12 Wilson, A.G. (1970) *Entropy in Urban and Regional Modeling*, London, Pion Ltd.
13 *Continuous Descent Approach, Implementation Guidance Information*, Eurocontrol (2007).

## 7 Monitoring of aircraft noise

1 Mann, R.E. (1980) 'The design of environmental monitoring systems,' *Prog. Phys. Geogr*, **4**(4), 567–76.
2 Tokarev, V.I., Vorotyntsev, V.M. and Zaporozhets, A.I. (1990) 'Structure of aircraft noise monitoring system around the airports', *Problems of Acoustical Ecology*, **1**, Leningrad, Strojizdat, 54–9 [in Russian].
3 *Consultation on Proposals for Transposition and Implementation of Directive 2002/49/EC of the European Parliament and of the Council of 25 June 2002 Relating to the Assessment and Management of Environmental Noise* (2005) February, 126, London, Defra Publications.
4 Tokarev, V.I., Zaporozhets, O.I. and Straholes, V.A. (1990) *Noise Abatement for Passenger Airplanes in Operation*. Kyiv, Tehnika, 1990. – 127c.
5 Bassanino, M., Mussin, M., Deforza, P., Lunesu, D. and Telaro, B. (2004) 'Methodology of statistical model results verification in a high traffic airport', in *Transport Noise – 2004*', St Petersburg.
6 Elliff, T., Cavadini, L. and Fuller, I. (2000) 'Enhance: an evolutionary improvement to aircraft noise modeling', in *Internoise 2000*, Nice, France.
7 Thomann, G. (2007) 'Mess- und Berechnungsunsicherheit von Fluglärmbelastungen und ihre Konsequenzen', Dissertation, Zürich, ETHZ, 318.
8 Thomann, G. and Bütikofer, R. (2007) 'Quantification of uncertainties in aircraft noise calculations', in *Internoise 2007*, Istanbul, Turkey, August.
9 ISO/ENV 13005 (1995) *Guide to the Expression of Uncertainty in Measurement*, Geneva, International Organization for Standardization.
10 Krebs, W., Bütikofer, R., Plüss, S. and Thomann, G. (2004) 'Sound source data for aircraft noise simulation', *Acta Acustica/Acustica*, **90**, 91–100.
11 Storeheier, S.A., Randeberg, R.T., Granoien, I.L.N., Olsen, H. and Ustad, A. (2002) *Aircraft Noise Measurements at Gardermoen Airport, 2001*, Part 1: *Summary of Results*, SINTEF Telecom and Informatics, Norway Report No. STF40 A02032, Trondheim, 2002-06-05.
12 Cadoux, R.E. and White, S. (2003) *An Assessment of the Accuracy of Flight Path Data Used in the Noise and Track-Keeping System at Heathrow, Gatwick and Stansted Airports*, ERCD Report 0209, CAA, March, London, The Stationery Office.
13 *Relationship between Aircraft Noise Contour Area and Noise Levels at Certification Points* (2003) NASA/TM-2003-212649, Hampton, Virginia, Langley Research Center.

14 Cadoux, R.E. and Kelly, J.A. (2003) *Departure Noise Limits and Monitoring Arrangements at Heathrow, Gatwick and Stansted Airports*, ERCD Report 0207, CAA, March, London, The Stationery Office.
15 Harlow, C. (2003) *Development of an Aircraft Operation Classification System for Louisiana's Airports*, LTRC Project No. 95-8SS, State Project No. 736-99-0241, June 2003, Louisiana State University, 71 p.
16 Fan, Y., Tysoe, B., Sim, J., Mirkhani, K., Sinclair, A.N. *et al.* (2003) 'Nondestructive evaluation of explosively welded clad rods by resonance acoustic spectroscopy', *Ultrasonics*, **41**, 369–75.
17 Duquennoy, M., Ouaftouh, M. and Ourak, M. (1999) 'Ultrasonic evaluation of stress in orthotropic materials using Rayleigh waves', *NDT & E International*, **32**, 189–99.
18 Murino, V. (2001) Reconstruction and segmentation of underwater acoustic images combining confidence information in MRF models. *Pattern Recognition*, **34**(5), 981–97.
19 Zha, D. and Qiu, T. (2006) Underwater sources location in non-Gaussian impulsive noise environments. *Digital Signal Processing*, **16** (2), 149–63.
20 Benko, U., Petrovcic, J., Juricic, D., Tavcar, J., Rejec, J. and Stefanovska, A. (2004) Fault diagnosis of a vacuum cleaner motor by means of sound analysis. *J. Sound Vib.*, **276** (3-5), 781–806.
21 Wu, J.D. and Chuang, C.Q. (2005) Fault diagnosis of internal combustion engines using visual dot patterns of acoustic and vibration signals. *NDT & E International*, **38**, 605–14.
22 Zadeh, LA. (1973) 'Outline of a new approach to the analysis of complex systems and decision processes', in *IEEE Transactions on Systems, Man and Cybernetics*, vol. SMC-3, 28-44.
23 Tong, R. M., Gupta, M. M., Ragade, R. K., Yager, R. R. (1979) *'The construction and evaluation of fuzzy models in Advances in Fuzzy Set Theory and Applications'*, Eds. Amsterdam: North-Holland.
24 Yager R. R., Filev, D. P. (1994) *Essentials of Fuzzy Modeling and Control.* New York John Wiley & Sons, Inc.
25 Couvreur, C. and Bresler, Y. (1995) 'A statistical pattern recognition framework for noise recognition in an intelligent noise monitoring system', *Proc. Euro-Noise'95*, Lyon, France, pp. 1007–12.
26 Couvreur, C. (1996) 'Adaptive classification of environmental noise sources', *Proc. Forum Acusticum*, Antwerp, Belgium, S220.
27 Dufournet, D. (2002) 'Automatic noise source recognition', *Transport Noise and Vibration, 6th International Symposium*, June, St Petersburg, tn02_s7_08. - 6 p.
28 *Guidance on the Balanced Approach to Aircraft Noise Management* (2004) ICAO Doc. 9829, Montreal, ICAO.
29 Yuriy Medvedev. PEGAS 1.2: CAEPport Results and Modelling assumptions. CAEP 9 MDG 3 meeting, Austin, TX, USA, March 7–9, 2011.
30 Konovalova, O. and Zaporozhets, O. (2005) 'NoBel – tool for aircraft noise spectra assessment account of ground and shielding effects on noise propagation', *World Congress Proc.: 'Aviation in the XXI Century', Environment Protection Symposium*, September.
31 Sinilo, K. and Zaporozhets, O. (2005) 'PolEmiCa – tool for air pollution and aircraft engine emission assessment in airports', *World Congress Proc.: 'Aviation in the XXI Century', Environment Protection Symposium*, September.
32 Gosudarskaja, I. and Zaporozhets, O. (2005) '3PRisk – third party risk assessment around the airports', *World Congress Proc.: 'Aviation in the XXI Century', Environment Protection Symposium*, September.

33 Glyva, V. and Lukjanchikov, A. (2008) 'EMISource – tool for electro-magnetic fields assessment in airports', *World Congress Proc.: 'Aviation in the XXI Century', Environment Protection Symposium*, September.

34 *Emissions and Dispersion Modelling System (EDMS) Reference Manual*, FAAAEE-01-01, September, US Department of Transportation, Federal Aviation Administration Washington, DC, CSSI.

# Index